# SUPERSPACE AND SUPERGRAVITY

# SUPERSPACE AND SUPERGRAVITY

## PROCEEDINGS OF THE NUFFIELD WORKSHOP, CAMBRIDGE

June 16 - July 12, 1980

edited by

S. W. HAWKING & M. ROČEK

Department of Applied Mathematics and Theoretical Physics

University of Cambridge

CAMBRIDGE UNIVERSITY PRESS

Cambridge

London New York New Rochelle

Melbourne Sydney

Published by the Press Syndicate of the University of Cambridge

The Pitt Building, Trumpington Street, Cambridge CB2 1RP

32 East 57th Street, New York, NY 100022, USA

296 Beaconsfield Parade, Middle Park, Melbourne 3206, Australia



First published 1981

Printed in Great Britain at the University Press, Cambridge

British Library cataloguing in publication data

Superspace and Supergravity.
I. Supergravity - Congresses
I. Hawking, S.W. II. Roček, M.
531'.5 QC173.59.S/ 80-42091

ISBN 0 521 23908 7

Table of Contents

"ACHTBEIN THE OCTOPUS"

Achtbein's eight legs refer to the eight Superspace coordinates.

# PREFACE

The Workshop on Supergravity was held in the Department of Applied Mathematics and Theoretical Physics at the University of Cambridge from 16 June to 12 July, 1980. It featured a variety of topics, and was generally held to be an instructive way to spend four weeks. The outcome included a number of articles, a blackboard (displayed on the cover), and, of course, these proceedings. We would like to thank the Nuffield Foundation for financing the Workshop, and Judy Fella without whose efforts nothing would have happened. We would also like to thank Jacques Richer for late night help with a computerized typing system.

S.W.H.
M.R.

## LIST OF CONTRIBUTORS

ADAMS, J.F., D.P.M.M.S., University of Cambridge, 4 Mill Lane, Cambridge, England.

ARAGONE, C.A., Division of Physics & Mathematics, University of S. Bolivar, Apartado 80659, Caracas 1080A, Venezuela.

CREMMER, E., Laboratoire de Physique de l'Ecole Normale Supérieure, 24 rue Lhomond, Paris, Cedex 5, France.

de WIT, B., NIKHEF-H, P.O. Box 41882, 1009 DB Amsterdam, The Netherlands.

DUFF, M., Physics Department, Imperial College, London, SW7 2BZ, England.

FREEDMAN, D.Z., Department of Mathematics, M.I.T., Cambridge, Massachusetts 02139, USA.

GATES, S.J., Lauritsen Lab. of Physics, California Inst. Tech., Pasadena, CA 91125, USA.

GIBBONS, G.W., D.A.M.T.P., University of Cambridge, Silver Street, Cambridge, England.

GRISARU, M., Physics Department, Brandeis University, Waltham, Massachusetts 02254, USA.

HAWKING, S.W., D.A.M.T.P., University of Cambridge, Silver Street, Cambridge, England.

HOWE, P., TH-Division, CERN, Geneva 23, Switzerland.

JULIA, B., Laboratoire de Physique de l'Ecole Normale Supérieure, 24 rue Lhomond, Paris, Cedex 5, France.

KAKU, M., Physics Department, C.C.N.Y., 138th St. & Morningside Drive, New York, N.Y. 10031, USA.

KARLHEDE, A., University of Stockholm, Institute of Physics, Vanadisvagen 9, S-113 46 Stockholm, Sweden.

LINDSTROM, U., University of Stockholm, Institute of Physics, Vanadisvagen 9, S-113 46 Stockholm, Sweden.

MacDOWELL, S.W., Physics Department, Yale University, New Haven, CT 06520, USA.

PERRY, M.J., Physics Department, Princeton University, Princeton, New Jersey 08540, USA.

LIST OF CONTRIBUTORS

POPE, C.N., D.A.M.T.P., University of Cambridge, Silver Street, Cambridge, England.

ROČEK, M., D.A.M.T.P., University of Cambridge, Silver Street, Cambridge, England.

SIEGEL, W., Lauritsen Lab. of Physics, California Inst. Tech., Pasadena, CA 91125, USA.

SLAVNOV, A.A., V.A. Steklov Mathematical Institute, Moscow, USSR.

SOKATCHEV, E., TH-Division, CERN, Geneva 23, Switzerland.

SOHNIUS, M., Imperial College, Prince Consort Road, London, SW 7, England.

STELLE, K.S., TH-Division, CERN, Geneva 23, Switzerland.

TAYLOR, J., Department of Mathematics, King's College, London, England.

TEITELBOIM, C., School of Natural Sciences, Institute for Advanced Study, Princeton NJ 08540, USA.

TOWNSEND, P., TH-Division, CERN, Geneva 23, Switzerland.

THIERRY-MIEG, J., Lyman Laboratory, Harvard University, Mass. 02138, USA.

VAN HOLTEN, J.W., TH-Division, CERN, Geneva 23, Switzerland.

VAN NIEUWENHUIZEN, P., I.T.P., State University of New York, Stony Brook, Long Island, N.Y. 11794, USA.

WEST, P.C., Flat 4, 6 Embankment Gardens, London, SW3, England.

ZUMINO, B., TH-Division, CERN, Geneva 23, Switzerland.

## LIST OF PARTICIPANTS

ADAMS, J.F.
AVIS, S.J.
ARAGONE, C.A.
BATCHELOR, M.
BAUM, E.
BEDDING, S.
BRINK, L.
CAPPER, D.M.
CASTAGNINO, M.
CASTELLANI, L.
CREMMER, E.
de WIT, B.
DeWitt, B.S.
DOUGHTY, N.
DUFF, M.
DOWNES-MARTIN, S.
D'ADDA, A.
FAWCETT, M.
FREEDMAN, D.Z.
GATES, S.J.
GIBBONS, G.
GRISARU, M.
HAWKING, S.W.
HITCHIN, N.
HULL, C.
HOWE, P.
ISHAM, C.J.
JONES, D.R.T.
JULIA, B.
KAKU, M.
KARLHEDE, A.
KOSTELECKY, A.
LINDSTROM, U.
LAPEDES, A.
MacDOWELL, S.W.
MOSS, I.G.
NIELSEN, N.K.
PAGE, D.N.
PAYNE, P.
PERRY, M.J.
POPE, C.N.
RICHER, J.
RIVELLES, V.O.
ROČEK, M.
ROGERS, A.
RYAN, P.
SIEGEL, W.
SLAVNOV, A.A.
SOKATCHEV, E.
SOHNIUS, M.
STELLE, K.S.
TAYLOR, J.
TEITELBOIM, C.
TOWNSEND, P.
THIERRY-MIEG, J.
VAN HOLTEN, J.W.
VAN NIEUWENHUIZEN, P.
WARNER, N.
WEST, P.C.
WHITING, B.
YUILLE, A.
ZUMINO, B.

## INTRODUCTION

The Newtonian theory of gravity was very successful in predicting planetary and stellar orbits, but because it implied that gravitational effects propagated with infinite velocity, it was incompatible with the local validity of special relativity. This difficulty was overcome in 1916 with the formulation of General Relativity in which the gravitational field was represented by the metric of curved spacetime. Since that time, the predictions of General Relativity have been found to be in excellent agreement with observation. However, the theory is incomplete in at least two ways:

1. It does not relate gravity to interactions and to matter fields that occur in physical theories.

2. It is a purely classical theory, whereas all other fields seem to be quantized.

Einstein himself was unhappy about the first shortcoming, and spent most of the rest of his life looking for a Unified Field Theory. He failed, partly because not much was known at the time about the other interactions except electromagnetism, and partly because he ignored the second limitation of General Relativity listed above. The necessity to quantize the gravitational field has become more urgent because it has been shown that classical General Relativity inevitably predicts singularities of spacetime.

In recent years there has been considerable success in unifying the other interactions apart from gravity. The Salam-Weinberg model combined electromagnetism with the weak nuclear forces while a number of Grand Unified Theories (GUTs) have been proposed to unite these two interactions with the strong nuclear forces. In the GUTs, the unification takes place at an energy of the order of $10^{15}$ GEV, which is quite near the Planck energy of $10^{19}$ GEV at which quantum gravitational effects might be expected to become important. It therefore seems natural to try to extend these theories to include gravity. At the present, the best hope of doing this seems to lie in the various supergravity theories. In these, the graviton, the spin 2 particle that carries the gravitational interaction, is related to particles of lower integer and half integer spin by so called "supersymmetry" transformations. This has the great merit of not only unifying the "interactions", represented by integer spin fields, but also of ending the old dichotomy between them and the "matter", represented by half-integer spin fields.

Another great advantage of supergravity theories is that many of the divergences that plague quantum field theory seem to cancel between the integer and half integer spin particles. Pure gravity on its own is known to be finite at the one loop level in topologically trivial situations. However, it is not finite when coupled to matter, or in topologically nontrivial configurations, and there are indications that it may not be finite at the two-loop level in any situation. On the other hand, it has been shown that the extended supergravity theories are all finite at the one and two levels in topologically trivial situations, and that the "N=8" theory is finite at one and two loops in all situations (see the article by M. J. Duff in this volume). It has also been shown that the analogous N=4 supersymmetric Yang-Mills theory is finite to three loops. These results suggest that the N=8 theory may turn out to be finite to all orders (although there is a possible counterterm at seven loops, see the article by P. Howe and U. Lindström in this volume). It is to be hoped that this is indeed the case, because it can be shown that theories including gravity are either finite or non-renormalizable: if there are divergences, one would have to add an infinite sequence of counterterms with a corresponding infinite number of undetermined renormalization parameters.

Supergravity is a very recent and rapidly developing subject. It is therefore difficult to get a clear view of the field and to see what progress has been made. We hope that this volume will do something to improve the situation. It is based on the Workshop on Supergravity that was held in Cambridge from 16 June to 12 July, 1980, and was supported by the Nuffield Foundation. The volume is divided into six parts. Part I contains two sets of introductory lectures on supergravity. Those by P. van Nieuwenhuizen concentrate on the component approach in ordinary spacetime. This is the form in which supergravity was first formulated, and in which it has the most direct physical interpretation in terms of fields of different spin. However, the supersymmetry is more apparent in the superspace formulation which is described for the case of unextended supergravity in the second set of lectures by M. Roček. We hope that these two sets of lectures provide an introduction to supergravity which is comprehensible to readers with some knowledge of General Relativity and quantum field theory, and which will enable them to follow the remaining more specialized articles.

Part II contains five articles on the quantization and regularization of supergravity and supersymmetric gauge theories. The first article, by M. T. Grisaru, describes the background superfield method which enables the perturbation expansion to be written in a very compact and manifestly supercovariant form of supergraphs. These supergraphs have a number of gauge invariances which in general require the introduction of several generations of ghosts as described in the article by W. Siegel. They also require a manifestly supersymmetric regularization technique. The next two articles describe two such schemes. The first, by W. Siegel, P. K. Townsend and P. van Nieuwenhuizen, deals with dimensional regularization by dimensional reduction. This seems to work well but has unresolved problems about anomalies. The second, by A. A. Slavnov, uses additional higher derivative terms with gauge invariant Pauli-Villars regularization of the remaining one loop divergences. It is more complicated to apply, and has so far been worked out only for supersymmetric Yang-Mills theory, but not for supergravity. The final article in this part, by C. Teitelboim, deals with the Hamiltonian quantization of supergravity.

Part III contains nine articles dealing with extended supergravity and related topics. In simple supergravity, described in the lectures in part I, there is only one supersymmetry generator which is a Majorana spinor whereas in extended supergravity there are N such generators, with $N \leq 8$. The Lagrangians and supersymmetry transformations are known, but only in component form and without the auxiliary fields needed to close the algebra of transformations. The first two articles by E. Sokatchev and S. J. Gates, Jr., describe the progress that has been made towards superspace formulations of the N=2 theory. It seems we do not yet have an elegant formulation of the theory. The next article, by B. de Wit and co-workers, uses extended supersymmetric versions of the conformal group to clarify the structure of the component formulation of N=2 supergravity with auxiliary fields. The approach seems promising, although there are difficulties which appear to limit it to $N \leq 4$. The article by S. MacDowell attempts to formulate all the extended theories by giving covariant constraints in superspace. It unfortunately gives an implicit form of the theories which is difficult to use. In the article by E. Cremmer, the extended theories are formulated in their component form (without auxiliary fields) in five dimensions. This is a useful starting point for obtaining symmetry breaking by dimensional reduction to four dimensions, and for attempting to derive auxiliary fields by a Legendre transformation as described in the following article by M. Sohnius, K. S. Stelle and P. C. West. This deals primarily with supersymmetric Yang-Mills theories. There are problems with constraints which arise because of gauge invariance. The authors try to solve these by introducing an infinite number of Lagrange multiplier fields which effectively are the coefficients of a Taylor expansion in a fifth dimension. A systematic attempt, from the point of view of group theory, to relate theories in lower and higher dimensions is presented in the next article by B. Julia. Related group theoretical considerations follow in the contribution of J. Thierry-Mieg and B. Morel, and the final article in this part, by J. G. Taylor, finds linearized representations of extended supersymmetry using superspin projection operators.

Part IV deals with the maximally extended N=8 supergravity theory, which is the best candidate for a complete unified theory without divergences. In the article by M. J. Duff, it is shown that the one loop term is finite even in topologically nontrivial situations if one takes a proper account of the antisymmetric tensors that arise in the derivation of the theory by dimensional reduction from eleven dimensions. These antisymmetric tensors can also give rise to a cosmological term, as described in the article by A. Aurilia, H. Nicolai and P. K. Townsend. There is a possible counter term for the N=8 theory at seven loops, as is shown in the article by P. Howe and U. Lindström, but it is not known whether this divergence actually appears or whether it is absent because of some as yet unknown symmetry. The N=8 theory at it stands cannot describe the known physical fields, since its 28 spin one fields cannot include the SU(3) x SU(2) x U(1) gauge symmetry that seems to be present in particle physics. A possible solution to this problem is presented in the article by B. Zumino, where it is suggested that the elementary fields of the theory (apart from the graviton) are not observed directly, but that they form a supermultiplet of bound states invariant under an SU(8) gauge group. Most of these bound states acquire very large masses on the order of $10^{19}$ GEV and are not observed, but a subset of them form a SU(5) Grand Unified Theory with three generations of quarks and leptons. This seems to be what is observed. It is a very promising idea to reconcile supergravity to the observed particle spectrum. It is however, at the moment, only a scenario because nothing is known about the mechanisms that would confine these bound states. The final article in this part, by J. F. Adams, describes the special mathematical features of Spin(8), triality, $F_4$ and all that.

Part V contains two articles on Kähler manifolds. D. Z. Freedman uses the properties of Kähler spaces to prove results about the divergences of supersymmetric nonlinear sigma-models. C. N. Pope uses the existence of gauge covariantly constant spinors in Kähler-Einstein backgrounds to find relations between the eigen-value spectra of different spin wave operators. These relations are responsible for the remarkable cancellations between integer and half-integer spins in supersymmetric theories.

Part VI contains five articles which do not fit under any of the previous headings. The first, by M. Sohnius, describes techniques useful for simplifying constraints in superspace, for example, of the sort used by S. MacDowell above. The article by M. Perry deals with the instabilities that arise because the gravitational part of the action is not positive definite, as for example, in a black hole background field. The next article, by U. Lindström, A. Karlhede and M. Roček, gives the explicit relation between the superspace and component formulations of N=1 supergravity. M. Kaku describes a scheme for quantising gravity and supergravity on a lattice, and C. Aragone presents failed attempts to find consistent interacting theories for fields with spins higher than two.

We feel that the articles and the lectures described above give a general introduction to the field, and indicate most of the current areas of research. Other surveys and research articles can be found in Recent Developments in Gravitation Cargèse 1978, eds. M. Lévy and S. Deser, Plenum, in Supergravity, eds. P. van Nieuwenhuizen and D. Z. Freedman, North-Holland, and in a forthcoming issue of Physics Reports C by P. van Nieuwenhuizen. At the present time, we think that the outstanding problems are:

1. The construction of a complete formulation of extended supergravity with auxiliary fields, preferably in superspace.

2. The development of general arguments to determine whether or not the higher loops in the N=8 supergravity theory are finite. This might follow immediately from the solution of the first problem.

3. The reconciliation of supergravity with the observed phenomenology of particle physics. This might be along the lines suggested in the article by B. Zumino.

It is impossible to predict the timescales, but given the current interest in the field, we can hope that all three problems will be solved in the next few years.

S. W. Hawking

M. Roček

# I INTRODUCTION TO SUPERGRAVITY

# SIX LECTURES AT THE CAMBRIDGE WORKSHOP ON SUPERGRAVITY

by

P. van Nieuwenhuizen

Institute for Theoretical Physics
State University of New York at Stony Brook
Stony Brook, New York 11794

"Well, however it may be, I will submit"

Schubert, die schöne Müllerin
(the Conference Concert)

Dedicated to Joel Scherk

## INTRODUCTION

This set of six lectures was given at the Nuffield Workshop on Supergravity at Cambridge, England in June 1980. It deals with supergravity in ordinary spacetime and is the counterpart of Dr Roček's lectures on superspace. There is a need for lectures covering the research results of the past year. We have summarized here mainly results on extended supergravities and on the relation between group theory and quantum field theory.

For a detailed introduction into many other areas of supergravity, we refer to a forthcoming Physics Report by the author. It was the intention that Joel Scherk would write a sequel, dealing with extended supergravity and spontaneous symmetry breaking, but his death has prevented this. Any reader of these notes will see how much Joel contributed to this field. We dedicate this set of lectures to him.

## LECTURE I. PROPERTIES OF GLOBAL SUPERSYMMETRY

Supergravity is the gauge theory of supersymmetry, so let us begin by discussing the necessary aspects of supersymmetry. To be concrete, we consider the simplest model, the Wess-Zumino model, (Wess & Zumino, 1974a, b) which contains a real scalar field A, a real pseudoscalar B and a Majorana spinor $\lambda^\alpha$. As explained below, a Majorana spinor satisfies <u>in four dimensions</u>

$$(\bar{\lambda}^D)_\alpha \equiv (\lambda^\dagger)_\beta (\gamma_4)^\beta{}_\alpha = \lambda^\beta C_{\beta\alpha} \equiv (\bar{\lambda}^M)_\alpha \qquad (1)$$

where C is defined up to an overall constant by $C\gamma_m C^{-1} = -\gamma_m^T$. We will use four- instead of two-component spinors since we shall also consider spinors in five and eleven dimensions.

The action reads

$$\mathcal{L} = -\frac{1}{2} (\partial_\mu A \partial_\nu A + \partial_\mu B\, \partial_\nu B)\eta^{\mu\nu} - \frac{1}{2} \bar{\lambda}\partial^\mu \gamma_\mu \lambda \qquad (2)$$

Since there is a symmetry between bosons and fermions, there are as many states of each (since $\lambda^\dagger$ is related to $\lambda$ by (1), there are only two fermion states with helicities $\pm \frac{1}{2}$). One can give a different formulation for such a spin ($0^+$, $0^-$, $\frac{1}{2}$) system by representing the spin $0^+$ field by an antisymmetric tensor field $t_{\mu\nu}$

$$\mathcal{L} = \frac{1}{2} F_{\mu\nu\rho} F^{\mu\nu\rho} - \frac{1}{2} \partial_\mu B \partial^\mu B - \frac{1}{2} \bar{\lambda}\, \gamma^\mu \partial_\mu \lambda \qquad (3)$$

Since $F_{\mu\nu\rho} = \partial_\mu \tau_{\nu\rho}$ + cyclic terms satisfies on-shell $\partial^\mu F_{\mu\nu\rho} = 0$ and in four dimensions any antisymmetric $F_{\mu\nu\rho}$ is equal to $\varepsilon_{\mu\nu\rho\sigma}\psi^\sigma$ it follows that on-shell $\psi_\sigma = \partial_\sigma \psi$ describes a spin 0 field.

One can find transformation rules which leave (2) and (3) invariant (up to a total four-divergence). There is, however, a difference: in (3) one has equal numbers of bosonic and fermionic field components, whereas in (2) one lacks two bosonic fields. (There is the gauge invariance $\delta\tau_{\mu\nu} = \varepsilon_{\mu\nu\rho\sigma} \partial^\rho \Lambda^\sigma$ in (3), but $\Lambda_\sigma = \partial_\sigma \Lambda$ cancels from $\delta\tau_{\mu\nu}$. Hence $\tau_{\mu\nu}$ has 6-3=3 field components). The algebra of global suspersymmetry transformations, translations and Lorentz rotations closes for (3), but for (2) only when the fermion is on-shell ($\not\partial\lambda = 0$ with $\not\partial = \gamma^\mu \partial_\mu$ ).Only after adding two auxiliary fields F and G does also this algebra close off-shell. This we shall discuss in the next lecture.

We conjecture that one can also find in supergravity different field representations which lead to an off-shell closed algebra. (Different field representations do not lead to a different S-matrix (Sezgin & van Nieuwenhuizen, 1980), but lead to different anomalies (Duff & van Nieuwenhuizen, 1980).

We shall always use hermitian Dirac matrices $\gamma_1 \ldots \gamma_4$ so that $\gamma_5 = \gamma_1\gamma_2\gamma_3\gamma_4$ is also hermitian. Moreover, $(\gamma_\mu)^2 = +1$ for $\mu = 1, 2,3,4,5$. Most people use instead of $\gamma_4$ the matrix $\gamma_o = i\gamma_4$, but since we will usually contract indices it does not matter which convention one uses (We use $\eta_{\mu\nu} = + + + +$, and $\partial_4 = \partial/\partial\,(ict)$. This is the Pauli metric but, again, the results are convention independent).

As we shall show in a moment, the Lagrangian density $\mathcal{L}$ in (2) transforms under global (also called rigid) supersymmetry transformations into a total derivative, and hence the action is invariant if one considers topologically trivial spacetimes. Thus supersymmetry is not really an internal symmetry, but, just as general coordinate transformations, it transforms Lagrangian densities into a total four-derivative. In the superspace approach, the general coordinate transformations of ordinary space and the local supersymmetry transformations constitute together general supercoordinate transformations, but in these lectures we will follow the ordinary space approach (also called component approach). For superspace, see the lectures by Dr. Roček.

We must now first of all explain (1). In a general (but hermitian) representation of the $\gamma^\mu$, the spinors $\lambda^\alpha$ are not real, but are Majorana spinors. Since we will also need Majorana spinors in higher dimensions, let me explain in detail what a Majorana spinor is. This will take some time, but since this is a set of lectures I feel justified and compelled to do so. Afterwards I will come back to (2).

One begins by defining a Dirac - and a Majorana-conjugate spinor $\bar{\lambda}^D = \lambda^\dagger B$ and $\bar{\lambda}^M = \lambda^T C$ by requiring that $\bar{\lambda}^D\chi$ and $\bar{\lambda}^M\chi$ transform under Lorentz transformations as a scalar. From the

Lorentz transformation rules for the arbitrary spinor $\chi$

$$(\chi')^{\alpha} = (\exp\tfrac{1}{2}\lambda^{mn}\sigma_{mn})^{\alpha}{}_{\beta}\chi^{\beta}, \qquad \sigma_{mn} = \tfrac{1}{4}[\gamma_m, \gamma_n] \tag{4}$$

one easily derives that B is proportional to $\gamma_o$, while C must satisfy

$$C\sigma_{mn}C^{-1} = -(\sigma_{mn})^T \tag{5}$$

A Majorana spinor is a spinor whose Dirac conjugate is proportional to its Majorana conjugate. Since the scales of B and C are arbitrary, we choose the scale of C such that $\bar{\lambda}^D = \bar{\lambda}^M$ as in (1).

Suppose one has at a given time a Majorana spinor $\lambda^{\alpha}$, then, in order that for all time this spinor remains a Majorana spinor, one must require that $\bar{\lambda}^D$ and $\bar{\lambda}^M$ satisfy the same field equations. Let us consider as field equation the Dirac equation (for the spin 3/2 Rarita Schwinger equation the argument is analogous). One easily derives

$$\bar{\lambda}^D(\gamma^{\mu}\overleftarrow{\partial}_{\mu} - M) = 0, \quad \lambda^T C(C^{-1}\gamma^{\mu,T}\overleftarrow{\partial}_{\mu}C + M) = 0 \tag{6}$$

Thus for massive spinors we have the condition $C^{-1}\gamma^{\mu T}C = -\gamma_{\mu}$, but for massless spinors (the case considered below), we have a weaker condition

$$C\gamma^{\mu}C^{-1} = \alpha\gamma^{\mu,T}, \qquad \alpha = \pm 1 \tag{7}$$

That $\alpha^2$ is +1 follows from squaring (7). In fact, (7) implies (5).

In four dimensions, $\alpha$ equals -1, but in five dimensions there exists only a matrix C which satisfies (7) when $\alpha$ equals +1. To see this, note that in a Majorana representation where $\gamma^1$, $\gamma^2$, $\gamma^3$ are real and symmetric but $\gamma^0$ (or $\gamma^4$) is antisymmetric, the matrix C is proportional to $\gamma^0$ (or $\gamma^4$). In five dimensions, the Clifford algebra

$$\{\gamma^{\mu},\gamma^{\nu}\} = 2\eta^{\mu\nu}, \qquad \mu,\nu = 1,5 \tag{8}$$

is spanned by $\gamma^1,\gamma^2,\gamma^3,\gamma^0$ (or $\gamma^4$ if one uses in (8) $\delta^{\mu\nu}$ instead of $\eta^{\mu\nu} = (+++-+)$) and $\gamma_5 = \gamma^1\gamma^2\gamma^3\gamma^4$. It is now easy to see that in five dimensions (7) is satisfied if in a Majorana representation one chooses C to be proportional to $\gamma^0\gamma^5$ (or $\gamma^4\gamma^5$), and that $\alpha = +1$.

Most people stop at this point, but there is more in the life of a Majorana spinor than $\bar{\lambda}^D = \bar{\lambda}^M$. Namely, suppose we take the complex conjugate of $\bar{\lambda}^D = \bar{\lambda}^M$ and insert into the result $\bar{\lambda}^D = \bar{\lambda}^M$ once more. A little calculation shows that one finds a new requirement

$$CB^{-1}C^*(B^*)^{-1} = 1. \tag{9}$$

Any phase of C and B cancels, and in four dimensions (9) holds, while in five dimensions (9) does not hold! Thus in five dimensions one must produce an extra sign in (9), and this is done by considering spinors $\lambda^{\alpha}_{a}$ with an internal index a. One defines then

$$(\bar{\lambda}^D)^a{}_{\alpha} = (\lambda^{\beta}{}_{a})^* B^{\beta}{}_{\alpha} = (\lambda^{\dagger})^a{}_{\beta} B^{\beta}{}_{\alpha}$$
$$(\bar{\lambda}^M)^a{}_{\alpha} = \lambda^{\beta}{}_{b} C_{\beta\alpha}\Omega^{ab} \tag{10}$$

and requires again that $(\bar{\lambda}^D)^a_{\alpha}\chi^{\alpha}$ and $(\bar{\lambda}^M)^a_{\alpha}\chi^{\alpha}$ are Lorentz invariant and satisfy the same field equations. Again one finds (7), while B is again proportional to the gamma matrix corresponding to the time index.

However, in order that the definition of a Majorana spinor is also invariant under internal transformations, and not only under Lorentz rotations, one finds a new condition: if $\delta\lambda^a = M^a_b\lambda^b$, then

$$\delta(\bar{\lambda}^D)^a_{\alpha} = \delta(\bar{\lambda}^M)^a_{\alpha} \quad \text{leads to} \quad (M_a{}^d)^* = \Omega^{-1}{}_{cd}M_b{}^c\Omega^{ab}$$

where $\Omega^{-1}{}_{ab}\Omega^{bc} = \delta^c_a$. Since unitarity demands that M be antihermitian ($M_a{}^b + (M_b{}^a)^* = 0$), we deduce that the internal symmetry group is restricted to

$$\Omega M + M^T\Omega = 0. \tag{11}$$

Going through the same steps as above (9), we now find the condition

$$(CB^{-1}C^{*}B^{*-1})(\Omega\Omega^{*}) = 1 \tag{9a}$$

In the applications we consider, $\Omega$ is a real antisymmetric matrix (a symplectic metric), and (11) shows that the internal symmetry group is USp(N). Thus, in five dimensions, $\Omega$ is a $2k \times 2k$ matrix for 2k-extended supergravity.

In eleven dimensions, (7) is satisfied if $\alpha = -1$ and (9) is satisfied without an internal index. Thus, in $d = 11$ one can expect an $N = 1$ supergravity (a supergravity theory with $N = 1$ gravitinos), but in $d = 5$ the lowest theory has $N = 2$. This is indeed the case, no $N = 1$ theory exists in $d = 5$.

Back finally to the Wess-Zumino model. We now know that $\bar{\lambda}_\alpha$ in (1) stands for $\lambda^\beta C_{\beta\alpha}$, and all symbols are defined. We could have added to (1) mass terms and self-couplings, which are all separately globally supersymmetric, but to keep things simple we shall not do so.

The Wess-Zumino action is invariant under

$$\delta A = \tfrac{1}{2}\bar{\varepsilon}\lambda, \quad \delta B = -\frac{i}{2}\,\bar{\varepsilon}\gamma_5\lambda, \quad \delta\lambda = \tfrac{1}{2}(\partial\!\!\!/(A - i\gamma_5 B))\varepsilon \tag{12}$$

as we shall show in a moment. In general, fermion fields F and boson fields B rotate into each other as (generically, suppressing indices etc, compare with examples in (12))

$$\delta B \sim F\varepsilon, \qquad \delta F \sim (\partial_\mu B)\varepsilon \tag{13}$$

It is easy to see why these rules are as they are. Since bosons are commuting and fermions anticommuting, covariance of (12) and (13) requires that $\varepsilon^\alpha$ be an anticommuting number (a Grassmann variable. N. B. That Einstein's mathematical mentor Grosman is not the same as Grassmann, otherwise Einstein might have invented supergravity. As it is, he stopped at $N = 0$ supergravity, which is general relativity with no spin 3/2 gravitinos). Since the spin-statistics theorem tells us that bosons have integer spins and fermions have half-integer spins, covariance of (13) requires that the $\varepsilon^\alpha$ transform as spinors under Lorentz transformations. In principle they could have spin 3/2 or higher, but in that case it has been shown that one ends up with theories with negative energies (ghosts). Thus the simplest case, spin ½, is the only acceptable case. We have the anticommutation rules

$$\{\varepsilon^\alpha,\varepsilon^\beta\} = \{\varepsilon^\alpha,\lambda^\beta\} = [\varepsilon^\alpha,A] = 0. \tag{14}$$

In fact, one must also require that $[A,B] = 0$ and $\{\lambda^\alpha,\lambda^\beta\} = 0$ since if one would require that for real A one has instead $\{\partial_\nu A,\partial_\mu A\} = 0$, then the Klein-Gordon Lagrangian density would be a total derivative. Similarly, $[\lambda^\alpha,\lambda^\beta] = 0$ would lead for a Majorana spinor to a vanishing Dirac action. (In quantum supergravity the normalization $N = (\det \not{D})^{-\frac{1}{2}}$ of the weight function $\int db\, \exp(\bar{b}\not{D}b)$ with anticommuting Majorana fields b would yield a vanishing action if one were to exponentiate N by means of Majorana spinors. However, replacing $(\det \not{D})^{-\frac{1}{2}}$ by $(\det \not{D})^{\frac{1}{2}}(\det \not{D})^{-1}$, one has no problems: an anticommuting Majorana and a commuting Dirac ghost appear. Thus the statistics problem of Nielsen and Kallosh ghosts is trivially solved).

The most important property of (13) is the derivative. From $\delta B \sim F\varepsilon$ it follows that $\varepsilon$ has dimension $-\frac{1}{2}$. Indeed, since actions are dimensionless (if $h = c = 1$), boson fields have dimension +1 (if derivatives have dimension +1), while fermion fields have dimension 3/2. But consider now the relation $\delta F \sim B\varepsilon$. There is a gap of one unit of dimension, and in flat space without dimensional (gravitational coupling) constant and without mass parameters, the only object which can fill this gap is a derivative.

The commutator of two consecutive fermi-bose rotations of a boson, for example, leads thus back to a boson (always to the same boson) but with a derivative . In other words, two "internal" fermi-bose rotations lead to a translation, and one has here a unification of internal and spacetime symmetries. In every model of global supersymmetry one always finds the same relation

$$[\delta_{sup}(\varepsilon_1),\delta_{sup}(\varepsilon_2)]\phi = \tfrac{1}{2}(\bar{\varepsilon}_2\gamma^\mu\varepsilon_1)\partial_\mu\phi \tag{15}$$

where $\phi$ is any field. We will demonstrate this general result for the Wess-Zumino model (as well as the invariance of the action under (12)) presently, but first we go on to derive a technical detail which we will need for that purpose.

Recalling that spinors anticommute and $C\gamma^{\mu}C^{-1} = \alpha\gamma^{\mu,T}$ (where $\gamma^{\mu}$ are the generators of a Clifford algebra in d-dimensions) it is easy to derive that

$$\bar{\chi}\gamma^{\mu_1}\cdots\gamma^{\mu_n}\lambda = (\alpha)^n\bar{\lambda}\gamma^{\mu_n}\cdots\gamma^{\mu_1}\chi \tag{16}$$

The only thing one needs is that C is always antisymmetric in 4, 5, and 11 dimensions. If there are internal indices, we use (10) and $\lambda^a \equiv \Omega^{ab}\lambda_b$, $\bar{\lambda}_b \equiv \lambda_b C$, to prove that instead of (16) one has

$$\bar{\chi}^a\gamma^{\mu_1}\cdots\gamma^{\mu_n}\lambda_b = (\alpha)^n\bar{\lambda}_b\gamma^{\mu_n}\cdots\gamma^{\mu_1}\chi^a. \tag{16a}$$

We are now in a position to conclude this first lecture on global supersymmetry by proving the invariance of the action and the algebraic relation in (15). Varying the field A according to $A = \frac{1}{2}\bar{\varepsilon}\lambda$, $\delta\mathcal{L}$(Klein-Gordon) yields

$$(-\partial^{\mu}A)(\tfrac{1}{2}\bar{\varepsilon}\partial_{\mu}\lambda) = \tfrac{1}{2}(\Box A)(\bar{\varepsilon}\lambda) - \tfrac{1}{2}\partial_{\mu}(\bar{\varepsilon}\lambda\partial^{\mu}A) \tag{17}$$

Varying $\mathcal{L}$ (Dirac), we note that with (16)

$$(\delta\bar{\lambda})\not\partial\lambda = (\delta\lambda^T)C\not\partial\lambda = \bar{\lambda}\not\partial\delta\lambda - \partial_{\mu}(\bar{\lambda}\gamma^{\mu}\delta\lambda). \tag{18}$$

Thus the variation of $\bar{\lambda}$ is simply equal to the variation of $\lambda$, except for a total derivative. Inserting $\delta\lambda = \frac{1}{2}(\not\partial A)\varepsilon$ it is clear that all A terms cancel in the action. Similarly all B-terms cancel. The Lagrangian density varies into a total derivative

$$\delta\mathcal{L} = \partial_{\mu}\kappa^{\mu}, \quad \kappa^{\mu} = -\tfrac{1}{2}\bar{\varepsilon}\lambda\partial^{\mu}A + \frac{i}{2}\bar{\varepsilon}\gamma_5\lambda\partial^{\mu}B + \tfrac{1}{4}\bar{\lambda}\gamma^{\mu}\not\partial(A - i\gamma_5 B)\varepsilon \tag{19}$$

Again we use (16) to rewrite this as

$$\kappa^{\mu} = \bar{\varepsilon}\gamma^{\mu}(\not\partial(A-i\gamma_5 B))\lambda \tag{20}$$

We will use this result in the next lecture.

To prove (15) for the case $\phi = A$, we note that

$$\delta(\varepsilon_1)\delta(\varepsilon_2)A - 1 \leftrightarrow 2 = \tfrac{1}{4}\bar{\varepsilon}_2 \not{\partial}(A - i\gamma_5 B)\varepsilon_1 - 1 \leftrightarrow 2 \tag{21}$$

Again with (16) it follows that $\bar{\varepsilon}_2\gamma^\mu\gamma_5\varepsilon_1 = \bar{\varepsilon}_1\gamma^\mu\gamma_5\varepsilon_2$ so that only the A-term survives in (21). For B (15) also is satisfied. For $\lambda$ it is only satisfied if the $\lambda$ field equation $\not{\partial}\lambda = 0$ holds, and this we will discuss in the next lecture.

LECTURE II. THE GAUGE ACTION OF SIMPLE SUPERGRAVITY

(Freedman, van Nieuwenhuizen, & Ferrara, 1976a and Deser & Zumino, 1976 ) In the last lecture we found a model with a global symmetry. Just as in the electromagnetic case, we expect to need gauge fields if we want to extend the global invariance of this matter system to a local gauge invariance. Which gauge fields do we expect?

First of all we expect the supersymmetry gauge field. Since gauge fields transform always into the derivative of the corresponding parameter plus more, we expect that the gauge field of supersymmetry is a vectorial spinor (or a spinorial vector) which transforms under local supersymmetry as

$$\delta\psi_\mu^\alpha = \partial_\mu \varepsilon^\alpha + \text{more} \tag{1}$$

This field contains spin 3/2 and is called the gravitino. However, for local $\varepsilon^\alpha(x)$, one would expect from (15) that in the commutator of two local supersymmetry transformations one finds a term $\frac{1}{2}\varepsilon_2(x)\gamma^\mu\varepsilon_1(x)\partial_\mu\phi$. This looks like a translation of $\partial_\mu$ but over distances $d^\mu(x) = \frac{1}{2}\varepsilon_2(x)\gamma^\mu\varepsilon_1(x)$ which differ from point to point. If one is sufficiently naive, one recognizes the notion of a general coordinate transformation, and one expects that curved space and hence gravity enter the stage. Of course this heuristic argument need not be correct (in curvilinear coordinates translations have also $d^\mu(x)$ but one can still be in flat space), but the correct end result confirms this expectation. Thus gravity is expected to be present, and since there are fermions in supergravity (it is, after all, the gauge theory of fermi-bose symmetry) we describe the gravitational field by vierbein fields $e_\mu^m$.

It is tempting to conjecture that the gravitino is the fermionic companion of the graviton. The graviton contains two states with helicities ±2, and the gravitino must thus also contain two states. Since its highest spin is 3/2

$$\text{spin } (\mu \otimes \alpha) = (1 + 0) \otimes \frac{1}{2} = \frac{3}{2} \oplus \frac{1}{2} \oplus \frac{1}{2} \tag{2}$$

it is likely that it contains two states with helicities ±3/2. The most general action for a Majorana gravitino which has only first order poles in its propagator and positive residue is unique, and given by

$$\mathcal{L} = -\frac{1}{2}\bar{\psi}_\mu \Gamma^{\mu\rho\sigma}\partial_\rho\psi_\sigma, \qquad \bar{\psi}_{\mu,\alpha} = \psi^\beta_\mu C_{\beta\alpha} \tag{3}$$

where $\Gamma^{\mu\rho\sigma}$ is a product of three Dirac matrices antisymmetrized in $\mu\rho\sigma$ with strength one. It is easy to show that (3) indeed leads to only two physical states with helicities $\pm 3/2$. The result in (3) holds in any dimension.

Since gravity is present, we also expect to have to deal with local Lorentz rotations. Thus we need spin connections. Which spin connection one takes is at this point not clear, any spin connection $\omega_\mu^{mn}$ which transforms under local Lorentz rotations as $\delta\omega_\mu{}^{mn} = -\partial_\mu\lambda^{mn}$ + more is acceptable. But one thing is clear: since we have already equal amounts of bosonic and fermionic states, the spin connection must not be propagating, since otherwise the balance would be upset. This rules out a gravitational action quadratic in curvatures, so that only the Hilbert action

$$R(e,\omega) = R_{\mu\nu}{}^{mn}(\omega)e_n{}^\mu e_m{}^\nu$$
$$R_{\mu\nu}{}^{mn} = (\partial_\mu\omega_\nu{}^{mn} + \omega_\mu{}^{mk}\omega_{\nu k}{}^n) - (\mu\leftrightarrow\nu) \tag{4}$$

can be used. (In conformal supergravity one finds instead of (3) a higher derivative action for the gravitino which contains more fermionic states than (3)).

Putting all these arguments together, and replacing ordinary derivatives by gravitationally covariant derivatives, we arrive at the following action

$$\mathcal{L} = -\frac{e}{2\kappa^2}R(e,\omega) - \frac{e}{2}\psi_\mu\Gamma^{\mu\rho\sigma}D_\rho\psi_\sigma$$
$$D_\rho\psi_\rho = \partial_\rho\psi_\rho + \frac{1}{2}\omega_\rho{}^{mn}\sigma_{mn}\psi_\sigma \tag{5}$$

Note that $\Gamma^{\mu\rho\sigma}$ depends on $e_\mu^m$ and that no Christoffel symbol is present in (5). (It need not be since only a curl appears in (5), but it might have been added). The spin connection is now expressed as a function of $e_\nu^n$ and $\psi_\mu^\alpha$ by solving its nonpropagating field equation (Palatini formalism). One finds (putting the gravitational constant $\kappa$ equal to 1)

$$\omega_\mu{}^{mn} = \omega_\mu{}^{mn}(e) + \tfrac{1}{4}(\bar{\psi}_\mu\gamma^m\psi^n - \bar{\psi}_\mu\gamma^n\psi^m + \bar{\psi}^m\gamma_\mu\psi^n) \tag{6}$$

Remarkably enough, the action in (5) is indeed the complete gauge action. It is invariant under

$$\delta\psi_\mu = D_\mu\varepsilon \equiv \partial_\mu\varepsilon + \tfrac{1}{2}\,\omega_\mu{}^{mn}\sigma_{mn}\varepsilon \tag{7}$$

$$\delta e^m{}_\mu = \tfrac{1}{2}\,\bar{\varepsilon}\,\gamma^m\psi_\mu \tag{8}$$

The result in (7) is the minimal extension of (1), but the result in (8) is not so obvious. It follows however, straight-forwardly from group theory (see lecture 6). A simpler argument for (8) is to try to prove the invariance of (5). Varying $\psi_\sigma$ one finds a commutator of two covariant derivatives, and hence a Riemann curvature

$$-\frac{e}{2}\,\bar{\psi}_\mu\Gamma^{\mu\rho\sigma}\,\tfrac{1}{4}(R_{\rho\sigma}{}^{mn}\sigma_{mn}\varepsilon) \tag{9}$$

On the other hand, varying $\bar{\psi}_\mu$ and partially integrating, one finds

$$+\frac{e}{2}\,\bar{\varepsilon}\,\Gamma^{\mu\rho\sigma}\,\tfrac{1}{4}(R_{\mu\rho}{}^{mn}\sigma_{mn}\psi_\sigma) \tag{10}$$

Using eq.(16) of lecture 1, it follows that the sum of (9) and (10) is equal to

$$+\frac{e}{8}\,\bar{\varepsilon}\,\{\Gamma^{\mu\rho\sigma},\sigma_{mn}\}\psi_\sigma R_{\mu\rho}{}^{mn} \quad . \tag{11}$$

In the anticommutator, only tensors with one and five gamma matrices remain, and the latter are proportional to torsion and hence of higher order in $\psi$ (as are, by the way, the terms due to varying the vierbeins in the spin 3/2 action and the tensors $D_\mu e^m_\nu - D_\nu e^m_\mu$ produced by partially integrating). All that remains is

$$-\tfrac{1}{2}\,e(\bar{\varepsilon}\gamma_m\psi_\mu)G^{m\mu} \tag{12}$$

where $G_{\mu\nu}$ is the Einstein tensor $R_{\mu\nu}-\tfrac{1}{2}g_{\mu\nu}R$. But if we vary the vierbeins in the Hilbert action we get, of course, also the Einstein tensor

$$\delta(-\frac{e}{2}\,R(e,\omega)) = e(G^{m\mu})(\delta e_{m\mu}) \tag{13}$$

It is now easy to see that the sum of (12) and (13) cancels if we choose the transformation law of the vierbein as in (8). This is thus a low-brow derivation of the vierbein transformation law.

We could stop here lecture II, because we have found the action of N=1 supergravity (supergravity with one gravitino, also called simple supergravity). All that remains to be done is to show that also the higher order terms in the variation of the action cancel. An enormous simplification occurs if one uses an observation which is so simple that it is often not understood. Namely, in the action one need not vary $\omega_\mu{}^{mn}(e,\psi)$ according to the chain rule, since this variation is always multiplied by $\delta I/\delta\omega_\mu{}^{mn} = 0$ (where I denotes the action). In other words, one can substitute for $\delta\omega_\mu{}^{mn}$ anything one likes, in particular $\delta\omega_\mu{}^{mn} = 0$. Note that what one really does in this 1.5 order formalism (as it is called) is to first replace everywhere in the action $\omega_\mu{}^{mn}$ by $\omega_\mu{}^{mn}(e,\psi)$, but not to expand in terms of e and $\psi$, but to leave the object $\omega_\mu{}^{mn}(e,\psi)$ as a black box in the action. Then, upon varying this black box, it is always multiplied by terms which are identically zero. So, one is still off-shell with respect to $e_\mu^m$ and $\psi_\mu{}^a$ .

We will not prove the invariance of the gauge action to all orders in $\psi_\mu{}^a$ . Suffice it to say that in the proof of Freedman, Ferrara and the author (1976a) a computer had to be used in order to show that 4000 terms cancelled, while in the simplifying reformulation of Deser and Zumino (1976), the field $\omega_\mu{}^{mn}$ was taken as an independent field whose transformation law was very complicated. The final reformulation based on 1.5 order formalism (Townsend & van Nieuwenhuizen (1977). Chemseddine and West, (1977) found $\delta\omega_\mu{}^{mn} = 0$ from group theory, but used first order formalism.) combines the virtues of both approaches, and makes the proof so simple that we leave it as an exercise. It is, however, in higher dimensions that 1.5 order formalism becomes indispensable as we shall see.

Let us now discuss the algebra. First of all the algebra of global supersymmetry. Going back to (12) of lecture 1, we derive that

$$[\delta(\varepsilon_1),\delta(\varepsilon_2)]\lambda = \tfrac{1}{2}\,\not\partial((\tfrac{1}{2}\bar{\varepsilon}_1\lambda) - i\gamma_5(-\tfrac{i}{2}\,\bar{\varepsilon}_1\gamma_5\lambda))\varepsilon_2 - 1\leftrightarrow 2$$

$$= \tfrac{1}{4}(\bar{\varepsilon}_1\partial_\mu\lambda)(\gamma^\mu\varepsilon_2) - \tfrac{1}{4}(\bar{\varepsilon}_1\gamma_5\partial_\mu\lambda)(\gamma^\mu\gamma_5\varepsilon_2) - 1\leftrightarrow 2 \tag{14}$$

Since, as we shall show below, for any spinors the following Fierz rearrangement formula holds in four dimensions

$$(\bar{\chi}M\lambda)(N\phi) = -\tfrac{1}{4}\sum_j(\bar{\chi}O_j\phi)(NO_jM\lambda)$$

$$O_j = \{1,\gamma_\mu,\ i\gamma_\mu\gamma_5,\gamma_5, 2\,i\sigma_{\mu\nu}\ \text{with}\ \mu>\nu\} \tag{15}$$

we can rewrite (14) as

$$-\frac{1}{16}\,(\bar{\varepsilon}_1 O_j\varepsilon_2 - \bar{\varepsilon}_2 O_j\varepsilon_1)(\gamma^\mu O_j\partial_\mu\lambda - \gamma^\mu\gamma_5 O_j\gamma_5\partial_\mu\lambda) \tag{16}$$

The first factor vanishes for $O_j = \{1,\gamma_5,\ i\gamma_\mu\gamma_5\}$, while in the second factor the terms with $O_j = \sigma_{\mu\nu}$ vanish, since $[\sigma_{\mu\nu},\ \gamma_5] = 0$. All that remains is

$$+\frac{1}{8}\,(\bar{\varepsilon}_2\gamma^\tau\varepsilon_1)(\gamma^\mu\gamma_\tau - \gamma^\mu\gamma_5\gamma_\tau\gamma_5)\partial_\mu\lambda$$

$$= \frac{1}{4}\,(\bar{\varepsilon}_2\gamma^\tau\varepsilon_1)(\gamma^\mu\gamma_\tau\partial_\mu\lambda) = \tfrac{1}{2}\,(\bar{\varepsilon}_2\gamma^\mu\varepsilon_1)\partial_\mu\lambda - \tfrac{1}{4}(\bar{\varepsilon}_2\gamma^\tau\varepsilon_1)(\gamma_\tau\not\partial\lambda) \tag{17}$$

Thus, as mentioned in lecture I, only when $\not\partial\lambda = 0$, is the algebra on $\lambda$ the same as on A and B. One can, however, add to the globally supersymmetric Wess-Zumino model in the action a term $\tfrac{1}{2}(F^2+G^2)$ and modify the transformation rules such that the commutator on A and B is not modified, but the $\not\partial\lambda$ terms in (17) disappear. One may check that

$\delta A$ and $\delta B$ unchanged

$$\delta\lambda = \not\partial(A - i\gamma_5 B)\varepsilon + (F+i\gamma_5 G)\varepsilon \tag{18}$$

$$\delta F = \bar{\varepsilon}\not\partial\lambda,\quad \delta G = i\bar{\varepsilon}\gamma_5\partial\lambda$$

does the trick. Since F=G=0 is a field equation, and field equations rotate into field equations, it is clear that $\delta F$ and $\delta G$ must be proportional to $\not{\partial}\lambda$. In fact, it is hard to miss the additions in the terms in (18). From the counting of field components, we know that we need two extra bosonic fields, and it is hard to think of something else than F and G. F and G are the auxiliary fields of the Wess-Zumino model. They do not contain physical modes so that the counting of states is unaltered, but it is only with auxiliary fields that one has equal numbers of bosonic and fermionic fields. In $\delta\lambda$ one can only allow a term with $\varepsilon$ or $\gamma_5\varepsilon$ or $\gamma_\mu\gamma_5\varepsilon$ in order that the algebra on A and B remains unchanged. We leave it to the reader to figure out for himself the rest of the arguments which lead to (18).

We now do the same thing for the gauge algebra, that is to say, we first count field components. There are $16e_\mu^m$-4 general coord. inv-6 local Lorentz = 6 bose fields and $16\psi_\mu^a$-4 local supersymmetry = 12 fermi fields. We do not count $\omega_\mu^{mn}$ since it is not an independent field. Hence, we need 6 more bosonic fields, and we expect again that the algebra closes on $e_\nu^m$ but not on $\psi_\mu^a$. This is indeed the case. (Freedman & van Nieuwenhuizen 1976b). To derive the algebra, we note that for the vierbein one gets a covariant translation

$$[\delta(\varepsilon_1),\, \delta(\varepsilon_2)]\; e_\mu^m = \tfrac{1}{2}\,\bar{\varepsilon}_2\gamma^m D_\mu\bar{\varepsilon}_1 - \tfrac{1}{2}\bar{\varepsilon}_1\gamma^m D_\mu\varepsilon_2$$

$$= \tfrac{1}{2}\,\partial_\mu(\bar{\varepsilon}_2\gamma^m\varepsilon_1) + \tfrac{1}{4}\bar{\varepsilon}_2[\gamma^m,\sigma_{pq}]\varepsilon_1\omega_\mu{}^{pq}(\varepsilon,\psi) = D_\mu\xi^m \qquad (19)$$

with $\xi^m = \tfrac{1}{2}\bar{\varepsilon}_2\gamma^m\varepsilon_1$ and $D_\mu\xi^m = \partial_\mu\xi^m + \omega_\mu{}^{mq}\xi q$ .

It is easy to rewrite (19) as follows:

$$D_\mu\xi^m = (\partial_\mu\xi^\nu)e^m{}_\nu + \xi^\nu\partial_\nu e^m{}_\mu$$

$$+\, \omega_\mu{}^{mn}\xi_n + \xi^\nu(\partial_\mu e^m{}_\nu - \partial_\nu e^m{}_\mu) \qquad (20)$$

Now $D_\mu e^m{}_\nu - D_\nu e^m{}_\mu = \tfrac{1}{2}\bar{\psi}_\mu\gamma^m\psi_\nu$ where $D_\mu e_\nu^m = \partial_\mu e_\nu^m + \omega_\mu{}^{mn}(e,\psi)e_{n\nu}$

To see this, note the well-known result that

$$D_\mu(\omega(e))e^m{}_\nu - D_\nu(\omega(e))e^m{}_\mu = 0 \qquad (21)$$

Hence, (20) can be rewritten as

$$(\partial_\mu\xi^\nu)e^m{}_\nu + \xi^\nu\partial_\nu e^m{}_\mu - \tfrac{1}{2}\,\xi^\nu\bar{\psi}_\nu\gamma^m\psi_\mu + \xi^\nu\omega_\nu{}^{mn}e_{n\mu} \qquad (22)$$

and one recognizes a sum of a general coordinate transformation, a local Lorentz rotation, and a supersymmetry transformation

$$[\delta_{\text{sup.}}(\varepsilon_1), \delta_{\text{sup.}}(\varepsilon_2)] = \delta_{\text{gen. coord.}}(\xi^\nu) + \delta_{\text{local Lorentz}}(\xi^\nu\omega_\nu^{mn}) + \delta_{\text{sup.}}(-\xi^\nu\psi_\nu) \qquad (23)$$

To find the same result for $\psi_\mu$ one must add again auxiliary fields. The simplest choice (the so-called minimal set of auxiliary fields) consists of a scalar S, a pseudoscalar P and an axial vector $A_m$. They appear as follows in the action in (5)

$$\mathcal{L} = \frac{e}{3}(S^2 + P^2 - A_m^2) \qquad (24)$$

The new transformation rules are

$\delta e^m{}_\mu$ unchanged

$$\delta\psi_\mu = D_\mu(\omega(e,\psi))\varepsilon + \frac{i}{2}A_\mu\gamma_5\varepsilon - \frac{1}{3}\gamma_\mu\eta\varepsilon$$

$$\eta = S - i\gamma_5 P - i\not{A}\gamma_5 \qquad (25)$$

$$\delta S = \tfrac{1}{2}\bar{\varepsilon}\gamma\cdot R^{cov}$$

$$\delta P = -\tfrac{1}{2}\bar{\varepsilon}\gamma_5\gamma\cdot R^{cov}$$

$$\delta A_m = \frac{3i}{2}\bar{\varepsilon}\gamma_5(R_m^{cov} - \frac{1}{3}\gamma_m\gamma\cdot R^{cov})$$

Again $\delta S$, $\delta P$ and $\delta A_m$ are proportional to the gravitino field equation. However, it follows from (23) that if one varies $\delta P$ and $\delta A_m$ itself one should not obtain terms with $\partial_\mu \varepsilon$. In other words, $\delta S$, $\delta P$, and $\delta A_m$ must be supercovariant. Indeed, if one acts on S with the three local symmetries on the right hand side of (23), one does not find terms with a derivative of $\xi^\mu$, while the left hand side contains the variation of $\delta S$.

A supercovariant object B rotates into terms without $\partial_\mu \varepsilon$. If, for example $\delta B = (\partial_\mu \bar{\varepsilon})C^\mu$ then one can make B supercovariant by adding a term $-\bar{\psi}_\mu C^\mu$. Indeed $\delta(B-\bar{\psi}_\mu C^\mu)$ now no longer contains $\partial_\mu \varepsilon$ terms. The object $R^{\alpha,\,cov}$ is obtained from $R^\alpha = \Gamma^{\alpha\rho\sigma} D_\rho \psi_\sigma$ by supercovariantization. We leave as an exercise to obtain $R^{\alpha,cov}$ from $\delta\psi_\mu$ in (25). Note that $\omega(e,\psi)$ in (6) is supercovariant.

Finally we come back to our promise to derive the Fierz rearrangement formula. We will do it in such a way that the result also holds in d dimensions. Consider

$$(\bar{\chi}_a M^\alpha{}_\beta \lambda^\beta)(N^\gamma{}_\delta \psi^\delta) \tag{26}$$

and expand the matrix $Z^\alpha{}_\delta = M^\alpha{}_\beta N^\gamma{}_\delta$ for fixed $\beta$ and $\gamma$ into a complete orthonormal set ($\mathrm{tr} O_j O_k = \alpha\delta_{jk}$)

$$M^\alpha{}_\beta N^\gamma{}_\delta = (O_j)^\alpha{}_\delta C^j{}_\beta{}^\gamma \tag{27}$$

The coefficients C follow by multiplying by $(O_k)^\delta{}_\alpha$ and using the orthonormality of the $O_j$. One finds

$$C^j{}_\beta{}^\gamma = \frac{1}{\alpha}(NO^jM)^\gamma{}_\beta \tag{28}$$

and this is sufficient to prove (15). Note that in 5 dimensions $O_j = [1, \gamma_\mu, 2i\sigma_{\mu\nu}]$ with $\mu < \nu$.

In this lecture we have obtained the gauge action of N = 1 supergravity, see (5). The spin connection is a function of tetrad and gravitinos, obtained by solving its own field equation $\delta I/\delta\omega_\mu{}^{mn} = 0$. Using 1.5 order formalism, the results for action (5) and transformation laws (7), (8) were very simple. Note that if one would have written the derivative in the spin 3/2 action in a different position (for example, $\bar{\psi}_\mu D_\rho \Gamma^{\mu\rho\sigma}\psi_\sigma$), then one needs extra four-fermion contact terms in the action.

We now consider theories with more than one gravitino and more than one local supersymmetry.

## LECTURE III. EXTENDED SUPERGRAVITIES

### 1. The N = 2, 3, 4 Models

The extended supergravities contain more than one gravitino. They are usually called N-extended supergravities where N is the number of real gravitinos. ($N = 0$ is then general relativity and $N = 1$ is simple supergravity). There are only eight viable supergravities, namely $1 \leq N \leq 8$ since for $N > 8$ one has particles with spins larger than two and more than one graviton (for $N = 9$ one has 10 gravitons, an "embarras de richesse"). Since it has been shown that the free field action for spin 5/2 is unique and cannot be coupled consistently to either gravity or to simple matter systems, it seems as if nature stops at spin 2, and supergravity at $N = 9$.

The simplest extended supergravity is the $N = 2$ model (Ferrara & van Nieuwenhuizen, 1976a). It was obtained by coupling the (2,3/2) gauge action to the (3/2,1) matter multiplet by means of the Noether method. It was only afterwards found that the model had a larger symmetry, namely (to begin with) a manifest O(2) invariance which rotates the two gravitinos into each other. This model unifies electromagnetism with gravity. The action reads

$$\mathcal{L} = -\frac{e}{2}R(e,\omega) - \frac{e}{2}\bar{\psi}^i_\mu\Gamma^{\mu\rho\sigma}D_\rho(\omega)\psi^i_\sigma - \frac{e}{4}F^2_{\mu\nu}$$
$$+ \frac{\kappa}{4\sqrt{2}}\bar{\psi}^i_\mu[e(F^{\mu\nu} + \hat{F}^{\mu\nu}) + \tfrac{1}{2}\gamma_5(\tilde{F}^{\mu\nu} + \hat{\tilde{F}}^{\mu\nu})]\psi_\nu\varepsilon^{ij} \tag{1}$$

and is invariant under general coordinate, local Lorentz and Maxwell transformations ($\delta A_\mu = \partial_\mu\Lambda$) as well as under the following <u>two</u> local supersymmetries

$$\delta e^m{}_\mu = \frac{\kappa}{2}\bar{\varepsilon}^i\gamma^m\psi^i_\mu, \quad \delta A_\mu = \frac{\kappa}{\sqrt{2}}\bar{\varepsilon}^i\psi^j_\mu\varepsilon^{ij}$$
$$\delta\psi^i_\mu = D_\mu(\omega)\varepsilon^i + \frac{\kappa}{2\sqrt{2}}\varepsilon^{ij}(\hat{F}_{\mu\lambda}\gamma^\lambda + \tfrac{1}{2}e\hat{\tilde{F}}_{\mu\lambda}\gamma^\lambda\gamma_5)\varepsilon^j \tag{2}$$

Indeed, each gravitino gauges one local supersymmetry as one sees from $\delta\psi_\mu{}^i = \partial_\mu\varepsilon^i$ + more. The symbol $\tilde{F}_{\mu\nu}$ equals $\varepsilon_{\mu\nu\rho\sigma}F^{\rho\sigma}$ and $\hat{F}_{\mu\nu}$ is the supercovariant photon curl

$$\hat{F}_{\mu\nu} = (\partial_\mu A_\nu - \frac{\kappa}{2\sqrt{2}} \bar{\psi}^i_\mu \psi^j_\nu \varepsilon^{ij}) - (\mu \leftrightarrow \nu) \tag{3}$$

One may always take the spin connection $\omega$ in the Hilbert action as that function of tetrads and gravitinos which solves $\delta I/\delta\omega = 0$ (1.5 order formalism). If one would have started in the Hilbert action with $\omega + \tau$ where $\tau$ is arbitrary, then one finds only $\tau^2 = \psi^4$ terms upon expanding about $\omega$. These $\psi^4$ terms are absorbed into the $\hat{F}_{\mu\nu}$ tensors, and it is easy to argue why this must be possible (as of course has been checked explicitly). Since fermionic field equations rotate into bosonic field equations and the former have only one derivative while the latter have two derivatives, no derivative can act on $\varepsilon$ (since otherwise there would no longer be two derivatives available for the bosonic field equations. Schematically: $\delta(\partial F) = (\partial\partial B)\varepsilon$ while $(\partial B)(\partial\varepsilon)$ cannot occur since $\partial B$ is not a field equation. This argument breaks down when there are nonpropagating (auxiliary) fields since then B is a field equation). The gravitino field equation is obtained by varying all $\psi$ fields, and since the terms with F have four $\psi$'s which appear symmetrically, one must take one half of the former in order that the field equation only contains the supercovariant $\hat{F}_{\mu\nu}$, but not, for example bare $F_{\mu\nu}$'s. Note that this argument also tells one that the spin connection in the gravitino field equation must be supercovariant. As it happens, $\omega$ is itself supercovariant, hence both in Hilbert and gravitino action one finds $\omega$. (In other cases, for example in d = 11 or d = 5 dimensions, one finds in the gravitino action $\frac{1}{2}(\omega + \hat{\omega})$ instead of $\omega$, but, based on our argument given before, one always chooses $\omega$ in the Hilbert action.)

The same kind of argument shows that if auxiliary fields are absent the gauge algebra must close on bosonic fields, and this is also a helpful criterion in constructing theories (It is difficult to credit one particular person, but certainly J. Scherk gave an important contribution).

This model was the model where finite quantum corrections were found for the first time (See Grisaru, van Nieuwenhuizen & Vermaseren (1976). Some chiral-dual symmetries were first found in another model which was not finite but displayed miraculous cancellations. (Vermaseren & van Nieuwenhuizen, 1976). For a complete treatment of dual-chiral symmetries, see the next paragraph. An explanation of this finiteness was given by using certain symmetries in this model which we now discuss.

According to the analysis of Haag, Lopuszanski and Sohnius (1975), the maximal symmetry group of the S-matrix of N = 2 Poincaré supergravity is U(2). In fact, the field theory has the global symmetry only on-shell (i.e., only the field equations have a U(2) global symmetry). Off-shell, only an SU(2) remains. These symmetries read (Ferrara, Scherk & Zumino, 1977a)

$$\begin{aligned} \delta\psi^{L} &= i\vec{\omega}\cdot\vec{\tau}\psi^{L}_{\mu} \\ \delta\psi^{R}_{\mu} &= -i\vec{\omega}\cdot\vec{\tau}\psi_{\mu} \end{aligned} \quad , \quad \psi_{\mu} = \begin{pmatrix} \psi^{1}_{\mu} \\ \psi^{2}_{\mu} \end{pmatrix} \quad , \quad \begin{aligned} \psi^{L}_{\mu} &= \tfrac{1}{2}(1+\gamma_5)\psi_{\mu} \\ \psi^{R}_{\mu} &= \tfrac{1}{2}(1-\gamma_5) \end{aligned}$$

$$\mathrm{U}(1):\quad \delta\psi_{\mu} = i\gamma_5\psi_{\mu} \text{ and } \delta\hat{F}_{\mu\nu} = ie\varepsilon_{\mu\nu\rho\sigma}\hat{F}^{\rho\sigma} \qquad (4)$$

Thus the SU(2) part rotates $\psi^{L}$ as $\underset{\sim}{(2)}$ and $\psi^{R}$ as $\underset{\sim}{(\bar{2})}$, while the U(1) part is a combined chirality-duality transformation. To prove the U(2) invariance, note that in the gravitino action $\psi_{L}$ couples to $\psi^{T}_{R}C$ so that these terms are SU(2) and of course chirally (U(1)) invariant. The torsion is separately U(2) invariant, and, finally, the remaining terms can be analyzed by eliminating the $F_{\mu\nu}$ kinetic terms through the equation of motion for $F_{\mu\nu}$. This cancels half of the Noether coupling with bare $F_{\mu\nu}$ and the result is that the action is the sum of the tetrad and gravitino actions plus the following term

$$\frac{\kappa}{4\sqrt{2}}\bar{\psi}^{i}_{\mu}\,[e\hat{F}^{\mu\nu} + \tfrac{1}{2}\gamma_5\hat{\tilde{F}}^{\mu\nu}]\psi^{j}_{\nu}\varepsilon^{ij} \qquad (5)$$

Clearly, also this last term is U(2) invariant. The SU(2) invariance holds off-shell, but for $N > 2$ only the O(N) invariance holds off-shell. The chiral-dual invariance never holds off-shell.

An interesting connection between the coupling of photons and the cosmological constant was found by Das and Freedman (1977a) and Fradkin & Vassiliev (1976). When one couples the photons minimally to the fermions one needs at the same time a cosmological constant and a mass-like term in the action. Thus it would seem that electromagnetism is due to the curvature of a de-Sitter universe. Actually, this cosmological constant is much too large, and can be eliminated altogether by spontaneous symmetry breaking as we shall discuss.

For the N = 2 model one can find this extension of the action by adding a cosmological constant to the action and finding the further modifications by the Noether method. The complete set of extra terms is given by

$$\mathcal{L}(\cos m) = 6eg^2 + 2e\bar{\psi}^i_\mu\sigma^{\mu\nu}\psi^i_\nu - \tfrac{1}{2}\bar{\psi}^i_\mu\Gamma^{\mu\rho\sigma}(D_\rho\psi^i_\sigma + g\varepsilon^{ik}A_\rho\psi^k_\sigma)$$

$$\delta\psi^i_\mu = D_\mu(\omega(e,\psi))\,\varepsilon^i + g\gamma_\mu\varepsilon^i + g\varepsilon^{ik}A_\mu\varepsilon^k \tag{6}$$

where we added the $D_\mu$ terms for comparison. Clearly, g is a dimensionless gauge coupling, but it is remarkable that also a mass-like term is needed (as well as a cosmological term) when one couples the photon to the gravitinos in an electromagnetic manner. Note that the theory still has the same number of local invariances. The mass term is actually needed in order that $\psi^i_\mu$ still be massless in de-Sitter space (Deser & Zumino, 1977), and for $g \neq 0$ there is only an $SO_2$ symmetry on and off-shell, since the mass term breaks the $SU_2$ invariance.

In fact, even in simple (N = 1) supergravity one needs a mass term if one adds a cosmological term even though here there is no photon (Townsend, 1977b). As the reader may easily verify, the action

$$\mathcal{L} = -\frac{e}{2} R(e,\omega) - \tfrac{1}{2} \bar{\psi}_\mu \Gamma^{\mu\rho\sigma} D_\rho \psi_\sigma + 6eg^2 + 2eg\bar{\psi}_\mu \sigma^{\mu\nu} \psi_\nu \tag{7}$$

is invariant under $\delta e^m_\mu = \tfrac{1}{2}\bar{\varepsilon}\gamma^m\psi_\mu$ and $\delta\psi_\mu = (D_\mu + g\gamma_\mu)\varepsilon$. This cosmological term plus mass term follows easily from the tensor calculus; it is simply the action belonging to the unit scalar multiplet $\Sigma = (1,0,0,0,0)$. This cosmological constant is positive because in the N = 1 gauge action one finds $-(1/3)S^2 + \alpha S$ with $\alpha$ arbitrary. On the other hand, in the Wess-Zumino model one finds a term $+ \tfrac{1}{2}F^2$ (with F the usual auxiliary field) and by coupling this model to supergravity, one can cancel the cosmological term (Cremmer, Julia, Scherk, Ferrara, Girardello & van Nieuwenhuizen, 1979a).

The next extended supergravity we discuss is the N = 3 model. It was obtained by Ferrara, Scherk & Zumino (1977b) and Freedman (1977). Here a new feature is that the Abelian photon gauge invariance turns into a non-Abelian Yang-Mills invariance in de-Sitter space. In other words, as in the N = 2 case, coupling the triplet of photons to the gravitinos, one needs at the same time a cosmological constant and a mass term. The action reads

$$\begin{aligned}\mathcal{L} &= \mathcal{L}(\text{kinetic terms for } e^m_\mu, \psi_\mu{}^i, A_\mu^i, \lambda \text{ with } i = 1,3) \\ &+ \tfrac{1}{2}\left(\mathcal{L}^{\text{Noether}}_{\text{bare}} + \mathcal{L}^{\text{Noether}}_{\text{supercov.}}\right) + \left(\text{eq(6) with } \varepsilon^{ik}A_\mu \to \varepsilon^{ijk}A^j_\mu\right)\end{aligned} \tag{8}$$

where the spin connection depends on $\psi_\mu{}^i$ <u>and $\lambda$</u> and solves $\delta I/\delta\omega = 0$. (We thus again use 1.5 order formalism). The bare (i.e. non-supercovariantized) Noether coupling is a sum of the Noether couplings of the (3/2,1) and (1,½) systems to the (2 , 3/2) system.

$$\begin{aligned}\mathcal{L}^{\text{Noether}}_{\text{bare}} &= -\frac{1}{2\sqrt{2}} \bar{\psi}^i_\mu (eF^{\mu\nu,j} + \tfrac{1}{2}\gamma_5 \tilde{F}^{\mu\nu,j})\psi^k_\nu \varepsilon^{ijk} \\ &+ \tfrac{1}{2}(\bar{\psi}^i_\mu \sigma^{\alpha\beta}\gamma^\mu\lambda)(F^i_{\alpha\beta})\end{aligned} \tag{9}$$

The supercovariantized Noether coupling follows from the local supersymmetry transformation rules, which read

$$\delta e^m_\mu = \tfrac{1}{2}\bar{\varepsilon}^i\gamma^m\psi^i_\mu\ , \qquad \delta\lambda = \tfrac{1}{2}(\sigma^{\mu\nu}\varepsilon^i)(F^i_{\mu\nu})$$

$$\delta A^i_\mu = \frac{1}{\sqrt{2}}\,\varepsilon^{ijk}\bar{\varepsilon}^j\psi^k_\mu - \tfrac{1}{2}\bar{\varepsilon}^i\gamma_\mu\lambda$$

$$\delta\psi^i_\mu = D_\mu(\omega(e,\psi,\lambda))\varepsilon^i + \frac{1}{2\sqrt{2}}\,\varepsilon^{ijk}(\sigma^{\rho\sigma}\gamma_\mu\varepsilon^k)(\hat{F}^j_{\rho\sigma}) \tag{10}$$

$$+ \frac{1}{4\sqrt{2}}\,\varepsilon^{ijk}[(\bar{\psi}^j_\mu\gamma_\rho\lambda)(\gamma^\rho\varepsilon^k) + (\bar{\psi}^j_\mu\gamma_\rho\gamma_5\lambda)(\gamma_5\gamma^\rho\lambda)\varepsilon^k)]$$

$$+ \frac{1}{8}(\lambda\gamma_5\gamma^\rho\lambda)(\gamma_\rho\gamma_\mu\gamma_5\varepsilon^i) + g\varepsilon^{ijk}A^j_\mu\varepsilon^k + g\gamma_\mu\varepsilon^i$$

Note that the expression for $\delta\psi_\mu$ is itself no longer supercovariant. The noncovariant terms involve the spin ½ fields, and this is a general (but not understood) feature: without spin ½ fields, all transformation rules are always supercovariant. This means that there must be extra terms in the commutator of an extra local Lorentz transformation with parameter

$$(2\sqrt{2})^{-1}\varepsilon^{ijk}(\bar{\varepsilon}^i_2\varepsilon^k_1\hat{F}^j_{mn} + \frac{e}{2}\,\bar{\varepsilon}^i_2\gamma_5\varepsilon^k_1\hat{\tilde{F}}^j_{mn} \tag{11}$$

and an extra local supersymmetry transformation with parameter

$$(2\sqrt{2})^{-1}\varepsilon^{ijk}(\bar{\varepsilon}^j_2\varepsilon^k_1\lambda - \bar{\varepsilon}^j_2\gamma_5\varepsilon^k_1\gamma_5\lambda) \tag{12}$$

and an O(3) Yang-Mills gauge transformation with parameter

$$\Lambda^i = (2\sqrt{2})^{-1}\varepsilon^{ijk}(\bar{\varepsilon}^j_2\varepsilon^k_1) - \tfrac{1}{2}\bar{\varepsilon}^j_2\gamma^\mu\varepsilon^j_1A^i_\mu$$

$$\delta A^i_\mu = \partial_\mu\Lambda^i + g\varepsilon^{ijk}A^j_\mu\Lambda^k, \qquad \delta e^m_\mu = 0 \tag{13}$$

$$\delta\psi^i_\mu = g\varepsilon^{ijk}\psi^j_\mu\Lambda^k, \qquad \delta\lambda = 0$$

Note that the parameter composition rules do not depend on g (or only through $F^j_{mn}$ as in (11)) although the transformation rules depend on g.

There is a global U(3) invariance for g = 0 on-shell. Off-shell only the O(3) invariance remains, since all other symmetries involve not only chiral transformations but also duality transformations. The precise rules follow by truncation from the U(4) group of the SO(4) version of the N = 4 model. (See below.) For $g \neq 0$, the off-shell O(3) invariance is local, but since the Yang-Mills coupling involves bare $A_\mu^i$ fields and the mass term breaks chiral invariance, the U(3) invariance is again lost for $g \neq 0$ on- and off-shell.

The gauge parameter in (13) contains a nonvanishing term in the global limit (when fields are set equal to zero). If one gauges an algebra, the commutator of two local symmetries with parameters $\varepsilon^i$ and $\eta^j$ is again a local symmetry, with parameter $\zeta^k = f^k{}_{ij}\eta^j\varepsilon^i$ where $f^k{}_{ij}$ are the structure constants $[X_i, X_j] = f_{ij}{}^k X_k$. Hence one finds on the right hand side of $\{Q_i^\alpha, Q_j^\beta\}$ a Maxwell gauge generator, which is a central charge (as one may verify directly) in the global limit with $g = 0$, but this charge does not act on physical states. Also in the super de-Sitter algebra (with $g \neq 0$) there is an "electric" charge in the {Q, Q} anticommutator, but now this charge is no longer a central charge, since it does not commute with supersymmetry. As a result $\delta(\text{gauge})\, \psi_\mu \neq 0$, so that for $g \neq 0$ the electric charge does act on physical states.

From the N = 3 theory (or any other theory) one can obtain other theories by consistent truncation. A truncation means putting certain fields equal to zero, and consistency requires that also their variations then vanish. Two consistent truncations are: $A_\mu^1 = A_\mu^2 = \psi_\mu^3 = \lambda = 0$ which yields the N = 2 model, while $A_\mu^2 = A_\mu^3 = \psi_\mu^2 = \psi_\mu^3 = 0$ yields the $(2, 3/2) + (1, \frac{1}{2})$ Maxwell-Einstein system (Ferrara, Scherk & van Nieuwenhuizen, 1976b).

We now discuss the N = 4 extended supergravity. New here is the occurrence of scalars, for example A, which appear in a non-polynomial way. This is possible since $\kappa A$ is dimensionless. For photons, $\kappa A_\mu$ is also dimensionless, but Maxwell gauge invariance forbids an infinite series in photons (Ferrara, Gliozzi, Scherk and van Nieuwenhuizen, 1977c) while one can also show that fermion fields can appear at most quartically.

There are actually two versions of the N = 4 model known, commonly called the SO (4) model ( Das, 1977b) with fields

$(e, \psi'^{i}, V_{\mu}^{ij}, \lambda'^{i}, A, B')$ and the $SU_4$ model (Cremmer, Ferrara & Scherk, 1978a) with fields $(e, \psi^{i}, A_{\mu}^{k}, B_{\mu}^{k}, \lambda^{i}, \phi, B)$. The six fields $V_{\mu}^{ij}$ are all vector fields, but the three $A_{\mu}^{k}$ are vector fields while the $B_{\mu}^{k}$ are axial fields. One can obtain the SU(4) model by reduction of the N = 1 model in d = 10 dimensions. Spinors in d = 10 can satisfy both the Majorana condition $\bar{\lambda}^{D} = \bar{\lambda}^{M}$ and the Weyl condition $\lambda = \frac{1}{2}(1 \pm \gamma_{11})\lambda$. Thus, in d = 10, $\psi_{\mu}$ has 16 components and becomes in d = 4 equal to four gravitinos. However, one finds in d = 4 matter multiplets, and splitting in d = 4 these matter couplings from the pure N = 4 gauge action is rather hard and has not been attempted. Rather, the SU(4) model was constructed directly. The appearance of an off-shell SU(4) symmetry is obvious since the Lorentz group O(9, 1) splits into the usual Lorentz group O(3,1) of d = 4 and an O(6) which is equivalent to SU(4). Since the pseudoscalar field B is obtained by a duality transformation on an antisymmetric tensor field (which can only appear polynomially in d = 10 and d = 4), B necessarily has opposite parity and appears polynomially; however, $\phi$ does appear nonpolynomially.

At the classical level the two theories are equivalent because one can find a transformation law of the fields which turns the one action into the other. For the axial and vector fields the correspondence is the following. Denoting by $G_{\mu\nu}^{ij}$ the tensor which gives the equation of motion of $V_{\mu}^{ij}$, hence $D^{\mu}G_{\mu\nu}^{ij} = 0$ (since only the curls $F_{\mu\nu}^{ij}$ of $V_{\mu}^{ij}$ appear in the action, $G_{\mu\nu}^{ij}$ can be defined) one has for the transformation laws of the curls of $A_{\mu}^{ij}$ and $B_{\mu}^{ij}$

$$\alpha^{ij}_{n} A^{n}_{\mu\nu} = \frac{1}{\sqrt{2}} \left(F^{ij}_{\mu\nu} + \tfrac{1}{2} F^{k\ell}_{\mu\nu} \varepsilon^{ij}{}_{kl}\right)$$

$$\beta^{ij}_{n} B^{n}_{\mu\nu} = \frac{1}{\sqrt{2}} \left(G^{ij}_{\rho\sigma} - \tfrac{1}{2} G^{k\ell}_{\rho\sigma} \varepsilon^{ij}{}_{kl}\right) \varepsilon_{\mu\nu}{}^{\rho\sigma} \qquad (14)$$

where $\alpha$ and $\beta$ are certain numerical matrices. The first equations can be carried over directly onto the gauge fields themselves, but the second equation together with the Bianchi identity $D^{\mu} \tilde{B}^{n}_{\mu\nu} = 0$ yields

$$(D^{\mu}G^{ij}_{\mu\nu}) - \tfrac{1}{2}\varepsilon^{ij}{}_{k\ell}(D^{\mu}G^{k\ell}_{\mu\nu}) = 0 \qquad (15)$$

Hence, one-half of the $V_\mu^{ij}$ field equations must be satisfied and one is partly on-shell. It is now also clear why in the SU(4) model there are three axial vector fields, namely due to the $\varepsilon_{\mu\nu}{}^{\rho\sigma}$ symbol in (14).

The SU(4) model is the simpler model. In the action one only finds nonpolynomiality in the scalar field $\phi$, and only by means of exponentials of $\phi$. In the SO(4) model, on the other hand, the scalar kinetic term reads

$(1-\kappa^2(A^2+B^2))^{-1}\ [(\partial_\mu A)^2 + (\partial_\mu B)^2]$, which, incidentally, limits

the range of $\kappa A$ and $\kappa B$ betwen -1 and +1.

In the SU(4) model there is a group of global SU(4) invariances off-shell, and a further global SU(1, 1) on-shell. The off-shell SU(4) transformations transform the fields $A_\mu^k$ and $B_\mu^k$ into each other and contain also chiral rotations of $\psi_\mu{}^i$, and of $\chi^i$, while $\phi$, and B and $e^m{}_\mu$ are inert.

The SU(1, 1) global symmetry group leaves the scalar kinetic terms invariant. These are given by $(\partial_\mu Y\ \partial^\mu\bar{Y})\ (Y + \bar{Y})^{-2}$ where $Y = \exp(-2\phi) - 2iB$ and the transformations $Y\to(\alpha Y+i\beta)(i\gamma Y+\delta)^{-1}$ with $\alpha\delta+\beta\gamma = 1$ leave it invariant. To see that these transformations are really SU(1, 1), note that the transformations

$$\delta(Z) = \delta\begin{pmatrix} Z_1 \\ Z_2 \end{pmatrix} = \begin{pmatrix} \alpha & i\beta \\ i\gamma & \delta \end{pmatrix}\begin{pmatrix} Z_1 \\ Z_2 \end{pmatrix}$$

can be written as $\exp(i\omega_1\tau_1 + \omega_2\tau_2 + \omega_3\tau_3)$ and leave $z^*\tau_1 z$ invariant. Since this noncompact group is realized nonlinearly it does not lead to ghosts. Again, this SU(1, 1) contains duality transformations of the form $\delta A_{\mu\nu}{}^k = \tilde{G}_{\mu\nu}{}^k(A)$ and idem for $B_\mu{}^k$.

We now turn to the SO(4) version of the N = 4 model. Here the global SU(4) consists of a manifest SO(4) symmetry plus transformations generated by symmetric antihermitian matrices $i\Lambda^{jk}$. More in detail

$$\delta\psi_\mu^{L,j} = i\Lambda^j{}_k \psi_\mu^{L,k}$$

$$\delta\chi^{L,j} = i(\Lambda^j{}_k - \delta^j_k \, tr\Lambda)\chi^k$$

$$\delta A = B tr\Lambda, \; \delta B = - A tr\Lambda \tag{16}$$

$$\delta F^{ij}_{\mu\nu} = \tfrac{1}{2} \Lambda^i{}_k \tilde{G}^{jk}_{\mu\nu} - i\leftrightarrow j$$

The tensor $D^\mu G_{\mu\nu}{}^{jk}$ is again the field equation of $V_\nu{}^{jk}$. All SU(4) symmetries except the O(4) hold only on-shell. The same is true for the N = 3 model, as one easily sees by reduction (i.e., A = B = 0, etc.). There does not seem to be a consistent truncation to N = 2 or N = 3 supergravity theories of the SU(4) version of the N = 4 theory.

One can gauge both the SO(4) and the SU(4) models. The SO(4) model was gauged by Das, Fischler and Rocek (1977c) and has a potential with an indifferent equilibrium. Of the SU(4) model, having only 6 spin 1 field, one can at best gauge an SU(2) x SU(2) subgroup and this was done by Freedman and Schwarz (1978). However, in this case the potential has no stationary point.

Let us now jump to the N = 8 theory. One might expect that this theory is prohibitively complicated - but one can formulate it in a very simple way in d = 11 dimensions as N = 1 (simple) supergravity. If one then uses dimensional reduction to d = 4 dimensions, the full theory emerges automatically.

2. The N = 8 Model in 11 Dimensions

(Cremmer, Julia & Scherk, 1978b). Dimensional reduction means that all fields are assumed to depend only on $x^1, \ldots x^4$ instead of $x^1 \ldots x^{11}$. A tetrad in d = 11 splits then up in d = 4 into one tetrad $e^m_\mu$(m = 1, 4), 7 vectors $e^a_\mu$(a = 5, 11) and 7 x 7 scalar fields $e^a_\alpha$($\alpha$ = 5, 11) which describe ½ 8 x 7 scalar particles (because the local O(7) symmetry, the residue of the O(10, 1) Lorentz symmetry, eliminates the antisymmetric parts of $e_\alpha^a$ ). Dimensional reduction has been extremely fruitful in supergravity (and in global supersymmetry).

We now explain why the simplest form of N = 8 model is in d = 11 dimensions, and then construct this theory in d = 11 in the remainder of this section. Dimensional reduction and spontaneous symmetry breaking will be discussed in following sections.

The example of the tetrad just given shows that by going to higher dimensions, supergravity theories can be reformulated in terms of fewer fields. The N = 8 theory in 4 dimensions has 8 gravitinos, hence one expects that its simplest form appears in d = 10 or d = 11 dimensions. Indeed, in d = 10 or d = 11, spinors have 32 components. The simplest version is the d = 11 theory, with only an 11 x 11 "elfbein" $e_\mu^m$, a 32 x 11 gravitino $\psi_\mu{}^\alpha$ and a "photon" $A_{\mu\nu\rho}$ (antisymmetric in $\mu\nu\rho$). That one needs $A_{\mu\nu\rho}$ follows from counting of states: $e_\mu^m$ transversal and traceless: $\frac{1}{2}\, 9 \times 10 - 1 = 44$, $\psi_\mu{}^a$ transversal in gauge $\gamma \cdot \psi = 0$: $(9 \times 32 - 32) \times \frac{1}{2} = 128$, $A_{\mu\nu\rho}$ transversal: $\binom{9}{3} = 84$.

A choice $A_{\mu_1 \cdots \mu_6}$ has also $\binom{9}{6} = \binom{9}{3} = 84$ states, but does not lead to a consistent theory,(Nicolai, Townsend & van Nieuwenhuizen, 1980). Thus there are indeed equal numbers of bosons and fermions.

Having decided upon 11 dimensions we note that the signature of this space must be (10, 1) in order that one does not find ghosts in the dimensionally reduced action. With more than one time coordinate, some of the scalars $e_\alpha^a$ would have kinetic terms with the wrong sign.

Since $A_{\mu\nu\rho}$ is transversal it must be a gauge field with gauge invariance $\delta A_{\mu\nu\rho} = \partial_\mu \Lambda_{\nu\rho} +$ cyclic terms, with $\Lambda_{\nu\rho} = -\Lambda_{\rho\nu}$. Hence, the action must start with

$$\mathcal{L} = -\frac{e}{2} R(e,\omega) - \frac{e}{2} \psi_\mu \Gamma^{\mu\rho\sigma} D_\rho(\omega) \psi_\sigma - \frac{e}{48} F^2_{\mu\nu\rho\sigma} \tag{1}$$

where $F_{\mu\nu\rho\sigma} = \partial_\mu A_{\nu\rho\sigma}$ plus 23 terms. (The symbol $\Gamma^{\mu\rho\sigma}$ has strength one, $\Gamma^{\mu\rho\sigma} = \Gamma^\mu\Gamma^\rho\Gamma^\sigma$ antisymmetrized).

The gravitino must be real, in order that the counting of states works, or rather it must satisfy a Majorana condition. (Since d = 11 is odd, we cannot invoke a Weyl condition to halve the number of complex fields.)

One may prove in general or by giving an explicit representation that in 10 + 1 dimensions there exists a real representation and that $C\gamma_\mu C^{-1} = -\gamma_\mu^{T}$. It follows that in d= 11 just as in d = 4 (because also in d = 11 is C antisymmetric)

$$(\bar{\lambda}\Gamma^{A_1} \dots \Gamma^{A_n}\chi) = (-)^n (\bar{\chi}\Gamma^{A_n} \dots \Gamma^{A_1}\lambda) \tag{2}$$

To construct the action, we begin by postulating

$$\delta e^m_\mu = \tfrac{1}{2}\bar{\varepsilon}\gamma^m\psi_\mu, \quad \delta\psi_\mu = D_\mu(\omega)\varepsilon \tag{3}$$

Using 1.5 order formalism, one finds for the variation of the Hilbert action

$$\tfrac{1}{2}e(\bar{\varepsilon}\gamma^\nu\psi^a)(R_{a\nu} - \tfrac{1}{2}\, e_{a\nu}R) \tag{4}$$

For variation of $\psi_\sigma$ in the gravitino action, we get

$$-\frac{e}{8}\,(\bar{\psi}_\mu\Gamma^{\mu\rho\sigma}\tfrac{1}{2}\Gamma^{mn}\varepsilon)R_{\rho\sigma mn}(\omega) \quad \text{where } \tfrac{1}{2}\,\Gamma^{mn} = \sigma^{mn} \tag{5}$$

On the other hand, varying $\bar{\psi}_\mu$ and partially integrating, neglecting torsion terms

$$D_\mu e^m_\rho - D_\rho \varepsilon^m_\mu$$

since these are of higher order in $\psi$, we find

$$+ \frac{e}{8} (\bar{\varepsilon}\Gamma^{\mu\rho\sigma} \tfrac{1}{2} \Gamma^{mn}\psi_\sigma) R_{\mu\rho mn}(\omega) \tag{6}$$

With the symmetry in (2), the sum of these two variations equals

$$\frac{e}{16} \bar{\varepsilon} \{\Gamma^{\mu\rho\sigma}, \Gamma^{mn}\}\psi_\sigma R_{\mu\rho mn} \tag{7}$$

In the anticommutator only the terms with five and one gamma survive but the three gamma terms cancel . The five gamma term leads to a complete antisymmetrization of $R_{\mu\rho mn}$ and hence we again neglect these terms, since they are of higher order in $\psi$ (without torsion the cyclic identity says that these terms vanish). The one-gamma terms are

$$2(\delta^{\sigma m}\delta^{\rho n}\Gamma^\mu - \delta^{\sigma m}\delta^{\mu n}\Gamma^\rho + \delta^{\mu m}\delta^{\rho n}\Gamma^\sigma) - (m \leftrightarrow n) \tag{8}$$

and one finally finds that these terms again lead to an Einstein tensor. In fact, all terms cancel! Thus, this justifies our choice for $\delta e_\mu^m$. This is a nontrivial result because it relies on the fact that no three gamma terms are present. If, for example, in d = 11 one would have $C\gamma_\mu C^{-1} = +\gamma_\mu^T$ , then one would have found a commutator $[\Gamma^{\mu\rho\sigma}, \Gamma^{mn}]$ and no invariance to order $\psi$ could have been achieved. (In 3 + 2 dimensions, $C\Gamma^A C^{-1} = + \Gamma^{A,T}$ and one finds a commutator. Hence, although one can define in 3 + 2 dimensions a Majorana spinor as usual, no supergravity exists. In 4 + 1 dimensions one still has $C\gamma^A C^{-1} = + \Gamma^{A,T}$ but here one defines a Majorana spinor with internal indices and one finds now an anticommutator).

To find the higher order terms, we recall that if $\omega$ is a solution of $\delta I(\text{total})/\delta\omega = 0$, then

$$I^{(2)}(\omega + \tau) + I^{(3/2)}(\omega+\tau) = I^{(2)}(\omega)+I^{(3/2)}(\omega) +\tau^2 \text{ terms} \tag{9}$$

(See Ferrara and the author, Phys.Rev. D20, 2079, 1979). Hence, without loss of generality, we can take the Hilbert action as $eR(e, \omega)$ but we must then use 1.5 order formalism (i.e.,

$\omega = \omega(e, \psi)$ is a solution of $\delta I/\delta\omega = 0$). The $\tau^2$ terms are of order $\psi^4$ and must be found later.

Taking the most general laws for $\delta A_{\mu\nu\rho} \sim \kappa\bar{\varepsilon}\psi$ and $\delta\psi \sim \kappa F\varepsilon$ and the most general terms for $\kappa\psi^2 F$ in the action, one finds a unique solution. A peculiar term bilinear in the Maxwell curls and linear in a bare photon field is needed as well. It is gauge invariant! (Just partially integrate the variation $\delta A \sim \partial\Lambda$ of the bare A field.) Thus one has the theory up to terms trilinear in $\psi$ in the variation laws and up to terms quartic in $\psi$ in the action. The remaining trilinear terms are fixed by requiring that $\delta\psi_\mu$ be supercovariant. Indeed, since in $[\delta(\varepsilon_1), \delta(\varepsilon_2)]\psi_\mu$ the $\partial_\mu\varepsilon$ terms cancel one must make $\delta\psi_\mu$ itself supercovariant. (In the commutator one finds a general coordinate contribution $(\partial_\mu\xi^\nu)\psi_\nu$ and a supersymmetry contribution $\partial_\mu(-\xi^\nu\psi_\nu)$ with $\xi^\nu = \frac{1}{2}\bar{\varepsilon}_2\gamma^\nu\varepsilon_1$. From the {Q,Q}anticommutator on vierbeins we know this.) This yields

$$\delta\psi_\mu = D_\mu(\hat{\omega})\varepsilon + (\text{gamma matrices})\ \varepsilon\hat{F}_{\alpha\beta\gamma\delta} \tag{10}$$

where $\hat{F}$ is supercovariant, while $\hat{\omega}$ is the supercovariant extension of $\omega(e)$, which is thus the same expression as in d = 4

$$\hat{\omega}_{\mu mn} = \omega_{\mu mn}(e) + \tfrac{1}{4}(\bar{\psi}_\mu\gamma_m\psi_n - \bar{\psi}_\mu\gamma_n\psi_m + \bar{\psi}_m\gamma_\mu\psi_n) \tag{11}$$

Finally the quartic terms in the action are fixed by requiring that the gravitino field equation be itself supercovariant.

The final result for the N = 8 action in d = 11 is very simple. It is polynomial since a power series in $\kappa A_{\mu\nu\rho}$ is not allowed (it would violate Maxwell gauge invariance)

$$\mathcal{L} = \frac{-e}{2} R(e,\omega) - \frac{e}{2}\bar{\psi}_\mu \Gamma^{\mu\rho\sigma} D_\rho\left(\frac{\omega+\hat{\omega}}{2}\right)\psi_\sigma - \frac{e}{48} F^2_{\mu\nu\rho\sigma}$$

$$-\frac{3D}{4} e\kappa[\bar{\psi}_\mu\Gamma^{\mu\alpha\beta\gamma\delta\nu}\psi_\nu + 12\bar{\psi}^\alpha\Gamma^{\beta\gamma}\psi^\delta][F_{\alpha\beta\gamma\delta} + \hat{F}_{\alpha\beta\gamma\delta}]$$

$$+ \, C e \kappa \varepsilon^{\mu_1 \dots \mu_{11}} F_{\mu_1 \dots \mu_4} F_{\mu_5 \dots \mu_8} A_{\mu_9 \mu_{10} \mu_{11}}$$

$$\delta e_\mu^m = \frac{\kappa}{2} \bar{\varepsilon} \gamma^m \psi_\mu$$

$$\delta \psi_\mu = \kappa^{-1} D_\mu (\hat{\omega}) \varepsilon + D(\Gamma^{\alpha\beta\gamma\delta}{}_\mu - 8\delta_\mu^\alpha \Gamma^{\beta\gamma\delta}) \varepsilon \hat{F}_{\alpha\beta\gamma\delta} \tag{12}$$

$$\delta A_{\mu\nu\rho} = E \bar{\varepsilon} \Gamma_{[\mu\nu} \psi_{\rho]}$$

Remarkably enough, the $\psi^4$ terms are obtained by putting in the gravitino action $\frac{1}{2}(\omega + \hat{\omega})$as connection and also taking $\frac{1}{2}(F + \hat{F})$ in the Pauli -type coupling. I do not know why this "minimal" solution yields all $\psi$ terms. (In N = 2 d = 4 supergravity, and also in the d = 4 Maxwell-Einstein system one finds the same features.) Roughly speaking, the covariantization terms in $\hat{F}$ have have twice as many $\psi$'s as the terms with F, so one needs half of them to obtain a supercovariant field equation, but unlike the N = 2 models,the four $\psi$'s in the F terms do not appear symmetrically, so that this argument is not complete.

The spin connection $\omega$ solves $\delta I/\delta\omega = 0$ and is given by

$$\omega_{\mu mn} = \hat{\omega}_{\mu mn} - \frac{1}{8}(\bar{\psi}^\alpha \Gamma_{\alpha\mu mn\beta} \psi^\beta) \tag{13}$$

In d = 4, $\omega = \hat{\omega}$, but here they differ. This illustrates once again that by going to more complicated theories one finds out which equalities are a coincidence and which are not.

The gauge algebra is the usual one except that the Lorentz parameter is

$$\lambda^{mn} = \xi^\nu \omega_\nu{}^{mn} + \bar{\varepsilon}_2 \left( \Gamma^{mn\alpha\beta\gamma\delta} - 24 e^{m\alpha} e^{n\beta} \Gamma^{\gamma\delta} \right) \hat{F}_{\alpha\beta\gamma\delta} \tag{14}$$

while two supersymmetry transformations on $A_{\mu\nu\rho}$ lead to a gauge transformation $\delta A_{\mu\nu\rho} = \partial_\mu \Lambda_{\nu\rho}$ + cyclic terms, where

$$\Lambda_{\mu\nu} = -\tfrac{1}{2}\bar{\varepsilon}_2 \Gamma_{\mu\nu}\varepsilon_1 - \xi^\sigma A_{\sigma\mu\nu} \tag{15}$$

Terms such as $-\xi^\sigma A_{\sigma\mu\nu}$ always come from converting P in $\{Q, Q\} = P$ + more, into

$$\delta_P(\xi) = \delta_{\text{gen.coord.}}(\xi^\nu) - \delta_{\text{gauge}}(\xi^\lambda h_\lambda{}^A) \tag{16}$$

The gauge algebra reads

$$\begin{aligned}[\delta_Q(\varepsilon_1), \delta_Q(\varepsilon_2)] &= \delta_{\text{gen.coord.}}(\xi^\nu) + \delta_Q(-\xi^\nu\psi_\nu) \\ &+ \delta_L(\lambda^{mn} \text{ in } (14)) + \delta_{\text{Maxwell}}(\Lambda_{\mu\nu} \text{ in } (15))\end{aligned} \tag{17}$$

and this suggest strongly that the N = 1 d = 11 model can be obtained by gauging OSp (1/32) since Sp (32) is spanned by $\Gamma^m$, $\Gamma^{mn}$, $\Gamma^{mnrst}$. (These are the matrices satisfying the definition of symplectic matrices $CM + M^TC = 0$ with antisymmetric C. We recall that $C\Gamma^A = -\Gamma^{A,T}C$). One would like to consider $e_\mu{}^m$ and $\omega_\mu{}^{mn}$ as the gauge fields of $\Gamma^m$, and $\Gamma^{mn}$. However, $A_{\mu\nu\rho}$ cannot be considered as the gauge field of $\Gamma^{mnrst}$ (it does not have enough indices and it is totally antisymmetric), and this is a problem for future research (Nicolai, Townsend & van Nieuwenhuizen, 1980). Because the N = 1 d = 11 model contains only gauge fields, one expects that it can be obtained by some kind of gauging of an algebra.

We now turn to a discussion of dimensional reduction. One considers a space with D + F coordinates and wraps the F coordinates into small circles (torus-geometry). Keeping only the lowest Fourier modes is a consistent truncation of the theory (it would not be for non-torus geometries), and the net result is that fields no longer depend on the F-coordinates. Finally one breaks the D + F dimensional indices of the fields up into D dimensional indices.

## 3. Ordinary Dimensional Reduction

In this section we discuss the important and pioneering work of Cremmer and Julia (1979b). As we saw, in ordinary dimensional reduction all fields depend only on $x^1, \cdots x^4$ instead of an all eleven coordinates. One then proceeds to compute the various quantities appearing in the action: inverse vierbein and connections. One writes the four-dimensional part of the elfbeins as $\delta^\gamma e^m{}_\mu$ where $\delta$ is the determinant of the 7×7 part of the elfbein, and by choosing $\gamma$ appropriately, the Einstein action of the reduced theory has the usual form.

In this section we discuss how the massless N = 8 model in d = 4 dimensions is obtained by ordinary (i.e. not accompanied by spontaneous symmetry breaking) dimensional reduction. This analysis will also lead us to deduce that the d = 4 theory has as internal symmetry an E(7) global × SU(8) local. However, the SU(8) local has not 63 gauge fields, so that this SU(8) cannot serve as a grand unification of SU(3) × SU(2) × U(1). Rather, the connections are nonlinear functions of other fields, in the same way as the spin connection $\omega(e)$ gauges SO(3, 1) but is not an independent gauge field.

In d = 11 dimensions, one has general coordinate transformations

$$\delta V_\Lambda = \xi^\Pi \partial_\Pi V_\Lambda + (\partial_\Lambda \xi^\Pi) V_\Pi , \qquad (\Lambda, \Pi = 1,11) \tag{1}$$

Splitting $\Lambda = (\lambda, \alpha)$ with $\lambda, \mu, \nu = 1, 4$ and $\alpha, \beta, \gamma = 5, 11$ one has under ordinary dimensional reduction with fields $\phi(x^1 \ldots x^{11})$ restricted to $(x^1, \ldots x^4)$ a global symmetry group generated by

$$\xi^\Pi = (\xi^\mu = 0, \xi^\alpha = M^\alpha{}_\beta\, x^\beta) \tag{2}$$

Indeed, $\delta V_\Lambda$ again depends only on $x^1 \ldots x^4$ as follows from

$$\delta V_\mu = 0, \quad \delta V_\alpha = (\partial_\alpha \xi^\beta) V_\beta = M^\beta{}_\alpha V_\beta \tag{3}$$

One must in fact require that $\partial_\Pi \xi^\Pi = M^\alpha{}_\alpha = 0$ in order that the reduced Hilbert action is still invariant under this special class of general coordinate transformations. To see this, note that $\delta(eR) = \partial_\Lambda(\xi^\Lambda eR)$ since the action is a scalar density. Usually one discards this total derivative since coordinates tend to infinity, but now 7 coordinates have become little circles and one must require that $\partial_\alpha \xi^\alpha = 0 = M^\alpha{}_\alpha$. Thus, in d=4 one expects a global group SL(7, R).

The Lorentz group O(10, 1) in d=11 acts on tetrads as

$$\delta e_\Lambda{}^M = \omega^M{}_N e_\Lambda{}^N \tag{4}$$

Splitting the flat index M into (m,a) with m, n, ⋯ = 1,4 and a, b, c ⋯ = 5, 11, we have

$$\delta e_\Lambda{}^m = \omega^m{}_n e_\Lambda{}^n + \omega^m{}_a e_\Lambda{}^a \tag{5}$$

$$\delta e_\Lambda{}^a = \omega^a{}_n e_\Lambda{}^n + \omega^a{}_b e_\Lambda{}^b$$

Thus we can fix the parameters $\omega^a{}_n = -\omega_n{}^a$ by putting $e^n_\alpha = 0$. In this case the vielbein has the form

$$e_\Lambda{}^m = \begin{pmatrix} e_\mu{}^m & e_\mu{}^a \\ 0 & e_\alpha{}^a \end{pmatrix} \tag{6}$$

The remaining symmetry group is now O(3, 1) × O(7), and one expects in d=4 a local group O(7).

Let us now do the reduction of the d=11 N=1 field down to d=4. If one reduces from d=11 to d=4, one finds for the bosonic fields (a, α, β, ⋯ internal indices)

$$e_{\Lambda\Pi}{}^m \to 1 \text{ tetrad, } 7 \text{ photons } e_\mu{}^a, \tfrac{1}{2}\, 7 \times 8 \text{ scalars } e_\alpha{}^a$$

$$A_{\Lambda\Pi\Sigma} \to A_{\mu\nu\rho},\ 21 \text{ photons } A_{\mu\alpha\beta},\ 7 \text{ scalars } A_{\mu\nu\alpha} \tag{7}$$

$$\text{and } \binom{7}{3} = 35 \text{ scalars } A_{\alpha\beta\gamma}$$

The field $A_{\mu\nu\rho}$ in d=4 is pure gauge, while $A_{\mu\nu\alpha}$ is a scalar since

$F_{\mu\nu\rho} = \varepsilon_{\mu\nu\rho\sigma}\psi^{\sigma}$ so that $\partial^{\mu}F_{\mu\nu\rho\sigma} = 0$ implies $\psi_{\sigma} = \partial_{\sigma}\psi$. This is indeed the bosonic sector of the N=8 d=4 model, but all bosons have only a manifest O(7) symmetry (from $O(10, 1) \to O(3, 1) \times O(7)$) while the O(8) invariance is hidden.

The fermionic sector is reduced as follows. One splits the d=11 gravitino $\psi^{A}_{\Lambda}$ into eight gravitinos with $\Lambda = \mu = 1, 4$ and $A = (1, 32) = ai$ with $a = 1, 4$ and $i = 1, 8$ and 56 spin ½ fields $\lambda^{i}_{a}$ with $\alpha = 5, 11$. One finds thus also the fermionic fields of N=8 d=4 theory in this way. The numbers (1, 8, 28, 56, 70) are the dimensions of antisymmetric tensor representations of SO(8) or U(8) or SU(8). From the work of Haag et. al one knows that in the Poincaré supersymmetry algebra with N supersymmetry changes $Q^{\alpha}_{i}$ the maximal symmetry group which can act on the indices i is U(8). However, the scalar fields are represented by $A^{ijkl}$ which are self-dual and the pseudoscalars by antiselfdual $B^{ijkl}$. These constraints are not U(8) but only SU(8) invariant: $(A + iB)^{*}_{ijkl} = \varepsilon_{ijklmnpq}(A + iB)^{mnpq}$ (Since the 8 dimensional $\varepsilon$ tensor is SU(8) but not U(8) invariant). Thus the maximal expected symmetry group is SU(8). According to theory of nonlinear realizations, the theory is expected to be invariant under a global group G times a local group H, such that H is a maximal compact subgroup of G (this eliminates ghosts!) while particles are in multiplets of the diagonal group $H \subset G$ times H. Hence we expect a local SU(8) invariance. This invariance is expected to be a symmetry on-shell (i.e. of the field equations) since the U(N) symmetries of the N = 1,2,3,4 models are only on-shell symmetries.

The first nontrivial extension of the expected SL(7, R) global × O(7) local occurred when it was found that there was a set of global scale invariances for the action in d=4 which extended SL(7, R) to GL(7, R). Secondly, it was found that the theory had a local SU(8) invariance, of which the local O(7) was a part. Upon reduction to the N=4 model (the SU(4) version of the N=4 model we have discussed) this gauge gets fixed and no SU(4) local is left.

General theory of nonlinear realization tells us that the scalars can be written as an exponent with in the exponential of the generators of a global group. Since one can eliminate 63 scalars by fixing the local SU(8), and there are 70 scalar particles in N=8 model, this global group must have 63 + 70 = 133 generators. To make these arguments more familiar, we compare with the usual tetrad $e^m_{\ \mu}$. It describes 10 particles (if we forget the general coordinate invariance for a moment). The local group is SO(3, 1) with 6 generators, and hence the global group has 10 + 6 = 16 generators. This group is of course GL(4, R).

Cremmer and Julia found the Lie algebra with 133 generators in Cartan's work: E(7)! This algebra is defined by the real matrices $\Lambda$ and $\Sigma$

$$\delta x^{ij} = \Lambda^i_{\ k} x^{kj} - \Lambda^j_{\ k} x^{ki} + \frac{1}{24}\,\varepsilon^{ij\alpha_1 \cdots \alpha_6}\Sigma_{\alpha_1 \cdots \alpha_4} y_{\alpha_5\alpha_6}$$

$$\delta y_{ij} = \Lambda_i^{\ k} y_{kj} - \Lambda_j^{\ k} y_{ki} + \Sigma_{ijkl} x^{kl} \tag{8}$$

with $\Lambda^i_{\ k} = -\Lambda_k^{\ i}$, $\Lambda^i_{\ i} = 0$. Further i = 1,8 and $\Sigma_{ijk\ell}$ is totally antisymmetric. The vectors $x^{ij} = -x^{ji}$ and $y^{ij} = -y^{ji}$ have each 28 components, and this 56 dimensional representation is in fact the fundamental representation of E(7). The 63 $\Lambda$'s generate SL(8, R) (and the previous GL(7, R) is thus contained in $E_7$) while the remaining $\binom{8}{4} = 70\Sigma$'s generate the rest of E(7). Also the expected SU(8) global can be found by considering $(x^{jk} + iy_{jk})$ (not to be confused with the SU(8) local).

The global E(7) symmetry acts on the curls of the photons, not on the photon fields themselves. It is an on-shell symmetry only, of the same kind as the dual-chiral symmetries in the N=2, 3, 4 models. In fact, truncating the N=8 d=4 model to the N=4 d=4 model, one finds the SU(4) × SU(1, 1) global symmetry of the so-called SU(4) model.

The reader may have wondered what happens with the general coordinate transformations whose parameters are in d=11 given by $\xi^{\alpha}(x^1, \cdots x^4)$ where $\alpha = 5, 11$. The answer is that they become Maxwell gauge parameters for the photons $A_{\mu}{}^{\alpha}$ which are defined by $e_{\mu}{}^{a} = A_{\mu}{}^{\alpha} e_{\alpha}{}^{a}$. As one easily checks, $\delta A_{\mu}{}^{\alpha} = \partial_{\mu}\xi^{\alpha}$, while $\delta e^{m}{}_{\mu} = \delta e^{a}_{\alpha} = 0$.

There are much more fascinating details. We regretfully refer to the literature for this massless N=8 d=4 model, and now turn to a massive version of it.

## LECTURE IV. SPONTANEOUS SYMMETRY BREAKING IN THE N=8 MODEL BY DIMENSIONAL REDUCTION

The work which we now discuss is due to Scherk and Schwarz (1979a). (For the massive N=8 model in d=5 see Cremmer, Scherk and Schwarz (1979c)). It solves a long-standing puzzle of how to obtain spontaneous symmetry breaking in the extended supergravities. For simple supergravity, spontaneous symmetry breaking can occur as soon as S or P are nonzero (which may or may not imply that the scalar fields A and B have nonzero vacuum expectation values). (Cremmer et al., 1979a). However, for N>1, many supergravity practitioners were afraid that no analogous results would be obtained. As usual, matters turned out to be better than expected.

A most interesting way to give masses to the particles of the N=8 model in d=4 dimensions is to first reduce the d=11 N=1 theory to the d=5 N=8 massless theory. Then, in d=5 dimensions, one writes fields as

$$\phi(x^1, \cdots x^5) = e^{Mx^5} \psi(x^1, \cdots x^4) \tag{1}$$

where M is in general a matrix; see below. In this way, for example, a massless scalar field in d=5 can become massive in d=4

$$\phi^* \nabla \phi = \psi^* (\Box - M^2)\psi = 0 \tag{2}$$

Note that this kind of symmetry breaking is not such that the theory itself determines the minimum of a Higgs potential, but, rather, it is we who (arbitrarily) prescribe that fields must have the factored form as in (1). Since coordinates are dimensional, the substitution in (1) introduces masses in a natural way.

In general, the procedure for construction of a D-dimensional theory with spontaneously broken symmetry is to first consider a theory with unbroken symmetry in D + E dimensions and to wrap the extra E dimensions into circles with such small radii that only the lowest Fourier coefficients have finite mass. If we write fields as in (1), then the Lagrangian density in D + E dimensions is independent of M (and hence remains gauge invariant in D + E and in D dimensions) as long as the transformation $\phi = \exp(Mx^5)\psi$

corresponds to a symmetry. In practice, the useful symmetries are the global symmetries.

How do we find global symmetries in D+E dimensions? A very interesting idea is to consider yet a higher dimension, say D + E + F, and to use local symmetries there to obtain global symmetries in D + E dimensions. For our applications we start with the N=1 model in D + E + F = 11 dimensions, and want to obtain a global symmetry of as high a rank as possible in D + E = 5 dimensions. The local symmetry we use is general coordinate invariance in d=11. (In order that the method works also for systems without fermions, we do not study whether local Lorentz invariance can be used.) As we discussed in the last section, the expected global symmetry group is SL(6,R). Now no extra scale transformations are present which extend this SL(6,R) to GL(6,R). (Note that $E_6$ does not contain GL(6,R) but that E(7) contains GL(7,R)). Basically the reason is that there are two types of terms in the action which cannot be made scale-invariant at the same time, namely $\delta^{\frac{1}{2}}(F_{\mu\nu}{}^{\alpha\beta})^2$ and $\varepsilon^{\mu\nu\rho\sigma\tau}\varepsilon_{abcdef}F_{\mu\nu}^{ab}F_{\rho\sigma}^{cd}A_{\tau}^{ef}$ . Thus the expected global symmetry is SL(6,R).

The Lorentz group SO(D + E + F - 1, 1) splits up into SO(D + E - 1,1)x SO(F) if we fix part of this local symmetry by casting the vielbein in triangular form. This we discussed in the previous section. Thus, naively we expect for the N=8 model in d=5 dimensions as internal symmetry group SO(6) local xSL(6,R) global.

However, things are more interesting. The particle spectrum in d=5 consists of irreducible representations of Sp(8) rather than of SO(6). This particle spectrum is: one tetrad $e_{\mu}^{m}$ (m, $\mu$ =1, 5) eight gravitinos $\psi_{\mu}^{a}$(a=1,4), 27 vector fields, 48 spin ½fields and 42 scalar fields. Indeed, all fields can be represented by completely antisymmetric tensors which are traceless with respect to the symplectic metric. For example, the photons are described by the 27-dimensional Sp(8) representation $A_{\mu}{}^{\alpha\beta} = -A_{\mu}{}^{\beta\alpha}$ with $\Omega_{\alpha\beta}A_{\mu}{}^{\alpha\beta}=0$

($\alpha,\beta$ = 1,8) which is not an irreducible SO(6) representation.

This particle spectrum is obtained from the fields $e_\Lambda{}^M$, $\psi_\Lambda{}^B$, $A_{\Lambda\Pi\Sigma}$ (M,A,$\Pi$,$\Sigma$ = 1,11 and B = 1,32) by ordinary dimensional reduction as follows:

$e_\Lambda^M \to e_\mu^m$, 6 vectors $e_\mu^a$, $\frac{1}{2}(6\times 7) = 21$ scalars $e_\alpha^a$
$\psi_\Lambda^B \to 8\psi_\mu^a$ and 48 spin $\frac{1}{2}\psi_\alpha^i$(i = 1,8 and $\alpha$ = 5,11)
$A_{\Lambda\pi\Sigma} \to$ 1 scalar $A_{\mu\nu\rho}$, 6 vectors $A_{\mu\nu\alpha}$, 15 vectors $A_{\mu\nu\beta}$,
20 scalars $A_{\alpha\beta\gamma}$ (3)

The field $A_{\mu\nu\rho}$ has as field equation $D^\mu F_{\mu\nu\rho\sigma} = 0$ and since $F_{\mu\nu\rho\sigma} = \partial_{[\mu}A_{\nu\rho\sigma]} = \varepsilon_{\mu\nu\rho\sigma\tau}\varphi^\tau$, it follows that $\varphi_\tau = \partial_\tau\varphi$ and $A_{\mu\nu\rho}$ describe vectors (since $\partial_{[\mu}A_{\nu\rho]\,a} = F_{\mu\nu\rho a} = \varepsilon_{\mu\nu\rho\sigma\tau}\varphi^{\sigma\tau}$ one has $\varphi_a^{\sigma\tau} = \partial^{[\sigma}\varphi_a{}^{\tau]}$. Inserting this into F yields a Maxwell action). As a check one may verify that, as in d = 11, there are 128 bosonic and 128 fermionic states (in d=5 the little group of a massless particle is SO(3) and hence a particle of spin J has 2J+1 polarizations).

Thus one expects that one can write the fields as completely antisymmetric and symplectic traceless tensors $e_\mu^m$, $\psi_\mu^a$, $A_\mu^{ab}$, $\chi^{abc}$, $\phi^{abcd}$ where $\Omega_{ab}A_{\mu\nu}^{ab} = 0$ and $\Omega = I_4 \times i\tau_2$ is the antisymmetric 8 × 8 metric of Sp(8).

The reality conditions of the bosonic fields are $(A_\mu^{ab})^* = A_{\mu ab}$ and idem for $\phi^{abcd}$ where indices are raised and lowered by $\Omega$. Note that under Sp(8) a vector $V^a$ transforms as $\delta V^a = \Lambda^a{}_b V^b$ and that $\delta(V^a\Omega_{ab}V^b) = 0$. Hence, $\delta V_a = -V_b\Lambda^b{}_a$ where $V_b = V^a\Omega_{ab}$. Only if $(V^a)^*$ transforms as $V_a$ can we impose the reality conditions mentioned above. Now $\delta(V^a)^* = (\Lambda^a{}_b)^*(V^b)^*$. Hence one must require that $(\Lambda^a{}_b)^* = -\Lambda^b{}_a$. Note that $\Lambda^b{}_a = (\Lambda^T)_a{}^b$. In other words, in addition to being symplectic, the matrices must be antihermitian. The local symmetry group is Usp(8), not Sp(8)!

Symplectic matrices satisfy $M^T\Omega + \Omega M = 0$. Multiplying by $\Omega$ one has also $\Omega M^T + M\Omega = 0$ since $\Omega^2 = -1$. Hence $M+M^T$ and $M-M^T$ form also a basis of Sp(N,R). The generators of Usp(N,C) differ from those of Sp(N,R) only by factors of i, and they have the same representations.

In order to discuss the reality conditions for the spinors, we want to define Majorana spinors. In d=5 one has $C\Gamma^A C^{-1} = +\Gamma^{A,T}$ with A=1,5. (In a Majorana representation $C = \gamma^4\gamma^5$. As always, C is antisymmetric.) Thus we define

$$(\bar{\lambda}^D)^a{}_\beta = (\lambda^\alpha_a)\ (\gamma^4)^\alpha{}_\beta,\quad (\bar{\lambda}^M)^a_\beta = \lambda^\alpha_b C_{\alpha\beta}\Omega^{ab} \tag{4}$$

since, as we discussed, $(\lambda_a)^*$ transforms as $\lambda^a$ under Usp(8), and since we can only equate tensors with indices in the same position. We now define a Majorana spinor by $\bar{\lambda}^D = \alpha\bar{\lambda}^M$ with $\alpha$ arbitrary. We leave as an exercise to show that taking the complex conjugate of this relation and reinserting in this result again that one finds indeed a consistent result for $|\alpha| = 1$. Crucial for this result is that $\Omega^*\Omega = -1$. (In fact, the most general solution of such $\Omega$ is probably equivalent to the symplectic metric we have a assumed so far.) Thus the reality conditions for the fermions are

$$(\chi^{abc})^* = \chi^T_{abc}C,\ \text{idem}\ \psi^a_\mu \tag{5}$$

We can now again use the theory of nonlinear representations to predict what the global symmetry group will be. Since USp(8) has ½(8×9)=36 generators and there are 42 scalars, one needs a Lie algebra with 78 generators: E(6).

The Lie algebra E(6) is most easily defined by giving its fundamental 27-dimensional representation

$$\delta z^{ij} = \Lambda^i{}_k z^{kj} - \Lambda^j{}_k z^{ki} + \Sigma^{ij}{}_{kl} z^{kl} \tag{6}$$

where $\Lambda^i{}_k$ is antihermitian and symplectic ($\Omega\Lambda + \Lambda^T\Omega = 0$ meaning that $\Lambda_{ij} = \Lambda_{ji}$) and $\Sigma_{ijkl}$ is totally antisymmetric, traceless with respect to $\Omega_{ij}$ and $(\Sigma_{ijkl})^* = \Sigma^{ijkl}$ The tetrad is of course a scalar under both groups (USp(8) local and E(6) global), but gravitinos are in the 8 representation of USp(8) (and denoted by $\psi_\mu^a$) while photons are in the 27 of E(6) and denoted by $A_\mu^{\alpha\beta}$ (and thus USp(8) scalars). Finally, the scalars are written as $\mathcal{V}_{\alpha\beta}{}^{ab}$ where the indices $\alpha,\beta$ are E(6) indices and a,b are USp(8) indices.

One writes $\mathcal{V}_{\alpha\beta}{}^{ab}$ as an exponential of E(6) generators

$$\mathcal{V} = [\exp(c^{de}\Lambda_{de} + C_{ijkl}\Sigma^{ijkl})]_{\alpha\beta}{}^{ab} \tag{7}$$

where c are coefficients and $\Lambda$, $\Sigma$ indicate the various generators which are matrices. For example

$$(\Lambda_{cd})_{\alpha\beta}{}^{ab} = (\Lambda_{cd})^{[a}_{[\alpha}\delta^{b]}_{\beta]} \ , \ (\Lambda_{cd})_\alpha{}^a = \Omega_{c\alpha}\delta^a_d - \Omega_{d\alpha}\delta^a_c \tag{8}$$

Since, in general, $\mathcal{V}^{-1}\partial_\mu\mathcal{V}$ lies in the algebra (it is <u>not</u> a group element), one has $\mathcal{V}^{-1}\partial_\mu\mathcal{V} = Q_\mu + P_\mu$ where $(Q_\mu)_{cd}{}^{ab} = Q_\mu{}^{[a}_{[c}\delta^{b]}_{d]}$ is a linear combination of USp(8) generators. Since

$$\delta Q_\mu{}^a{}_b = \partial_\mu\Lambda^a{}_b + \Lambda^a{}_c Q_\mu{}^c{}_b - Q_\mu{}^a{}_c\Lambda^c{}_b \tag{9}$$

Q is the USp(8) gauge field, and one can define USp(8) covariant derivatives. For example, $\mathcal{V}^{-1}D_\mu\mathcal{V} = P_\mu$ (Just as for the spin connection in d=4 N=1 theory, one can find the exact form of Q by solving its field equation $\delta I/\delta Q_\mu = 0$. This is again 1.5 order formalism).

We will not write down the complete action; for that we refer to the literature. However, let us now turn to the mass generation. One writes only the fields with curved indices in the form of (1), namely

$$A_\mu{}^{\alpha\beta}(x,x^5) = U^\alpha{}_{\alpha'}(x^5)U^\beta{}_{\beta'}(x^5)A_\mu{}^{\alpha'\beta'}(x) \tag{10}$$

where M is a USp(8) matrix, which can be taken as

$$M = \begin{pmatrix} 0 & m_1 & & & & \\ -m_1 & 0 & & & & \\ & & \ddots & & & \\ & & & & 0 & m_4 \\ & & & & -m_4 & 0 \end{pmatrix} \quad \text{and} \quad U^\alpha{}_\beta(x^5) = (\exp Mx^5)^\alpha{}_\beta \tag{11}$$

In principle any E(6) matrix could have been chosen as mass matrix, but in order to exclude tachyons (with negative $m^2$), one selects the maximal compact subgroup of E(6). As it happens, this is also a USp(8) group. (This is thus a global USp(8) and has nothing to do with the USp(8) local which extended the O(6) local.) Since USp(8) has rank 4, there are precisely four arbitrary mass parameters generated.

Inserting (10) into the Maxwell action, one finds as mass term for the photons

$$\mathcal{L}^{\text{mass}}(\text{photons}) = A^\mu_{\alpha\beta}\, A^{\alpha\beta}_\mu \left( (-)^\alpha m_{[\frac{\alpha}{2}]} + (-)^\beta m_{[\frac{\beta}{2}]} \right) \tag{12}$$

This leads in d=4 dimensions to 3 massless photons ($A_{12}$, $A_{34}$, $A_{56}$, and $A_{78}$, satisfying $\Omega^{\alpha\beta}A_{\alpha\beta} = 0$) and 24 massive vector fields with masses $(m_i \pm m_j)^2$ where i,j = 1,4 (each mass is doubly degenerate). A fourth massless photon is obtained by the reduction of the vielbein from 5 to 4 dimensions. Usually one takes $m_1 = m_2 = m_3$ in which case one finds six more massless photons, yielding a total of 10. These represent the 8 gluons, the ordinary photon and the $Z^o$ or the antigraviton (Scherk, 1979b).

For the fermions, masses are due to the USp(8) connections $Q_5$ in the covariant derivatives, while the scalars obtain masses through $P_5$. We recall that

$$\mathcal{V}_{\alpha\beta}{}^{ab}(x,x^5) = \mathcal{V}_{\gamma\delta}{}^{ab}(x)\left(U^{-1}(x^5)\right)^\gamma{}_\alpha \left(U^{-1}(x^5)\right)^\delta{}_\beta \tag{13}$$

so that one finds for $\mathcal{V}^{-1}\partial_5\mathcal{V}$ the result

$$\mathcal{V}^{-1}{}_{cd}{}^{\alpha'\beta'}(2M^{\alpha}{}_{\alpha'}\delta^{\beta}{}_{\beta'})\mathcal{V}_{\alpha\beta}{}^{ab} = e^{-c\Sigma}M\cdot\Lambda e^{c\Sigma} \tag{14}$$

$$= M.\Lambda + [M\cdot\Lambda, c\Sigma] + \cdots$$

in the special Sp(8) gauge where there are no $\Lambda$ generators in the exponent of $\mathcal{V}$. Clearly, $Q_5$ starts with $M.\Lambda$ and hence $Q_5$ yields masses to the fermions.

The masses of the scalars come from $P_5$ because the scalar kinetic matrix is $\mathrm{tr}D_\mu\mathcal{V}^{-1}D^\mu\mathcal{V} = -\mathrm{tr}P_\mu P^\mu$ and because $P_5$ is linear in masses <u>and</u> scalar fields: $P_5 = [M.\Lambda, c\Sigma] + \cdots$ In fact, the E(6) algebra has a $Z_2$ grading: $[\Lambda, \Lambda] \sim \Lambda$, $[\Lambda, \Sigma] \sim \Sigma$, $[\Sigma, \Sigma] \sim \Lambda$. (This is thus a Cartan grading of an ordinary Lie algebra. Superalgebras have also a $Z_2$ grading, but the bracket relation of two odd elements is for superalgebras symmetric whereas it is here still, as usual, antisymmetric.) Thus the Q are even in the number of scalar fields c, while the P are odd.

This concludes our discussion of the way in which one can introduce masses into the N=8 model in four dimensions. One obtains a massive action without cosmological constant (which previously was always disastrously large) and with a potential which is bounded from below. It is a beautiful method, but is may not be the last word. Hence (in the author's opinion at least) it is not impossible that one can also construct different mass spectra. Perhaps not everything we need for physical applications descends upon us from higher dimensions. Therefore we refer the interested reader to a result in the literature which gives the most general spontaneous symmetry breaking which can take place in the N=1 model in 4 dimensions.

As shown by Cremmer, Julia, Scherk, Ferrara, Girardello and the author (1979a), one can indeed break supersymmetry spontaneously (at least in N=1 supergravity) such that the super-Higgs effect occurs, the potential is bounded from below, there is no cosmological constant (this fixes only part of the freedom in the action) and a mass formula $(2J + 1) (-)^J (m_J)^2 = 0$ is found. All masses are further totally free, and the model still contains an arbitrary function of two variables. Clearly, local supersymmetry is much less restrivitive than was thought in the beginning of supergravity. A sub class of these results was previously obtained by Das, Fischler and Rocek, 1977d, who used the Noether method instead of the tensor calculus.

Finally, we mention one of the outstanding problems for the N=8 model: in how far are the quantum corrections still finite if one adds masses? It has been shown that all one-loop divergences in the corrections to the cosmological constant vanish, leaving only a finite constant correction. However, whether also physical processes are finite is an open question. If they are, one might call the symmetry breaking discussed in this section spontaneous, since for spontaneously broken theories the theory behaves in the ultraviolet region as if there are no masses. Let us conclude with an optimistic note. Perhaps the N=8 model with masses is the ultimate finite quantum theory of gravity, and it will be shown that all particles of nature appear as bound states of this model.

ACKNOWLEDGEMENT: I am indebted to E. Sezgin for explaining to me some of the aspects of mass generation.

LECTURE V. GAUGING OF SUPERALGEBRAS

Superalgebras are algebras with a $Z_2$-grading which divide the generators $X_A$ into even and odd elements. Moreover, the bracket relation of two odd elements is symmetric, while in all other cases it is antisymmetric. Finally, the super Jacobi identities must be satisfied

$$[X_A, [X_B, X_C\} = \left[[X_A, X_B\}, X_C\right\} - \left[[X_A, X_C\}, X_B\right\}(-)^{bc} \tag{1}$$

where $b = 0$ if B is even and $b = 1$ if B is odd. Thus superalgebras are graded algebras, but not every graded algebra is a superalgebra. (Counterexample: The rotation and boost generators form a $Z_2$-grading of the Lorentz algebra which is not a superalgebra).

There are two main families of superalgebras, the orthosymplect algebras $Osp(N|2M)$ and the superunitary algebras $SU(N|M)$. Consider the line element

$$\ell^2 = x^i g_{ij} y^j + \theta^\alpha C_{\alpha\beta}\eta^\beta \quad (i,j = 1,N;\ \alpha,\beta = 1,2M) \tag{2}$$

where $g_{ij}$ is symmetric and $C_{\alpha\beta}$ is antisymmetric and all entries are ordinary and real numbers. Clearly $O(N)$ and $Sp(2M)$ are subalgebras which leave $\ell^2$ invariant. However, one cannot define $Osp(N|2M)$ as those transformations which rotate x and $\theta$ into each other and in the same way y and $\eta$ into each other. This would require that

$$\delta x = M\theta,\ \delta\theta = Nx,\ \delta y = M\eta,\ \delta\eta = Ny$$
$$M^T g + CN = 0 \ \underline{\text{and}}\ gM + N^T C = 0 \tag{3}$$

which is not possible if C is antisymmetric. Rather than finding modified transformation rules for $(y,\eta)$ we note that if we let $\theta$ and $\eta$ be anticommuting numbers, then all is consistent. (In this case $M\theta = -\theta M$). The composition rule involves only commutators. One can replace the Grassmann variables in M and N by ordinary numbers, but then the composition rule of two (M,N) transformations becomes an anticommutator.

The superunitary algebras leave invariant the complex line element

$$\ell^2 = (x^i)^* g_{ij} y^j + (\theta^\alpha)^* C_{\alpha\beta}\eta^\beta \tag{4}$$

One can always diagonalize C by a complex similarity transformation, so that one can also write

$$\ell^2 = (x^i)^* g_{ij} y^j + (\theta^\alpha)^* g_{\alpha\beta}\eta^\beta \tag{5}$$

Only if $\theta^\alpha$ and $\eta^\beta$ are Grassmann variables does one get the SU(N|M), otherwise one simply gets SU(N + M).

One can also use diagonal line elements. Typical examples are:

$$\begin{aligned} &\mathrm{Osp}(1/4):\quad \ell^2 = x^2 + \bar{\theta}^M\theta \quad (\bar{\theta}^M \text{ is the Majorana conjugate}) \\ &\mathrm{Osp}(2/4):\quad \ell^2 = x^2 + y^2 + \bar{\theta}^M\theta \\ &\mathrm{SU}(1/4):\quad \ell^2 = x^* x + \bar{\theta}^D\theta \quad (\bar{\theta}^D \text{ is the Dirac conjugate}) \end{aligned} \tag{6}$$

The field theories derived below from these algebras are: simple supergravity, N = 2 extended supergravity and simple conformal supergravity.

A nice physical way, due to Berezin, to introduce the superalgebra Osp(N,N|2M) is to consider supercanonical transformations between N fermion absorption operators $C_i$, N fermion creation operators $C_i^\dagger$, M boson absorption operators $a_i$ and M boson creation operators $a_i^\dagger$. Requiring that $\hat{a}_i^\dagger$, $\hat{a}_i$, $\hat{C}_i$ and $\hat{C}_i^\dagger$ have again canonical transformation rules, where

$$\hat{C}_i = C_i + F_{ik}C_k + P_{ik}C_k^\dagger + E_{ik}a_k + H_{ik}A_k^\dagger$$

$$\hat{C}_i^\dagger = C_i^\dagger + (F_{ik})^* C_k^t + (P_{ik})^* C_k + (E_{ik})^* a_k^\dagger + (H_{ik})^* a_k$$

$$\hat{a}_i = a_i + \phi_{ik}a_k + \psi_{ik}a^\dagger{}_k + \varepsilon_{ik}C_k + \eta_{ik}C_k^\dagger \qquad (7)$$

$$\hat{a}_i^\dagger = a_i^\dagger + (\phi_{ik})^* A_k^\dagger + (\psi_{ik})^* a_k - (\varepsilon_{ik})^* C_k^\dagger - (\eta_{ik})^* C_k$$

it is a straightforward algebra to show that

$$\begin{pmatrix} A & B \\ C & D \end{pmatrix}\begin{pmatrix} J & O \\ O & S \end{pmatrix} + \begin{pmatrix} J & O \\ O & S \end{pmatrix}\begin{pmatrix} A & B \\ C & D \end{pmatrix}^T = 0, \quad \begin{pmatrix} A & B \\ C & D \end{pmatrix}^T = \begin{pmatrix} A^T & C^T \\ -B^T & D^T \end{pmatrix}$$

$$A = \begin{pmatrix} \phi & \psi \\ \psi^* & \phi^* \end{pmatrix}, \quad B = \begin{pmatrix} \varepsilon & \eta \\ -\eta^* & -\varepsilon^* \end{pmatrix}, \quad C = \begin{pmatrix} E & H \\ H^* & E^* \end{pmatrix}, \quad D = \begin{pmatrix} F & P \\ P^* & F^* \end{pmatrix}$$

$$J = \begin{pmatrix} 0 & I \\ -I & 0 \end{pmatrix}, \quad S = \begin{pmatrix} 0 & I \\ I & 0 \end{pmatrix} \qquad (8)$$

The purely bosonic sector is of course Sp(2M) while the purely fermionic sector is O(N,N). (Note that one can diagonalize S by means of $(\tau^1 + \tau^3) \otimes 1_N$ into a diagonal matrix with N entries + 1 and N entries -1). The interesting thing is that transformations mixing bosons and fermions have Grassmann elements. Clearly supercanonical transformations form a group, involving only commutators, but if one formally replaces the Grassmann variables by ordinary numbers and at the same time replaces the commutator of two odd elements by anticommutator, one again obtains a closed algebraic system. This system is Osp(N,N|2M).

General formalism. Consider a general Lie superalgebra with bracket relations

$$[X_A, X_B\} = f_{AB}{}^C X_C \qquad (9)$$

We restrict our attention to ordinary and real numbers $f_{AB}{}^C$.

Associated with every generator $X_A$ is a gauge field $h_\mu{}^A$ and a gauge parameter $\varepsilon^A$. The parameters $\varepsilon^A$ transform in the adjoint representation, $[X_A, \varepsilon^B X_B] = 0$. The gauge fields transform for constant $\varepsilon^A$ also in the adjoint representation, but for local gauge transformation one has an extra term $\partial_\mu \varepsilon^A$

$$\delta h_\mu{}^A = \partial_\mu \varepsilon^A + h_\mu^B \varepsilon^C f_{CB}{}^A \equiv (D_\mu \varepsilon)^A$$
$$D_\mu = \partial_\mu + h_\mu^A X_A \quad \delta h_\mu{}^A \equiv [h_\mu{}^A, \varepsilon^B X_B] \tag{10}$$

Curvatures are defined by $[D_\mu, D_\nu] = R_{\mu\nu}{}^A X_A$ and one finds

$$R_{\mu\nu}{}^A = \partial_\mu h_\nu^A - \partial_\nu h_\mu{}^A + h_\mu{}^B h_\nu{}^C f_{CB}{}^A \tag{11}$$

Under local gauge transformations curvatures transform homogeneously

$$\delta R_{\mu\nu}{}^A = R_{\mu\nu}{}^B \varepsilon^C f_{CB}{}^A \tag{12}$$

while under arbitrary variations $h_\mu{}^A \to h_\mu{}^A + \delta h_\mu{}^A$ one has

$$\delta R_{\mu\nu}{}^A = (D_\mu \delta h_\nu)^A - (D_\nu \delta h_\mu)^A \tag{13}$$

This is the usual general set-up. We have been careful to contract indices in such a way that all expressions are manifestly covariant. Moreover, our results hold also when one has anticommuting base manifold indices $\mu$ and $\nu$. (In the second terms in (11) and (13) one needs an extra factor $(-)^{\sigma(\mu)\sigma(\nu)}$ where $\sigma(\mu) = +1$ if $\mu$ is fermionic). We now will treat three subjects in the next lecture. First we will discuss the gauge algebra and show how constraints on curvatures are needed in order that the commutator of two local supersymmetry transformations closes. Then we will discuss two approaches to the construction of actions: one approach which starts with Yang-Mills type of actions in ordinary spacetime and another approach which starts with actions linear in curvatures in a much bigger manifold, the group manifold.

## LECTURE VI. CONSTRAINTS AND ACTIONS. CLOSURE OF THE GAUGE ALGEBRA

In the cases of interest (the super de-Sitter algebra, its contraction the super Poincaré algebra and the super conformal algebra) one has the anticommutator

$$\{Q^{\alpha}, Q^{\beta}\} = (\gamma^{m}C^{-1})^{\alpha\beta}P_{m} + \text{possibly more} \tag{1}$$

P-gauge transformations are not desired as symmetries of the theory (phrased differently: although we start out with them as part of the tangent group, we do not want to leave them there), but rather we want general coordinate transformations as symmetries of still-to-be-constructed actions. The relation between both is well-known; it is a matter of elementary algebra to prove that

$$\delta_{gc}(\xi)h_{\mu}{}^{A} = (D_{\mu}\xi\cdot h)^{A} + \xi^{\nu}R_{\nu\mu}{}^{A} \tag{2}$$

In other words, general coordinate transformations of a gauge field $h_{\mu}^{A}$ are a sum of <u>all</u> gauge transformations (with parameters $\varepsilon^{B} = \xi^{\nu}h_{\nu}{}^{B}$ including P-gauge transformations with parameter $\xi^{\nu}e_{\nu}^{m}$) plus a term proportional to the curvature of the field $h_{\mu}^{A}$. Thus the P-gauge transformation is a sum of symmetries of the final theory minus a curvature term which is not a symmetry of the final theory. How does the theory manage to get rid of this curvature term?

There are basically two mechanisms possible

(i) The curvature term itself vanishes. This constitutes a constraint on curvatures. In this case {Q,Q} is indeed equal to P, but P is a symmetry of the theory.

(ii) Some fields are expressed in terms of other fields (as a result of solving the constraints under (i)) so that the Q-gauge transformation of these dependent fields is no longer given by $\delta h_{\mu}{}^{A} = (D_{\mu}\varepsilon)^{A}$ but rather by the chain rule. The extra terms in $\delta h_{\mu}{}^{A}$ must then cancel the curvature term in (1).

In both cases one ends up with (see (2))
$[\delta_Q(\varepsilon_1),\delta_Q(\varepsilon_2)] = \delta_{gc}(\bar{\varepsilon}_2\gamma^\mu\varepsilon_1) + \Sigma\delta_{gauge}(-\bar{\varepsilon}_2\gamma^\mu\varepsilon_1 h^A_\mu)$ where the sum $\Sigma$ runs over all symmetries of the super algebra except P.

To illustrate this general phenomenon, consider first the vierbein field $e^m_\mu$. Under Q it rotates always into the gravitino, since $\{Q,X\}$ = P+ more can only occur if X = Q. Since we want to keep the gravitino as a physical field and do not want to express it in terms of other fields we conclude: <u>in all theories the P-curvature $R^m_{\mu\nu}$ (P) must vanish</u>. This is indeed the case. One can solve $\omega_\mu^{mn}$ from $R^m_{\mu\nu}(P) = 0$. In the super-Poincaré and super de-Sitter cases the result is the same as if one had solved $\omega_\mu{}^{mn}$ from its own nonpropagating field equation, and thus here we see how 1.5 order formalism appears in the context of group theory. However, in conformal supergravity, the spin connection is propagating and one could not have eliminated it from its field equation.

Let us now see how the arguments apply to the gravitino. Inspection of $R_{\mu\nu}(Q)$ reveals that in the de-Sitter and Poincaré cases there is no gauge field which can solve $R_{\mu\nu}(Q) = 0$. Thus mechanism (ii) must come into effect. Since we do not want to eliminate the gravitino as an independent field we must change the transformation rule(s) of the fields into which the gravitino varies. Under Q the gravitino transforms into $h_\mu{}^X$ with $[Q,X] \sim Q$. Hence X = P,M or X equals the chiral charge A and the dilaton charge D in the super conformal case. From the constraint $R_{\mu\nu}(P) = 0$ we already eliminated $\omega_\mu{}^{mn}$ (the M-gauge field) as an independent field, and it is easy to see that this suffices to close the $\{Q,Q\}$ commutator. The extra term in the transformation rule for $\omega_\mu{}^{mn}$ is indeed proportional to the Q-curvature. To see this note that

$$\delta_Q(\text{gauge})R^m_{\mu\nu}(P) - \delta_Q(\text{extra})(\omega_\mu{}^m{}_\nu - \omega_\nu{}^m{}_\mu) \equiv 0 \qquad (3)$$

whereas $\delta_Q$ (gauge) $R(P) \sim R(Q)$. So we see that the theory kills two birds in one stroke ($R_{\mu\nu}{}^{m}(P) = 0$).

Just for fun, let us consider what happens with $\{Q,Q\}A_\mu$ in superconformal gravity. From the preceding arguments it is clear that the transformation rule of $A_\mu$ under Q is unchanged. Under Q $A_\mu$ rotates into the conformal supersymmetry S gauge field $\bar{\phi}_{\mu\alpha}$ and either we must impose that $R_{\mu\nu}(A) = 0$ or otherwise $\bar{\phi}_{\mu,\alpha}$ must be expressed in terms of other fields in such a way that there is an extra term in its transformation rule proportional to $R_{\mu\nu}(A)$. Again there is no possibility to solve $R_{\mu\nu}(A) = 0$, so we need to solve $\bar{\phi}_{\mu,\alpha}$ from a constraint. This is only possible if in some curvature we find $\phi_\mu^\alpha$ times $e_\nu^m$, and since $[S, P] \sim Q$ we need a constraint on $R(O)$. There is indeed such a constraint (first found by requiring invariance of the action) and it states that $R(Q)$ contains two terms: one with three dotted indices and totally symmetric, the other with three undotted indices and totally symmetric.

We leave as an exercise to argue that we now have a consistency test on the dilaton field (which rotates under Q into $\phi_{\mu,\alpha}$), since from the preceding arguments we deduce that it still must transform according to the group.

Construction of Yang-Mills type actions.

To construct actions quadratic in the curvatures of some super algebra, one writes

$$I = \int d^4x \left[ R_{\mu\nu}{}^{A} R_{\rho\sigma}{}^{B} Q_{BA}{}^{\mu\nu\rho\sigma} \right] \tag{4}$$

One chooses a subgroup under which Q is an invariant tensor (usually the Lorentz group). For example, for N = 1 supergravity, one chooses the super-de-Sitter algebra Osp(1/4) and writes

$$I = \int d^4x\; R_{\mu\nu}^{mn}(M) R_{\rho\sigma}^{pq}(M) \varepsilon_{mnpq} + R_{\mu\nu,\alpha}(Q) R_{\rho\sigma,\beta}(Q) (\gamma_5 C^{-1})^{\alpha\beta} \Big] \varepsilon^{\mu\nu\rho\sigma} \tag{5}$$

where $R_{\mu\nu}{}^{mn}(M)$ is the Lorentz curvature and $R_{\mu\nu,\alpha}(Q) = \partial_\mu \bar{\psi}_{\nu,\alpha} + \ldots$

the supersymmetry curvature. The action is by construction invariant under general coordinate and local Lorentz rotation, but using general formulae of how curvatures rotate under gauge transformations, one discovers that it is only invariant under local supersymmetry provided $R^m_{\mu\nu}(P) = 0$. Hence, again 1.5 order formalism is found, now from the action.

It is interesting to see how the higher derivative terms in the action cancel. Since

$$R^{mn}_{\mu\nu}(M) = R^{mn}_{\mu\nu}(\omega) + (e^m{}_\mu e^n{}_\nu - e^m{}_\nu e^n{}_\mu) + \bar{\psi}_\mu \sigma^{mn} \psi_\nu$$

$$R_{\mu\nu}(Q) = D_\mu \bar{\psi}_\nu - D_\nu \bar{\psi}_\mu + \bar{\psi}_\mu \gamma_\nu - \bar{\psi}_\nu \gamma_\mu \qquad (6)$$

one finds that the squares of the Riemann tensor are multiplied by two ε-symbols and yield the Gauss-Bonnet invariant. Similarly, the squares of $(D_\mu \psi_\nu - D_\nu \psi_\mu)$ cancel after partial integration the $R(\omega)\bar{\psi}\sigma\psi$ cross term. The $R(\omega)ee$ cross terms yield the Hilbert action, while the squares of the ee terms yield the cosmological constant and the $\bar{\psi}\sigma\psi$ terms yield the mass-term which always accompanies the cosmological constant.

For N = 2 supergravity one can do similar things. The action is essentially the same except that one sums over two R(Q)R(Q) terms (since there are two gravitini) and one adds the photon curvatures squared. All aspects of the N = 2 model are recovered: the action (including the gauged photon, cosmological terms, mass term) and also all transformation rules come out exactly right, except that in $\delta\psi_\mu{}^i$ one has to add by hand one term of the form $\delta\psi_\mu \sim F_{\mu\nu}(A)\varepsilon$. The origin of this term has not yet been explained group-theoretically. Also N = 1 conformal supergravity was obtained in this way.

The tetrad law $\delta e^m_\mu$ is $\frac{1}{2}\bar{\varepsilon}\gamma^m\psi_\mu$, and in lecture II we mentioned that group theory predicts this form correctly.

Construction of Einstein-type actions.

Another approach to the gauging of groups G is to use the

group manifold: a manifold with as many coordinates as there are generators. One starts with an action (using wedge -notation)

$$I = \int_M R^A \wedge h^{K_1} \wedge \cdots\cdots \wedge h^{K_n} C_{AK_1 \cdots K_n} \tag{7}$$

where C is an invariant tensor of a subgroup H, $K_i \varepsilon G/H$ and $A \varepsilon G$. Moreover one requires that $h^{K_1} \wedge \ldots . C_{A \cdots K_n}$ transforms as a covariant vector under G (not only H!) when all curvatures vanish. The constants C follow from requiring that flat space be a solution of the field equations. Flat space being defined by $R^A = 0$, one must solve the C coefficients from

$$D \wedge (h^{K_1} \wedge \ldots C_{A \cdots K_n}) = 0 \tag{8}$$

for the case that $h^{K_i}$ are pure gauge and according to the Torino group--this can be done and there are always only a finite number of solutions. Depending on the number of indices C, one finds what the dimensions of the hypersurface M must be over which one integrates. One uses a generalized variational principle: one varies not only the fields but also the curved hypersurface M, so that the field equations hold in each point in the group manifold.

For simple supergravity one can now start directly (if one wishes) with the super Poincaré algebra. The group manifold has 4 + 4 + 6 coordinates, M is four-dimensional, the C's are of the form of $\varepsilon^{ijkl}$ and $\gamma_5$ etc., H is the Lorentz group, and $K \sim P + Q$.

From the field equations one deduces

(i) horizontality: curvatures $R_{AB}{}^C$ (with flat A, B) are zero if A and/or B lies in H. This allows one to deduce that fields do not depend on the H-coordinates and that one can forget about the gauge fields $h_A{}^B$ when $A \varepsilon H$. One arrives at this point at superspace.

(ii) rheonomic symmetry: curvatures $R_{AB}{}^C$ with A and/or B in G/H-M (in the Poincaré case M = P and G/H - M = Q) can be expressed in terms of curvatures $R_{MN}{}^{C}$.

One can now show that general coordinate transformations in the group manifold with parameters $\xi^{\Lambda}$ with $\Lambda\varepsilon$ G/H become local symmetries of the final theory. (In the Poincaré case this is just ordinary general coordinate transformations and local supersymmetry).

This approach states that ordinary spacetime (M) is more fundamental for supergravity than superspace. It starts with a closed algebra in the group manifold with a large number of auxiliary fields ("the maximal number"). One might try to find out which subset of auxiliary fields can be eliminated such that one ends up in ordinary spacetime with an invariant action and still with a closed gauge algebra. This is not yet fully understood. Einstein general relativity and N = 1 supergravity in d = 4 dimensions and N = 2 supergravity in d = 5 dimensions have been recovered in this way.

We close by giving some literature on the gauging of groups. Yang-Mills gauging: S.W. MacDowell and F.Mansouri, Phys. Rev. Lett.38, 739 (1977), P. K. Townsend and P. Van Nieuwenhuizen, Phys. Lett. 67B, 439 (1977), A. H. Chemseddine and P.C. West, Nucl.Phys.B 129, 39 (1977)

Einstein gauging: Y. Ne'eman and T. Regge, Rivista del Nuovo Cimento, 1, 5, 1 (1978) and several Torino preprints by A.D'Adda, R.D'Auria, P. Fre', T. Regge. See the article by these four authors in Proceedings of the Supergravity Workshop at Stony Brook. North-Holland, 1979, P. Van Nieuwenhuizen and D.Z.Freedman editor's.

## OUTLOOK

Supergravity is the best candidate for a true unification of all interactions. However, the mechanism of spontaneous symmetry breaking should be further developed to generate the different strengths of weak, electromagnetic and strong interactions. Moreover, it may be the ultimate quantum theory of gravity. The next crucial test is the 3-loop finiteness of the N = 8 model and I believe it will be finite (whether a candidate counter term exists or not is not relevant).

There are only eight modelsof supergravity, because couplings of higher spin are inconsistent. This was first shown for the coupling of spin 5/2 to gravity and matter by Berends, de Wit, van Holten and the author (J. Phys.A 13 , 1643,1980). Since then various attempts to improve these results have failed. Thus we seem stuck in supergravity with the N = 8 model. Let us hope that Nature is aware of our efforts.

REFERENCES

Cremmer, E., Ferrara, S. & Scherk, J. (1978a). SU(4) invariant supergravity theory. Phys. Lett. 74B, pp. 61-64.

Cremmer, E., Julia, B. & Scherk, J. (1978b). Supergravity theory in 11 dimensions. Phys. Lett. 76B, pp. 409-412.

Cremmer, E., Julia, B., Scherk, J., Ferrara, S., Girardello, L. & van Nieuwenhuizen, P. (1979a). Spontaneous symmetry breaking and Higgs effect in supergravity without cosmological constant. Nucl. Phys. B 147, pp. 105-131.

Cremmer, E. & Julia, B. (1979b) The SO(8) supergravity. Nucl. Phys. B 159, pp. 141-212.

Cremmer, E. Scherk, J. & Schwarz, J. H. (1979c) Spontaneously broken N = 8 supergravity. Phys. Lett. 84B, pp. 83-86.

Chemseddine, A. H. & West, P. C. (1977) Supergravity as a gauge theory of supersymmetry. Nucl. Phys. B 129, pp. 39-44.

Das, A. & Freedman, D. Z. (1977a). Gauge internal symmetry in extended supergravity. Nucl. Phys. B 120, pp. 221-230.

Das, A. (1977b). SO(4) invariant extended supergravity. Phys. Rev. D 15, pp. 2805-2809.

Das, A., Fischler, M. & Rocek, M. (1977c) Massive self-interacting scalar multiplet coupled to supergravity. Phys. Lett. 69B, pp. 186-188.

Das, A., Fischler, M. & Roček, M. (1977d). Super-Higgs effect in a new class of scalar models and a model of super QED. Phys. Rev. D 16, pp. 3427-3436.

Deser, S. & Zumino, B. (1976). Consistent supergravity Phys. Lett. 62B, pp. 335-337.

Deser, S. & Zumino, B. (1977). Broken supersymmetry and supergravity. Phys. Rev. Lett. 38, pp. 1433-1436.

Duff, M. J. & van Nieuwenhuizen, P. (1980). Quantum inequivalence of different field representations. Phys. Lett. 94B, 179.

Fradkin, E. S. & Vassiliev, M. A. (1976). SO(2) Supergravity with minimal electromagnetic interaction, 1.5 order formalism and explicit form of the structure coefficient. Lebedev Institute preprint N 197.

Ferrara, S. & van Nieuwenhuizen, P. (1976a). Consistent supergravity with complex spin - $\frac{3}{2}$ gauge fields. Phys. Rev. Lett. 37, pp. 1669-1671.

Ferrara, S., Scherk, J. & van Nieuwenhuizen, P. (1976b). Locally supersymmetric Maxwell-Einstein theory. Phys. Rev. Lett. 37 pp. 1035-1037.

Ferrara, S., Scherk, J. & Zumino, B. (1977a). Algebraic properties of extended supergravity theories. Nucl. Phys. B 121, pp. 393-402.

Ferrara, S., Scherk, J. & Zumino, B. (1977b). Supergravity and local extended supersymmetry. Phys. Lett. 66B, pp. 35-38.

Ferrara, S., Gliozzi, F., Scherk, J. & van Nieuwenhuizen, P. (1977c). Matter couplings in supergravity theory. Nucl. Phys. B 117, pp. 333-335.

Freedman, D. Z., van Nieuwenhuizen, P. & Ferrara, S. (1976a). Progress toward a theory of supergravity. Phys. Rev. D 13 pp. 3214-3218.

Freedman, D. Z. & van Nieuwenhuizen, P. (1976b). Properties of supergravity theory. Phys. Rev. D 14, pp. 912-916.

Freedman, D. Z. (1977). SO(3) invariant extended supergravity. Phys. Rev. Lett. 38, pp. 105-108.

Freedman, D. Z. & Schwarz, J. H. (1978). N = 4 supergravity with local SU(2) × SU(2) invariance. Nucl. Phys. B 137, pp. 333-339.

Grisaru, M. T., van Nieuwenhuizen, P. & Vermaseren, J. A. M. (1976) One-loop renormalizibility of pure supergravity and of Maxwell-Einstein theory in extended supergravity. Phys. Rev. Lett. 37, pp. 1662-1666.

Haag, R., Lopuszanski, J. & Sohnius, M. (1975). All possible generators of supersymmetries of the S-matrix. Nucl. Phys. B 88, pp. 257-274.

Nicolai, H., Townsend, P. K. & van Nieuwenhuizen, P. (1980). Comments on 11-dimensional supergravity. To be published in Nuovo Cim. Letts.

Scherk, J. & Schwarz, J. H. (1979a). How to get masses from extra dimensions. Nucl. Phys. B 153, pp. 61-88.

Scherk, J. (1979b). From supergravity to antigravity. In Supergravity: Proceedings of the Supergravity Workshop at Stony Brook, eds. P. van Nieuwenhuizen and D. Z. Freedman, pp. 43-51. North-Holland Publishing Company, Amsterdam.

Sezgin, E. & van Nieuwenhuizen, P. (1980). Renormalizibility properties of antisymmetric tensor fields coupled to gravity. Phys. Rev.D 22, 301(1980)

Townsend, P. K. (1977a) Cosmological constant in supergravity. Phys. Rev. D 15, pp. 2802-2804.

Townsend, P. K. & van Nieuwenhuizen, P. (1977b). Geometrical interpretation of extended supergravity. Phys. Lett. 67B, pp. 439-442.

Vermaseren, J. A. M. & van Nieuwenhuizen, P. (1976). One-loop divergences in quantum theory of supergravity. Phys. Lett. 65B, pp. 263-266.

Wess, J. & Zumino, B. (1974a). A Lagrangian model invariant under supergauge transformations. Phys. Lett. 49B, pp. 52-54.

Wess, J. & Zumino, B. (1974b). Supergauge transformations in four dimensions. Nucl. Phys. B 70, pp. 39-50.

# AN INTRODUCTION TO SUPERSPACE AND SUPERGRAVITY

M. Roček

University of Cambridge

## INTRODUCTION

In these lectures, I shall discuss some of the formal aspects of supersymmetry and supergravity. I hope to convince you that the formalism is both natural and elegant, though, for supergravity at least, there is plenty of room for improvement; but more importantly, the formalism is very useful for actual calculations, and I want to make these techniques accessible.

### Motivation

The physical aspects of supersymmetry will not be stressed here (for a physically motivated approach to the subject see P. van Nieuwenhuizen's lectures in this volume); however, a few comments are in order. Supersymmetry is a symmetry which relates bosons and fermions; it is indeed the only such symmetry compatible with relativistic quantum field theory (see Haag, Lopuszanski & Sohnius 1975). As a result, it offers hope for a new unification: the merging of spacetime symmetries such as Lorentz rotations with internal symmetries such as gauge and isospin invariance (supersymmetry circumvents the no-go theorem of Coleman & Mandula 1967). To be honest I must add that to date there is no experimental evidence for such a unification of fundamental interactions (in nuclear physics the situation is quite different; see Iachello 1980). Supersymmetric theories are typically less divergent in perturbation theory than their more conventional kin, and they may teach us new things about quantum field theory itself.

All the features noted above become particularly interesting in the context of gravitation, since both the unification of gravity with other forces and the divergences of quantum gravity are venerable and formidable problems. Since gravity is the gauge theory of spacetime symmetries, supersymmetry seems particularly well suited to unifying gravity with particle physics; and the divergence taming powers of supersymmetry have already manifested themselves in quantum gravity through the finiteness of the two-loop supergravity S-matrix.

Finally, I want to emphasize the pragmatic reason for learning the formalism described here: supergraph techniques allow the practical calculation of quantities which, though in principle calculable, in practice are inaccessible in conventional quantum field theory. Indeed, even the unrealized possibility that some features of these techniques could be extended to non-supersymmetric theories is sufficient reason to study the formalism.

## HISTORY

I merely want to note that supersymmetry was invented in a long overlooked paper by Gol'fand & Likhtman (1971). My lectures draw most heavily on Salam & Strathdee (1974, 1974a, 1975), Ferrara, Wess & Zumino (1974), Wess (1977), Siegel (1977, 1978), Siegel & Gates (1979, 1979a), and Ogievetsky & Sokatchev (1978, 1978a). Useful review articles in the field include Wess (1977), Fayet & Ferrara (1977), Zumino (1978), and van Nieuwenhuizen (1980). A bibliography can be found in Freedman & van Nieuwenhuizen (1979).

## Preliminaries

A comment. In quantum field theory, it is convenient to represent "classical" fermi fields by anticommuting Grassmann elements (Berezin 1973); in supersymmetric theories, this continues to be convenient. It is not, however, essential (see Nicolai 1980, 1980a).

Notation. I will use two-component spinor notation, that is, left(right)handed spinors will bear undotted (dotted) indices respectively:

$$\tfrac{1}{2}(1+\gamma_5)\psi \leftrightarrow \psi^\alpha, \quad \tfrac{1}{2}(1-\gamma_5)\psi \leftrightarrow \psi_{\dot\alpha}.$$

For a left(right)handed Weyl spinor, $\psi^{\dot\alpha}(\psi^\alpha) = 0$ respectively; for a Majorana spinor,

$$\psi^{\dot\alpha} = \bar\psi^{\dot\alpha} := (\psi^\alpha)^\dagger.$$

Spinor indices can be raised and lowered by the alternating tensor $\varepsilon_{\alpha\beta} = -\varepsilon_{\beta\alpha}$:

$$\psi_\alpha = \psi^\beta\varepsilon_{\beta\alpha}, \quad \psi^\alpha = \varepsilon^{\alpha\beta}\psi_\beta,$$

$$\psi_{\dot\alpha} = \psi^{\dot\beta}\varepsilon_{\dot\beta\dot\alpha}, \quad \psi^{\dot\alpha} = \varepsilon^{\dot\alpha\dot\beta}\psi_{\dot\beta}.$$

I adopt the convention that

$$\psi^\alpha\psi_\alpha + \bar\psi^{\dot\alpha}\bar\psi_{\dot\alpha}$$

is Hermitian, which, since $\Psi$ is anticommuting, implies

$$(\varepsilon_{\alpha\beta})^{\dagger} = -\varepsilon_{\dot\alpha\dot\beta}, \text{ and } \bar\Psi_{\dot\alpha} = -(\Psi_\alpha)^{\dagger}.$$

The fundamental two-component identity is simply the result that any antisymmetric tensor must be proportional to $\varepsilon$:

$$T_{\alpha\beta} - T_{\beta\alpha} = \varepsilon_{\alpha\beta}\, T_\gamma{}^\gamma = -\varepsilon_{\alpha\beta}\, T^\gamma{}_\gamma,$$

$$T_{\dot\alpha\dot\beta} - T_{\dot\beta\dot\alpha} = \varepsilon_{\dot\alpha\dot\beta}\, T_{\dot\gamma}{}^{\dot\gamma} = -\varepsilon_{\dot\alpha\dot\beta}\, T^{\dot\gamma}{}_{\dot\gamma}.$$

Finally, a useful identity gives the product of two $\varepsilon$'s:

$$\varepsilon^{\alpha\beta}\, \varepsilon_{\gamma\beta} = \delta^\alpha{}_\gamma, \quad \varepsilon^{\dot\alpha\dot\beta}\, \varepsilon_{\dot\gamma\dot\beta} = \delta^{\dot\alpha}{}_{\dot\gamma}.$$

Vector indices are related to spinor indices by the Pauli matrices:

$$k_{\alpha\dot\alpha} = \sigma^a_{\alpha\dot\alpha}\, k_a$$

$$\sigma^a_{\alpha\dot\alpha}\, \sigma^{\beta\dot\beta}_a = 2\, \delta_\alpha{}^\beta\, \delta_{\dot\alpha}{}^{\dot\beta}, \quad \sigma^a_{\alpha\dot\alpha}\, \sigma^{\alpha\dot\alpha}_b = 2\, \delta_b{}^a, \text{ and}$$

$$\eta_{ab} = \mathrm{diag}(++++).$$

Hermitian conjugation of derivatives is defined by:

$$(\zeta^a\partial_a\, \phi)^{\dagger} = (\phi^{\dagger}\, \overleftarrow{\partial}_a \zeta^a) = (\zeta^a \partial_a\, \phi^{\dagger}),$$

which implies

$$[\zeta^a\partial_a, \phi]^{\dagger} = +[\zeta^a\partial_a, \phi^{\dagger}].$$

Sometimes it is convenient to consider indices that can take either spinor or vector values; they are represented by capital letters, e.g.,

$$V^A W_A = V^a W_a + V^\alpha W_\alpha + V^{\dot\alpha} W_{\dot\alpha} = W_a V^a - W_\alpha V^\alpha - W_{\dot\alpha} V^{\dot\alpha}$$

$$= (-1)^a\, W_A V^A.$$

Note that spinor components anticommute whereas vector components commute. The symbol $(-1)^a = (+1, -1, -1)$ for $A = (a, \alpha, \dot\alpha)$ respectively. Finally, when it is necessary to distinguish tangent space (flat) indices from world (curved) indices, the former are taken from the beginning of the alphabet ($A$, $a$, $\alpha$, etc.) and the latter from the middle of the alphabet (M, m, $\mu$, etc.).

## LIE ALGEBRAS, SUPERALGEBRAS, AND THEIR REPRESENTATIONS

### Introduction

In field theory, one deals with objects, e.g. fields $\phi$, that transform under some representation of a Lie algebra (or Lie group). The simplest situation arises when the algebra acts by matrix multiplication; in that case, I shall write

$$\delta(\Lambda)\phi := i\Lambda\phi. \tag{0}$$

Here $\Lambda$ is an element of a Lie algebra and can be written as $\Lambda = \Lambda^i T_i$ where the $T_i$ are a matrix representation of the generators of the Lie algebra, and $\phi$ is an element of a vector space on which the $T_i$ act. The Lie algebra equips the generators with a Lie bracket, which, for a matrix representation, is simply the commutator:

$$i[T_i, T_j] = i\ (T_i\ T_j - T_j\ T_i) = f_{ij}{}^k\ T_k$$

The $f_{ij}{}^k$ are the structure constants of the algebra, and satisfy Jacobi identities which follow from:

$$[T_i,[T_j,T_k]] + [T_j,[T_k,T_i]] + [T_k,[T_i,T_j]] \equiv 0 \tag{1}$$

The fields $\phi$ transform under a representation of the Lie algebra in the following sense: the commutator of two transformations with parameters

$$\Lambda = \Lambda^i T_i \text{ and } \Omega = \Omega^i T_i$$

is a third transformation with parameter

$$i[\Lambda, \Omega] = (\Omega^j \Lambda^i f_{ij}{}^k) T_k.$$

Thus:

$$\begin{aligned}
[\delta(\Lambda), \delta(\Omega)]\phi &= \delta(\Lambda)\delta(\Omega)\phi - \delta(\Omega)\delta(\Lambda)\phi \\
&= \delta(\Lambda)(i\Omega\phi) - \delta(\Omega)(i\Lambda\phi) \\
&= (i\Lambda)(i\Omega\phi) - (i\Omega)(i\Lambda\phi) \\
&= i(i[\Lambda, \Omega])\phi \\
&= \delta(i[\Lambda, \Omega])\phi.
\end{aligned} \tag{2}$$

The Lie algebra can be exponentiated to a Lie group; in that case, I shall write

$$g(\Lambda)\phi := e^{i\Lambda}\phi \text{ or simply } \phi' = e^{i\Lambda}\phi.$$

Here $e^{i\Lambda}$ is an element of a matrix representation of the Lie group. If $\Lambda$ is Hermitian, $\Lambda = \Lambda^{\dagger}$, then $e^{i\Lambda}$ is unitary:

$$[e^{i\Lambda}]^{\dagger} = e^{-i\Lambda}.$$

A particular case of interest arises when the fields $\phi$ are themselves elements of the Lie algebra, i.e., the vector space to which the $\phi$'s belong is spanned by the $T_i$'s; in that case, the $\phi$'s are said to be in the adjoint representation, and are conveniently written as matrices $\phi = \phi^i T_i$ on which the Lie algebra acts by

$$\delta(\Lambda)\phi = i[\Lambda,\phi].$$

Equation (2) now becomes a consequence of the Jacobi identities (1):

$$\begin{aligned}[\delta(\Lambda),\delta(\Omega)]\phi &= i[\Lambda,i[\Omega,\phi]] - i[\Omega,i[\Lambda,\phi]] \\ &= i[i[\Lambda,\Omega],\phi] \\ &= \delta(i[\Lambda,\Omega])\phi.\end{aligned}$$

The exponentiated or finite transformation on the matrix $\phi$ is written

$$g(\Lambda)\phi = e^{i\Lambda}\phi e^{-i\Lambda} \text{ or simply } \phi' = e^{i\Lambda}\phi e^{-i\Lambda}.$$

<u>Spacetime symmetries: the Poincaré group</u>

The preceding example of a matrix representation is typical of an internal symmetry; spacetime symmetries need to be implemented differently. Notions of spacetime, fields, translations and rotations are so familiar that they are usually taken for granted; they can, however, be derived in an elegant albeit formal way by considering induced representations of the Poincaré group ISO(3,1). I shall present this derivation in detail, as it generalizes immediately to give superspace, superfields, and supersymmetry transformations.

The Poincaré algebra has generators $P_a$ and $J_{ab}$ with Lie brackets:

$$i[P_a,P_b] = 0,$$

$$i[J_{ab},P_c] = \tfrac{1}{2}(\eta_{bc}\,P_a - \eta_{ac}\,P_b) \iff i[\omega^{ab}\,J_{ab},P_c] = \omega^a{}_c\,P_a$$

$$i[J_{ab},J_{cd}] = \tfrac{1}{2}(\eta_{bc}\,J_{ad} + \eta_{bd}\,J_{ca} - \eta_{ac}\,J_{bd} - \eta_{ad}\,J_{cd}) \iff$$

$$i[\omega^{ab}\,J_{ab},J_{cd}] = \omega^a{}_c\,J_{ad} + \omega^a{}_d\,J_{ca}. \qquad (3)$$

In two-component notation,

$$J_{\alpha\dot\alpha\beta\dot\beta} = \sigma^a_{\alpha\dot\alpha}\,\sigma^b_{\beta\dot\beta}\,J_{ab} = \varepsilon_{\alpha\beta}\,J_{\dot\alpha\dot\beta} + \varepsilon_{\dot\alpha\dot\beta}\,J_{\alpha\beta}, \text{ where}$$

$$J_{\alpha\beta} := \tfrac{1}{2}J_{\alpha\dot\alpha\beta}{}^{\dot\alpha} \text{ and } J_{\dot\alpha\dot\beta} := \tfrac{1}{2}J_{\alpha\dot\alpha}{}^{\alpha}{}_{\dot\beta};$$

$$\omega^{ab}\,J_{ab} = \tfrac{1}{2}(\omega^{\alpha\beta}\,J_{\alpha\beta} + \bar\omega^{\dot\alpha\dot\beta}\,J_{\dot\alpha\dot\beta}).$$

(Every antisymmetric tensor $T_{ab} = -\,T_{ba}$ can be decomposed into a selfdual part $T_{\alpha\beta} = T_{\beta\alpha}$ and antiselfdual part $T_{\dot\alpha\dot\beta} = T_{\dot\beta\dot\alpha}$).Equations (3) become:

$$i[P_{\alpha\dot\alpha},P_{\beta\dot\beta}] = 0,$$

$$\tfrac{1}{2}i[\omega^{\alpha\beta}\,J_{\alpha\beta} + \bar\omega^{\dot\alpha\dot\beta}\,J_{\dot\alpha\dot\beta},P_{\gamma\dot\gamma}] = -\tfrac{1}{2}(\omega_\gamma{}^\alpha\,P_{\alpha\dot\gamma} + \bar\omega_{\dot\gamma}{}^{\dot\alpha}\,P_{\gamma\dot\alpha}),$$

$$i[J_{\alpha\beta},J_{\dot\gamma\dot\delta}] = 0,$$

$$\tfrac{1}{2}i[\omega^{\alpha\beta}\,J_{\alpha\beta},J_{\gamma\delta}] = -\tfrac{1}{2}(\omega_\gamma{}^\alpha\,J_{\alpha\delta} + \omega_\delta{}^\alpha\,J_\gamma{}^\alpha), \text{ etc.} \qquad (4)$$

The Lorentz group SO(3,1) is a subgroup of the Poincaré group ISO(3,1) with only the J's (but not the P's) as generators. Consider now the coset space ISO(3,1)/SO(3,1), that is, the space of orbits that the Lorentz group SO(3,1) sweeps out in the Poincaré group ISO(3,1). One can choose an origin in this space, and coordinatize its neighborhood by exponentiating the coordinates in the tangent space at that point, i.e., write a point in the coset space as:

$$h(x) = e^{ix^aP_a}\,\mathbb{1}. \qquad (5)$$

The group action is defined by left multiplication:

$$h(gx) = h(x') := g\,h(x) \bmod SO(3,1);$$

thus, for $g = e^{i\xi^aP_a}$,

$$h(x') = e^{i\xi^aP_a}\,e^{ix^bP_b}\,\mathbb{1} = e^{i(\xi+x)^aP_a}\,\mathbb{1}, \text{ so}$$

$$g(\xi)x^a = x'^a = x^a + \xi^a, \quad \text{or} \quad \delta(\xi)x^a = \xi^a; \tag{6}$$

for $g = e^{i\omega^{ab}J_{ab}}$,

$$h(x') = e^{i\omega^{ab}J_{ab}}\, e^{ix^cP_c}\, 1\!\!1 \mod SO(3,1)$$

$$= e^{i\omega^{ab}J_{ab}}\, e^{ix^cP_c}\, e^{-i\omega^{ab}J_{ab}}\, 1\!\!1 \mod SO(3,1)$$

$$= e^{\left(e^{i\omega^{ab}J_{ab}}\,(ix^cP_c)\,e^{-i\omega^{ab}J_{ab}}\right)}\, 1\!\!1 \mod SO(3,1)$$

$$= e^{ix^cP_c + ix^c(i[\omega^{ab}J_{ab},P_c]) + o(\omega^2)}\, 1\!\!1$$

$$= e^{i(x^c + \omega^c{}_b\, x^b)P_c + o(\omega^2)}\, 1\!\!1, \text{ so}$$

$$g(\omega)x^a = x'^a = x^a + \omega^a{}_b\, x^b + o(\omega^2), \text{ or } \delta(\omega)x^a = \omega^a{}_b\, x^b. \tag{7}$$

Fields can now be defined as functions from the x's into a vector space on which the Lorentz group SO(3,1) acts by ordinary matrix multiplication; for example, in the case of a scalar field $\phi(x)$, the matrix part of the SO(3,1) transformation is trivial, and one has:

$$g\phi(x) = \phi(g^{-1}x), \text{ or } \phi'(x') = \phi(x). \tag{8}$$

For $g = e^{i\xi P}$,

$$\phi'(x + \xi) = \phi(x), \text{ or } \delta(\xi)\phi = -\xi^a\, \partial_a\, \phi, \tag{9}$$

whereas for $g = e^{i\omega J}$,

$$\phi'(x + \omega x + o(\omega^2)) = \phi(x), \text{ or } \delta(\omega)\phi = -\omega^b{}_a\, x^a\, \partial_b\, \phi. \tag{10}$$

Thus the generators of the Poincaré group are represented as differential operators acting on fields $\phi(x)$:

$$P_a = i\partial_a,$$

$$J_{ab} = \tfrac{1}{2}i(x_b\partial_a - x_a\partial_b). \tag{11}$$

A more complicated representation arises when the matrix part of the SO(3,1) transformation is nontrivial; for a vector $V_a(x)$,

$$\delta(\omega)V_a = -\omega^b{}_c\, x^c\, \partial_b\, V_a + \omega_a{}^b\, V_b; \tag{12}$$

for a spinor $\Psi_\alpha(x)$,

$$\delta(\omega)\Psi_\alpha = -\omega^b{}_c\, x^c\, \partial_b\, \Psi_\alpha + \omega_\alpha{}^\beta\, \Psi_\beta. \tag{13}$$

For any representation, it follows that

$$[\delta(\omega),\delta(\Omega)]\phi = \delta(\omega_a{}^b\, \Omega_b{}^c - \Omega_a{}^b\, \omega_b{}^c)\phi. \tag{14}$$

In addition to transforming under a representation of the Poincaré group, a field may of course transform under the action of some internal symmetry group as well; however, the no-go theorem of Coleman & Mandula (1967) states that all the generators of the internal symmetry group must commute with all the generators of the Poincaré group.

## Superalgebras

A superalgebra differs from an ordinary Lie algebra in that certain Lie brackets are symmetric, i.e., a matrix representation has anticommutators as well as commutators. The generators of a superalgebra are graded: some, $\{E_i\}$, are called even, and others, $\{Q_s\}$, are called odd. The Lie bracket is antisymmetric for two even or an odd and an even generator:

$$i[E_i,E_j] = f_{ij}{}^k\, E_k \text{ and}$$

$$i[Q_s,E_j] = f_{sj}{}^t\, Q_t, \tag{15}$$

but is symmetric for two odd generators:

$$\{Q_s,Q_t\} = f_{st}{}^k\, E_k. \tag{16}$$

The structure constants obey graded Jacobi identities, that is , Jacobi identities with an extra factor of -1 whenever two Q's are interchanged. To keep track of signs, one saturates the indices of the generators with graded parameters $\zeta_i$ and $\varepsilon_s$ which obey

$$[\zeta_i,\xi_j] = [\zeta_i,\varepsilon_s] = \{\varepsilon_s,\eta_t\} = 0 \text{ and}$$

$$[\zeta_i,E_j] = [\zeta_i,Q_s] = [\varepsilon_s,E_j] = \{\varepsilon_s,Q_t\} = 0.$$

Then equations (15) and (16) become:

$$i[\zeta^i\, E_i,\xi^j\, E_j] = \zeta^i\, \xi^j\, f_{ij}S^k\, E_k,$$

$$i[\varepsilon^s\, Q_s,\zeta^j\, E_j] = \varepsilon^s\, \zeta^j\, f_{sj}{}^t\, Q_t, \text{ and}$$

$$i[\varepsilon^s Q_s, \eta^t Q_t] = i\, \eta^t \varepsilon^s f_{st}{}^k E_k.$$

The graded Jacobi identities now follow from e.g.,

$$[\varepsilon^s Q_s,[\eta^t Q_t,\zeta^i E_i]] + [\eta^t Q_t,[\zeta^i E_i,\varepsilon^s Q_s]] + [\zeta^i E_i,[\varepsilon^s Q_s,\eta^t Q_t]] \equiv 0,$$

$$[\varepsilon^s Q_s,[\xi^j E_j,\zeta^i E_i]] + [\xi^j E_j,[\zeta^i E_i,\varepsilon^s Q_s]] + [\zeta^i E_i,[\varepsilon^s Q_s,\xi^j E_j]] \equiv 0, \text{ etc.} \tag{17}$$

The even generators $\{E_i\}$ form an ordinary Lie algebra which acts on the odd generators $\{Q_s\}$ through a matrix representation

$$T(E_i)\;_s{}^t = i\, f_{is}{}^t.$$

As a consequence of the Jacobi identities (17), the matrices $T_i := T(E_i)$ obey:

$$i[T_i,T_j] = f_{ij}{}^k\, T_k.$$

Super-Poincaré algebra

The superalgebra on which I shall focus is the super-Poincaré algebra; it has, for its even part, the ordinary Poincaré algebra (eqs. (3) and (4)), and in addition, a single odd Majorana charge

$$Q_\alpha = -(Q_{\dot\alpha})^\dagger$$

with four independent components. The super-Lie brackets are those for the Poincaré algebra (eqs. (3) and (4)), and in addition:

$$\tfrac{1}{2} i[\omega^{\alpha\beta} J_{\alpha\beta} + \bar\omega^{\dot\alpha\dot\beta} J_{\dot\alpha\dot\beta}, Q_\gamma] = -\tfrac{1}{2}\, \omega_\gamma{}^\alpha\, Q_\alpha, \text{ or}$$

$$i[J_{\alpha\beta},Q_\gamma] = \tfrac{1}{2}(\varepsilon_{\gamma\beta}\, Q_\alpha + \varepsilon_{\gamma\beta}\, Q_\beta) \text{ and } i[J_{\dot\alpha\dot\beta},Q_\gamma] = 0; \tag{18}$$

analogous relations hold for $[J,\bar Q_{\dot\alpha}]$, and finally,

$$\{Q_\alpha,Q_\beta\} = \{\bar Q_{\dot\alpha},\bar Q_{\dot\beta}\} = i[P_a,Q_\alpha] = i[P_a,\bar Q_{\dot\alpha}] = 0,$$

$$\{Q_\alpha,\bar Q_{\dot\beta}\} = -2\, P_{\alpha\dot\beta} = -2\, \sigma^a_{\alpha\dot\beta}\, P_a. \tag{19}$$

Equations (3), (18), and (19) are the structure equations of the super-Poincaré algebra. This is the relevant algebra for what is called N=1 or simple supersymmetry. Before discussing its representations, however, I want to describe briefly some other superalgebras of physical interest, or at least, of interest to physicists.

Extended super-Poincaré algebras, etc.

The N=1 supersymmetry algebra described above can be easily extended: the number of odd charges is increased to N $Q_{\alpha i}$, and even internal symmetry charges $T_q$ which rotate the $Q_{\alpha i}$ amongst themselves are added; also (even) central charges $Z_{ij}$ which commute with all generators may be added. Thus eq. (3) is unmodified, but (18) and (19) become:

$$i[J_{\alpha\beta},Q_{\gamma i}] = \tfrac{1}{2}(\varepsilon_{\gamma\beta}\, Q_{\alpha i} + \varepsilon_{\gamma\alpha}\, Q_{\beta i}),$$

$$\{Q_{\alpha i},\bar{Q}_{\dot{\beta}}{}^{j}\} = -2\,\delta_i{}^j\, P_{\alpha\dot{\beta}},$$

$$\{Q_{\alpha i},Q_{\beta j}\} = \varepsilon_{\alpha\beta}\, Z_{ij},$$

$$i[T_q,Q_{\alpha i}] = (M_q)_i{}^j\, Q_{\alpha j},$$

$$i[T_q,T_r] = f_{qr}{}^s\, T_s,\ \text{etc.} \tag{20}$$

The matrices $(M_q)_i{}^j$ are a representation of the generators $T_q$. The algebra generated by the $T_q$'s is not uniquely determined: for example, in the absence of central charges ($Z_{ij}$ = 0), the algebra may be as large as U(N), where i = 1,...,N; if $Z_{ij} = \Omega_{ij}\, Z$, where $\Omega_{ij}$ is a symplectic form, e.g., a nondegenerate real antisymmetric matrix, the algebra must preserve $\Omega_{ij}$ as a consequence of the Jacobi identities, and hence is restricted to the symplectic algebra USp(2K) (sometimes called Sp(K) in the literature) where N = 2K or N = 2K + 1; for details, see Haag et al. (1975). Note that in contrast with the ungraded case, internal and spacetime generators mix in nontrivial ways.

In addition to the various super-Poincaré algebras, there exist both simple and extended superversions of the conformal algebra, the anti-de Sitter algebra SO(3,2), and the Poincaré algebra in spacetime dimensions other than four (see Nahm 1978 for a complete list). These will not be discussed here; interested readers are referred to Kaku 1979 and references therein, Ferrara 1977, and the work of Joel Scherk, e.g., Scherk 1978 and references therein.

## Representations of N=1 super-Poincaré algebra

To find representations of the simple super-Poincaré algebra, one simply repeats the analysis of the Poincaré algebra, taking in this case the coset space super-ISO(3,1)/SO(3,1), that is, the space of orbits that the ordinary Lorentz group SO(3,1) sweeps out in the super-Poincaré group super-ISO(3,1). A point in this space can be parametrized in analogy with eq. (5):

$$h(x,\theta,\bar{\theta}) = e^{(ix^a P_a - \theta^\alpha Q_\alpha - \bar{\theta}^{\dot{\alpha}}\bar{Q}_{\dot{\alpha}})}\, 1\!\!1. \tag{21}$$

Here $\theta$ and $\bar{\theta}$ are odd anticommuting Majorana coordinates in the coset space; $x$, $\theta$ and $\bar{\theta}$ will be referred to collectively as $z$, and are the coordinates of superspace. The group action is again defined by left multiplication

$$h(gz) = h(z') := g\, h(z) \bmod SO(3,1);$$

thus, for $g = e^{i\xi P}$, one finds eq. (6) as before, and in addition $\delta(\xi)\theta = \delta(\xi)\bar{\theta} = 0$. For $g = e^{-(\varepsilon Q + \bar{\varepsilon}\bar{Q})}$,

$$\begin{aligned} h(z') &= e^{-(\varepsilon Q + \bar{\varepsilon}\bar{Q})}\, e^{(ixP - \theta Q - \bar{\theta}\bar{Q})}\, 1\!\!1 \\ &= e^{(ixP - (\varepsilon+\theta)Q - (\bar{\varepsilon}+\bar{\theta})\bar{Q} + \frac{1}{2}[\varepsilon Q+\bar{\varepsilon}\bar{Q},\, \theta Q+\bar{\theta}\bar{Q}])}\, 1\!\!1 \\ &= e^{(ix^a P_a - (\varepsilon+\theta)^\alpha Q_\alpha - (\bar{\varepsilon}+\bar{\theta})^{\dot{\alpha}}\bar{Q}_{\dot{\alpha}} + (\varepsilon^\alpha\bar{\theta}^{\dot{\alpha}}+\bar{\varepsilon}^{\dot{\alpha}}\theta^\alpha)P_{\alpha\dot{\alpha}})}\, 1\!\!1 \\ &= e^{(i(x^a - i(\varepsilon^\alpha\bar{\theta}^{\dot{\alpha}} + \bar{\varepsilon}^{\dot{\alpha}}\theta^\alpha)\sigma^a_{\alpha\dot{\alpha}})P_a} \\ &\qquad {}^{- (\varepsilon^\alpha + \theta^\alpha)Q_\alpha - (\bar{\varepsilon}^{\dot{\alpha}} + \bar{\theta}^{\dot{\alpha}})\bar{Q}_{\dot{\alpha}})}\, 1\!\!1 \end{aligned} \tag{22}$$

Hence, for supersymmetry transformations,

$$\begin{aligned} x'^a &= x^a - i(\varepsilon^\alpha\bar{\theta}^{\dot{\alpha}} + \bar{\varepsilon}^{\dot{\alpha}}\theta^\alpha)\sigma^a_{\alpha\dot{\alpha}}, \\ \theta'^\alpha &= \theta^\alpha + \varepsilon^\alpha, \quad \bar{\theta}'^{\dot{\alpha}} = \bar{\theta}^{\dot{\alpha}} + \bar{\varepsilon}^{\dot{\alpha}}, \text{ or} \\ \delta(\varepsilon)x^a &= -i(\varepsilon^\alpha\bar{\theta}^{\dot{\alpha}} + \bar{\varepsilon}^{\dot{\alpha}}\theta^\alpha)\sigma^a_{\alpha\dot{\alpha}}, \\ \delta(\varepsilon)\theta^\alpha &= \varepsilon^\alpha, \quad \delta(\varepsilon)\bar{\theta}^{\dot{\alpha}} = \bar{\varepsilon}^{\dot{\alpha}}. \end{aligned} \tag{23}$$

For $g = e^{i\omega J}$, one finds in addition to eq. (7),

$$\delta(\omega)\theta^\alpha = -\tfrac{1}{2}\omega^\alpha{}_\beta\, \theta^\beta \text{ and } \delta(\bar{\omega})\bar{\theta}^{\dot{\alpha}} = -\tfrac{1}{2}\bar{\omega}^{\dot{\alpha}}{}_{\dot{\beta}}\, \bar{\theta}^{\dot{\beta}}.$$

Superfields can now be defined as Grassmann-valued functions from the z's into a vector space on which the Lorentz group SO(3,1) acts by ordinary matrix multiplication; for example, the scalar superfield $\phi(z)$ transforms analogously to eq. (8):

$$g\phi(z) = \phi(g^{-1}z), \text{ or } \phi'(z') = \phi(z).$$

The analogs of equations (9) and (10) are:

$$\delta(\xi)\phi = -\xi^a \partial_a \phi,$$

$$\delta(\varepsilon)\phi = -(\varepsilon^\alpha \partial_\alpha + \bar{\varepsilon}^{\dot\alpha} \bar{\partial}_{\dot\alpha} - i(\varepsilon^\alpha \bar{\theta}^{\dot\alpha} + \bar{\varepsilon}^{\dot\alpha}\theta^\alpha) \partial_{\alpha\dot\alpha}) \phi, \tag{24}$$

where

$$\partial_\alpha = \frac{\partial}{\partial\theta^\alpha} \text{ and } \bar{\partial}_{\dot\alpha} = \frac{\partial}{\partial\bar{\theta}^{\dot\alpha}} \Rightarrow (\partial_\alpha \theta^\alpha) = \delta_\alpha{}^\beta \text{ and } (\bar{\partial}_{\dot\alpha} \bar{\theta}^{\dot\beta}) = \delta_{\dot\alpha}{}^{\dot\beta},$$

and hence

$$(\partial_\alpha)^\dagger = -\bar{\partial}_{\dot\alpha}, \quad (\bar{\partial}_{\dot\alpha})^\dagger = -\partial_\alpha.$$

Thus the generators of the super-Poincaré algebra are represented as differential operators acting on superfields $\phi(z)$ in analogy with eq. (11):

$$P_a = i\partial_a,$$

$$J_{ab} = \tfrac{1}{2}i(x_b\partial_a - x_a\partial_b) + \tfrac{1}{2}i\Sigma_{ab}^{\alpha\beta}(\theta_\alpha\partial_\beta + \theta_\beta\partial_\alpha)$$
$$+ \tfrac{1}{2}i\bar{\Sigma}_{ab}^{\dot\alpha\dot\beta}(\bar{\theta}_{\dot\alpha}\bar{\partial}_{\dot\beta} + \bar{\theta}_{\dot\beta}\bar{\partial}_{\dot\alpha})$$

$$Q_\alpha = \partial_\alpha - i\bar{\theta}^{\dot\alpha}\partial_{\alpha\dot\alpha}, \text{ and}$$

$$\bar{Q}_{\dot\alpha} = \bar{\partial}_{\dot\alpha} - i\theta^\alpha\partial_{\alpha\dot\alpha}, \tag{25}$$

where

$$\Sigma_{ab}^{\alpha\beta} = \tfrac{1}{4}\,\sigma_a{}^\alpha{}_{\dot\alpha}\,\sigma_b{}^{\beta\dot\alpha} \text{ and } \bar{\Sigma}_{ab}^{\dot\alpha\dot\beta} = \tfrac{1}{4}\,\sigma_{a\alpha}{}^{\dot\alpha}\,\sigma_b{}^{\alpha\dot\beta}.$$

Equations (12), (13), and (14) generalize to the graded case with no further effort. In addition to transforming under a representation of the super-Poincaré group, a superfield may also transform under a further internal symmetry group which commutes with the super-Poincaré group (as in the nonsupersymmetric case; but superalgebras do unify spacetime and internal symmetries).

At this point, some new features without analog in the nonsupersymmetric case enter; because the odd generators have nonvanishing anticommutator, a different (conjugate) representation can be found by using right multiplication rather than left to define the group action; however, because the odd generators Q and $\bar{Q}$ do not rotate the orbit within super-ISO(3,1), that is, their action on the elements of the coset space is identical to their action on the points in the whole supergroup (this is not the case for the Lorentz generators J), one can define new operators $D_\alpha$ and $\bar{D}_{\dot{\alpha}}$ by

$$e^{2i(\zeta^\alpha D_\alpha - \bar{\zeta}^{\dot{\alpha}}\bar{D}_{\dot{\alpha}})}\, e^{(ix^a P_a - \theta^\alpha Q_\alpha - \bar{\theta}^{\dot{\alpha}}\bar{Q}_{\dot{\alpha}})}\, 1\!\!1$$

$$:= e^{(ix^a P_a - \theta^\alpha Q_\alpha - \bar{\theta}^{\dot{\alpha}}\bar{Q}_{\dot{\alpha}})}\, e^{-(\zeta^\alpha Q_\alpha + \bar{\zeta}^{\dot{\alpha}}\bar{Q}_{\dot{\alpha}})}\, 1\!\!1$$

$$= e^{-(\zeta^\alpha Q_\alpha + \bar{\zeta}^{\dot{\alpha}}\bar{Q}_{\dot{\alpha}}) + [\theta^\alpha Q_\alpha + \bar{\theta}^{\dot{\alpha}}\bar{Q}_{\dot{\alpha}},\, \zeta^\alpha Q_\alpha + \bar{\zeta}^{\dot{\alpha}}\bar{Q}_{\dot{\alpha}}]}$$

$$\times\, e^{(ix^a P_a - \theta^\alpha Q_\alpha - \bar{\theta}^{\dot{\alpha}}\bar{Q}_{\dot{\alpha}})}\, 1\!\!1$$

$$= e^{-(\zeta^\alpha Q_\alpha + \bar{\zeta}^{\dot{\alpha}}\bar{Q}_{\dot{\alpha}}) - 2(\zeta^\alpha\bar{\theta}^{\dot{\alpha}} + \bar{\zeta}^{\dot{\alpha}}\theta^\alpha)P_{\alpha\dot{\alpha}}}$$

$$\times\, e^{(ix^a P_a - \theta^\alpha Q_\alpha - \bar{\theta}^{\dot{\alpha}}\bar{Q}_{\dot{\alpha}})}\, 1\!\!1$$

$$= e^{-(\zeta^\alpha(Q_\alpha + 2\bar{\theta}^{\dot{\alpha}}P_{\alpha\dot{\alpha}}) + \bar{\zeta}^{\dot{\alpha}}(\bar{Q}_{\dot{\alpha}} + 2\theta^\alpha P_{\alpha\dot{\alpha}}))}$$

$$\times\, e^{(ix^a P_a - \theta^\alpha Q_\alpha - \bar{\theta}^{\dot{\alpha}}\bar{Q}_{\dot{\alpha}})}\, 1\!\!1$$

$$\Rightarrow D_\alpha = \tfrac{1}{2}i(Q_\alpha + 2\bar{\theta}^{\dot{\alpha}}P_{\alpha\dot{\alpha}}) = \tfrac{1}{2}i(\partial_\alpha + i\bar{\theta}^{\dot{\alpha}}\partial_{\alpha\dot{\alpha}}),\text{ and}$$

$$\bar{D}_{\dot{\alpha}} = -\tfrac{1}{2}i(\bar{Q}_{\dot{\alpha}} + 2\theta^\alpha P_{\alpha\dot{\alpha}}) = -\tfrac{1}{2}i(\bar{\partial}_{\dot{\alpha}} + i\theta^\alpha\partial_{\alpha\dot{\alpha}}). \qquad (26)$$

Now note that group associativity has interesting consequences:

$$\left(e^{-(\varepsilon Q + \bar{\varepsilon}\bar{Q})}\, h(z)\right) e^{-(\zeta Q + \bar{\zeta}\bar{Q})} = e^{-(\varepsilon Q + \bar{\varepsilon}\bar{Q})}\left(h(z)\, e^{-(\zeta Q + \bar{\zeta}\bar{Q})}\right)$$

$$\Rightarrow e^{2i(\zeta D + \bar{\zeta}\bar{D})}\, e^{-(\varepsilon Q + \bar{\varepsilon}\bar{Q})}\, h(z)$$

$$= e^{-(\varepsilon Q + \bar{\varepsilon}\bar{Q})}\, e^{2i(\zeta D + \bar{\zeta}\bar{D})}\, h(z)$$

$$\Rightarrow \{D,Q\} = \{\bar{D},Q\} = \{D,\bar{Q}\} = \{\bar{D},\bar{Q}\} = 0.$$

Thus right multiplication leads to supersymmetrically invariant Majorana operators D and $\bar{D}$. Finally, let me note that the algebra of the D's is, up to the normalization factors, conjugate to the algebra of the Q's:

$$\{D_\alpha, \bar{D}_{\dot\alpha}\} = +\tfrac{1}{2} i \partial_{\alpha\dot\alpha}. \tag{27}$$

Chiral representations and coordinates

There is an essential difference between the nonsupersymmetric and the supersymmetric cases in the action of the factor group SO(3,1) on the coordinates of the coset space: although the spacetime coordinates are rotated into each other, the fermionic and the bosonic coordinates do not mix; this reducibility of the action of the Lorentz group on superspace leads to many interesting consequences.

For example, one has much greater freedom in parametrizing the coset space; instead of (21), one could use as well

$$h_+(x,\theta,\bar\theta) = e^{ixP - \theta Q} e^{-\bar\theta\bar Q} \mathbb{1}, \tag{28}$$

or instead,

$$h_-(x,\theta,\bar\theta) = e^{ixP - \bar\theta\bar Q} e^{-\theta Q} \mathbb{1}. \tag{29}$$

One then finds, instead of (25) and (26),

$$Q_{\alpha+} = \partial_\alpha, \quad \bar Q_{\dot\alpha+} = \bar\partial_{\dot\alpha} - 2i\theta^\alpha \partial_{\alpha\dot\alpha},$$

$$D_{\alpha+} = \tfrac{1}{2} i(\partial_\alpha + 2i\bar\theta^{\dot\alpha}\partial_{\alpha\dot\alpha}), \quad \bar D_{\alpha+} = -\tfrac{1}{2} i\bar\partial_{\dot\alpha},$$

$$Q_{\alpha-} = \partial_\alpha - 2i\bar\theta^{\dot\alpha}\partial_{\alpha\dot\alpha}, \quad \bar Q_{\dot\alpha-} = \bar\partial_\alpha,$$

$$D_{\alpha-} = \tfrac{1}{2} i\partial_\alpha, \quad \bar D_{\dot\alpha-} = -\tfrac{1}{2} i(\bar\partial_{\dot\alpha} + 2i\theta^\alpha\partial_{\alpha\dot\alpha}). \tag{30}$$

Thus one arrives at three equivalent representations of the super-Poincaré algebra; the equivalence can be made explicit by defining

$$Q_A := (P_a, Q_\alpha, \bar Q_{\dot\alpha}) \quad \text{and} \quad D_A := (\partial_a, D_\alpha, \bar D_{\dot\alpha}).$$

A quick calculation reveals that for

$$U_o := 2\theta^\alpha\bar\theta^{\dot\alpha}\partial_{\alpha\dot\alpha},$$

$$Q_{A+} = e^{-\frac{1}{2}iU_o} Q_A e^{\frac{1}{2}iU_o}, \quad D_{A+} = e^{-\frac{1}{2}iU_o} D_A e^{\frac{1}{2}iU_o},$$

$$Q_{A-} = e^{\frac{1}{2}iU_o} Q_A e^{-\frac{1}{2}iU_o}, \quad D_{A-} = e^{\frac{1}{2}iU_o} D_A e^{-\frac{1}{2}iU_o}. \tag{31}$$

The transformations (31) can be thought of as similarity transformations or equivalently as coordinate transformations to complex coordinates:

$$z^A := e^{\pm\frac{1}{2}iU_o} z^A$$

$$\Rightarrow \theta_\pm = \theta, \quad \bar{\theta}_\pm = \bar{\theta}, \quad x_\pm^{\alpha\dot{\alpha}} = x^{\alpha\dot{\alpha}} \pm i\theta^\alpha\bar{\theta}^{\dot{\alpha}}. \tag{32}$$

The coordinates $z_\pm$ are called chiral (antichiral) coordinates; the operators $Q_+$ and $\bar{D}_+$ are said to be in the chiral (antichiral) representation. Of course other representations can be defined; but these two, and the original symmetric or vector representation are the most convenient.

## Chiral superfields

A consequence of the reducibility discussed in the previous section is that complex scalar superfields are not irreducible representations of the super-Poincaré group; explicitly, one can impose the invariant constraint $\bar{D}\phi(z) = 0$ and find nontrivial solutions. Indeed, in the chiral representation (28), $\bar{D}_+ = -\frac{1}{2}i\bar{\partial}$ (see (30)), and hence the constraint is solved by any superfield $\phi_+(z)$ independent of $\bar{\theta}$:

$$\phi_+(z) = \phi_+(x,\theta) \Rightarrow \phi(z) = (e^{\frac{1}{2}iU_o}\phi_+(x,\theta)) = \phi(x_+,\theta).$$

Similarly, there exist antichiral superfields; the Hermitian conjugate of a chiral field is automatically antichiral:

$$\bar{D}\,\phi(z) = 0 \Leftrightarrow D\,\bar{\phi}(z) = 0;$$

in antichiral representation, an antichiral field is independent of $\theta$:

$$\bar{\phi}_-(z) = \bar{\phi}_-(x,\bar{\theta}) \Rightarrow \bar{\phi}(z) = (e^{-\frac{1}{2}iU_o}\bar{\phi}_-(x,\bar{\theta})) = \bar{\phi}(x_-,\bar{\theta}).$$

A real superfield cannot be chiral, as Hermitian conjugation would imply it is antichiral as well, and then (see (27)):

$$D\phi = \bar{D}\phi = 0 \Rightarrow \{D_\alpha,\bar{D}_{\dot{\alpha}}\}\phi = \tfrac{1}{2}i\partial_{\alpha\dot{\alpha}}\phi = 0 \Rightarrow \phi \text{ is constant.}$$

# COMPONENTS OF SUPERFIELDS AND ACTIONS IN SUPERSPACE

## Components of superfields

The formalism presented above, though elegant, has little obvious relation to conventional field theory in ordinary four dimensional spacetime; to find such a relation, one needs to express superfields in terms of their conventional field components. I shall present two equivalent procedures establishing the relation.

The θ-expansion. Because the θ's anticommute, any expression containing a sufficiently high power as a factor vanishes. Thus:

$$(\theta)^3 = (\bar{\theta})^3 = 0. \tag{33}$$

Similarly,

$$(\partial_\alpha)^3 = (\bar{\partial}_{\dot{\alpha}})^3 = (D_\alpha)^3 = (\bar{D}_{\dot{\alpha}})^3 = 0. \tag{34}$$

This implies that any superfield has a formal expansion in θ and $\bar{\theta}$, with conventional spacetime dependent components as coefficients; for example, a chiral field has the expansion:

$$\phi_+(x,\theta) = A(x) - \theta^\alpha\chi_\alpha(x) + \theta^2F(x).$$

Equations (30) now allow one to find the action of the supersymmetry algebra on the components:

$$\begin{aligned}\delta(\varepsilon)\phi_+ &= (-\varepsilon^\alpha Q_{\alpha+} - \bar{\varepsilon}^{\dot{\alpha}}\bar{Q}_{\dot{\alpha}+})\phi_+ = -(\varepsilon^\alpha\partial_\alpha + 2i\theta^\alpha\bar{\varepsilon}^{\dot{\alpha}}\partial_{\alpha\dot{\alpha}})\phi_+ \\ &= \varepsilon^\alpha\chi_\alpha(x) - 2\theta^\alpha(\varepsilon_\alpha F(x) + i\bar{\varepsilon}^{\dot{\alpha}}\partial_{\alpha\dot{\alpha}}A(x)) + \theta^2(i\bar{\varepsilon}^{\dot{\alpha}}\partial_{\alpha\dot{\alpha}}\chi^\alpha(x)) \\ &:= \delta A(x) - \theta^\alpha\delta\chi_\alpha + \theta^2\delta F(x).\end{aligned}$$

Hence

$$\begin{aligned}\delta(\varepsilon)A(x) &= \varepsilon^\alpha\chi_\alpha(x), \\ \delta(\varepsilon)\chi_\alpha(x) &= 2(\varepsilon_\alpha F(x) + i\bar{\varepsilon}^{\dot{\alpha}}\partial_{\alpha\dot{\alpha}}A(x)), \\ \delta(\varepsilon)F(x) &= i\bar{\varepsilon}^{\dot{\alpha}}\partial_{\alpha\dot{\alpha}}\chi^\alpha(x).\end{aligned} \tag{35}$$

Thus one discovers that a chiral superfield is a multiplet of complex component fields $(A(x), \chi_\alpha(x), F(x))$. Similarly, one can expand an unrestricted superfield as a polynomial in θ and $\bar{\theta}$, and interpret it as a multiplet of components. Thus for a real superfield V(z):

$$\begin{aligned}V(x,\theta,\bar{\theta}) &= C(x) - \theta^\alpha\psi_\alpha(x) - \bar{\theta}^{\dot{\alpha}}\bar{\psi}_{\dot{\alpha}}(x) + \tfrac{1}{2}\theta^2H(x) + \tfrac{1}{2}\bar{\theta}^2\bar{H}(x) \\ &\quad + \theta^\alpha\bar{\theta}^{\dot{\alpha}}W_{\alpha\dot{\alpha}} - \tfrac{1}{2}\theta^2\bar{\theta}^{\dot{\alpha}}(\bar{\lambda}_{\dot{\alpha}} + i\partial_{\alpha\dot{\alpha}}\psi^\alpha) \\ &\quad - \tfrac{1}{2}\bar{\theta}^2\theta^\alpha(\lambda_\alpha + i\partial_{\alpha\dot{\alpha}}\bar{\psi}^{\dot{\alpha}}) + \tfrac{1}{4}\theta^2\bar{\theta}^2(D + \Box C),\end{aligned}$$

and

$$\delta(\varepsilon)C(x) = \varepsilon^{\alpha}\psi_{\alpha} + \bar{\varepsilon}^{\dot{\alpha}}\bar{\psi}_{\dot{\alpha}}(x),$$

$$\delta(\varepsilon)\psi_{\alpha}(x) = \varepsilon_{\alpha}H(x) + i\bar{\varepsilon}^{\dot{\alpha}}(\partial_{\alpha\dot{\alpha}}C(x) - iW_{\alpha\dot{\alpha}}(x)),$$

$$\delta(\varepsilon)\bar{\psi}_{\dot{\alpha}}(x) = \bar{\varepsilon}_{\dot{\alpha}}\bar{H}(x) + i\varepsilon^{\alpha}(\partial_{\alpha\dot{\alpha}}C(x) + iW_{\alpha\dot{\alpha}}(x))$$

$$\delta(\varepsilon)H(x) = \bar{\varepsilon}^{\dot{\alpha}}(\bar{\lambda}_{\dot{\alpha}}(x) + 2i\partial_{\alpha\dot{\alpha}}\psi^{\alpha}(x)),$$

$$\delta(\varepsilon)W_{\alpha\dot{\alpha}}(x) = \bar{\varepsilon}_{\dot{\alpha}}\lambda_{\alpha}(x) - \varepsilon_{\alpha}\bar{\lambda}_{\dot{\alpha}}(x) + \partial_{\alpha\dot{\alpha}}(i\varepsilon^{\beta}\psi_{\beta}(x) - i\bar{\varepsilon}^{\dot{\beta}}\bar{\psi}_{\dot{\beta}}(x)),$$

$$\delta(\varepsilon)\lambda_{\alpha}(x) = \varepsilon_{\alpha}D(x) + 2i\varepsilon^{\beta}F_{\beta\alpha}(x),$$

$$\delta(\varepsilon)D(x) = i(\bar{\varepsilon}^{\dot{\alpha}}\partial_{\alpha\dot{\alpha}}\lambda^{\alpha}(x) + \varepsilon^{\alpha}\partial_{\alpha\dot{\alpha}}\bar{\lambda}^{\dot{\alpha}}(x)), \text{ etc.,} \tag{36}$$

where

$$F_{\beta\alpha}(x) = \tfrac{1}{2}(\partial_{\beta\dot{\alpha}}W_{\alpha}{}^{\dot{\alpha}}(x) + \partial_{\alpha\dot{\alpha}}W_{\beta}{}^{\dot{\alpha}}(x))$$

is the selfdual part of the field strength tensor for the vector field $W_{\alpha\dot{\beta}}$.

Components by projection. For some applications, explicit θ-expansions are inconvenient; an alternative way of defining components is to project out the θ-independent sector of various derivatives of the superfield. Such a procedure has the virtue of being representation independent. Explicitly, for a chiral superfield, one defines:

$$A(x) := \phi(z)\,|_{\theta=\bar{\theta}=0},$$

$$\chi_{\alpha}(x) := 2iD_{\alpha}\,\phi(z)\,|_{\theta=\bar{\theta}=0},$$

$$F(x) := D^{\alpha}D_{\alpha}\,\phi(z)\,|_{\theta=\bar{\theta}=0}.$$

Now noting:

$$Q_{\alpha}|_{\theta=\bar{\theta}=0} = -2iD_{\alpha}|_{\theta=\bar{\theta}=0},$$

$$\bar{Q}_{\dot{\alpha}}|_{\theta=\bar{\theta}=0} = +2i\bar{D}_{\dot{\alpha}}|_{\theta=\bar{\theta}=0},$$

one finds

$$\delta(\varepsilon)A(x) = -(\varepsilon^{\alpha}Q_{\alpha} + \bar{\varepsilon}^{\dot{\alpha}}\bar{Q}_{\dot{\alpha}})\phi(z)\,|_{\theta=\bar{\theta}=0}$$

$$= 2i(\varepsilon^{\alpha}D_{\alpha} - \bar{\varepsilon}^{\dot{\alpha}}\bar{D}_{\dot{\alpha}})\phi(z)\ |_{\theta=\bar{\theta}=0}$$

$$= \varepsilon^{\alpha}\chi_{\alpha}(x)$$

in agreement with (35). Similarly,

$$\delta(\varepsilon)\chi_{\alpha}(x) = -2iD_{\alpha}\ (\varepsilon^{\beta}Q_{\beta} + \bar{\varepsilon}^{\dot{\beta}}\bar{Q}_{\dot{\beta}})\ \phi(z)\ |_{\theta=\bar{\theta}=0}$$

$$= -2i(\varepsilon^{\beta}Q_{\beta} + \bar{\varepsilon}^{\dot{\beta}}\bar{Q}_{\dot{\beta}})\ D_{\alpha}\ \phi(z)\ |_{\theta=\bar{\theta}=0}$$

$$= 4(\bar{\varepsilon}^{\dot{\beta}}\bar{D}_{\dot{\beta}} - \varepsilon^{\beta}D_{\beta})\ D_{\alpha}\ \phi(z)\ |_{\theta=\bar{\theta}=0}$$

$$= 2(i\bar{\varepsilon}^{\dot{\beta}}\partial_{\alpha\dot{\beta}} + \varepsilon_{\alpha}D^{\beta}D_{\beta})\ \phi(z)\ |_{\theta=\bar{\theta}=0}$$

$$= 2(i\bar{\varepsilon}^{\dot{\beta}}\partial_{\alpha\dot{\beta}}\ A(x) + \varepsilon_{\alpha}\ F(x)),\ \text{and}$$

$$\delta(\varepsilon)F(x) = -(\varepsilon Q + \bar{\varepsilon}\bar{Q})\ D^2\ \phi(z)\ |_{\theta=\bar{\theta}=0}$$

$$= 2i(\varepsilon D - \bar{\varepsilon}\bar{D})\ D^2\ \phi(z)\ |_{\theta=\bar{\theta}=0}$$

$$= -2i\ [\bar{\varepsilon}\bar{D}, D^2]\ \phi(z)\ |_{\theta=\bar{\theta}=0}$$

$$= -2i\bar{\varepsilon}^{\dot{\alpha}}\partial_{\alpha\dot{\alpha}}\ (-iD^{\alpha})\ \phi(z)\ |_{\theta=\bar{\theta}=0}$$

$$= +i\bar{\varepsilon}^{\dot{\alpha}}\partial_{\alpha\dot{\alpha}}\ \chi^{\alpha}(x),$$

reproducing the remainder of (35). Likewise, one can define the components of a real superfield by projection, and reproduce (36) using only (27).

## A few notes

Before continuing, I'd like to note a few useful facts about superfields.

Products of superfields. Because $D$ and $\bar{D}$ obey the Leibnitz rule, the product of two (anti)chiral superfields is again (anti)chiral. This can also be seen by using the (anti)chiral representation: since (anti)chiral fields are independent of $(\theta)\bar{\theta}$, it is clear their products will be as well. The product of a chiral superfield and its Hermitian conjugate (which is antichiral) is a real superfield.

Chiral superfields from unrestricted superfields. Because of (34), applying $(D^2)\bar{D}^2$ to an unrestricted superfield results in an (anti)chiral superfield.

Dimensions. Due to (27), and because

$$\partial_{\alpha\dot{\alpha}} \sim M$$

one finds

$$D \sim \bar{D} \sim M^{\frac{1}{2}}, \quad \theta \sim \bar{\theta} \sim M^{-\frac{1}{2}}.$$

Now a chiral superfield contains two spin zero components, and it may not be immediately clear which should be taken to have physical dimension; however, it contains only one spin $\frac{1}{2}$ component, and since

$$\chi_\alpha \sim M^{\frac{3}{2}},$$

one concludes

$$\phi \sim \theta^\alpha \chi_\alpha \sim M \quad \Rightarrow \quad A \sim M, \ F \sim M^2.$$

Similarly, for a real superfield the vector component is alone identified unambiguously, and so

$$V \sim \theta^\alpha \bar{\theta}^{\dot{\alpha}} W_{\alpha\dot{\alpha}} \sim 1 \quad \Rightarrow \quad C\sim 1, \ \Psi\sim M^{\frac{1}{2}}, \ldots, \ \lambda\sim M^{\frac{3}{2}}, \ D\sim M^2.$$

## Supersymmetric actions

Inspecting (35) and (36), one discovers that the F or $\theta^2$ component of a chiral field and the D or $\theta^2\bar{\theta}^2$ component of a real field transform as total derivatives; as a result, they are candidates for supersymmetric Lagrange densities, and

$$\int d^4x \ F(x) \text{ and } \int d^4x \ D(x) \tag{37}$$

are indeed supersymmetrically invariant actions. (Here F(x) and D(x) are meant as the $\theta^2$ and $\theta^2\bar{\theta}^2$ components of some chiral or real superfield which will in general be a product of elementary superfields). However, (37) is counter to the whole spirit of superspace; one would expect expressions of the following form:

$$\int d^4x d^2\theta \ \phi(x,\theta) \quad \text{and} \int d^8z \ V(z). \tag{38}$$

To make sense of such expressions, one needs an interpretation of integration with respect to anticommuting numbers. Such a thing is known (see Berezin 1973), and is in any case easily derived. Consider a one dimensional anticommuting space; then any function can be written as

$$f(x,\theta) = f_0 + f_1\theta.$$

If one requires linearity, then:

$$\int d\theta \ f(x,\theta) = \int d\theta(f_0 + f_1\theta) = I_0 \ f_0 + I_1 \ f_1,$$

where $I_0$ and $I_1$ are universal constants. Imposing in addition translation invariance, one finds:

$$\int d(\theta+\varepsilon)\ f(x,\theta+\varepsilon) = \int d\theta(f_0 + f_1\varepsilon + f_1\theta) = I_0(f_0 + f_1\varepsilon) + I_1 f_1$$

$$\Rightarrow I_0 = 0,$$

and hence:

$$\int d\theta(f_0 + f_1\theta) = I_1 f_1 \quad \Leftrightarrow \quad \int d\theta \propto \frac{\partial}{\partial\theta}. \tag{39}$$

$I_1$ is an arbitrary normalization constant; a useful choice is

$$\int d^2\theta\ \theta^2 = 1 \quad \Rightarrow \quad \int d^2\theta = -\tfrac{1}{4}\partial^\alpha\partial_\alpha,$$

$$\int d^2\bar\theta\ \bar\theta^2 = 1 \quad \Rightarrow \quad \int d^2\bar\theta = -\tfrac{1}{4}\bar\partial^{\dot\alpha}\bar\partial_{\dot\alpha}. \tag{40}$$

As a result of the Berezin integral, the action formulae (38) are in fact the same as (37). Note that

$$\int d\theta \sim M^{\frac{1}{2}} \quad \Rightarrow \quad \int d^4x d^2\theta \sim M^{-3} \quad \text{and} \int d^8z \sim M^{-2}. \tag{41}$$

Action for chiral superfields

The only local dimensionless action quadratic in chiral superfields and not involving a dimensional constant is:

$$I = \int d^8z\ \bar\phi\phi = \int d^8z\ \phi\bar\phi = \int d^8z(e^{\frac{1}{2}iU_0}\ \phi_+)(e^{-\frac{1}{2}iU_0}\ \bar\phi_-)$$

$$= \int d^8z\ \phi_+\ e^{-iU_0}\ \bar\phi_-. \tag{42}$$

(In the last line of (42), I have performed an integration by parts). The easiest way to express the action (42) in terms of components is by a $\theta$-expansion; however it is pedagogically useful to use the projection technique. To do so, one needs to observe that up to total derivatives (see (40)):

$$\int d^8z = \int d^4x\ D^2\ \bar D^2 = \int d^4x\ \bar D^2\ D^2, \text{ etc.} \tag{43}$$

Then

$$I = \int d^8z\ \phi\bar\phi = \int d^4x\ D^2\ \bar D^2(\phi\bar\phi) = \int d^4x\ D^2(\phi\bar D^2\bar\phi)$$

$$= \int d^4x(\phi D^2\bar D^2\bar\phi + 2(D^\alpha\phi)(D_\alpha\bar D^2\bar\phi) + (D^2\phi)(\bar D^2\bar\phi)).$$

Repeated use of (27) leads to:

$$D^2\bar{D}^2\bar{\phi} = \Box\bar{\phi} \quad \text{and} \quad D_\alpha \bar{D}^2\bar{\phi} = -i\partial_{\alpha\dot{\alpha}}\bar{D}^{\dot{\alpha}}\bar{\phi}. \tag{44}$$

Then

$$I = \int d^4x(\phi\Box\bar{\phi} - \tfrac{1}{2}i(2iD^\alpha\phi)\partial_{\alpha\dot{\alpha}}(-2i\bar{D}^{\dot{\alpha}}\bar{\phi}) + (D^2\phi)(\bar{D}^2\bar{\phi}))$$

$$\Rightarrow \quad I|_{\theta=\bar{\theta}=0} = \int d^4x(A\Box\bar{A} - \tfrac{1}{2}i\chi^\alpha\partial_{\alpha\dot{\alpha}}\bar{\chi}^{\dot{\alpha}} + F\bar{F}) \tag{45}$$

This is just the free massless action for a complex spin zero field A, a Majorana fermi field $\chi$, and a nonpropagating complex auxiliary field F. Nonderivative interactions can be added to the free action (42):

$$\begin{aligned} I_{int} &= \int d^4x d^2\theta\, V(\phi) + \int d^4x d^2\bar{\theta}\, \bar{V}(\bar{\phi}) \\ &= \int d^4x\, D^2\, V(\phi) + \int d^4x\, \bar{D}^2\, \bar{V}(\bar{\phi}) \\ &= \int d^4x(V'(\phi)D^2\phi + 2V''(\phi)(D^\alpha\phi)(D_\alpha\phi) \\ &\quad + \bar{V}'(\bar{\phi})\bar{D}^2\bar{\phi} + 2\bar{V}''(\bar{\phi})(\bar{D}^{\dot{\alpha}}\bar{D}_{\dot{\alpha}}\bar{\phi})) \end{aligned}$$

$$\begin{aligned} \Rightarrow \quad I_{int}|_{\theta=\bar{\theta}=0} &= \int d^4x(V'(A)F + \bar{V}'(\bar{A})\bar{F} \\ &\quad - \tfrac{1}{2}(V''(A)\chi^\alpha\chi_\alpha + \bar{V}''(\bar{A})\bar{\chi}^{\dot{\alpha}}\bar{\chi}_{\dot{\alpha}})). \end{aligned}$$

Eliminating the auxiliary field F finally leads to an action:

$$\begin{aligned} (I + I_{int})|_{\theta=\bar{\theta}=0} &= \int d^4x(A\Box\bar{A} - \tfrac{1}{2}i\chi^\alpha\partial_{\alpha\dot{\alpha}}\bar{\chi}^{\dot{\alpha}} - V'(A)\bar{V}'(\bar{A}) \\ &\quad - \tfrac{1}{2}(V''(A)\chi^2 + \bar{V}''(\bar{A})\bar{\chi}^2)), \end{aligned}$$

that is, an action for a complex spin zero field A which interacts with itself via an interaction potential $|V'(A)|^2$ and interacts with a Majorana spinor field $\chi$ via a generalized Yukawa interaction $\mathrm{Re}(V''(A)\chi^2)$. For a renormalizable theory (in four dimensions),

$$V(\phi) = \tfrac{1}{2}m\phi^2 + \frac{\lambda}{3!}\phi^3.$$

## SUPERSYMMETRIC YANG-MILLS THEORIES

### Rigid gauge invariance

If one considers an action (42) for several chiral fields, that is, if one takes $\phi$ in a vector space, then one can consider the action of a Lie algebra on the space (see eq. (0)):

$$\delta(\Lambda)\phi = i\Lambda\phi \text{ or } g(\Lambda)\phi = e^{i\Lambda}\phi. \tag{46}$$

In order to preserve the chirality of $\phi$, one requires that $\Lambda$ be chiral as well:

$$\bar{D}\phi = 0 \Rightarrow \bar{D}(\delta\phi) = 0 \Rightarrow \bar{D}\Lambda = 0. \tag{47}$$

Hermitian conjugation yields:

$$\delta(\Lambda)\bar{\phi} = -i\bar{\phi}\bar{\Lambda} \text{ or } g(\Lambda)\bar{\phi} = \bar{\phi}e^{-i\bar{\Lambda}}, \tag{48}$$

where

$$D\bar{\phi} = D\bar{\Lambda} = 0. \tag{49}$$

The action (42) will be invariant under (46) and (48) only if

$$\Lambda = \bar{\Lambda}. \tag{50}$$

However, as was noted earlier, the only real chiral superfield is a constant, and thus one finds that the action (42) has a rigid, or spacetime independent, gauge invariance.

Local gauge invariance

To promote the rigid, or global, gauge invariance above to a local superspace dependent invariance, one must relax either condition (50) or conditions (47) and (49); both choices lead to the same theory, and I will start by considering the former, that is:

$$\bar{D}\phi = \bar{D}\Lambda = D\bar{\phi} = D\bar{\Lambda} = 0 \tag{51}$$

as before, but

$$\Lambda \neq \bar{\Lambda}. \tag{52}$$

Now the action is no longer gauge invariant:

$$I' = \int d^8z\, \bar{\phi}'\phi' = \int d^8z\, \bar{\phi}\, e^{-i\bar{\Lambda}}\, e^{i\Lambda}\, \phi \neq I.$$

Evidently, one needs a new dynamical field which acts as a $\Lambda$ to $\bar{\Lambda}$ "converter"; it is convenient to parametrize it in the following way:

$$g(\Lambda)e^V = (e^V)' := e^{i\bar{\Lambda}}e^V e^{-i\Lambda} \text{ or } \delta(\Lambda)e^V = i(\bar{\Lambda}e^V - e^V\Lambda), \tag{53}$$

where

$$V = V^i\, T_i = V^\dagger$$

is an Hermitian superfield in a particular (Hermitian) matrix representation of the Lie algebra. Recall that an Hermitian superfield contains a vector component, so it is sensible to use V as a gauge "potential"; further, since earlier it was shown that such a V is dimensionless, it is sensible to write expressions such as (53) involving exponentials of V.

The new action

$$I_{inv} = \int d^8z\ \bar{\phi}e^V\phi \tag{54}$$

is invariant under the transformations (47), (49), and (53), subject to (51) and (52), with chiral superspace dependent parameters $\Lambda$. Because this representation of the Yang-Mills transformations involves a chiral parameter, it is often called a chiral representation; such chiral representations of gauge invariance must not be confused with the chiral representation of the supersymmetry algebra discussed earlier; indeed, I shall mostly work in the vector or symmetric representation of supersymmetry in this section, and all reference to chiral representations, unless otherwise explicitly indicated, will be to the gauge algebra. Supersymmetry transformations of the new field V are just the same as for any other field (36):

$$\delta(\varepsilon)V = -(\varepsilon Q + \bar{\varepsilon}\bar{Q})V \quad \Longleftrightarrow \quad \delta(\varepsilon)e^V = -(\varepsilon Q + \bar{\varepsilon}\bar{Q})e^V. \tag{55}$$

I now want to examine the transformations (53) more carefully; adopting the notation

$$L_V X := [V,X]$$

one finds

$$e^V X e^{-V} = e^{L_V}X;$$

then

$$[V,e^V] = 0 \Rightarrow (\delta V)e^V + V(\delta e^V) - e^V(\delta V) - (\delta e^V)V = 0$$

$$\Rightarrow e^{-\frac{1}{2}V}(\delta V)e^{\frac{1}{2}V} - e^{\frac{1}{2}V}(\delta V)e^{-\frac{1}{2}V} + e^{-\frac{1}{2}V}[V,\delta e^V]e^{-\frac{1}{2}V} = 0$$

$$\Rightarrow 2\sinh\tfrac{1}{2}L_V(\delta V) = e^{-\frac{1}{2}V}L_V(\delta e^V)e^{-\frac{1}{2}V}$$

$$= iL_V(e^{-\frac{1}{2}V}(\bar{\Lambda}e^V - e^V\Lambda)e^{-\frac{1}{2}V})$$

$$= iL_V(e^{-\frac{1}{2}L_V}\bar{\Lambda} - e^{\frac{1}{2}L_V}\Lambda)$$

$$= iL_V(\cosh\tfrac{1}{2}L_V(\bar{\Lambda}-\Lambda) - \sinh\tfrac{1}{2}L_V(\bar{\Lambda}+\Lambda))$$

$$\Rightarrow \delta V = -\tfrac{1}{2}iL_V(\bar{\Lambda}+\Lambda + \coth\tfrac{1}{2}L_V(\Lambda-\bar{\Lambda})). \tag{56}$$

Expanding, one finds:

$$\delta V = i(\bar{\Lambda}-\Lambda) - \tfrac{1}{2}i[V,\bar{\Lambda}+\Lambda] + o(V^2), \tag{57}$$

where

$$\Lambda(z) = e^{\frac{1}{2}iU_0}\,\Lambda_+(x,\theta) \text{ and } \bar{\Lambda}(z) = e^{-\frac{1}{2}iU_0}\bar{\Lambda}_-(x,\bar{\theta}).$$

If one writes

$$\Lambda_+(x,\theta) = \Lambda_0(x) - \theta^\alpha\Lambda_\alpha(x) + \tfrac{1}{2}\theta^2\Lambda_2(x)$$

one finds for the components of V (eq.(36)):

$$\delta C = i(\bar{\Lambda}_0-\Lambda_0) + o(V)$$

$$\delta\Psi_\alpha = -i\Lambda_\alpha + o(V)$$

$$\delta\bar{\Psi}_{\dot{\alpha}} = i\bar{\Lambda}_{\dot{\alpha}} + o(V)$$

$$\delta H = -i\Lambda_2 + o(V)$$

$$\delta\bar{H} = +i\bar{\Lambda}_2 + o(V)$$

$$\delta W_{\alpha\dot{\alpha}} = \partial_{\alpha\dot{\alpha}}(\Lambda_0+\bar{\Lambda}_0) - \tfrac{1}{2}i[W_{\alpha\dot{\alpha}},\Lambda_0+\bar{\Lambda}_0] + o(V^2)$$

$$\delta\lambda_\alpha = -\tfrac{1}{2}i[\lambda_\alpha,\Lambda_0+\bar{\Lambda}_0] + o(V^2)$$

$$\delta\bar{\lambda}_{\dot{\alpha}} = -\tfrac{1}{2}i[\bar{\lambda}_{\dot{\alpha}},\Lambda_0+\bar{\Lambda}_0] + o(V^2)$$

$$\delta D = -\tfrac{1}{2}i[D,\Lambda_0+\bar{\Lambda}_0] + o(V^2). \tag{58}$$

From equations (58) it follows that there exists a <u>nonsupersymmetric</u> gauge choice $C=\Psi=\bar{\Psi}=H=\bar{H}=0$, preserved by <u>nonsupersymmetric</u> gauge transformations with parameters $\Lambda(z)=\bar{\Lambda}(z)$, and hence

$$\bar{\Lambda}_0(x) = \Lambda_0(x),\ \Lambda_\alpha = \bar{\Lambda}_{\dot{\alpha}} = \Lambda_2 = \bar{\Lambda}_2 = 0. \tag{59}$$

In this gauge, commonly called Wess-Zumino gauge, V takes the simple form:

$$V_{W-Z}(z) = \theta^\alpha\bar{\theta}^{\dot{\alpha}}W_{\alpha\dot{\alpha}}(x) - \tfrac{1}{2}\theta^2\bar{\theta}^{\dot{\alpha}}\bar{\lambda}_{\dot{\alpha}}(x) - \tfrac{1}{2}\bar{\theta}^2\theta^\alpha\lambda_\alpha(x) + \tfrac{1}{4}\theta^2\bar{\theta}^2D(x). \tag{60}$$

In such a gauge, the exponential terminates after $o(V^2)$, and the action (54) becomes polynomial; indeed, using (60), one can show that the action (54) is just the gauge covariantization of (45), with extra interactions of the form $\bar{A}\chi\lambda$, $A\bar{\chi}\bar{\lambda}$, and $\bar{A}AD$. Despite the manifest violation of supersymmetry in (59) and (60), all traces of supersymmetry are not lost in Wess-Zumino gauge; one can define combined supersymmetry and gauge transformations (53) and (55) which preserve the Wess-Zumino gauge condition (60). To do this, one performs a supersymmetry transformation of V(z), and then adds in a compensating gauge transformation to restore (60). Thus the gauge parameter is:

$$\Lambda_+(x,\theta) = (-i)(-\varepsilon Q - \bar{\varepsilon}\bar{Q})V|_{\bar{\theta}=0} = +i\bar{\varepsilon}^{\dot{\alpha}}\bar{\partial}_{\dot{\alpha}}V|_{\bar{\theta}=0}$$

$$= -i\bar{\varepsilon}^{\dot{\alpha}}(\theta^{\alpha}W_{\alpha\dot{\alpha}}(x)+\tfrac{1}{2}\theta^2\bar{\lambda}_{\dot{\alpha}}(x)) := +i\bar{\varepsilon}^{\dot{\alpha}}\bar{\Lambda}_{\dot{\alpha}+}(x,\theta),$$

$$\bar{\Lambda}_-(x,\theta) = (+i)(-\varepsilon Q - \bar{\varepsilon}\bar{Q})V|_{\theta=0} = -i\varepsilon^{\alpha}\partial_{\alpha}V|_{\theta=0}$$

$$= -i\varepsilon^{\alpha}(\bar{\theta}^{\dot{\alpha}}W_{\alpha\dot{\alpha}}(x)-\tfrac{1}{2}\bar{\theta}^2\lambda_{\alpha}(x)) := -i\varepsilon^{\alpha}\Lambda_{\alpha-}(x,\bar{\theta}). \tag{61}$$

The combined transformations acting on the gauge field V(z) are now:

$$\delta V = -(\varepsilon Q+\bar{\varepsilon}\bar{Q})V + i(\bar{\Lambda}-\Lambda) - \tfrac{1}{2}i[V,\bar{\Lambda}+\Lambda]$$

$$= (\bar{\varepsilon}^{\dot{\alpha}}\bar{\Lambda}_{\dot{\alpha}}+\varepsilon^{\alpha}\Lambda_{\alpha}) - [\varepsilon^{\alpha}(Q_{\alpha}-\tfrac{1}{2}\Lambda_{\alpha})+\bar{\varepsilon}^{\dot{\alpha}}(\bar{Q}_{\dot{\alpha}}+\tfrac{1}{2}\bar{\Lambda}_{\dot{\alpha}}),V]. \tag{62}$$

On matter fields, one finds:

$$\delta\phi_+(x,\theta) = -(\varepsilon Q_+ + \bar{\varepsilon}\bar{Q}_+ - i\Lambda_+(x,\theta))\phi_+(x,\theta)$$

$$= -(\varepsilon^{\alpha}\partial_{\alpha} + \bar{\varepsilon}^{\dot{\alpha}}(\bar{Q}_{\dot{\alpha}+} + \bar{\Lambda}_{\dot{\alpha}+}(x,\theta))\phi_+(x,\theta). \tag{63}$$

Both (62) and (63) are suggestive of some sort of gauge covariantized supersymmetry transformation with spinorial connections $\Lambda_{\alpha}$ and $\bar{\Lambda}_{\dot{\alpha}}$; indeed, examining in detail the action of (62) and (63) on component fields, one finds just the gauge covariantizations of (35) and (36) (see de Wit & Freedman 1975).

## Gauge covariant derivatives

Having studied the action of gauge transformations on matter fields in superspace, and the relation of such transformations to components, I now want to turn to an investigation which will result in a super-Lagrange density for the gauge field V itself. To this end, I construct gauge covariant derivatives

$$\nabla_A = (\nabla_{\alpha\dot{\alpha}}, \nabla_{\alpha}, \nabla_{\dot{\alpha}}).$$

By a gauge covariant derivative, I mean covariant with respect to $\Lambda(z)$ transformations :

$$g(\Lambda)\phi = e^{i\Lambda}\phi \Rightarrow g(\Lambda)(\nabla_A\phi) := e^{i\Lambda}(\nabla_A\phi)$$

$$\Rightarrow g(\Lambda)\nabla_A = \nabla_A' = e^{i\Lambda}\nabla_A e^{-i\Lambda}. \tag{64}$$

To construct such an operator, one uses the chirality of $\Lambda$; thus one finds:

$$\bar{D}_{\dot{\alpha}}\Lambda = 0 \Rightarrow \bar{D}_{\dot{\alpha}} = (\bar{D}_{\dot{\alpha}})' = e^{i\Lambda}\bar{D}_{\dot{\alpha}}e^{-i\Lambda}. \tag{65}$$

That is very encouraging; one can simply use the uncovariantized $\bar{D}$ without any further effort:

$$\nabla_{\dot{\alpha}} := \bar{D}_{\dot{\alpha}}. \tag{66}$$

To find the undotted component, one needs to work a little harder: $D\Lambda \neq 0$ but $D\bar{\Lambda} = 0$. So one needs to convert $\Lambda$ to $\bar{\Lambda}$; fortunately, V was devised precisely for that purpose. Thus one finds:

$$(e^{-V}D_\alpha e^V)' = (e^{i\Lambda}e^{-V}e^{-i\bar{\Lambda}})D_\alpha(e^{i\bar{\Lambda}}e^{V}e^{-i\Lambda})$$

$$= e^{i\Lambda}(e^{-V}D_\alpha e^V)e^{-i\Lambda}, \tag{67}$$

and hence one can choose

$$\nabla_\alpha := e^{-V}D_\alpha e^V. \tag{68}$$

To do the rest is now no effort at all: one simply covariantizes (27):

$$\nabla_{\alpha\dot{\alpha}} := -2i\,\{\nabla_\alpha, \nabla_{\dot{\alpha}}\}, \text{ or}$$

$$\nabla_a = -i\sigma_a^{\alpha\dot{\alpha}}\,\{\nabla_\alpha, \nabla_{\dot{\alpha}}\}. \tag{69}$$

Note that $\nabla_A \neq \nabla_A^\dagger$; this is a feature of the chiral representation. Indeed, $\bar{\nabla}_A$ is not covariant with respect to $\Lambda$, but with respect to $\bar{\Lambda}$.

## Field strengths and the gauge action

Equipped with gauge covariant derivatives, one can calculate field strengths by commuting them; define

$$[\nabla_A,\nabla_B\} := \nabla_A\nabla_B - (-1)^{ab}\nabla_B\nabla_A$$

$$:= T_{AB}{}^C\nabla_C + F_{AB}, \qquad (70)$$

where $T_{AB}{}^C$ is the fixed torsion of superspace which is present as a consequence of (27) and (69), and is simply:

$$T_{\alpha\dot\beta}{}^a = T_{\dot\beta\alpha}{}^a = \tfrac{1}{2}i\sigma^a_{\alpha\dot\beta}, \qquad (71)$$

and all other components zero. Calculating field strengths, one finds

$$F_{\alpha\beta} = F_{\dot\alpha\dot\beta} = F_{\alpha\dot\beta} = 0, \qquad (72)$$

and the first interesting component is:

$$F_{\dot\alpha b} = [\nabla_{\dot\alpha},\nabla_b],$$

or equivalently,

$$F_{\dot\alpha,\beta\dot\beta} = [\nabla_{\dot\alpha},\nabla_{\beta\dot\beta}] = -2i[\nabla_{\dot\alpha},\{\nabla_\beta,\nabla_{\dot\beta}\}]$$

$$= -2i[\bar D_{\dot\alpha},\{e^{-V}D_\alpha e^V,\bar D_{\dot\beta}\}]$$

$$= -2i\,\bar D_{\dot\alpha}\bar D_{\dot\beta}(e^{-V}D_\alpha e^V)$$

$$= +i\varepsilon_{\dot\alpha\dot\beta}\bar D^2(e^{-V}D_\alpha e^V)$$

$$:= +i\varepsilon_{\dot\alpha\dot\beta}\,W_\alpha. \qquad (73)$$

Here $W_\alpha$ is a gauge covariant field strength spinor; as it contains three spinor derivatives acting on V, it has dimension $M^{\frac{3}{2}}$, that is, the canonical dimension for a spinor. It is manifestly chiral. All the other nonvanishing field strength components can be expressed in terms of it:

$$F_{\alpha b} \sim W_{\dot\alpha} = e^{-V}(\bar W_{\dot\alpha})e^V, \text{ and}$$

$$F_{ab} \sim \varepsilon_{\alpha\beta}(\nabla_{\dot\alpha}W_{\dot\beta} + \nabla_{\dot\beta}W_{\dot\alpha}) + \varepsilon_{\dot\alpha\dot\beta}(\nabla_\alpha W_\beta + \nabla_\beta W_\alpha). \qquad (74)$$

Thus the unique action, up to total derivatives, quadratic in field strengths, manifestly super and gauge invariant, and requiring no dimensional parameters, is:

$$I_{Gauge} = -\frac{tr}{g^2}\int d^4x\,d^2\theta\; W^\alpha W_\alpha \qquad (75)$$

(one can show using the Bianchi identities that up to total derivatives, this action is Hermitian). I leave it as an exercise for the reader to show that in Wess-Zumino gauge this action becomes explicitly polynomial, and that in components it is nothing other than an ordinary gauge action for the vector field at the $\theta\bar{\theta}$ level in $V(z)$, a gauge invariant action for the Majorana spinor at the $\theta^2\bar{\theta}$ and $\bar{\theta}^2\theta$ levels in $V(z)$, and a nonpropagating $D^2(x)$ term for the $\theta^2\bar{\theta}^2$ component (see (60)). All the component fields in the superfield $V(z)$ are of course in the adjoint representation of the gauge group. This constitutes a complete description of classical supersymmetric Yang-Mills theory in superfield language, and a detailed description of its relation to component formulations; but what of eqs.(51) and (52)?

## The road not taken: vector representation

I want to indicate briefly an alternative and entirely equivalent superfield formulation of super Yang-Mills theory which arises when one insists on an Hermitian representation of the gauge group. Rather than deriving such a formulation from the beginning, I shall take the one I have already shown you and generalize it to include an Hermitian representation; then I shall show how one could have derived such a formulation from first principles.

Suppose one wants a chiral field to transform under some Hermitian representation, call it $K(z)$, of the same gauge group considered earlier and represented by $\Lambda(z)$; that is, for

$$K(z) = K^i\, T_i = K^\dagger(z), \tag{76}$$

one demands

$$g(K)\phi_{(V)} = e^{iK}\phi_{(V)} \text{ and } g(K)\bar{\phi}_{(V)} = \bar{\phi}_{(V)}e^{-iK}. \tag{77}$$

Here the subscript "(V)" indicates that $\phi$ is in a vector representation of the Yang-Mills group (remember this bears no relation to the representation of the supersymmetry group). Now the original action (42) with $\phi_{(V)}$ instead of $\phi$ has local K gauge invariance. The obvious way to "replace" (46) and (48) with (77) is to invent a $\Lambda$ to K "converter", just as V was introduced as a $\Lambda$ to $\bar{\Lambda}$ converter (see (53)); thus one defines

$$W := W^i\, T_i,$$

$$g(K,\Lambda)e^W = (e^W)' := e^{i\bar{\Lambda}}e^W e^{-iK}. \tag{78}$$

Manifestly W cannot be Hermitian, so one has as well:

$$\bar{W} = \bar{W}^i\, T_i,$$

$$g(K,\Lambda)e^{\bar{W}} = (e^{\bar{W}})' = e^{iK}e^{\bar{W}}e^{-i\Lambda}. \tag{79}$$

Since one wants to introduce the minimum number of new fields, one imposes the covariant constraint (see (78), (79), and (53)):

$$e^V = e^W e^{\bar{W}}. \tag{80}$$

Thus $e^W$ is a sort of "square root" of $e^V$, in the same sense that the vierbein is the "square root" of the metric (Marc Grisaru has suggested that W should be called an einbein). Now one can define quantities covariant under K type transformations from quantities covariant under $\Lambda$ or $\bar{\Lambda}$ type transformations simply by using $\bar{W}$ and W respectively; thus

$$\phi_{(V)} = e^{\bar{W}}\phi, \quad \bar{\phi}_{(V)} = \bar{\phi}e^W,$$

$$\nabla_{A(V)} = e^{\bar{W}}\, \nabla_A\, e^{-\bar{W}}, \quad W_{\alpha(V)} = e^{\bar{W}}\, W_\alpha\, e^{-\bar{W}}, \text{ etc.} \tag{81}$$

To see that this achieves the desired result, use (78) and (79) in conjunction with (46), (48), (64), etc. Note that the new K invariance can be used to choose a gauge

$$e^W = e^{\bar{W}} = e^{\frac{1}{2}V}, \tag{82}$$

so in a sense there is no content in the new formulation; however, the new vector quantities <u>are</u> explicitly Hermitian. Calculating, one finds

$$\begin{aligned}\nabla_{A(V)} &= e^{\bar{W}}\, \nabla_A\, e^{-\bar{W}} \\ &= (e^{\bar{W}}\, D_{\dot{\alpha}}\, e^{-\bar{W}},\ e^{-W}\, D_\alpha\, e^W,\ -2i\{\nabla_{\alpha(V)},\nabla_{\dot{\alpha}(V)}\}) \\ &= \nabla^\dagger_{A(V)}.\end{aligned} \tag{83}$$

(Strictly speaking one needs to be careful to specify what sort of a representation $\nabla$ acts on; on any Hermitian representation, (83) is true).

How could one arrive at such a result without first finding the chiral representation? Clearly, (76) and (77) imply that $\phi_{(V)}$ cannot be taken to be chiral anymore; it must be <u>gauge covariantly</u> chiral:

$$\nabla_{\dot{\alpha}(V)}\phi_{(V)} = 0, \tag{84}$$

where one has simply postulated the existence of a gauge covariant derivative such that

$$g(K)\nabla_{A(V)} = e^{iK}\ \nabla_{A(V)}\ e^{-iK}. \qquad (85)$$

Now (84) implies:

$$\nabla_{\dot{\beta}}(V)\nabla_{\dot{\alpha}}(V)\Phi(V) = 0 \ \Rightarrow\ F_{\dot{\alpha}\dot{\beta}}(V) = 0. \qquad (86)$$

Similarly, one finds the gauge covariant constraint on the field strengths

$$F_{\alpha\beta}(V) = 0. \qquad (87)$$

Finally, given a covariant $\nabla_\alpha$ and $\nabla_{\dot{\alpha}}$, one can always define a vector derivative by (69), and hence take

$$F_{\alpha\dot{\beta}}(V) = 0. \qquad (88)$$

The constraints (86), (87), and (88) are precisely what the chiral approach yielded as a result (72). They have as solution (83). Examining (83), one finds a new, hidden gauge invariance of the covariant derivatives

$$g(\Lambda)\nabla_{A(V)} = \nabla_{A(V)} \qquad (89)$$

generated precisely by the chiral transformations $\Lambda$ acting as in (78) and (79).

For a more detailed description of the constrained approach see Wess (1977), Gates & Siegel (1979), Gates (1979), Stelle & West (1979), and Gates, Stelle & West (1980); for the utility of mixing chiral and vector representations, see Grisaru, Siegel & Roček (1979), and the contribution of M. Grisaru to this volume. At any rate, both the chiral and the constrained approaches contain aspects that are more "art" than "science"; in the chiral approach, one needs to know the art of constructing covariant derivatives (which was actually very simple in this case); in the constrained approach, one needs the art of postulating constraints and further the art of solving them; for some applications, the Bianchi identities provide a systematic procedure for the latter problem (see, e.g., Wess 1977), but for quantum calculations as described in the final lecture, an explicit solution is necessary.

## SUPERFIELD SUPERGRAVITY

### Introduction

This approach to constructing a superfield theory of supergravity is based on the work of Siegel & Gates (1979); that work itself is based on Siegel (1977, 1978), and is very closely related to the work of Ogievetsky & Sokatchev (1978, 1978a, and subsequent papers). The basic idea is to mimic as closely as possible the chiral approach to Yang-Mills theory, taking, in a certain sense to be made precise below, instead of Yang-Mills generators $T_i$, the superderivative

$$\partial_M := (\partial_m, \partial_\mu, \bar{\partial}_{\dot{\mu}}). \tag{90}$$

One might expect to use $Q_A$ or $D_A$ (see (30) and (31)), but in a general curved superspace, $Q_A$ and $D_A$ have no particular property that makes them natural or interesting, whereas $\partial_M$ is at least familiar as a local basis for vector fields (in the differential geometric sense). One might prefer to use covariant derivatives $\nabla_A$, but I want to construct such an object, and not postulate it _ab initio_; otherwise, I would be forced to adopt the constrained approach, and as I have said, that is another art.

## Curved superspace

Before I begin the actual construction of a superfield theory of supergravity, I shall need to discuss some of the general features of curved superspace (see Wess 1977 and Zumino 1978). The basic geometric objects in curved superspace are the achtbein (or, in general, the vielbein)

$$E_M{}^A, \tag{91}$$

and the Lorentz connection

$$\Omega_{MA}{}^B := (\Omega_{M\alpha\dot{\alpha}}{}^{\beta\dot{\beta}}, \Omega_{M\alpha}{}^\beta, \Omega_{M\dot{\alpha}}{}^{\dot{\beta}}) \tag{92}$$

where

$$\Omega_{Ma}{}^b \leftrightarrow \Omega_{M\alpha\dot{\alpha}}{}^{\beta\dot{\beta}} = \delta_\alpha{}^\beta\, \Omega_{M\dot{\alpha}}{}^{\dot{\beta}} + \delta_{\dot{\alpha}}{}^{\dot{\beta}}\, \Omega_{M\alpha}{}^\beta.$$

Eq. (92) implies that the Lorentz group acts reducibly on the tangent space of curved superspace: vectors rotate into vectors but not into spinors, and spinors into spinors but not into vectors, and both are rotated with the same Lorentz rotation. Indeed, a Lorentz rotation acts on a supervector $V_A$ by

$$g(L)V_A = V'_A = L_A{}^B V_B,$$

where

$$L_A{}^B = (L_a{}^b, L_\alpha{}^\beta, L_{\dot{\alpha}}{}^{\dot{\beta}}) \text{ and } L_a{}^b \leftrightarrow L_{\alpha\dot{\alpha}}{}^{\beta\dot{\beta}} = L_\alpha{}^\beta L_{\dot{\alpha}}{}^{\dot{\beta}}$$

or, infinitesimally,

$$\delta(\omega)V_A = \tfrac{1}{2}\omega_A{}^B V_B, \quad \omega_A{}^B = (\omega_{\alpha\dot\alpha}{}^{\beta\dot\beta},\ \omega_\alpha{}^\beta,\ \omega_{\dot\alpha}{}^{\dot\beta}), \text{ and}$$

$$\omega_a{}^b \leftrightarrow \omega_{\alpha\dot\alpha}{}^{\beta\dot\beta} = \delta_\alpha{}^\beta\ \omega_{\dot\alpha}{}^{\dot\beta} + \delta_{\dot\alpha}{}^{\dot\beta}\ \omega_\alpha{}^\beta.$$

The achtbein (91) is invertible; there exists an inverse achtbein $E_A{}^M$ such that:

$$E_M{}^A E_A{}^N = \delta_M{}^N \quad \text{and} \quad E_A{}^M E_M{}^B = \delta_A{}^B, \tag{93}$$

where

$$\delta_M{}^N = (\delta_m{}^n,\ \delta_\mu{}^\nu,\ \delta_{\dot\mu}{}^{\dot\nu}) \quad \text{and} \quad \delta_A{}^B = (\delta_a{}^b,\ \delta_\alpha{}^\beta,\ \delta_{\dot\alpha}{}^{\dot\beta}),$$

$$\delta_a{}^b \leftrightarrow \delta_{\alpha\dot\alpha}{}^{\beta\dot\beta} = 2\delta_\alpha{}^\beta \delta_{\dot\alpha}{}^{\dot\beta}.$$

Often it will be useful to write the inverse achtbein as a vector field:

$$\underline{E}_A = E_A{}^M \partial_M. \tag{94}$$

(Throughout this section, any vector field or differential operator not written with an explicit $\partial_M$ will be underlined).

In addition to tangent space Lorentz rotations, superdiffeomorphisms or general coordinate transformations act naturally on curved superspace; I shall consider those that can be written as the exponential of a vector field (see De Witt 1978). Thus I shall write:

$$g(\Lambda)z := z' := (e^{\underline{\Lambda}}z) = e^{\underline{\Lambda}}ze^{-\underline{\Lambda}} \tag{95}$$

where $\underline{\Lambda} := \Lambda^M\partial_m$ is a vector field, and $(e^{\underline{\Lambda}}z)$ is the "number part" of the differential operator $e^{\underline{\Lambda}}z$; because $\underline{\Lambda}$ obeys the Leibnitz rule $e^{\underline{\Lambda}}ze^{-\underline{\Lambda}}$ is a number and not a differential operator. Infinitesimally,

$$\delta(\Lambda)z = (\underline{\Lambda}z) = [\underline{\Lambda}, z] \tag{96}$$

Eq. (95) implies

$$(\partial_M)' = e^{\underline{\Lambda}}\partial_M e^{-\underline{\Lambda}}. \tag{97}$$

For a scalar field:

$$\phi(z) = \phi'(z') = \phi'(e^{\underline{\Lambda}}z) = (e^{\underline{\Lambda}}\phi'(z)) = e^{\underline{\Lambda}}\phi'(z)e^{-\underline{\Lambda}} \tag{98}$$

and hence

$$\delta(\Lambda)\phi = -(\underline{\Lambda}\phi) = [\phi,\underline{\Lambda}]. \tag{99}$$

On a density of weight w, diffeomorphisms act slightly differently; defining

$$\overleftarrow{\Lambda} := \Lambda^M\overleftarrow{\partial}_M = (-1)^M\overleftarrow{\partial}_M\Lambda^M + (1\cdot\overleftarrow{\Lambda}), \tag{100}$$

where

$$(1\cdot\overleftarrow{\Lambda}) = (\partial_m\Lambda^m) - (\partial_{\dot\mu}\Lambda^{\dot\mu}) - (\partial_\mu\Lambda^\mu), \tag{101}$$

one has, for w = 1,

$$\delta(\Lambda)\phi = -(\phi\overleftarrow{\Lambda}) = -((\underline{\Lambda} + (1\cdot\overleftarrow{\Lambda}))\phi) = [\phi,\underline{\Lambda}] - (1\cdot\overleftarrow{\Lambda})\phi. \tag{102}$$

For a finite transformation:

$$\phi(z) = e^{(\underline{\Lambda}+(1\cdot\overleftarrow{\Lambda}))}\phi'(z) = \phi'(z)e^{\overleftarrow{\Lambda}}. \tag{103}$$

However, it is well known that (103) can be written:

$$\phi(z) = \mathrm{sdet}|\partial_N z'^M|\phi'(z') = \mathrm{sdet}|\partial_N z'^M|(e^{\underline{\Lambda}}\phi'(z))$$

and hence it follows that

$$\mathrm{sdet}|\partial_N z'^M| = e^{(\underline{\Lambda}+(1\cdot\overleftarrow{\Lambda}))}e^{-\underline{\Lambda}} = (1\cdot e^{\overleftarrow{\Lambda}}) \tag{104}$$

(For a discussion of the superdeterminant, see the Appendix). For a density of arbitrary weight one has, instead of (102),

$$\delta(\Lambda)\phi = [\phi,\underline{\Lambda}] - w(1\cdot\overleftarrow{\Lambda})\phi. \tag{105}$$

Lastly, on vector fields, diffeormorphisms act by

$$V'^M(z') = V^N(z)(\partial_N z'^M), \tag{106}$$

or, for $\underline{V} = V^M\partial_M$, $\underline{V}' = (V')^M\partial_M$,

$$\delta(\Lambda)\underline{V} = [\underline{V},\underline{\Lambda}] \leftrightarrow \underline{V}(z) = e^{\underline{\Lambda}}\underline{V}'(z)e^{-\underline{\Lambda}}, \tag{107}$$

Finally, in addition to diffeomorphisms, one sometimes wants to consider Weyl transformations; these will be written

$$g(\Sigma)\phi = e^{\Sigma}\phi, \quad \text{or} \quad \delta(\Sigma)\phi = \Sigma\phi. \tag{108}$$

Note that the transformation of a density (102) can be interpreted as a diffeomorphism acting on an ordinary scalar (99) combined with a Weyl transformation (108) with parameter

$\Sigma = -(1 \cdot \overset{+}{\Lambda})$.

## Supergravity

I shall develop a formulation of supergravity modeled closely on the formulation of supersymmetric Yang-Mills theories described in previous sections (see eq. (46)-(89)). Thus I shall start with fundamental unconstrained objects and construct out of them covariantly transforming quantities. Central to the construction, as indeed to N=1 supersymmetry in general, is the notion of a chiral superfield; however, in curved superspace without an achtbein (and since I want to construct covariantly transforming quantities out of only unconstrained objects, I cannot begin with a given achtbein), a chiral superfield is necessarily noncovariant. Thus I define a chiral superfield by

$$\bar{\partial}_{\dot{\mu}}\phi(z) = 0. \tag{109}$$

The particular complex diffeomorphisms which preserve this constraint are generated by (see(99)):

$$\delta(\Lambda)\phi = -(\underline{\Lambda}\phi) = -((\Lambda^m\partial_m + \Lambda^\mu\partial_\mu + \Lambda^{\dot{\mu}}\bar{\partial}_{\dot{\mu}})\phi)$$

$$= -((\Lambda^m\partial_m + \Lambda^\mu\partial_\mu)\phi)$$

$$\bar{\partial}_{\dot{\mu}}(\delta\phi) = 0 \Rightarrow (\bar{\partial}_{\dot{\mu}}(\Lambda^m\partial_m + \Lambda^\mu\partial_\mu)\phi) = 0$$

$$\Rightarrow \bar{\partial}_{\dot{\mu}}\Lambda^m = \bar{\partial}_{\dot{\mu}}\Lambda^\mu = 0. \tag{110}$$

Note that $\Lambda^{\dot{\mu}}$ is unconstrained, since $\Lambda^{\dot{\mu}}\bar{\partial}_{\dot{\mu}}\phi=0$ as a consequence of (109). By Hermitian conjugation, one finds antichiral fields

$$\partial_\mu\bar{\phi}(z) = 0,$$

transforming as

$$\delta(\Lambda)\bar{\phi} = -(\bar{\underline{\Lambda}}\bar{\phi}) = -(\bar{\Lambda}^m\partial_m + \bar{\Lambda}^{\dot{\mu}}\bar{\partial}_{\dot{\mu}} + \bar{\Lambda}^\mu\partial_\mu)\bar{\phi}$$

and thus the constraints

$$\partial_\mu\bar{\Lambda}^m = \partial_\mu\bar{\Lambda}^{\dot{\mu}} = 0 \tag{111}$$

with $\bar{\Lambda}^\mu$ unconstrained. By analogy with (53), I introduce a $\underline{\Lambda}$ to $\bar{\underline{\Lambda}}$ converter; in this case it is an Hermitian vector field:

$$\underline{U} = U^M\partial_M, \quad U^M = (U^\dagger)^M, \tag{112}$$

transforming as

$$e^{i\underline{U}(z)} = e^{\bar{\underline{\Lambda}}}\,(e^{i\underline{U}'(z)})\;e^{-\underline{\Lambda}}, \text{ or}$$

$$\delta(\Lambda)e^{i\underline{U}} = -\bar{\underline{\Lambda}}\, e^{i\underline{U}} + e^{i\underline{U}}\, \underline{\Lambda}. \tag{113}$$

It is easy to verify that the transformations (113) preserve the Hermiticity of $\underline{U}$ (112). One can solve (113) to find the transformation of $\underline{U}$ itself in precise analogy with (56):

$$\delta(\Lambda)\underline{U} = +\tfrac{1}{2}L_{\underline{U}}(\underline{\Lambda}+\bar{\underline{\Lambda}} + i\cot\tfrac{1}{2}L_{\underline{U}}(\bar{\underline{\Lambda}}-\underline{\Lambda})); \tag{114}$$

expanding in powers of $\underline{U}$, one finds:

$$\delta(\Lambda)\underline{U} = i(\bar{\underline{\Lambda}}-\underline{\Lambda}) + \tfrac{1}{2}[\underline{U},\underline{\Lambda}+\bar{\underline{\Lambda}}] + o(\underline{U}^2). \tag{115}$$

Formulae (113), (114), and (115) can be checked by substituting $V\to i\underline{U}$, $\Lambda\to +i\underline{\Lambda}$, and $\bar{\Lambda}\to +i\bar{\underline{\Lambda}}$ in (53),(56), and (57). The transformation (115) can be written more explicitly as:

$$\delta(\Lambda)U^M = i(\bar{\Lambda}-\Lambda)^M + \tfrac{1}{2}(\underline{U}(\Lambda+\bar{\Lambda})^M) - \tfrac{1}{2}((\underline{\Lambda}+\bar{\underline{\Lambda}})U^M) + o(\underline{U}^2). \tag{116}$$

Because $\Lambda^{\dot{\mu}}$ and $\bar{\Lambda}^{\mu}$ are unconstrained (see (110) and (111) above), $\bar{U}^{\dot{\mu}}$ and $U^{\mu}$ undergo arbitrary displacements, and can be gauged to zero; such a gauge is preserved by transformations with

$$\Lambda^{\dot{\mu}} = e^{-i\underline{U}}\, \bar{\Lambda}^{\dot{\mu}}\, e^{i\underline{U}}, \quad \text{and} \quad \bar{\Lambda}^{\mu} = e^{i\underline{U}}\, \Lambda^{\mu}\, e^{-i\underline{U}}, \tag{117}$$

as can be verified by examining (113) directly:

$$\begin{aligned}
\delta(\Lambda)e^{i\underline{U}} &= \delta e^{iU^m\partial_m} \\
&= -\bar{\Lambda}^n\partial_n e^{i\underline{U}} + e^{i\underline{U}}\Lambda^n\partial_n \\
&\quad -\bar{\Lambda}^{\mu}\partial_{\mu} e^{i\underline{U}} + e^{i\underline{U}}\Lambda^{\mu}\partial_{\mu} \\
&\quad -\bar{\Lambda}^{\dot{\mu}}\bar{\partial}_{\dot{\mu}} e^{i\underline{U}} + e^{i\underline{U}}\Lambda^{\dot{\mu}}\bar{\partial}_{\dot{\mu}} \\
&= -\bar{\Lambda}^n\partial_n e^{i\underline{U}} + e^{i\underline{U}}\Lambda^n\partial_n \\
&\quad -\bar{\Lambda}^{\mu}\,[\partial_{\mu},e^{i\underline{U}}] - (\bar{\Lambda}^{\mu}e^{i\underline{U}} - e^{i\underline{U}}\Lambda^{\mu})\partial_{\mu} \\
&\quad -\bar{\Lambda}^{\dot{\mu}}\,[\bar{\partial}_{\dot{\mu}},e^{i\underline{U}}] - (\bar{\Lambda}^{\dot{\mu}}e^{i\underline{U}} - e^{i\underline{U}}\Lambda^{\dot{\mu}})\bar{\partial}_{\dot{\mu}} \\
&= -\bar{\Lambda}^n\partial_n e^{i\underline{U}} + e^{i\underline{U}}\Lambda^n\partial_n \\
&\quad - e^{i\underline{U}}\Lambda^{\mu}e^{-i\underline{U}}\,[\partial_{\mu},e^{i\underline{U}}] + 0 \\
&\quad - \bar{\Lambda}^{\dot{\mu}}\,[\bar{\partial}_{\dot{\mu}},e^{i\underline{U}}] + 0.
\end{aligned} \tag{118}$$

It is clear that (118) preserves $U^\mu=\bar{U}^{\dot\mu}=0$. For many purposes, e.g., quantization (see Grisaru and Siegel, this volume), this is a very useful gauge; indeed, Ogievetsky and Sokatchev work exclusively in this gauge. As before, one can solve for the transformation of $\underline{U}=U^m\partial_m$:

$$\delta(\Lambda)U^m\partial_m = \tfrac{1}{2}L_{\underline{U}}((\Lambda+\bar\Lambda)^n\partial_n - \Lambda^\mu e^{-i\underline{U}}[\partial_\mu, e^{i\underline{U}}] - \bar\Lambda^{\dot\mu}e^{i\underline{U}}[\bar\partial_{\dot\mu}, e^{-i\underline{U}}]$$

$$+ \; i\cot\tfrac{1}{2}L_{\underline{U}}((\bar\Lambda-\Lambda)^n\partial_n + \Lambda^\mu e^{-i\underline{U}}[\partial_\mu, e^{i\underline{U}}]$$

$$- \bar\Lambda^{\dot\mu}e^{i\underline{U}}[\bar\partial_{\dot\mu}, e^{-i\underline{U}}]))$$

$$= i(\bar\Lambda-\Lambda)^n\partial_n + (\Lambda^\mu[\underline{U},\partial_\mu] + \bar\Lambda^{\dot\mu}[\underline{U},\bar\partial_{\dot\mu}])$$

$$- \tfrac{1}{12}i[\underline{U},[\underline{U},(\bar\Lambda-\Lambda)^n\partial_n]]$$

$$+ \tfrac{1}{2}[\underline{U},(\bar\Lambda+\Lambda)^n\partial_n + i(\Lambda^\mu[\underline{U},\partial_\mu] - \bar\Lambda^{\dot\mu}[\underline{U},\bar\partial_{\dot\mu}])]$$

$$+ \tfrac{1}{2}i(\bar\Lambda^{\dot\mu}[\underline{U},[\underline{U},\bar\partial_{\dot\mu}]] - \Lambda^\mu[\underline{U},[\underline{U},\partial_\mu]]) + o(\underline{U}^3)$$

$$= i(\bar\Lambda-\Lambda)^n\partial_n + \tfrac{1}{2}[\underline{U},(\bar\Lambda+\Lambda)^n\partial_n] - \tfrac{1}{12}i[\underline{U},[\underline{U},(\bar\Lambda-\Lambda)^n\partial_n]]$$

$$+ (\Lambda^\mu + \tfrac{1}{2}i[\underline{U},\Lambda^\mu])[\underline{U},\partial_\mu] + (\bar\Lambda^{\dot\mu} - \tfrac{1}{2}i[\underline{U},\bar\Lambda^{\dot\mu}])[\underline{U},\bar\partial_{\dot\mu}]$$

$$+ o(\underline{U}^3). \tag{119}$$

To derive (119) I have used the identity

$$0 \equiv [\bar\partial_{\dot\mu},1] \equiv [\bar\partial_{\dot\mu}, e^{i\underline{U}}\, e^{-i\underline{U}}] \equiv e^{i\underline{U}}[\bar\partial_{\dot\mu}, e^{-i\underline{U}}] + [\bar\partial_{\dot\mu}, e^{i\underline{U}}]e^{-i\underline{U}}.$$

In precise analogy to (58), one can use the chiral parameter $\Lambda^m$ and the antichiral parameter $\bar\Lambda^m$ to gauge away the chiral and antichiral pieces of $U^m$. Thus one finds a Wess-Zumino gauge

$$U^\mu = \bar{U}^{\dot\mu} = 0$$

$$U^m = 2e_{\alpha\dot\alpha}{}^m(\theta^\alpha\bar\theta^{\dot\alpha} + \tfrac{1}{2}\theta^2\bar\theta_{\dot\beta}\bar\psi^{\dot\alpha\alpha\dot\beta} + \tfrac{1}{2}\bar\theta^2\theta_\beta\psi^{\alpha\beta\dot\alpha}$$

$$+ \tfrac{1}{4}\theta^2\bar\theta^2(A^{\alpha\dot\alpha} + \ldots)) \tag{120}$$

where the terms indicated by (...) are functions of the basic fields

$$(e_{\alpha\dot\alpha}{}^m, \quad \bar\psi^{\dot\alpha\alpha\dot\beta} = \bar\psi_m^{\dot\alpha}(\sigma^m)^{\alpha\dot\beta}, \quad \psi^{\alpha\beta\dot\alpha} = \psi_m^\alpha(\sigma^m)^{\beta\dot\alpha})$$

needed only if the fields are to be identified with the conventional component fields of supergravity (see, for example, van Nieuwenhuizen this volume, and also Lindström, Karlhede and Roček, this volume). The transformations which preserve (120) are generated by parameters

$$\xi^m(x) = \tfrac{1}{2}(\Lambda^m + \bar{\Lambda}^m)|_{\theta=\bar{\theta}=0},$$

$$\Lambda^\mu,\ \bar{\partial}_{\dot{\mu}}\Lambda^\mu = 0 \Rightarrow \Lambda^\mu = \varepsilon^\mu - \theta^\nu(\delta_\nu{}^\mu\, \sigma + \omega_\nu{}^\mu) + \theta^2\zeta^\mu,$$

$$\bar{\Lambda}^{\dot{\mu}},\ \partial_\mu\bar{\Lambda}^{\dot{\mu}} = 0 \Rightarrow \bar{\Lambda}^{\dot{\mu}} = \bar{\varepsilon}^{\dot{\mu}} - \bar{\theta}^{\dot{\nu}}(\delta_{\dot{\nu}}{}^{\dot{\mu}}\, \bar{\sigma} + \bar{\omega}_{\dot{\nu}}{}^{\dot{\mu}}) + \bar{\theta}^2\bar{\zeta}^{\dot{\mu}} \qquad (121)$$

One might expect, by analogy to (59), to set $\Lambda^m = \bar{\Lambda}^m$; however, the transformations generated by $\Lambda^\mu$ do not preserve (120), and one needs compensating transformations (see 61) with parameters:

$$\Lambda^m_{comp} = -i\bar{\Lambda}^{\dot{\mu}}(\bar{\partial}_{\dot{\mu}}U^m)|_{\bar{\theta}=0} = 2i\, e_{\alpha\dot{\alpha}}{}^m(\bar{\varepsilon}^{\dot{\alpha}}\theta^\alpha - \tfrac{1}{2}\theta^2\bar{\varepsilon}_{\dot{\beta}}\bar{\Psi}^{\dot{\alpha}\alpha\dot{\beta}})$$

$$\bar{\Lambda}^m_{comp} = +i\Lambda^\mu(\partial_\mu U^m)|_{\theta=0} = 2i\, e_{\alpha\dot{\alpha}}{}^m(\varepsilon^\alpha\bar{\theta}^{\dot{\alpha}} + \tfrac{1}{2}\bar{\theta}^2\varepsilon_\beta\Psi^{\alpha\beta\dot{\alpha}}). \qquad (122)$$

Then the residual symmetry can be found by combining (121), (122), and (119); the $\xi^m(x)$ transformations are readily identifiable as conventional spacetime dependent coordinate transformations:

$$\delta(\xi)U^m = U^n(\partial_n\xi^m) - \xi^n(\partial_n U^m) \qquad (123)$$

Similarly, the $\varepsilon^\mu(x)$ transformations are identifiable as supersymmetry transformations, and the $\omega_\nu{}^\mu(x)$ transformations as Lorentz rotations of the component fields; the transformations generated by $\sigma(x)$ and $\zeta^\mu(x)$ are, however, unfamiliar and unwanted. The knowledgable reader may recognize them as superconformal transformations, but in any case, the objects that contain them:

$$\partial_\mu\Lambda^\mu = 2(-\sigma(x) + \theta_\mu\zeta^\mu) \quad \text{and}$$

$$\bar{\partial}_{\dot{\mu}}\bar{\Lambda}^{\dot{\mu}} = 2(-\bar{\sigma}(x) + \bar{\theta}_{\dot{\mu}}\bar{\zeta}^{\dot{\mu}}), \qquad (124)$$

must somehow be restricted or eliminated. One can impose a constraint directly on (124); but it is more convenient and more covariant to introduce a compensating field $\mathcal{E}$ to absorb the extra gauge freedoms (124). There is some freedom in the way this is done, leading to different though apparently physically equivalent theories; the simplest, and most natural procedure leads to a theory with a minimal number of field components. One takes $\mathcal{E}$ to be a chiral density, that is, one demands

$$\int d^4x d^2\theta \mathcal{E} \tag{125}$$

be invariant; then

$$\delta(\Lambda)\mathcal{E} = -(\underline{\Lambda}\,\mathcal{E}) - (\partial_m\Lambda^m - \partial_\mu\Lambda^\mu)\mathcal{E}. \tag{126}$$

Note the absence of a term $(\bar{\partial}_{\dot{\mu}}\bar{\Lambda}^{\dot{\mu}})$ in (126). For (125) to be a θ-independent object, one assumes $\mathcal{E}$ is chiral:

$$\bar{\partial}_{\dot{\mu}}\mathcal{E} = 0; \tag{127}$$

Clearly, the transformation (126) preserves the chirality of $\mathcal{E}$ (127). In a θ-expansion, one can write:

$$\mathcal{E} = Q - \theta^\nu\rho_\nu + \theta^2 S;$$

Then one finds that the transformation (126) leads to:

$$\delta Q = \varepsilon^\nu\rho_\nu - 2\sigma(x)Q = (\varepsilon^\nu\rho_\nu Q^{-1} - 2\sigma(x))\,Q$$

$$\delta\rho_\nu = +2Q\zeta_\nu + \ldots$$

If one assumes that Q is nonsingular, $Q(x) \neq 0$, then $\sigma(x)$ can be used to scale Q to an arbitrary value, e.g., $Q = e = \det|e^a{}_m|$, where $e^a{}_m$ is the ordinary spacetime vierbein (see 120); further, $\zeta_\nu$ can be used to gauge $\rho_\nu$ to an arbitrary value, e.g., $\rho_\nu = 0$ or $\rho_\nu = e_{\nu\dot{\nu}}{}^m\bar{\Psi}_m^{\dot{\nu}}$. Such gauge choices completely determine σ and ζ, eliminating the unwanted gauge freedom; for any particular choice, compensating σ and ζ transformations analogous to (122) will be needed. The complex spin-zero field S remains as one of the auxiliary fields of supergravity.

Thus the basic unconstrained objects available for the description of supergravity and arising naturally out of considerations of chiral fields are the vector field $\underline{U}$ (see (112)-(120)) and a chiral density $\mathcal{E}$ (see (125)-(127)). These transform under gauge transformations generated by complex chirality preserving vector fields $\underline{\Lambda}$ (110) and (111). I shall now attempt to construct, by analogy to the Yang-Mills case (66)-(69), achtbeins and connections (91)-(93) that transform covariantly with respect to the transformations generated by $\underline{\Lambda}$.

## The achtbein

If one were to proceed blindly in analogy with the Yang-Mills case, one would hope that $\bar{\partial}_{\dot{\mu}}$ is already covariant under chirality preserving diffeomorphisms (see(65) and (66)); this, however, is not the case:

$$\delta\bar{\partial}_{\dot{\alpha}} = 0 \neq [\bar{\partial}_{\dot{\alpha}}, \underline{\Lambda}]$$

(see (107)). Indeed, one finds:

$$\delta\bar{\partial}_{\dot{\alpha}} = 0 = [\bar{\partial}_{\dot{\alpha}}, \underline{\Lambda}] - (\bar{\partial}_{\dot{\alpha}}\Lambda^{\dot{\beta}})\bar{\partial}_{\dot{\beta}}.$$

The last, noncovariant term, can be rewritten to give:

$$\delta\bar{\partial}_{\dot{\alpha}} = [\bar{\partial}_{\dot{\alpha}}, \underline{\Lambda}] - \tfrac{1}{2}(\bar{\partial}_{\dot{\alpha}}\Lambda^{\dot{\beta}} + \bar{\partial}^{\dot{\beta}}\Lambda_{\dot{\alpha}})\bar{\partial}_{\dot{\beta}} - \tfrac{1}{2}(\bar{\partial}_{\dot{\beta}}\Lambda^{\dot{\beta}})\bar{\partial}_{\dot{\alpha}}$$

$$:= [\bar{\partial}_{\dot{\alpha}}, \underline{\Lambda}] + \tfrac{1}{2}\omega_{\dot{\alpha}}{}^{\dot{\beta}}\bar{\partial}_{\dot{\beta}} - \tfrac{1}{2}(\bar{\partial}_{\dot{\beta}}\Lambda^{\dot{\beta}})\bar{\partial}_{\dot{\alpha}}, \tag{128}$$

where I have defined

$$\omega_{\dot{\alpha}}{}^{\dot{\beta}} := -(\bar{\partial}_{\dot{\alpha}}\Lambda^{\dot{\beta}} + \bar{\partial}^{\dot{\beta}}\Lambda_{\dot{\alpha}}). \tag{129}$$

Note that the term involving $\omega$ is harmless; it is merely a tangent space Lorentz rotation, and is perfectly covariant. The last term in (128), on the other hand, is noncovariant; it is some sort of a peculiar scale transformation. However, its form is such that terms needed to make it a density transformation (see (105) and (101)) are precisely the terms in the transformation of $\mathcal{E}$ (see (126)). Thus one can define a covariant tensor density of weight $-\frac{1}{2}$:

$$\underline{\tilde{E}}_{\dot{\alpha}} := \mathcal{E}^{-\frac{1}{2}}\partial_{\dot{\alpha}}$$

$$\delta\underline{\tilde{E}}_{\dot{\alpha}} = [\underline{\tilde{E}}_{\dot{\alpha}}, \underline{\Lambda}] + \tfrac{1}{2}\omega_{\dot{\alpha}}{}^{\dot{\beta}}\underline{\tilde{E}}_{\dot{\beta}} - \tfrac{1}{2}(\bar{\partial}_{\dot{\beta}}\Lambda^{\dot{\beta}})\underline{\tilde{E}}_{\dot{\alpha}} + \tfrac{1}{2}(\partial_m\Lambda^m - \partial_\mu\Lambda^\mu)\underline{\tilde{E}}_{\dot{\alpha}}$$

$$= [\underline{\tilde{E}}_{\dot{\alpha}}, \underline{\Lambda}] + \tfrac{1}{2}\omega_{\dot{\alpha}}{}^{\dot{\beta}}\underline{\tilde{E}}_{\dot{\beta}} + \tfrac{1}{2}(1 \cdot \overleftarrow{\Lambda})\underline{\tilde{E}}_{\dot{\alpha}}. \tag{130}$$

Proceeding with the analogy to Yang-Mills theory (see (67) and (68)), one would next be inclined to consider for an undotted spinor component of the achtbein:

$$e^{-i\underline{U}}\ \bar{\mathcal{E}}^{-\frac{1}{2}}\partial_\alpha e^{i\underline{U}} \tag{131}$$

where $\bar{\mathcal{E}} = \mathcal{E}^\dagger$, and transforms as (see(126)):

$$\delta(\Lambda)\bar{\mathcal{E}} = -\underline{\bar{\Lambda}}\bar{\mathcal{E}} - (\partial_m\bar{\Lambda}^m - \bar{\partial}_{\dot{\mu}}\bar{\Lambda}^{\dot{\mu}})\bar{\mathcal{E}}\ ; \tag{132}$$

it is antichiral:

$$\partial_\mu\bar{\mathcal{E}} = 0. \tag{133}$$

One can recognize (131) as a diffeomorphism with parameter $-i\underline{U}$ acting on $\bar{\mathcal{E}}^{-\frac{1}{2}}\partial_\alpha$ (see(97) and (107)). However, $\bar{\mathcal{E}}^{-\frac{1}{2}}\partial_\alpha$ is a density of weight $-\frac{1}{2}$ (with respect to $\bar{\Lambda}$ transformations). Therefore, one should not expect (131) to be covariant (a quick calculation reveals it is not), but rather, including the correct density factor (see (100) to (105)):

$$\tilde{\underline{E}}_\alpha := (\text{sdet}\,|\partial_N(e^{-i\underline{U}}z^M)|)^{-\frac{1}{2}}\ e^{-i\underline{U}}\ \bar{\mathcal{E}}^{-\frac{1}{2}}\partial_\alpha e^{i\underline{U}}$$

$$= (1\cdot e^{-i\overleftarrow{U}})^{-\frac{1}{2}}\ e^{-i\underline{U}}\ \bar{\mathcal{E}}^{-\frac{1}{2}}\partial_\alpha e^{i\underline{U}}. \qquad (134)$$

Here

$$(1\cdot e^{-i\overleftarrow{U}}) = (e^{-i(\underline{U}+(1\cdot\overleftarrow{U}))}\cdot 1),\quad \overleftarrow{U} = U^M\overleftarrow{\partial}_M. \qquad (135)$$

The expression (134) is indeed a covariantly transforming density of weight $-\frac{1}{2}$, as I shall verify by explicit calculation. One can show

$$\delta e^{-i(\underline{U}+(1\cdot\overleftarrow{U}))} = -(\underline{\Lambda} + (1\cdot\overleftarrow{\Lambda}))\ e^{-i(\underline{U}+(1\cdot\overleftarrow{U}))}$$

$$+\ e^{-i(\underline{U}+(1\cdot\overleftarrow{U}))}(\bar{\underline{\Lambda}} + (1\cdot\overleftarrow{\bar{\Lambda}})), \qquad (136)$$

using the observation that, for arbitrary test fields $\phi$ and $\chi$,

$$\int\phi\ e^{i\underline{U}}\ \chi = \int\chi\ e^{-i(\underline{U}+(1\cdot\overleftarrow{U}))}\ \phi \qquad (137)$$

which follows from integration by parts. As a result of (136), it follows that

$$\delta(1\cdot e^{-i\overleftarrow{U}}) = -(\underline{\Lambda}(1\cdot e^{-i\overleftarrow{U}})) - (1\cdot\overleftarrow{\Lambda})(1\cdot e^{-i\overleftarrow{U}})$$

$$+\ (1\cdot e^{-i\overleftarrow{U}})(e^{-i\underline{U}}(1\cdot\overleftarrow{\bar{\Lambda}})), \qquad (138)$$

where I have used (see (104)):

$$(1\cdot e^{-i\overleftarrow{U}}) = e^{-i(\underline{U}\,+\,(1\cdot\overleftarrow{U}))}\ e^{i\underline{U}} \qquad (139)$$

Combining (113), (132), and (138), one finds

$$\delta\tilde{\underline{E}}_\alpha = [\tilde{\underline{E}}_\alpha, \underline{\Lambda}] + \tfrac{1}{2}\omega_\alpha{}^\beta\tilde{\underline{E}}_\beta + \tfrac{1}{2}(1\cdot\overleftarrow{\Lambda})\tilde{\underline{E}}_\alpha \qquad (140)$$

where

$$\omega_\alpha{}^\beta = -e^{-i\underline{U}}\ (\partial_\alpha\bar{\Lambda}^\beta + \partial^\beta\bar{\Lambda}_\alpha)\ e^{i\underline{U}}. \qquad (141)$$

Thus $\tilde{\underline{E}}_\alpha$ is a density of weight $-\frac{1}{2}$, and transforms covariantly with a Lorentz rotation (note that $\omega_\alpha{}^\beta \neq (\omega_{\dot\alpha}{}^{\dot\beta})^\dagger$; this is a consequence of working in a chiral representation).

Continuing the analogy with the Yang-Mills case, one can define (see eq. (69)),

$$\underline{\tilde{E}}^{(NC)}_{\alpha\dot{\alpha}} := -2i\{\underline{\tilde{E}}_{\alpha},\underline{\tilde{E}}_{\dot{\alpha}}\}; \tag{142}$$

unfortunately, $\underline{\tilde{E}}^{(NC)}_{\alpha\dot{\alpha}}$ is not covariant, as the right hand side of eq. (142) is missing connections for the Lorentz rotations and for the density transformations in eqs. (130) and (140). There is a straightforward procedure for finding such connection pieces, which undoubtedly has a geometric significance that, I'm afraid, I do not know. For convenience, I rewrite eqs. (130) and (140) as:

$$\delta\underline{\tilde{E}}_{\dot{\alpha}} = [\underline{\tilde{E}}_{\dot{\alpha}},\underline{\Lambda}] + \tfrac{1}{2}X_{\dot{\alpha}}{}^{\dot{\beta}}\underline{\tilde{E}}_{\dot{\beta}}, \quad X_{\dot{\alpha}}{}^{\dot{\beta}} = \omega_{\dot{\alpha}}{}^{\dot{\beta}} + (1\cdot\overleftarrow{\Lambda})\delta_{\dot{\alpha}}{}^{\dot{\beta}}$$

$$\delta\underline{\tilde{E}}_{\alpha} = [\underline{\tilde{E}}_{\alpha},\underline{\Lambda}] + \tfrac{1}{2}X_{\alpha}{}^{\beta}\underline{\tilde{E}}_{\beta}, \quad X_{\alpha}{}^{\beta} = \omega_{\alpha}{}^{\beta} + (1\cdot\overleftarrow{\Lambda})\delta_{\alpha}{}^{\beta}. \tag{143}$$

Then (142) implies:

$$\delta\underline{\tilde{E}}^{(NC)}_{\alpha\dot{\alpha}} = [\underline{\tilde{E}}_{\alpha\dot{\alpha}},\underline{\Lambda}] + \tfrac{1}{2}X_{\alpha}{}^{\beta}\underline{\tilde{E}}_{\beta\dot{\alpha}} + \tfrac{1}{2}X_{\dot{\alpha}}{}^{\dot{\beta}}\underline{\tilde{E}}_{\alpha\dot{\beta}}$$
$$- i(\underline{\tilde{E}}_{\alpha}X_{\dot{\alpha}}{}^{\dot{\beta}})\underline{\tilde{E}}_{\dot{\beta}} - i(\underline{\tilde{E}}_{\dot{\alpha}}X_{\alpha}{}^{\beta})\underline{\tilde{E}}_{\beta}, \tag{144}$$

and, for compactness, I define

$$\delta\underline{\tilde{E}}^{(NC)}_{A} := [\underline{\tilde{E}}^{(NC)}_{A},\underline{\Lambda}] + \tfrac{1}{2}X_{A}{}^{B}\underline{\tilde{E}}^{(NC)}_{B},$$

$$X_{\alpha}{}^{B} := X_{\alpha}{}^{\beta}\delta_{\beta}{}^{B}, \quad X_{\dot{\alpha}}{}^{B} := X_{\dot{\alpha}}{}^{\dot{\beta}}\delta_{\dot{\beta}}{}^{B},$$

$$X_{\alpha\dot{\alpha}}{}^{B} := (X_{\alpha}{}^{\beta}\delta_{\dot{\alpha}}{}^{\dot{\beta}} + X_{\dot{\alpha}}{}^{\dot{\beta}}\delta_{\alpha}{}^{\beta})\delta_{\beta\dot{\beta}}{}^{B} - 2i(\underline{\tilde{E}}_{\alpha}X_{\dot{\alpha}}{}^{\dot{\beta}})\delta_{\dot{\beta}}{}^{B}$$
$$- 2i(\underline{\tilde{E}}_{\dot{\alpha}}X_{\alpha}{}^{\beta})\delta_{\beta}{}^{B}. \tag{145}$$

Now consider the anholonomy coefficients $\tilde{C}_{AB}{}^{C}$:

$$[\underline{\tilde{E}}^{(NC)}_{A},\underline{\tilde{E}}^{(NC)}_{B}\} := \tilde{C}_{AB}{}^{C}\,\underline{\tilde{E}}^{(NC)}_{C}. \tag{146}$$

They transform as:

$$\delta\tilde{C}_{AB}{}^{C} = -(\underline{\Lambda}\,\tilde{C}_{AB}{}^{C}) - \tfrac{1}{2}\tilde{C}_{AB}{}^{D}\,X_{D}{}^{C}$$
$$+ \tfrac{1}{2}(X_{A}{}^{D}\,\tilde{C}_{DB}{}^{C} + (\underline{\tilde{E}}^{(NC)}_{A}\,X_{B}{}^{C}) - (-1)^{ab}(A\leftrightarrow B)). \tag{147}$$

Some useful properties of the anholonomy coefficients which follow from (130), (134), and (142) are:

$$\tilde{C}_{\alpha\beta}{}^{C} = \tilde{C}_{\alpha\beta}{}^{\gamma}\,\delta_{\gamma}{}^{C}, \quad \tilde{C}_{\dot{\alpha}\dot{\beta}}{}^{C} = \tilde{C}_{\dot{\alpha}\dot{\beta}}{}^{\dot{\gamma}}\,\delta_{\dot{\gamma}}{}^{C},$$

$$\text{and} \quad \tilde{C}_{\alpha\dot{\beta}}{}^{C} = \tfrac{1}{2}i\,\delta_{\alpha}{}^{\gamma}\,\delta_{\dot{\beta}}{}^{\dot{\gamma}}\,\delta_{\gamma\dot{\gamma}}{}^{C}. \tag{148}$$

The particular components that contain the connections transform as:

$$\begin{aligned}\delta\tilde{C}_{\alpha,\beta\dot{\beta}}{}^{\gamma\dot{\gamma}} &= -(\underline{\Lambda}\,\tilde{C}_{\alpha,\beta\dot{\beta}}{}^{\gamma\dot{\gamma}}) + \tfrac{1}{2}(X_{\alpha}{}^{\delta}\,\tilde{C}_{\delta,\beta\dot{\beta}}{}^{\gamma\dot{\gamma}} + X_{\beta}{}^{\delta}\,\tilde{C}_{\alpha,\delta\dot{\beta}}{}^{\gamma\dot{\gamma}} \\ &\quad + X_{\dot{\beta}}{}^{\dot{\delta}}\,\tilde{C}_{\alpha,\beta\dot{\delta}}{}^{\gamma\dot{\gamma}} - \tilde{C}_{\alpha,\beta\dot{\beta}}{}^{\gamma\dot{\delta}}\,X_{\dot{\delta}}{}^{\dot{\gamma}} - \tilde{C}_{\alpha,\beta\dot{\beta}}{}^{\delta\dot{\gamma}}\,X_{\delta}{}^{\gamma}) \\ &\quad + \tfrac{1}{2}((\underline{\tilde{E}}_{\alpha}X_{\beta}{}^{\gamma})\delta_{\dot{\beta}}{}^{\dot{\gamma}} + (\underline{\tilde{E}}_{\alpha}X_{\dot{\beta}}{}^{\dot{\gamma}})\delta_{\beta}{}^{\gamma} - (\underline{\tilde{E}}_{\beta}X_{\dot{\beta}}{}^{\dot{\gamma}})\delta_{\alpha}{}^{\gamma}) \\ &= -(\underline{\Lambda}\tilde{C}) + \tfrac{1}{2}(X\tilde{C}) \\ &\quad + \tfrac{1}{2}((\underline{\tilde{E}}_{\alpha}X_{\beta}{}^{\gamma})\delta_{\dot{\beta}}{}^{\dot{\gamma}} - (\underline{\tilde{E}}^{\gamma}X_{\dot{\beta}}{}^{\dot{\gamma}})\varepsilon_{\alpha\beta}),\end{aligned} \tag{149}$$

where $(\underline{\Lambda}\tilde{C})$ and $\frac{1}{2}(X\tilde{C})$ are the covariant pieces of the transformations, and of course the complex conjugate components contain the conjugate connections. The density connection is easily extracted; it is simply $\tilde{C}_{\alpha,\beta\dot{\beta}}{}^{\beta\dot{\beta}}$, and transforms as:

$$\begin{aligned}\delta\tilde{C}_{\alpha,\beta\dot{\beta}}{}^{\beta\dot{\beta}} := \delta\tilde{C}_{\alpha} &= -(\underline{\Lambda}\tilde{C}_{\alpha}) + \tfrac{1}{2}(X_{\alpha}{}^{\beta}\tilde{C}_{\beta}) \\ &\quad + (\underline{\tilde{E}}_{\alpha}X_{\beta}{}^{\beta}) + \tfrac{1}{2}(\underline{\tilde{E}}_{\alpha}X_{\dot{\beta}}{}^{\dot{\beta}}) \\ &= -(\underline{\Lambda}\tilde{C}_{\alpha}) + \tfrac{1}{2}(X_{\alpha}{}^{\beta}\tilde{C}_{\beta}) + 3\underline{\tilde{E}}_{\alpha}(1\cdot\bar{\Lambda}) \\ &= -(\underline{\Lambda}\tilde{C}_{\alpha}) + \tfrac{1}{2}(X_{\alpha}{}^{\beta}\tilde{C}_{\beta}) + \tfrac{3}{2}(\underline{\tilde{E}}_{\alpha}X_{\beta}{}^{\beta}).\end{aligned} \tag{150}$$

The connection needed to covariantize (142), (see (144)), is now:

$$\Gamma_{\alpha\dot{\beta}}{}^{\dot{\gamma}} = \tfrac{1}{3}\tilde{C}_{\alpha}\delta_{\dot{\beta}}{}^{\dot{\gamma}} - \tilde{C}_{\alpha,\beta\dot{\beta}}{}^{\beta\dot{\gamma}},$$

which transforms as

$$\begin{aligned}\delta\Gamma_{\alpha\dot{\beta}}{}^{\dot{\gamma}} &= -(\underline{\Lambda}\Gamma_{\alpha\dot{\beta}}{}^{\dot{\gamma}}) + \tfrac{1}{2}(X_{\alpha}{}^{\beta}\Gamma_{\beta\dot{\beta}}{}^{\dot{\gamma}} + X_{\dot{\beta}}{}^{\dot{\delta}}\Gamma_{\alpha\dot{\delta}}{}^{\dot{\gamma}} - \Gamma_{\alpha\beta}{}^{\dot{\delta}}X_{\dot{\delta}}{}^{\dot{\gamma}}) \\ &\quad - \tfrac{1}{2}(\underline{\tilde{E}}_{\alpha}X_{\dot{\beta}}{}^{\dot{\gamma}}).\end{aligned} \tag{151}$$

Then a truly covariant $\underline{\tilde{E}}_{\alpha\dot{\alpha}}$ can be defined:

$$\underline{\tilde{E}}_{\alpha\dot{\alpha}} = \underline{\tilde{E}}^{(NC)}_{\alpha\dot{\alpha}} - 2i(\Gamma_{\alpha\dot{\alpha}}{}^{\dot{\beta}}\underline{\tilde{E}}_{\dot{\beta}} + \Gamma_{\alpha\dot{\alpha}}{}^{\beta}\underline{\tilde{E}}_{\beta})$$

$$= -2i[(\tilde{\underline{E}}_{\alpha}\delta_{\dot{\alpha}}{}^{\dot{\beta}} + \Gamma_{\alpha\dot{\alpha}}{}^{\dot{\beta}})\tilde{\underline{E}}_{\dot{\beta}} + (\tilde{\underline{E}}_{\dot{\alpha}}\delta_{\alpha}{}^{\beta} + \Gamma_{\dot{\alpha}\alpha}{}^{\beta})\tilde{\underline{E}}_{\beta}]. \tag{152}$$

The new $\tilde{\underline{E}}_{\alpha\dot{\alpha}}$ is a density of weight -1, transforming as:

$$\delta\tilde{\underline{E}}_{\alpha\dot{\alpha}} = [\tilde{\underline{E}}_{\alpha\dot{\alpha}}, \underline{\Lambda}] + \tfrac{1}{2}\omega_{\alpha}{}^{\beta}\tilde{\underline{E}}_{\beta\dot{\alpha}} + \tfrac{1}{2}\omega_{\dot{\alpha}}{}^{\dot{\beta}}\tilde{\underline{E}}_{\alpha\dot{\beta}} + (1\cdot\overleftarrow{\Lambda})\tilde{\underline{E}}_{\alpha\dot{\alpha}}. \tag{153}$$

The superdeterminant of $\tilde{E}_A{}^M$ is a density of weight (see the Appendix):

$$- 1 + 4(\text{weight of } \tilde{E}_{\alpha\dot{\alpha}}{}^M) - 2(\text{weight of } \tilde{E}_{\alpha}{}^M)$$
$$- 2(\text{weight of } \tilde{E}_{\dot{\alpha}}{}^M)$$

$$= - 1 - 4 + 1 + 1 = -3.$$

Defining $E = (\text{sdet } \tilde{E}_A{}^M)^{-\frac{1}{3}}$ = a true scalar density of weight +1, it is easy to find the true inverse achtbein which is an ordinary vector field and not a density:

$$\underline{E}_{\dot{\alpha}} = E^{\frac{1}{2}}\,\tilde{\underline{E}}_{\dot{\alpha}}, \quad \underline{E} = E^{\frac{1}{2}}\,\tilde{\underline{E}}_{\alpha}, \quad \text{and}$$

$$\underline{E}_{\alpha\dot{\alpha}} = E\,\tilde{\underline{E}}_{\alpha\dot{\alpha}}. \tag{154}$$

The only local, gauge invariant action of the correct dimensions for supergravity is now:

$$I_{SG} \propto \frac{1}{\kappa^2}\int d^4x d^4\theta\; E(U^M, \mathcal{E}), \tag{155}$$

where $\kappa$ is the Planck length and E is a function of the unconstrained superfields $U^M$ and $\mathcal{E}$. Tangent space Lorentz rotations are easily included by generalizing (154) to:

$$\underline{E}_{\dot{\alpha}} = N_{\dot{\alpha}}{}^{\dot{\beta}}\,E^{\frac{1}{2}}\,\tilde{\underline{E}}_{\dot{\alpha}}, \quad \underline{E}_{\alpha} = N_{\alpha}{}^{\beta}\,E^{\frac{1}{2}}\,\tilde{\underline{E}}_{\alpha},$$

$$\underline{E}_{\alpha\dot{\alpha}} = N_{\alpha}{}^{\beta}\,N_{\dot{\alpha}}{}^{\dot{\beta}}\,E\,\tilde{\underline{E}}_{\alpha\dot{\alpha}}, \tag{156}$$

where $N_{\dot{\alpha}}{}^{\dot{\beta}}$ is an arbitrary Lorentz rotation satisfying the modified Hermiticity condition (analogous to (129) and (141)):

$$N_{\alpha}{}^{\beta} = e^{-i\underline{U}}(N_{\dot{\alpha}}{}^{\dot{\beta}})^{\dagger}e^{i\underline{U}}. \tag{157}$$

Ordinary Lorentz connections $\Omega_{AB}{}^C$ (see(92)) can be determined in terms of anholonomy coefficients $C_{AB}{}^C$ defined now in terms of the achtbein (156):

$[\underline{E}_A, \underline{E}_B\} = C_{AB}{}^C \, \underline{E}_C.$

I'll spare you the details of the calculation, and give you the result

$$\Omega_{\alpha\dot{\beta}}{}^{\dot{\gamma}} = -C_{\alpha\dot{\beta}}{}^{\dot{\gamma}}, \quad \Omega_{\dot{\alpha}\beta}{}^{\gamma} = -C_{\dot{\alpha}\beta}{}^{\gamma},$$

$$\Omega_{\alpha\beta}{}^{\gamma} = -\tfrac{1}{2}(C_{\alpha\beta}{}^{\gamma} + C^{\gamma}{}_{\alpha\beta} - C^{\gamma}{}_{\beta\alpha}),$$

$$\Omega_{\dot{\alpha}\dot{\beta}}{}^{\dot{\gamma}} = -\tfrac{1}{2}(C_{\dot{\alpha}\dot{\beta}}{}^{\dot{\gamma}} + C^{\dot{\gamma}}{}_{\dot{\alpha}\dot{\beta}} - C^{\dot{\gamma}}{}_{\dot{\beta}\dot{\alpha}}), \quad \text{and}$$

$$\Omega_{ab}{}^{c} = -\tfrac{1}{2}(C_{ab}{}^{c} + C^{c}{}_{ab} + C^{c}{}_{ba}). \tag{158}$$

Of course, these connections can be modified by the addition of any true tensor with the correct symmetry properties. Coupling to matter is now straightforward: derivatives are covariantized in the normal fashion, and chiral or general Lagrangians are integrated with

$$\int d^4x d^2\theta \, \mathcal{E} \quad \text{or} \quad \int d^4x d^4\theta \, E. \tag{159}$$

Torsions and curvatures are defined by (see(70)):

$$[\nabla_A, \nabla_B\} := T_{AB}{}^C \nabla_C + R_{AB}(J), \tag{160}$$

where

$$\nabla_A := \underline{E}_A + \Omega_A(J). \tag{161}$$

This section has been rather heavy going, so I will summarize briefly what all those calculations were about. Starting with unconstrained objects $U^M$ and $\mathcal{E}$, I constructed the inverse achtbein and connections, and the gauge action for supergravity. I worked in close analogy with the Yang-Mills case. For the actual construction it was convenient to construct directly a density $\tilde{\underline{E}}_A$ rather than the inverse achtbein $\underline{E}_A$ itself. $\tilde{\underline{E}}_{\dot{\alpha}}$ was formed simply out of the chiral volume element $\mathcal{E}$ (see (130)); $\tilde{\underline{E}}_{\alpha}$ was simply a coordinate transform of the complex conjugate of $\tilde{\underline{E}}_{\dot{\alpha}}$ (see (134)). $\tilde{\underline{E}}_{\alpha\dot{\alpha}}$ was just the anticommutator of $\tilde{\underline{E}}_{\dot{\alpha}}$ and $\tilde{\underline{E}}_{\alpha}$ with extra connection pieces formed out of the anholonomy coefficients $\tilde{C}_{AB}{}^C$ (see (146-152)).

## Vector representation

Just as in the Yang-Mills case, one can go to a vector representation by introducing a new dynamical field, in this case a non-

Hermitian vector field $\underline{W} = W^M \partial_M \neq (\underline{W})^\dagger$ which converts the complex $\Lambda$ transformations into real K transformations:

$$\underline{K} = K^M \partial_M, \quad K^M = (K^\dagger)^M,$$

$$e^{i\underline{W}} := e^{\underline{\bar{\Lambda}}} (e^{i\underline{W}'}) e^{-\underline{K}},$$

or

$$\delta(\Lambda,K) e^{i\underline{W}} = -\underline{\bar{\Lambda}} e^{i\underline{W}} + e^{i\underline{W}}\underline{K} \tag{162}$$

(see (78) and (113)). The K transformations are precisely ordinary superdiffeomorphisms in superspace; $\underline{K}$ is Hermitian but satisfies no differential constraint. Just as in the Yang-Mills case, $e^{i\underline{W}}$ and its Hermitian conjugate $e^{-i\underline{\bar{W}}}$ can be used to convert all $\Lambda$'s to K's, that is, to define objects which are inert under $\Lambda$ transformations and transform under K transformations. Thus, for example, one can define a scalar field (see (81))

$$\Phi_{(V)} := (e^{i\underline{\bar{W}}}\Phi) = e^{i\underline{\bar{W}}}\ \Phi\ e^{-i\underline{\bar{W}}} \tag{163}$$

which transforms as

$$\Phi_{(V)} = (e^{\underline{K}}\Phi_{(V)}) \tag{164}$$

(see (98)). Just as in the Yang-Mills case, one could have started with only K transformations, found constraints, solved them, and arrived at the same point; and just as in the approach I have taken above, supergravity was considerably more complex and subtle than Yang-Mills theory, so in the constrained approach both the constraints themselves and their solutions are much more difficult to find in supergravity than in the Yang-Mills theory. And now for something completely different.

## SUPERGRAPHS

### Introduction

Up to now, I have considered only classical supersymmetry -- first in flat superspace, and then in curved superspace; it is, however, in the quantum theory that much of the utility of superfield techniques becomes evident (see, for example, Abbot & Grisaru 1979, Grisaru, Roček & Siegel 1980, and Roček & Townsend 1980). I shall try to present the groundwork for the quantization of globally supersymmetric theories in flat superspace; further material can be found in Grisaru , Siegel & Roček 1979, and in the contributions of Grisaru and Siegel to this volume. For the purposes of quantization, I shall no longer try to treat the θ's and the x's rigidly on an equal footing; for example, I shall work in momentum space, but I shall not Fourier transform the θ's (since $D_\alpha$ contains explicit θ's, this would be of little use), but rather I shall treat the θ's as a continuous internal symmetry index. I assume familiarity with standard modern path integral quantization techniques.

The first theory I want to consider is not physically interesting, but is a simple toy model. Consider a supersymmetric theory with the following action:

$$I = \int d^8z\{-\tfrac{1}{2}V(z)(\Box-m^2)V(z) + U(V(z)) + V(z)J(z)\} \tag{165}$$

$$\sim \int d^4x \sum_\theta \{-\tfrac{1}{2}V_\theta(x)(\Box-m^2)V_\theta(x)+U(V_\theta(x))+V_\theta(x)J_\theta(x)\}, \tag{166}$$

where $V(z)$ is a real unconstrained superfield (see (36)), $J(z)$ is a real unconstrained superfield acting as a source and $U$ is some interaction potential. Eq. (166) is meant merely as a reminder that the θ's are to be thought of as an internal symmetry index. The free part of the action (165) is:

$$I_0 = \int d^8z\{-\tfrac{1}{2}V(z)(\Box-m^2)V(z) + V(z)J(z)\}, \tag{167}$$

and the free generating functional is:

$$\begin{aligned} Z_0[J] &= N\int(dV)\ e^{I_0[J,V]} \\ &= N\int(dV)\ \exp\int d^8z\{-\tfrac{1}{2}V(z)(\Box-m^2)V(z) + J(z)\}. \end{aligned} \tag{168}$$

Shifting the variable to complete the square, and performing the gaussian integration, one finds, in the usual fashion:

$$\begin{aligned} Z_0[J] &= N\int(dV)\ \exp\int d^8z\{-\tfrac{1}{2}(V - J\frac{1}{\Box-m^2})(\Box-m^2)(V - \frac{1}{\Box-m^2}J) \\ &\qquad + \tfrac{1}{2}J\frac{1}{\Box-m^2}J\} \\ &= N'\ \exp(\tfrac{1}{2}\int d^8z\ J(z)\frac{1}{\Box-m^2}J(z)) \end{aligned} \tag{169}$$

$$\sim N'\ \exp(\tfrac{1}{2}\int d^4x \sum_\theta J_\theta(x)\frac{1}{\Box-m^2}J_\theta(x)). \tag{170}$$

Thus the free propagator is

$$\langle T(V(-k,\theta_1)V(k,\theta_2))\rangle = -\frac{\delta^4(\theta_1-\theta_2)}{k^2 + m^2}, \tag{171}$$

where $-k^2$ is the Fourier transform of $\Box$. (Throughout this section, I shall work with Euclidean Feynman rules, and thus avoid factors of i). Usually, I shall use the notation

$\delta_{12} := \delta^4(\theta_1-\theta_2)$.

To find interactions, one needs the full generating functional:

$$Z[J] = \exp(\int d^8z\ U(\frac{\delta}{\delta J(z)}))\ Z_0[J], \tag{172}$$

and

$$\frac{\delta}{\delta J(z_2)} J(z_1) = \delta^8(z_1-z_2) := \delta^4(x_1-x_2)\ \delta_{12}. \tag{173}$$

Thus the Feynman rules for the generating functional $\Gamma$ of one particle irreducible graphs

$$\Gamma[V(z)] := \ell nZ[J[V(z)]] - \int d^8z\ V(z)J[V(z)], \quad \text{where}$$

$$V(z) := \frac{\delta \ell nZ[J]}{\delta J(z)} \Rightarrow J[V(z)] \tag{174}$$

are:

1. For every propagator, a factor of $-\frac{\delta_{12}}{k^2+m^2}$ (see 171)
2. For every n-point vertex, a factor of $\int d^4\theta\ U^{(n)}\delta^4(\sum_{i=1}^{n} k_i)$, where $U(x) = \sum_{n=3}^{\infty} \frac{U^{(n)}}{n!} x^n$ is the interaction potential, and $\delta^4(\sum_{i=1}^{n} k_i)$ is the momentum conserving $\delta$-function.
3. For every loop, a factor of $\int\frac{d^4k}{(2\pi)^4}$, where k is the loop momentum; for dimensional regularization by dimensional reduction (see Siegel, Townsend & van Nieuwenhuizen, this volume, and references therein), use instead a factor of $\int\frac{d^Dk}{(2\pi)^D}$.
4. For every external line, a factor $\int\frac{d^4k}{(2\pi)^4}V(-k,\theta)$, where k is the momentum flowing out of the graph along that line.
5. Usual symmetry factors: $\frac{1}{n!}$ for each set of n indistinguishable lines.

As a simple check of these rules, one can verify that the interaction piece of $\Gamma[V]$ at zero loops is just

$$\Gamma_{int}[V]_{0\text{-Loop}} = \int d^8z\ U(V(z)).$$

This is just a sum of terms, each with one vertex and varying numbers of external lines; since all the external lines are indistinguishable, the symmetry factor for n external lines attached to the same vertex is $\frac{1}{n!}$. Thus

$$\Gamma_{int}[V]_{0\text{-Loop}}$$

$$= \sum_{n=3}^{\infty} \frac{1}{n!} \int d^4\theta \; U^{(n)} \left( \prod_{i=1}^{n} \int \frac{d^4k_i}{(2\pi)^4} V(-k_i,\theta) \right) \; \delta(\sum_{i=1}^{n} k_i) \tag{175}$$

$$= \sum_{n=3}^{\infty} \frac{1}{n!} \int d^4\theta d^4x \; U^{(n)} (V(z))^n \tag{176}$$

$$= \int d^8z \; U(V(z))$$

where (175) is recognizable as the Fourier transform of (176).

"Realistic" theories

Preliminaries: Before quantizing physically more interesting theories, e.g., super Yang-Mills theory and chiral superfields, it is convenient to collect a number of formulae. As noted in eq. (43), up to total derivative terms,

$$\int d^8z = \int d^4x d^2\theta \; \bar{D}^2 = \int d^4x d^2\bar{\theta} D^2. \tag{177}$$

The basic anticommutation relation (27) allows one to define superspin projection operators (see eq. (44)):

$$\Pi_+ := \frac{\bar{D}^2 D^2}{\Box}, \qquad \Pi_- := \frac{D^2 \bar{D}^2}{\Box}, \qquad \Pi_0 := \Pi_+ + \Pi_-,$$

$$\Pi_{\frac{1}{2}} := 1 - \Pi_0 = -2\frac{D^\alpha \bar{D}^2 D_\alpha}{\Box} = -2\frac{\bar{D}^{\dot{\alpha}} D^2 \bar{D}_{\dot{\alpha}}}{\Box}. \tag{178}$$

Being projection operators, these of course obey:

$$(\Pi_i)^2 = \Pi_i$$

$$\Pi_i \Pi_j = 0 \quad i \neq j; \quad i,j = \tfrac{1}{2},+,- \tag{179}$$

Further, as a consequence of eq. (34),

$$D_\alpha \Pi_- = \bar{D}_{\dot{\alpha}} \Pi_+ = 0, \tag{180}$$

that is, $\Pi_+$ and $\Pi_-$ are chiral and antichiral projectors, respectively. In particular, it follows that on chiral (antichiral) superfields $\phi(\bar{\phi})$

$$\Pi_+ \phi(z) = \phi(z), \qquad \Pi_- \bar{\phi}(z) = \bar{\phi}(z), \tag{181}$$

and hence, for arbitrary chiral superfields $\phi(z)$ and $J(z)$,

$$\begin{aligned}\int d^4x d^2\theta \ \phi J &= \int d^4x d^2\theta \ \phi \ (\Pi_+ J) \\ &= \int d^4x d^2\theta \ \phi \ (\frac{\bar{D}^2 D^2}{\Box} J) \\ &= \int d^4x d^2\theta \ \bar{D}^2 \ (\phi(\frac{D^2}{\Box} J)) \\ &= \int d^8z \ \phi(\frac{D^2}{\Box} J). \end{aligned} \tag{182}$$

Thus (177) and (182) allow one to convert general $\theta$-integrals into chiral integrals, and vice-versa.

Super Yang-Mills theory: The action for supersymmetric Yang-Mills theory was found earlier (see eqs.(73) and (75)):

$$\begin{aligned} I_{Gauge} &= -\frac{tr}{g^2}\int d^4x d^2\theta \ W^\alpha W_\alpha \\ &= -\frac{tr}{g^2}\int d^4x d^2\theta \ (\bar{D}^2 \ e^{-gV} \ D^\alpha \ e^{gV})(\bar{D}^2 \ e^{-gV} \ D_\alpha \ e^{gV}), \end{aligned} \tag{183}$$

where, relative to (73), a factor of the coupling constant $g$ has been scaled in: $V \to gV$. The action (183) can be rewritten as a general $\theta$-integral using (177):

$$\begin{aligned} I_{Gauge} &= -\frac{tr}{g^2}\int d^8z \ (e^{-gV} \ D^\alpha \ e^{gV})\bar{D}^2(e^{-gV} \ D_\alpha \ e^{gV}) \\ &:= I_0 + I_{int} \end{aligned} \tag{184}$$

where

$$\begin{aligned} I_0 &= -tr\int d^8z \ (D^\alpha V)(\bar{D}^2 D_\alpha V) \\ &= \ tr\int d^8z \ V(D^\alpha \bar{D}^2 D_\alpha V) \\ &= \ tr\int d^8z \ (-\tfrac{1}{2} V\Box\Pi_{\frac{1}{2}} V) \end{aligned} \tag{185}$$

(I have integrated by parts with respect to $\theta$ and used (178) to obtain (185)), and

$$I_{int} = g \ tr\int d^8z \ V\{D^\alpha V, \bar{D}^2 D_\alpha V\} + o(g^2). \tag{186}$$

The projection operator in (185) appears because the theory is gauge invariant, and it makes the kinetic term (185) degenerate; to find a propagator, one must, in the normal way, add a gauge breaking term. To make (185) nondegenerate, a term of the form

$$I_{G.B.} = \mathrm{tr}\int d^8z\, (-\tfrac{1}{2\alpha} V \Box \Pi_0 V) \tag{187}$$

is desirable; using (178) and integrating by parts, (187) can be rewritten as

$$I_{G.B.} = \mathrm{tr}\int d^8z\, (-\tfrac{1}{2\alpha} V (D^2\bar{D}^2 + \bar{D}^2 D^2) V$$

$$= \mathrm{tr}\int d^8z\, (-\tfrac{1}{\alpha}(D^2V)(\bar{D}^2V)). \tag{188}$$

The gauge breaking term is now recognizable: it can be obtained by using the standard Faddeev-Popov prescription and averaging the gauge fixing δ-functional

$$\delta(D^2V - \bar{a})\ \delta(\bar{D}^2V - a)$$

with a Gaussian smearing fuctional

$$N\int (da)(d\bar{a})\ e^{-\frac{1}{\alpha}\int d^8z\ a\bar{a}}.$$

(Details can be found in Grisaru, Siegel & Roček 1979, and in the contributions of M. Grisaru and W. Siegel to this volume). To avoid infrared singularities, one must work in a Fermi-Feynman gauge $\alpha$ = +1; then

$$I_{kin} := I_0 + I_{G.B.} = \mathrm{tr}\int d^8z\, (-\tfrac{1}{2} V \Box V); \tag{189}$$

in this gauge, the propagator for super Yang-Mills theory is the same as for the toy model discussed above (in the m=0 case). Indeed, only the interactions differ: (186) contains spinor derivatives, whereas (165) does not. The spinor derivatives modify the Feynman rules; one now has:

2'. In addition to the usual factors, derivatives at a vertex act on the propagator or external field attached.

For practical calculation the Feynman rules must be augmented with a procedure for evaluating θ-integrals around a loop; the following is a straightforward approach: a propagator is chosen at random; if it has spinor derivatives, they are integrated by parts onto adjoining propagators or external lines until a pure δ-function in θ-space remains. Then the θ-integral at one end is performed, entirely contracting that propagator to a point in θ-space. This procedure of clearing propagators of all spinor derivatives, and then contracting them to a point in θ-space, is continued until only two propagators remain in a loop; this is simplified using the following identities:

$$\begin{aligned}
\delta_{12}\delta_{12} &= 0, \\
\delta_{12}D\delta_{12} &= \delta_{12}\bar{D}\delta_{12} = 0, \\
\delta_{12}D^2\delta_{12} &= \delta_{12}\bar{D}^2\delta_{12} = \delta_{12}D\bar{D}\delta_{12} = 0, \\
\delta_{12}D^2\bar{D}\delta_{12} &= \delta_{12}\bar{D}^2D\delta_{12} = 0, \\
\delta_{12}D^2\bar{D}^2\delta_{12} &= \delta_{12}\bar{D}^2D^2\delta_{12} = \delta_{12}D^\alpha\bar{D}^2D_\alpha\delta_{12} = \delta_{12}.
\end{aligned} \tag{190}$$

One continues until the whole graph is reduced to a single θ-integral. The identities (190) follow from the explicit representation of the δ-function (see eq. (40)):

$$\delta_{12} = (\theta_1-\theta_2)^2(\bar{\theta}_1-\bar{\theta}_2)^2. \tag{191}$$

A very useful identity which can be derived by explicit calculation using (191) allows one to transfer a derivative across a propagator:

$$D_\alpha(k,\theta_1)\delta_{12} = -D_\alpha(-k,\theta_2)\delta_{12}. \tag{192}$$

Note that the identities (190) imply that any graph containing a closed loop with less than 2 $D_\alpha$'s and 2 $\bar{D}_{\dot{\alpha}}$'s vanishes. In particular, the toy model considered above, lacking spinor derivatives in its interactions, receives no quantum corrections!

Power counting for the super Yang-Mills theory is simple. Because of (27), as noted earlier, $D \sim \bar{D} \sim M^{\frac{1}{2}}$; because every loop cancels four D's (see (190)), a loop integration contributes only 4-2=2 powers of momentum; each vertex, having 4 D's (see (184)), contributes 4/2=2 powers of momentum, and each propagator, as usual, contributes -2 powers of momentum. Hence the superficial degree of divergence is

$$d = 2L + 2V - 2P. \tag{193}$$

Noting that after all the θ-integrals are simplified, one remains, and that each vertex contributes a θ-integral, each propagator contributes a counter-balancing δ-function, and each loop cancels a δ-function (see(190)), one finds:

$$1 = V - P + L, \tag{194}$$

and hence,

$$d = 2. \tag{195}$$

Thus all graphs in super Yang-Mills theory (ignoring ghosts) have superficial degree of divergence two. Renormalizability of the theory can be proven using the Slavnov-Taylor identities (see

Ferrara & Piguet 1975).

Chiral superfields: Recall (see (54)) that the action for a chiral superfield coupled to super Yang-Mills theory is:

$$I_{inv} = \int d^8z\ \bar{\phi}e^{gV}\phi := I_{CH} + I', \tag{196}$$

where

$$I_{CH} = \int d^8z\ \bar{\phi}\phi, \quad I' = g\int d^8z\ \bar{\phi}V\phi + o(g^2). \tag{197}$$

A chiral self-interaction term can also be added (for simplicity, I consider the massless case):

$$\begin{aligned} I'' &= \frac{\lambda}{3!}\int d^4x(d^2\theta\ \phi^3 + \text{h.c.}) \\ &= \frac{\lambda}{3!}\int d^8z(\phi^2(\frac{D^2}{\Box}\phi) + \text{h.c.}) \end{aligned} \tag{198}$$

where I have used (182) to write I" as a general θ-integral. The sources for a chiral superfield are themselves chiral, and so the free part of the action is

$$\begin{aligned} I_0 &= \int d^8z\ \bar{\phi}\phi + \int d^4x(d^2\theta\ \phi J + \text{h.c.}) \\ &= \int d^8z(\bar{\phi}\phi + \phi(\frac{D^2}{\Box}J) + \bar{\phi}(\frac{\bar{D}^2}{\Box}\bar{J})). \end{aligned} \tag{199}$$

As in (168-169), one finds the free generating functional:

$$\begin{aligned} Z_0[J] &= N\int(d\phi)(d\bar{\phi})\ \exp\int d^8z(\bar{\phi}\phi + \phi(\frac{D^2}{\Box}J) + \bar{\phi}(\frac{\bar{D}^2}{\Box}\bar{J})) \\ &= N\int(d\phi)(d\bar{\phi})\ \exp\int d^8z\{(\bar{\phi} + \frac{D^2}{\Box}J)(\phi + \frac{\bar{D}^2}{\Box}\bar{J}) - (\frac{D^2}{\Box}J)(\frac{\bar{D}^2}{\Box}\bar{J})\} \\ &= N'\ \exp\{-\int d^8z\ (\frac{D^2}{\Box}J)(\frac{\bar{D}^2}{\Box}\bar{J})\} \\ &= N'\ \exp\{-\int d^8z\ J(\frac{D^2}{\Box})(\frac{\bar{D}^2}{\Box})\bar{J}\} \\ &= N'\ \exp\{-\int d^8z\ J(\frac{\Pi_-}{\Box}\bar{J})\} \\ &= N'\ \exp\{-\int d^8z\ J\ \frac{1}{\Box}\ \bar{J}\}, \end{aligned} \tag{200}$$

where (178) and (181) have been used to find the final form of (200). Thus the propagator for a massless chiral superfield is the same as (171) up to an overall minus sign:

$$\langle T(\phi(-k,\theta_1)\bar{\phi}(k,\theta_2))\rangle = +\ \frac{\delta_{12}}{k^2} \tag{201}$$

To find the interactions for the theory, one needs to know how to functionally differentiate with respect to chiral sources. In a chiral representation (see (30)), chiral fields are independent of $\bar{\theta}$, and it is clear that:

$$\frac{\delta}{\delta J_+(x_2,\theta_2)}\int d^4x_1 d^2\theta_1\; J_+(x_1,\theta_1)\phi_+(x_1,\theta_1) = \phi_+(x_2,\theta_2). \qquad (202)$$

Requiring (202) to be representation independent, and using (182), one finds:

$$\begin{aligned}\phi(z_2) &= \int d^4x_1 d^2\theta_1 \left(\frac{\delta}{\delta J(z_2)} J(z_1)\right)\phi(z_1)\\ &= \int d^8z_1 \left(\frac{\delta}{\delta J(z_2)} J(z_1)\right)\frac{D_1^2}{\Box_1}\phi(z_1)\ ;\end{aligned} \qquad (203)$$

but,

$$\begin{aligned}\phi(z_2) &= \int d^8z_1\; \delta^8(z_2-z_1)\; \frac{\bar{D}_1^2 D_1^2}{\Box_1}\phi(z_1)\\ &= \int d^8z_1\; (\bar{D}_1^2\delta^8(z_2-z_1))\; \frac{D_1^2}{\Box_1}\phi(z_1)\end{aligned} \qquad (204)$$

and so, for chiral sources $\bar{D}J = 0$,

$$\frac{\delta}{\delta J(z_2)} J(z_1) = \bar{D}_1^2\delta^8(z_2-z_1). \qquad (205)$$

Thus the Feynman rules for interactions involving chiral superfields are:

2a. Differentiate every chiral line with a factor $\bar{D}^2$, and every antichiral line with a factor $D^2$.

Note that the $\phi^3$ self-coupling interaction (198) has a factor $\frac{D^2}{\Box}$ on one line when written as a general $\theta$-integral; this cancels one factor of $\bar{D}^2$, leaving precisely four D's, just as in the $\bar{\phi}V\phi$ vertex (197) and the $V^3$ self-interactions (186). Finally, external chiral lines come with a factor of $\frac{D^2}{\Box}$ (see (199)), but because of (204), this just serves to cancel the factor of $\bar{D}^2$ at the vertex, and hence one has:

4a. For every external chiral line, drop a factor of $\bar{D}^2$ at the vertex to which the line attaches; similarly, for external antichiral lines, drop a factor of $D^2$.

These rules can be checked most easily by calculating $\Gamma_{int}[\phi]_{0\text{-Loop}}$.

Power counting proceeds as before, since, as noted above, all vertices still have dependence momentum²; however, external chiral lines, because of rule 4a above, have dependence momentum⁻¹, so the superficial degree of divergence with $E_{CH}$ external chiral plus antichiral lines is

$$d = 2 - E_{CH}. \tag{206}$$

Hence it follows that <u>any graph with three or more external chiral (antichiral) lines is superficially convergent</u>.

<u>Sample calculations:</u> As an example, I shall consider a model with some real interest. This is the N=4 super Yang-Mills theory (see Grisaru, Siegel & Roček 1979 and Grisaru, Roček & Siegel 1980); it has an action, in Fermi-Feynman gauge:

$$I = \int d^8z \, (\bar\phi_i^a \phi_i^a - \tfrac{1}{2}V^a \Box V^a + g(if_{abc})\bar\phi_i^a V^b \phi_i^c)$$
$$+ \int d^4x \, \{d^2\theta \, (\tfrac{g}{3!})(i\epsilon^{ijk})(if_{abc})\phi_i^a \phi_j^b \phi_k^c + \text{h.c.}\} + \ldots \tag{207}$$

where i=1,2,3, and $f_{abc}$ are the structure constants for an arbitrary group, such that

$$f_{abc} \, f_{dbc} = \kappa \, \delta_{ad}. \tag{208}$$

I shall calculate the one-loop contribution to the chiral superfield self energy $\langle T(\bar\phi_i^a \phi_j^b)\rangle$. This receives two contributions shown in figure 1. (All momentum integrations which can be done by virtue of momentum conserving δ-functions have been done).

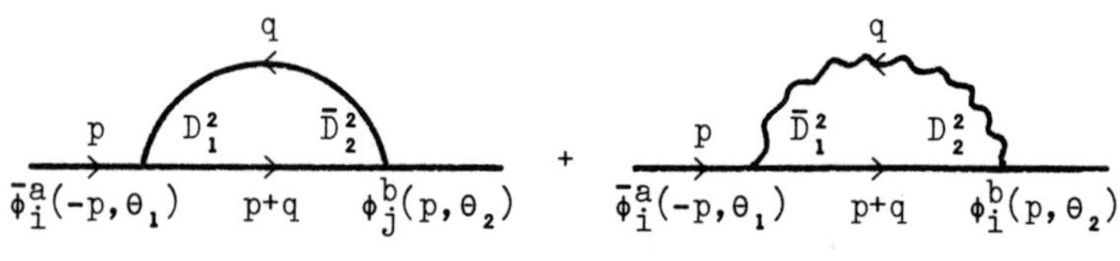


Fig. 1 Contributions to $\langle T(\bar\phi_i^a \phi^b{}_j)\rangle$

The first graph contributes:

$$\tfrac{1}{2}\{g^2(i\epsilon_{ikl})(i\epsilon_{jlk})(if_{acd})(if_{bdc})\}\int \frac{d^4p}{(2\pi)^4}\frac{d^Dq}{(2\pi)^D}\,(d^4\theta_1 d^4\theta_2)$$

S V E L V

$$\times\{\bar{\phi}^a_i(-p,\theta_1)\phi^b_j(p,\theta_2)\}\ \frac{\delta_{12}}{q^2}(D^2_1\bar{D}^2_2)\ \frac{\delta_{12}}{(p+q)^2} \tag{209}$$

E P V P

where the labels S, V, E, L, P, underneath the various factors indicate their origins: symmetry, vertex, external line, loop, and propagator respectively. The numerical factors in front of the integral reduce to $g^2\kappa$; the spinor derivatives and $\delta$-functions, after using (192) and (190), reduce to $\delta_{12}$; performing the $\theta_2$-integral, one is left with:

$$g^2\kappa\int\frac{d^4p}{(2\pi)^4}\frac{d^Dq}{(2\pi)^D}\,d^4\theta_1\ \bar{\phi}^a_i(-p,\theta_1)\phi^a_i(p,\theta_1)\ \frac{1}{q^2(p+q)^2}. \tag{210}$$

The second graph contributes:

$$\{g^2(if_{acd})(if_{bdc})\}\int\frac{d^4p}{(2\pi)^4}\frac{d^Dq}{(2\pi)^D}\,(d^4\theta_1 d^4\theta_2)$$

V E L V

$$\times\{\bar{\phi}^a_i(-p,\theta_1)\phi^b_i(p,\theta_2)\}\ -\frac{\delta_{12}}{q^2}(\bar{D}^2_1D^2_2)\ \frac{\delta_{12}}{(p+q)^2} \tag{211}$$

E P V P

which reduces to

$$-g^2\kappa\int\frac{d^4p}{(2\pi)^4}\frac{d^Dq}{(2\pi)^D}\,d^4\theta_1\ \bar{\phi}^a_i(-p,\theta_1)\phi^a_i(p,\theta_1)\ \frac{1}{q^2(p+q)^2}. \tag{212}$$

precisely cancelling (210) to give a final contribution zero.

These techniques have been used to calculate 3-loop processes in a number of models <u>without</u> the aid of a computer (see the references given at the beginning of this lecture). The methods are powerful and elegant, and I hope that I have made them more accessible. They can be applied to the formulation of supergravity given earlier in these lectures, and may possibly allow the resolution of the problem concerning the absence or presence of the 3-loop counterterm by explicit calculation.

ACKNOWLEDGEMENTS

I would like to thank Professor Hawking for encouraging me first to give, and then to write these lectures; all the participants of the Nuffield workshop on "Progress in supergravity" for many useful comments, discussions, and corrections; Jacques Richer, the control engineering department, and Judy Fella for their astonishing endurance in helping me type the manuscript; and Sean and Anne O'Cathasaigh for tolerating me during the typing.

APPENDIX

The achtbein is an example of a supermatrix which, in general, has structure:

$$\left(\begin{array}{c|c} A & B \\ \hline C & D \end{array}\right) \tag{A.1}$$

where A and D are ordinary square matrices with commuting entries, and B and C are in general rectangular matrices with anticommuting entries. The essential property that the superdeterminant (sdet) or Berezinian of a matrix has is that the sdet of the product of two matrices is the product of the sdets of each. Because supermatrices contain anticommuting elements, the conventional determinant will not work. In the case of an ordinary matrix with structure (A.1), for A and D nondegenerate matrices, the determinant can be found by noting

$$\begin{aligned}
\det\left(\begin{array}{c|c} A & B \\ \hline C & D \end{array}\right) &= \det\left(\begin{array}{c|c} A & B \\ \hline C & D \end{array}\right)\cdot\left(\begin{array}{c|c} 1 & -A^{-1}B \\ \hline 0 & 1 \end{array}\right) \\
&= \det\left(\begin{array}{c|c} A & 0 \\ \hline C & D-CA^{-1}B \end{array}\right) \\
&= (\det A)(\det(D-CA^{-1}B)) \\
&= (\det(A-BD^{-1}C))(\det D).
\end{aligned} \tag{A.2}$$

The superdeterminant has a similar expression:

$$\begin{aligned}
\operatorname{sdet}\left(\begin{array}{c|c} A & B \\ \hline C & D \end{array}\right) &= (\det A)(\det(D-CA^{-1}B))^{-1} \\
&= (\det(A-BD^{-1}C))(\det D)^{-1}.
\end{aligned} \tag{A.3}$$

As a check, consider the sdet of the product of two particularly simple supermatrices:

$$\begin{aligned}
\operatorname{sdet}\begin{pmatrix} 1 & a \\ 0 & 1 \end{pmatrix}\begin{pmatrix} 1 & 0 \\ b & 1 \end{pmatrix} &= \operatorname{sdet}\begin{pmatrix} 1 & a \\ 0 & 1 \end{pmatrix}\operatorname{sdet}\begin{pmatrix} 1 & 0 \\ b & 1 \end{pmatrix} = 1 \\
&= \operatorname{sdet}\begin{pmatrix} 1+ab & a \\ b & 1 \end{pmatrix}
\end{aligned}$$

$$= (1+ab)(1-b(1+ab)^{-1}a)^{-1}$$

$$= (1+ab)(1-ba)^{-1} = (1+ab)(1+ab)^{-1} = 1$$

where $ab + ba = 0$ implies $a^2 = b^2 = 0$. This definition of the sdet has the property that the logarithm of the sdet of a matrix is the supertrace (str) of the logarithm of the matrix, where the supertrace is defined by

$$\mathrm{str}\left(\begin{array}{c|c} A & B \\ \hline C & D \end{array}\right) = \mathrm{tr}A - \mathrm{tr}D \tag{A.4}$$

and has the cyclicity property $\mathrm{str}(XY) = \mathrm{str}(YX)$ for X and Y two supermatrices. As a consequence of the definition of sdet (A.3), a supermatrix has the following scaling property:

$$\mathrm{sdet}\left(\begin{array}{c|c} \alpha A & \alpha B \\ \hline \beta C & \beta D \end{array}\right) = \alpha^n \beta^{-m}\, \mathrm{sdet}\left(\begin{array}{c|c} A & B \\ \hline C & D \end{array}\right), \tag{A.5}$$

n and m being the dimensions of A and D respectively.

REFERENCES

Abbott, L.F. & Grisaru, M.T. (1979) The 3 loop β-function for the Wess-Zumino model, Nucl. Phys. B, to be published.

Berezin, F.A. (1973) The Method of Second Quantization, NY, Academic Press (1966).

Coleman, S. & Mandula, J. (1967) All Possible symmetries of the S matrix, Phys. Rev., 159, p.1251.

de Wit, B. & Freedman, D.Z. (1975) Combined Supersymmetric and gauge-invariant Field Theories, Phys. Rev., D12, pp. 2286-2297.

De Witt, B.S. (1978) The Formal Structure of Quantum Gravity, in Recent Developments in Gravitation Cargèse 1978, eds. M. Lévy & S. Deser, NY, Plenum Press, pp. 275-322.

Fayet, P. & Ferrara, S. (1977) Supersymmetry, Physics Reports, 32C, pp. 1-250.

Ferrara, S. (1977) Algebraic properties of Extended Supergravity Theories in de Sitter space, Phys. Lett., 69B, p. 481.

Ferrara, S. & Piguet, O. (1975) Perturbation theory and renormalization of supersymmetric Yang-Mills theories, Nucl. Phys., B93, pp. 261-302.

Ferrara, S., Wess, J. & Zumino, B. (1974) Supergauge multiplets and superfields, Phys. Lett., 51B, p. 239.

Freedman, D.Z. & van Nieuwenhuizen, P. (1979) Supergravity, Amsterdam, North-Holland Publishing Co.

Gates , S.J. (1979) "Superconformal" symmetry breakdown as a guide to supergravity constraints, in Supergravity, Freedman & van Nieuwenhuizen (1979), pp. 215-220.

Gates, S.J. & Siegel, W. (1979) Understanding Constraints in Superspace formulations of Supergravity, Nucl. Phys., B163, pp. 519-545. (1980).

Gates, S.J., Stelle, K.S. & West, P.C. (1980) Algebraic Origins of Superspace Constraints in Supergravity. Nucl. Phys., B169, p. 347.

Gol'fand, Y.A. & Likhtman, E.P. (1971) JETP Letters, 13, p. 323.

Grisaru, M.T., Roček, M. & Siegel, W. (1980) Zero value of the 3-loop β-function in N=4 Super Yang-Mills Theory, Phys. Rev. Lett., 45, p. 1063.

Grisaru, M.T., Siegel, W. & Roček, M. (1979) Improved methods for Supergraphs, Nucl. Phys., B159, pp. 429-450.

Haag, R., Lopuszanski, J.T. & Sohnius, M. (1975) All possible generators of Supersymmetries of the S matrix, Nucl. Phys., B88, p. 257.

Iachello (1980) Dynamical supersymmetries in nuclei, Phys. Rev. Lett., 44, p. 772.

Kaku, M. (1979) Conformal Supergravity, in Supergravity, Freedman & van Nieuwenhuizen (1979), pp. 93-102.

Nahm, W. (1978) Supersymmetries and their Representations, Nucl. Phys., B135, p. 149.

Nicolai, H. (1980) On a new characterization of scalar Supersymmetric theories, Phys. Lett., 89B, p. 341.

Nicolai, H. (1980a) On a new characterization of Supersymmetric theories II, CERN preprint Ref. TH. 2811 unpublished.

Ogievetsky, V. & Sokatchev, E. (1978) Structure of the Supergravity Group, Phys. Lett., B79, p. 222.

Ogievetsky, V. & Sokatchev, E. (1978a) Supergravity as a theory of an axial Superfield (Talk presented at the Seminar on Quantum Gravity, Moscow, Dec. 5-7, 1978).

Roček, M. & Townsend, P.K. (1980) Three Loop Finiteness of the N=4 supersymmetric non-linear σ-model, CERN Preprint Ref. TH. 2914, to appear in Phys. Lett. B.

Salam, A. & Strathdee, J. (1974) Superguage transformations, Nucl. Phys., B79, p. 477.

Salam, A. & Strathdee, J. (1974a) Super-symmetry and non-abelian gauges, Phys. Lett., 51B, p. 353.

Salam, A. & Strathdee, J. (1975) Feynman Rules for Superfields, Nucl. Phys., B86, p. 142.

Scherk, J. (1978) Extended Supersymmetry and Extended Supergravity theories, in Recent Developments in Gravitation Cargèse 1978, eds. M. Lévy & S. Deser, NY, Plenum Press, pp. 479-517.

Siegel, W. (1977) Supergravity Superfields without a Supermetric, Harvard Preprint, HUTP-77/A068, unpublished.
The Superfield Supergravity Action, Harvard Preprint, HUTP-77/A080, unpublished.

Siegel, W. (1978) Solution to Constraints in the Wess-Zumino Supergravity formalism, Nucl. Phys., B142, pp. 301-305.

Siegel, W. & Gates, S.J. (1979) Superfield Supergravity, Nucl. Phys., B147, pp. 77-104.

Siegel, W. & Gates, S.J. (1979a) see Gates & Siegel (1979).

Stelle, K.S. & West, P.C. (1979) Algebraic understanding of constraints, in Supergravity, Freedman & van Nieuwenhuizen (1979), pp. 132-142.

van Nieuwenhuizen, P. (1980) Supergravity, to appear in Physics Reports C.

Wess, J. (1977) Supersymmetry-Supergravity, Lectures given at VIII GIFT International Seminar on Theo. Phys., Salamanca June 1977; in Phys., 77, Springer-Verlag.

Zumino, B. (1978) Supergravity and Superspace, in Recent Developments in Gravitation Cargèse 1978, eds. M. Lévy & S. Deser, NY, Plenum Press, pp. 405-459.

The following additional references may be of interest:

Breitenlohner, P. (1979) Auxiliary fields and tensor calculus for N=2 extended Supergravity, in Supergravity, Freedman & van Nieuwenhuizen (1979), pp. 123-132.

Gates, S.J. (1977) Geometry of Superspace and Local Supersymmetry, Phys. Rev., D17, p. 3188 (1977).

Gates, S.J. (1978) A note on the Geometry of Local Supersymmetry, Harvard Preprint HUTP-78/A001, unpublished.
Geometry of Superspace and Local Supersymmetry II, Harvard Preprint HUTP-78/A028, unpublished.

Gates, S.J. & Shapiro, J. A. (1978) Local Supersymmetry in Superspace, Phys. Rev., D18, p. 2768.

Grimm, R., Wess, J. & Zumino, B. (1978) Consistency checks on the Superspace formulation of Supergravity, Phys. Lett., 79B, p. 394.

Grisaru, M.T., Roček, M. & Siegel, W. (1980) Superloops 3, Beta 0 A calculation in N=4 Yang-Mills Theory, Brandeis Prepreint, submitted to Nucl. Phys. B.

Howe, P. (1979) The N=8 Supergravity on the mass shell in Superspace, in Supergravity, Freedman & van Nieuwenhuizen (1979), pp. 11-18.

Howe, P. & Tucker, R. (1978) Scale invariance in Superspace, _Phys. Lett._, 80B, pp. 138-141.

Siegel, W. (1979) Superconformal invariance of Superspace with nonminimal auxiliary fields, _Phys. Lett._, 80B, pp. 224-227.

Wess, J. (1979) Constraints for Supergravity, in _Supergravity_, Freedman & van Nieuwenhuizen (1979), pp. 103-112.

# II QUANTIZATION

# BACKGROUND SUPERFIELD METHOD

M. T. Grisaru

Physics Department
Brandeis University

## 1. INTRODUCTION

This is an introduction (a gentle one I hope) to the first part of "Supergraphity" (Grisaru & Siegel, 1980) on the background field quantization of superfield supergravity in the formalism of Gates & Siegel (1979). The second part will be described by W. Siegel (1980).

In that work we have attempted to present a fairly complete and self contained approach to the problem of computing quantum effects (loops) in simple supergravity, and supergravity coupled to supersymmetric matter, using superfield Feynman diagrams (Grisaru, Roček & Siegel, 1979) so as to maintain manifest global supersymmetry, and the background field method ('t Hooft, 1976) so as to maintain manifest local supersymmetry. We discuss the background-quantum splitting of the fields and the quantization (gauge fixing and ghost Lagrangians) and give a sample calculation using the resulting Feynman rules. Effective techniques of calculation using superfield Feynman diagrams and various examples can be found elsewhere (Grisaru et al, 1979; Abbott & Grisaru, 1980).

There are several reasons for doing superfield calculations in supergravity. First, we expect them to be easier than component field calculations, an expectation supported by our experience with globally supersymmetric theories. Second, we hope that they might shed some light on issues such as the structure of anomalies in supergravity. Third, we do not find it inconceivable that a calculation to settle the presence or absence of the three-loop divergence in supergravity could be carried out. Finally, we believe that many theoretical insights can be gained in the course of doing such calculations.

We have chosen to do background field quantization because it maintains manifest local supersymmetry. Ordinary quantization breaks it, because of gauge fixing. (Of course the Feynman rules for ordinary quantization can be easily obtained from ours simply by setting the background fields equal to their flat space value.)

We have restricted ourselves to on-shell background fields, although our formalism was chosen so that this restriction can be easily removed (at the expense of extra algebra). Since most off-shell quantities are gauge dependent (although covariant) and we are primarily interested in gauge independent ones we do not believe that anything important has been lost.

We present the rules which are adequate for one-loop calculations. The rules for higher loops can be obtained in a straightforward manner by means of rather tedious algebra and we will work them out when we need them. The one-loop rules owe their simplicity to the fact that they involve only quantum-background interactions (quadratic in the quantum fields) rather than quantum interactions (cubic and higher).

I shall discuss the following topics:

a) Background field formalism for ordinary (gauge) fields.
b) Background field formalism for Yang-Mills superfields.
c) The Siegel-Gates superfield formalism for supergravity.
d) The background-quantum splitting in supergravity.

The choice of gauge fixing terms, ghosts and resulting quantum action will be described by W. Siegel (1980).

## 2. THE BACKGROUND FIELD FORMALISM

For ordinary gauge theories a good discussion can be found in 't Hooft's lectures ('t Hooft, 1976). There are several issues: First, setting up a procedure in gauge theories (for theories without local symmetries some of the problems are trivial) which allows one to compute a set of Green's functions which are covariant (but in general do depend on gauge fixing parameters), even though the gauge fixing procedure has destroyed the gauge invariance of the fields which appear in loops. Second, giving a prescription for calculating physical quantities from those Green's functions. Third, modifying the procedure for cases where the fields transform nonlinearly.

The first issue is well known and I shall review it shortly. For the second one, see 't Hooft's lectures. There are some subtleties associated with the question of gauge invariant sources and with higher loop calculations which are discussed there. The third one has been discussed by Kazakov, Pervushin & Pushkin (1977) for ordinary theories with nonlinear global invariances and by Grisaru et al (1979) for superfield Yang-Mills theories. I shall come back to it presently.

In a typical gauge theory, with gauge field $A_\mu$ and transformation $\delta A_\mu \to -\partial_\mu \Lambda + \Lambda \times A_\mu$ one considers the splitting into a classical and a quantum field

BACKGROUND SUPERFIELD METHOD

$$A_\mu = C_\mu + Q_\mu \tag{1}$$

and a generating functional

$$\Omega(C) = \int [dQ]\, e^{\int \mathcal{L}(C+Q) - \mathcal{L}(C) - \mathcal{L}'(C)Q} \quad . \tag{2}$$

$\mathcal{L}(C+Q)$ is invariant under

$$\delta(C_\mu + Q_\mu) = -\partial_\mu \Lambda + \Lambda \times (C_\mu + Q_\mu) \tag{3}$$

which can be interpreted as either

a) Quantum transformation

$$\delta C_\mu = 0 \tag{4a}$$

$$\delta Q_\mu = -D_\mu \Lambda + \Lambda \times Q_\mu \quad , \quad D_\mu \Lambda = \partial_\mu \Lambda + C_\mu \times \Lambda \tag{4b}$$

or

b) Background transformation

$$\delta C_\mu = -\partial_\mu \Lambda + \Lambda \times C_\mu \tag{5a}$$

$$\delta Q_\mu = \Lambda \times Q_\mu \tag{5b}$$

Note that under background transformations $Q_\mu$ transforms like a global isovector, homogeneously and linearly. It follows that not only is the whole $\Omega$ invariant under the gauge transformation of $C_\mu$, eq. (5a) [obviously $\mathcal{L}(C)$ and $\mathcal{L}'(C)Q$ in the exponential are] but also this invariance will be preserved if $\mathcal{L}(C + Q)$ is expanded in powers of Q and the series is truncated at any point. In other words, in a loop expansion of $\Omega(C)$ the n-loop contribution will be separately invariant. (Furthermore the invariance is maintained if one chooses gauge fixing terms for the invariance eq. (4b) which are background covariant, e.g. $(D_\mu Q_\mu)^2 = [\partial_\mu Q_\mu + C_\mu \times Q_\mu]^2$ .

In theories invariant under a set of nonlinear transformations the invariance of the loop expansion may be lost if one makes the splitting A = C + Q. The resulting problems have been discussed by DeWit & Grisaru (1979) and the correct procedure has been described by Kazakov et al (1977) and Grisaru et al (1979). The essential point is that if the invariance transformation is nonlinear the splitting should not be A = C + Q but rather a splitting which is correlated to the nonlinear transformation law. Here is a trivial example:

Consider

$$\mathcal{L}(A) = \frac{(\partial_\mu A)^2}{(1 + A^2)^2} \tag{6}$$

which is invariant under

$$A \rightarrow \frac{\alpha + A}{1 - \alpha A} \tag{7}$$

The group multiplication law is

$$\gamma = \alpha \oplus \beta = \frac{\alpha + \beta}{1 - \alpha\beta} \tag{8}$$

If one makes the splitting A = C + Q the resulting Lagrangian

$$\mathcal{L} = \frac{[\partial_\mu (C + Q)]^2}{[1 + (C + Q)^2]^2} \tag{9}$$

will be invariant under

$$C + Q \rightarrow \frac{\alpha + C + Q}{1 - \alpha\ (C + Q)} \tag{10}$$

and this can be interpreted, for example, as a background transformation

$$C \rightarrow \frac{\alpha + C}{1 - \alpha C} \tag{11a}$$

$$Q \rightarrow \frac{(1 + \alpha^2)\ Q}{[1 - \alpha(C + Q)][1 - \alpha C]} \tag{11b}$$

However it is clear that individual terms in a power series expansion of (9) will not be invariant.

By contrast the splitting suggested by eq. (7)

$$A = C \oplus Q = \frac{C + Q}{1 - CQ} \tag{12}$$

leads to the transformation law

$$\frac{C + Q}{1 - CQ} \rightarrow \frac{\alpha + \dfrac{C + Q}{1 - CQ}}{1 - \alpha\ \dfrac{C + Q}{1 - CQ}} \tag{13}$$

which can be interpreted as the background transformation

$$C \rightarrow \frac{\alpha + C}{1 - \alpha C} \tag{14a}$$

$$Q \rightarrow Q \tag{14b}$$

Obviously

$$\mathcal{L}\,(C \oplus Q) = \left[\frac{\partial_\mu C(1 + Q^2) + \partial_\mu Q(1 + C^2)}{(1 + C^2)(1 + Q^2)}\right]^2 \tag{15}$$

$$= \left[\frac{\partial_\mu C}{1 + C^2} + \frac{\partial_\mu Q}{1 + Q^2}\right]^2$$

and every term in its Q power series expansion will be invariant.

As we shall see, in both supersymmetric Yang-Mills theory and in supergravity the superfield transformations are nonlinear and finding the proper splitting, and the proper choice of fields to make the splitting easy, are essential problems.

## 3. BACKGROUND FIELD METHOD IN SUPERFIELD YANG-MILLS THEORY

The usual discussion is in terms of the chiral representation gauge field $V = V^i G_i$ [$G_i$ = generators of the gauge group] with the gauge transformation $\Lambda = \Lambda^i G_i$ [$\Lambda^i$ = chiral superfield].

$$e^V \to e^{V'} = e^{i\bar{\Lambda}}\, e^V\, e^{-i\Lambda} \tag{16}$$

Chiral fields transform as

$$\phi \to \phi' = e^{i\Lambda}\, \phi \tag{17}$$

and $\Lambda$ was chosen to be chiral in order to maintain the chirality of $\phi$. A typical invariant action is

$$S = \frac{1}{64}\, \mathrm{tr} \int d^4 x\, d^2\theta W^\alpha W_\alpha + \int d^4 x\, d^4\theta\, \bar{\phi}\, e^V \phi \tag{18}$$

where the field strength $W_\alpha$ is

$$W_\alpha = \bar{D}^2 [e^{-V} D_\alpha\, e^V] \tag{19}$$

In this representation covariant derivatives are

$$\nabla_\alpha = e^{-V} D_\alpha e^V \quad , \quad \bar{\nabla}_{\dot{\alpha}} = \bar{D}_{\dot{\alpha}} \qquad \nabla_a = -\frac{i}{4}\, \sigma_a{}^{\alpha\dot{\beta}} \{\nabla_\alpha , \bar{\nabla}_{\dot{\beta}}\} \,. \tag{20}$$

V is Hermitean, $V^\dagger = V$ but the hermiticity condition on the covariant derivatives involves V:

$$(i\, \nabla_A)^\dagger = e^V\, [(-1)^a\, i\, \nabla_A]\, e^{-V} \tag{21}$$

The gauge transformation is nonlinear. Note that if instead of eq. (16) we had $e^V \rightarrow e^{iL} e^V e^{-iL}$ (e.g., if $\Lambda$ were real) we could immediately deduce $V \rightarrow e^{iL} V e^{-iL}$. This is what we would like to have for the quantum field. In fact what we want to achieve is a splitting $V = f(V_C, V_Q)$ such that $V_C$ transforms according to eq. (16) while $V_Q \rightarrow MV_Q M^{-1}$. The splitting

$$e^V = e^{V_C} e^{V_Q} \tag{22}$$

with the transformation

$$e^{V_C} e^{V_Q} \rightarrow e^{i\bar{\Lambda}} e^{V_C} e^{V_Q} e^{-i\Lambda} = (e^{i\bar{\Lambda}} e^{V_C} e^{-i\Lambda})(e^{i\Lambda} e^{V_Q} e^{-i\Lambda}) \tag{23}$$

looks likes it might do the job, except that we cannot maintain the hermiticity condition $V = V^\dagger$, $V_C = V_C^\dagger$ except in the Abelian case. Other splittings, e.g., $e^V \rightarrow e^{V_Q} e^{V_C} e^{V_Q}$ do not work either.

It turns out that the way to proceed is to introduce the "square root" (einbein?) of V by

$$e^V = e^W e^{\bar{W}} \tag{24}$$

where W is a complex general superfield, and to do the splitting on W. This procedure increases the invariance group (as it does when one goes from $g_{\mu\nu}$ to $e^a{}_\mu$) to

$$e^W \rightarrow e^{i\bar{\Lambda}} e^W e^{-iK} \tag{25}$$

with $K = K^i G_i = K^\dagger$. Furthermore the derivatives

$$\nabla_\alpha = e^{-\bar{W}} e^{-W} D_\alpha e^W e^{\bar{W}} \ , \quad \bar{\nabla}_{\dot{\alpha}} = \bar{D}_{\dot{\alpha}} \tag{26}$$

look more symmetrical if one goes to the vector representation by multiplying from left and right by $e^{\bar{W}}$, $e^{-\bar{W}}$ respectively. For example

$$\nabla_\alpha = e^{-W} D_\alpha e^W \ , \quad \bar{\nabla}_{\dot{\alpha}} = e^{\bar{W}} \bar{D}_{\dot{\alpha}} e^{-\bar{W}} \ , \tag{27}$$

with transformation properties

$$\nabla_\alpha \rightarrow e^{iK} \nabla_\alpha e^{-iK} \ , \quad \bar{\nabla}_{\dot{\alpha}} = e^{-iK} \bar{\nabla}_{\dot{\alpha}} e^{iK} \tag{28}$$

The background-quantum splitting is now almost obvious:

If we define

$$e^W = e^{W_C} e^{W_Q} \tag{29}$$

the transformation law

$$(e^{W_C}\, e^{W_Q}) \rightarrow e^{i\bar{\Lambda}}(e^{W_C}\, e^{W_Q})e^{-iK} = e^{W_C}(e^{-W_C}\, e^{i\bar{\Lambda}}\, e^{W_C})e^{W_Q}\, e^{-iK} \tag{30a}$$

$$= (e^{i\bar{\Lambda}}\, e^{W_C}\, e^{-iK})(e^{iK}\, e^{W_Q}\, e^{-iK}) \tag{30b}$$

can be interpreted in two ways:

a) Quantum: $W_C \rightarrow W_C$ (31a)

$$e^{W_Q} \rightarrow e^{i\bar{\tilde{\Lambda}}}\, e^{W_Q}\, e^{-iK} \quad , \quad \tilde{\Lambda} = e^{\bar{W}_C}\Lambda e^{-\bar{W}_C} \tag{31b}$$

b) Background: $e^{W_C} \rightarrow e^{i\bar{\Lambda}}\, e^{W_C} e^{-iK}$ (32a)

$$W_Q \rightarrow e^{iK}\, W_Q\, e^{-iK} \tag{32b}$$

which is what we want. Exponentials make life simple! An invariant action will be separately invariant under these two transformations. Note the similarity to eqs. (4, 5). Also the covariant derivative becomes

$$\nabla_\alpha = e^{-W_Q}\, e^{-W_C}\, D_\alpha\, e^{W_C}\, e^{W_Q} = e^{-W_Q}\mathcal{D}_\alpha\, e^{W_Q} \tag{33}$$

where we have introduced the <u>background covariant derivative</u> $\mathcal{D}_\alpha$ (same transformation properties).

Finally, we note that if we define

$$e^{V_Q} = e^{W_Q}\, e^{\bar{W}_Q} \tag{34}$$

the transformations (31, 32) become

$$(a')\quad W_C \rightarrow W_C \ , \quad e^{V_Q} = e^{i\bar{\tilde{\Lambda}}}\, e^{V_Q}\, e^{-i\tilde{\Lambda}} \tag{35a}$$

$$(b')\quad e^{W_C} \rightarrow e^{i\bar{\Lambda}}\, e^{W_C}\, e^{-iK}, \quad V_Q \rightarrow e^{iK}\, V_Q\, e^{-iK} \tag{35b}$$

After all is said and done, it is clear that what we have done is to adopt a splitting

$$e^V = e^{W_C}\, e^{V_Q}\, e^{\bar{W}_C}\ . \tag{36}$$

If we compare this to the gauge transformation eq. (16) we see that it follows a pattern very similar to that of the trivial example; just compare eqs. (7) and (12) with eqs. (16) and (36).

It turns out that in all quantities of interest one can express everything in terms of $V_Q$, provided one goes to yet another representation, a background vector-quantum chiral representation by multiplying from left and right by $e^{-\bar{W}Q}$ and $e^{\bar{W}Q}$. In particular the covariant derivatives become

$$\hat{\nabla}_\alpha = e^{-V_Q}\mathcal{D}_\alpha e^{V_Q} \quad , \quad \hat{\nabla}_{\dot{\alpha}} = \bar{\mathcal{D}}_{\dot{\alpha}} \tag{37}$$

This should be compared to eq. (20): One has replaced the "flat" covariant derivatives by the background covariant derivatives.

Finally we work out the vector action by substituting, for example, the expression of eq. (36) into eq. (18) (after replacing one $-\bar{D}^2/4$ by a $\int d^2\bar{\theta}$)

$$S = -\frac{1}{16}\,\mathrm{tr}\int d^4x\, d^4\theta\, (e^{-V_Q}\mathcal{D}^\alpha e^{V_Q})\bar{\mathcal{D}}^2(e^{-V_Q}\mathcal{D}_\alpha e^{V_Q}) \tag{38}$$

It is invariant under (35b) and also under (35a) which requires gauge fixing. A background covariant gauge fixing term can be used:

$$S_{GB} = -\frac{1}{16}\,\mathrm{tr}\int d^4x\, d^4\theta (\bar{\mathcal{D}}^2 V_Q)(\mathcal{D}^2 V_Q) \tag{39}$$

as a generalization of the flat space term $D^2V\bar{D}^2V$. The particular number in front will lead to propagators in Feynman gauge and other simplifications. The ghost Lagrangian will be discussed by W. Siegel (1980).

One can expand now $S + S_{GB}$ in powers of $V_Q$, dropping the constant and linear terms. One finds remarkable simplifications for the quadratic term (which is sufficient for doing one-loop calculations):

$$S_2 = -\frac{1}{2}\,\mathrm{tr}\int d^4x\, d^4\theta\, V_Q[\mathcal{D}^a\mathcal{D}_a + \frac{1}{8}\mathcal{W}^\alpha\mathcal{D}_\alpha - \frac{1}{8}\bar{\mathcal{W}}^{\dot{\alpha}}\bar{\mathcal{D}}_{\dot{\alpha}}]\, V_Q \tag{40}$$

$$\mathcal{D}_a = -\frac{i}{4}\sigma_a{}^{\alpha\dot{\alpha}}\{\mathcal{D}_\alpha, \bar{\mathcal{D}}_{\dot{\alpha}}\} \tag{41}$$

and $\mathcal{W}^\alpha$ is the background field strength (as in eq. (19)). Propagators can be obtained from the free part of $S_2$:

$$S_2^{(0)} = -\frac{1}{2}\,\mathrm{tr}\int d^4x\, d^4\theta\, V_Q \,\square\, V_Q \tag{42}$$

while vertices are obtained from the remainder. An interesting fact is that these vertices contain at most one spinor derivative (from $\mathcal{D}_\alpha = D_\alpha + \ldots$). Since as discussed by M. Roček (1980) at least two D's and two $\bar{D}$'s are needed in a supergraph loop to obtain non zero contributions, the quantum field $V_Q$ gives one loop contributions only to the four- and higher-point functions (the ghosts are responsible for the full contribution to two- and three-point functions). We note that this is, among other things, a consequence of our having chosen a particular background covariant gauge fixing term. A similar effect will be seen in supergravity.

## 4. SUPERFIELD SUPERGRAVITY

In the Yang-Mills case the background-quantum splitting was greatly facilitated by the fact that the basic field appeared in the exponential. Since in the Siegel-Gates formalism (Gates & Siegel, 1979) the Ogievetsky-Sokatchev axial vector field (Ogievetsky & Sokatchev, 1977) also appears in an exponential, one might expect that again matters will be very simple. Here is a quick summary of the formalism. For more details see M. Roček's lectures (Roček 1980).

The starting point is again the statement that general super-coordinate transformations act on chiral fields $\eta$ $(\bar{\partial}_{\dot{\alpha}}\eta = 0)$ as

$$\eta \to e^{i\Lambda}\,\eta \tag{43}$$

$$\Lambda = \Lambda^{M} i\partial_{M} \tag{44}$$

with $\bar{\partial}_{\dot{\alpha}}\Lambda^{m} = \bar{\partial}_{\dot{\alpha}}\Lambda^{\mu} = 0$ so as to maintain the chirality of $\eta$. The corresponding gauge field $U^{M}$ is introduced with the transformation laws

$$e^{2U} \to e^{i\bar{\Lambda}}\, e^{2U}\, e^{-i\Lambda} \tag{45}$$

$$U = U^{M}\, i\, \partial_{M} \qquad\qquad M = (m,\ \mu,\ \dot{\mu}) \tag{46}$$

It turns out that one can use $\Lambda$ to gauge away algebraically $U^{\mu}$, $U^{\dot{\mu}}$. $U^{m}$ contains then the physical fields of supergravity, e.g.,

$$U^{m}(x, \theta, \bar{\theta}) = \ldots + \theta\sigma^{a}\bar{\theta}e_{a}^{\ m} + \bar{\theta}^{2}\theta^{\alpha}\psi_{\alpha}^{\ m} + \ldots \tag{47}$$

The flat space value of U (all fields are zero except $e_{a}^{\ m} = \delta_{a}^{\ m}$) is

$$U_{o} = i\,\theta^{\alpha}\sigma^{a}_{\alpha\dot{\alpha}}\bar{\theta}^{\dot{\alpha}}\,\partial_{a} \tag{48}$$

We note that the flat <u>superspace</u> chiral representation covariant derivatives can be written as

$$D_{\alpha} = e^{-2U_{o}}\,\partial_{\alpha}e^{2U_{o}}\ ,\ \bar{D}_{\dot{\alpha}} = \bar{\partial}_{\dot{\alpha}} \qquad D_{a} = \partial_{a} = -\frac{i}{4}\,\sigma_{a}^{\alpha\dot{\alpha}}\,\{D_{\alpha}, \bar{D}_{\dot{\alpha}}\} \tag{49}$$

and it is natural to try a generalization

$$\hat{E}_\alpha = e^{-2U} \partial_\alpha e^{2U} \ , \quad \hat{E}_{\dot\alpha} = \bar\partial_{\dot\alpha} \ , \quad \hat{E}_a = - \frac{i}{4} \sigma_a^{\alpha\dot\alpha} \{\hat{E}_\alpha, \hat{E}_{\dot\alpha}\} \tag{50}$$

Although these derivatives are not covariant they are useful intermediate objects in the construction of covariant derivatives. In particular the action can be written in terms of them.

$$S = - \frac{6}{\kappa^2} \int d^4 x \, d^4\theta \, (1 \cdot e^{-2\overleftarrow{U}})^{1/3} \hat{E}^{-1/3} \phi \, e^{-2U} \bar\phi \tag{51}$$

where $\overleftarrow{U} = U^M i \overleftarrow{\partial}_M$ and $\phi$ is a "compensating" chiral scalar with transformation properties

$$\phi \to \exp[i\Lambda - \frac{1}{3} (\partial_m \Lambda^m - \partial_\mu \Lambda^\mu)]\phi \tag{52}$$

It can be gauged away, but it is preferable to keep it. For quantum calculations it facilitates choosing a good gauge for the axial vector superfield. Finally

$$\hat{E} = \det \hat{E}_A{}^M \tag{53}$$

with

$$\hat{E}_A = (\hat{E}_\alpha, \hat{E}_{\dot\alpha}, \hat{E}_a) = \hat{E}_A^M \partial_M \tag{54}$$

Also for quantum calculations, the analogue of the splitting $e_a{}^m = \delta_a{}^m + \frac{1}{2} h_a{}^m$ is here

$$e^{2U} = e^{U_o} e^H e^{U_o} \tag{55a}$$

$$\phi = e^{-U_o} (1 + \chi) \tag{55b}$$

with

$$H = \bar{H} = H^A \, i \, D_A \ , \qquad \bar{D}_{\dot\alpha} \chi = 0 \tag{56}$$

Such a splitting is not unique (others amount to field redefinitions) but very convenient: If one goes to a new representation by multiplying from the left by $e^{U_o}$ and from the right (where relevant) by $e^{-U_o}$ the various quantities we have introduced become

$$\hat{E}_\alpha = e^{-H} D_\alpha e^H \ , \quad \hat{E}_{\dot\alpha} = \bar{D}_{\dot\alpha} \qquad \phi = 1 + \chi \tag{57}$$

$$\hat{E}_A = (\hat{E}_\alpha, \hat{E}_{\dot\alpha}, \hat{E}_a) = \hat{E}_A{}^M D_M$$

and everything is manifestly globally supersymmetric since spinor derivatives appear only in the covariant form $D_\alpha$, $\bar{D}_{\dot\alpha}$. This form should be compared to eq. (20). The transformation laws can be written

$$e^H \to e^{i\bar{\Lambda}} e^H e^{i\Lambda} \quad (58a)$$

$$\chi \to \exp[i\Lambda - \frac{1}{3}(\partial_a \Lambda^a - D_\alpha \Lambda^\alpha)](1 + \chi) \quad (58b)$$

$$\Lambda = \Lambda^A \, i \, D_A \quad (59)$$

and the action is

$$S = -\frac{6}{\kappa^2} \int d^4 x \, d^4 \theta \, (1.e^{-\overleftarrow{H}})^{1/3} \hat{E}^{-1/3} (1 + \chi) \, e^{-H} (1 + \bar{\chi}) \quad (60)$$

This is the starting point of quantum calculations in terms of H and $\chi$ (Siegel, 1979). However we wish to do a background-quantum splitting. We expect to introduce background supergravity fields, and quantum fields such as $H_Q = H_Q{}^M \, i\mathcal{D}_M$ where $\mathcal{D}_M$ is a background covariant derivative (cf. eq. (33)). We must first describe the construction of such derivatives: The covariant derivatives have the general form (M = Lorentz generators)

$$\nabla_A = E_A + \frac{1}{2} \phi_{AB}{}^C M_C{}^B \quad (61)$$

$$E_A = E_A{}^M \partial_M = E_A{}^B D_B \quad (62)$$

$$\phi_{A\alpha}{}^\beta = \frac{1}{4} \sigma^a{}_{\alpha\dot{\gamma}} \sigma_b{}^{\beta\dot{\gamma}} \phi_{Aa}{}^b \quad , \text{ etc.} \quad (63)$$

and must be constructed out of H (or U) in such a way that they transform covariantly and satisfy the Wess-Zumino constraints (Wess & Zumino, 1977) on the torsion $T_{AB}{}^C$ defined by

$$[\nabla_A, \nabla_B\} = T_{AB}{}^C \nabla_C + \frac{1}{2} R_{ABc}{}^d M_d{}^c \quad (64)$$

Their construction in terms of the exponential has been given by Siegel (1978). It is somewhat involved so what I shall give is a sequence of steps which allows them to be explicitly calculated to any order in $\phi$, H and its derivatives (in a particular, locally sypersymmetric gauge):

1) Calculate

$$\psi = \phi^{1/2} \bar{\phi}^{-1} (1.e^{-2\overleftarrow{U}})^{-1/3} \hat{E}^{-1/6} \quad (65)$$

where $\hat{E}$ was defined, and is calculable from eq. (57).

2) Define and compute $\hat{C}_{AB}{}^C$ from

$$[\hat{E}_A, \hat{E}_B\} = \hat{C}_{AB}{}^C \hat{E}_C \quad (66)$$

3) Define and compute the supervierbein from

$$E_\alpha = \psi \hat{E}_\alpha , \quad E_{\dot\alpha} = \bar\psi \hat{E}_{\dot\alpha}$$
$$E_a = -\frac{i}{4} [\sigma_a{}^{\alpha\dot\alpha}\{\psi\hat{E}_\alpha, \bar\psi\hat{E}_{\dot\alpha}\} - \frac{1}{2} \psi\bar\psi \sigma^{\alpha\dot\alpha}_{[a} (\hat{C}_\alpha{}^b{}_{b]}\hat{E}_{\dot\alpha} + \hat{C}_{\dot\alpha}{}^b{}_{b]} \hat{E}_\alpha)] \tag{67}$$

4) Define and compute $C_{AB}{}^C$ from

$$[E_A, E_B\} = C_{AB}{}^C E_C \tag{68}$$

5) Compute the superconnections $\phi_{Abc}$ from

$$\phi_{\alpha bc} = - C_{\alpha[bc]}$$
$$\phi_{abc} = -\frac{i}{4} \sigma_a{}^{\alpha\dot\alpha} (E_\alpha\phi_{\dot\alpha bc} + E_{\dot\alpha}\phi_{\alpha bc} + \phi_{\dot\alpha[b}{}^d\phi_{\dot\alpha dc]}) \tag{69}$$

and eq. (63).

These five steps, with eq. (61), give the covariant derivatives. Torsions and curvatures can then be obtained from eq. (64). It is known (Grimm, Wess & Zumino, 1978) that the Bianchi identities show them to be expressible in terms of three superfields R, $G_{\alpha\dot\beta}$, $W_{\alpha\beta\gamma}$ (and their conjugates). One finds

$$\{\nabla_\alpha, \nabla_\beta\} = -\bar{R}\, M_{\alpha\beta}$$
$$[\nabla^\alpha, 2i\nabla_{\beta\dot\gamma}] = \frac{1}{2} \delta^\alpha{}_\beta\, [\bar{R}\, \bar\nabla_{\dot\gamma} - G^\delta{}_{\dot\gamma}\nabla_\delta - (\nabla^\delta G^\varepsilon{}_{\dot\gamma})\, M_{\delta\varepsilon} - \bar{W}_{\dot\gamma}{}^{\dot\delta\dot\varepsilon}\, \bar{M}_{\dot\delta\dot\varepsilon}] - \frac{1}{2} (\bar\nabla_{\dot\gamma}\, \bar{R})\, M^\alpha{}_\beta \tag{70}$$

where

$$\nabla_{\alpha\dot\beta} = \sigma^a{}_{\alpha\dot\beta}\nabla_a, \quad M_\alpha{}^\beta = \frac{1}{4} \sigma^a{}_{\alpha\dot\gamma}\sigma_b{}^{\beta\dot\gamma} M_a{}^b \tag{71}$$

These equations allow one to express R, G, W in terms of the basic fields H and $\phi$.

## 5. BACKGROUND SUPERGRAVITY

The similarity to the Yang-Mills case strongly suggests that the proper approach is via a splitting of $e^{2U}$ into its "square roots". We do it in fact from eq. (55a) by writing

$$e^H = e^W e^{\bar W} \tag{72}$$

with

$$W = W^A\, i\, D_A , \quad \bar W = \bar W^A\, i\, D_A \tag{73}$$

and the transformation properties

$$e^{W} \to e^{i\bar{\Lambda}} e^{W} e^{-i\hat{K}} \tag{74}$$

$$(\hat{K} = \hat{K}^{M}\, iD_{M} = \bar{\hat{K}} \quad ; \Lambda \text{ as in eq. (59) })$$

Finally we make the Q-B splitting:

$$e^{W} = e^{W_C} e^{W_Q} \tag{75}$$

with

$$W_C = W_C{}^{A}\, iD_A \tag{76a}$$

$$W_Q = W_Q{}^{A}\, \mathcal{E}_A \tag{76b}$$

as well as

$$\phi = \phi_C \phi_Q \tag{76c}$$

This particular form may need some explanation: $W_C$ will play the role of background field, and we construct a background U out of it as in eqs. (55) and (72). Now proceed and construct background quantities, following the steps in eqs. (65-71), expressing everything now in terms of $W_C$, $\bar{W}_C$, $\phi_C$. In particular we have the background supervierbein $\mathcal{E}_A$ (cf. eqs. (62, 67) and we choose to expand $W_Q$ in terms of it. The more natural choice $W_Q{}^{A} D_A$ is not possible since then eq. (75) could not be true.

With this splitting we return now to eqs. (61-65) and express the covariant derivative $\nabla_A$ in terms of the background and quantum fields, and do the same for the action in eq. (51). We discover that only the quantity $H_Q^A$ defined by

$$e^{H_Q^A i\mathcal{E}_A} = e^{W_Q} e^{\bar{W}_Q} \tag{77}$$

enters and the action is

$$S = -\frac{6}{\kappa^2} \int d^4x\, d^4\theta\, \mathcal{E}^{-1}\, \hat{E}_Q^{-1/3} (1.\, e^{-\overleftarrow{H}_Q})^{1/3} \phi_Q e^{-H_Q} \bar{\phi}_Q \tag{78}$$

where

$$H_Q = H_Q^A\, i\mathcal{D}_A \tag{79}$$

(expressed in terms of the background covariant derivative). $\hat{E}_Q$ is computed as in eq. (53).

The original invariance of the action has its counterpart in a set of two invariances under

a) Quantum transformations:

$$\mathcal{D}_A \to \mathcal{D}_A \tag{80a}$$

$$e^H \to \langle e^{i\bar{\Lambda}} e^H e^{-i\Lambda} \rangle \tag{80b}$$

$$\phi \to \exp[i\Lambda - \frac{1}{3} (\mathcal{D}_a \Lambda^a - \mathcal{D}_\alpha \Lambda^\alpha - \frac{i}{4} G_a \Lambda^a)] \tag{80c}$$

(here $\Lambda = \Lambda^A \, i\mathcal{D}_A$ and < > means to drop all terms which cannot be written in this form; $G_a = -\frac{1}{4} \sigma_a^{\alpha\dot{\beta}} G_{\alpha\dot{\beta}}$ is defined in eq. (70).

b) Background transformations:

$$e^{W_B} \to e^{i\bar{\Lambda}} e^{W_B} e^{-iK} \tag{81a}$$

$$\mathcal{D}_A \to e^{iK} \mathcal{D}_A e^{-iK} \tag{81b}$$

$$H_Q \to e^{iK} H_Q e^{-iK} \tag{81c}$$

$$\phi_Q \to e^{iK} \phi_Q \tag{81d}$$

Eqs. (81c, d) show that our goal has been achieved.

The rest is sheer torture: The action must be expanded in powers of the quantum fields $H_Q$ and $\chi_Q = \phi_Q - 1$. We can work in a gauge $H_Q^\alpha$, $H_Q^{\dot{\alpha}} = 0$ so that only the axial vector superfield $H^a$ appears. The final result can contain the background fields only through covariant derivatives and the field strengths R, $G_{\alpha\dot{\beta}}$, $W_{\alpha\beta\gamma}$. We find for example, the quadratic part of the action, sufficient for one loop calculations

$$S = \frac{-6}{\kappa^2} \int d^4 x\, d^4\theta \, [\chi\bar{\chi} - \frac{i}{3}(\chi-\bar{\chi})\mathcal{D}\cdot H + \frac{1}{12} H \cdot \Box H + \frac{1}{12}(\mathcal{D}\cdot H)^2$$

$$+ \frac{1}{576}\left([\bar{\mathcal{D}}_{\dot\beta}, \mathcal{D}_\alpha]H^{\alpha\dot\beta}\right)^2 - \frac{1}{96}[(\bar{\mathcal{D}}^2 + \tfrac{3}{2}R)H] \cdot [(\mathcal{D}^2 + \tfrac{3}{2}\bar{R})H]$$

$$+ \frac{1}{12}(\chi + \bar{\chi})\, G \cdot H + \frac{1}{288}(G\cdot H)^2 + \frac{1}{48} R\bar{R}\, H^2$$

$$+ \frac{1}{128}(\mathcal{D}^2 R + \bar{\mathcal{D}}^2\bar{R})H^2 - \frac{1}{192} H \cdot G^{\alpha\dot\beta}[\bar{\mathcal{D}}_{\dot\beta}, \mathcal{D}_\alpha]H$$

$$- \frac{1}{288}(G\cdot H)\, [\bar{\mathcal{D}}_{\dot\beta}, \mathcal{D}_\alpha]H^{\alpha\dot\beta} \tag{82}$$

$$+ \frac{1}{384} H^{\alpha\dot\beta}[(\mathcal{D}_{(\alpha} G_{\gamma)}{}^{\dot\sigma})\bar{\mathcal{D}}_{\dot\sigma} H^{\gamma}{}_{\dot\beta} - (\bar{\mathcal{D}}_{(\dot\beta} G^{\sigma}{}_{\dot\gamma)})\mathcal{D}_\sigma H_\alpha{}^{\dot\gamma}]$$

$$- \frac{1}{192} H^{\alpha\dot\beta}(W_\alpha{}^{\gamma\sigma}\mathcal{D}_\gamma H_{\sigma\dot\beta} + \bar{W}_{\dot\beta}{}^{\dot\gamma\dot\sigma}\bar{\mathcal{D}}_{\dot\gamma} H_{\alpha\dot\sigma}) .$$

After a suitable gauge fixing term has been added as discussed by Siegel (1980) the action simplifies considerably if the background fields are on shell:

$$S = \frac{-6}{\kappa^2}\int d^4 x\, d^4\theta\, \mathcal{E}^{-1}\, [\, -\tfrac{1}{2} H^a \mathcal{D}^b \mathcal{D}_b H_a - \frac{18}{5}\chi(e^{2U}\chi)^\dagger$$

$$+ \frac{1}{32} H^{\alpha\dot\beta}[W_\alpha{}^{\gamma\sigma}\mathcal{D}_\gamma H_{\sigma\dot\beta} + \bar{W}_{\dot\beta}{}^{\dot\gamma\dot\sigma}\bar{\mathcal{D}}_{\dot\gamma} H_{\alpha\dot\sigma}] \tag{83}$$

The above form is quite similar to that of eq. (40). Again the H field gives no contributions to a loop with fewer than four external lines. Contributions to such loops come from the chiral field $\chi$ and from ghosts. The ghost Lagrangian will be discussed by W. Siegel.

REFERENCES

Abbott, L.F. & Grisaru, M.T. (1980). The 3-loop β-function for the Wess-Zumino model. Nucl. Phys. B169, pp. 415-429.

DeWit, B. & Grisaru, M.T. (1979). On-shell counterterms and non-linear invariances. Phys. Rev. D, 8, pp. 2082-5.

Gates, S.J. & Siegel, W. (1979). Superfield supergravity. Nucl. Phys. B147, pp. 77-104.

Grimm, R., Wess, J. & Zumino, B. (1978). Consistency checks on the superspace formulation of supergravity. Phys. Lett. 73B, pp. 415-417.

Grisaru, M. T., Roček, M. & Siegel, W. (1979). Improved methods for supergraphs. Nucl. Phys. B159, pp. 429-450.

Grisaru, M.T. & Siegel, W. (1980). Supergraphity. (in preparation).

Kazakov, D.I., Pervushin, V.N. & Pushkin, S.V. (1977). Invariant renormalization for theories with nonlinear symmetry. Theor. Math. Phys. 31, pp. 389-394.

Roček, M. (1980). (this volume).

Ogievetsky, V. & Sokatchev, E. (1977). On a vector superfield generated by the supercurrent. Nucl. Phys. B124, pp. 309-316.

Siegel, W. (1978). Solution to constraints in Wess-Zumino supergravity. Nucl. Phys. B142, pp. 301-305.

Siegel, W. (1979). Supergravity supergraphs. Phys. Lett. 84B, pp. 197-200.

Siegel, W. (1980). The wraiths of graphs. (this volume).

't Hooft, G. (1976). The background field method in gauge field theories. In Acta Universitatis Wratislavensis No. 368, XIIth Winter School of Theoretical Physics in Karpacz Feb.-March 1975.

Wess, J. & Zumino, B. (1977). Superspace formulation of supergravity. Phys. Lett. 66B, pp. 361-364.

# THE WRAITHS OF GRAPHS

W. Siegel

Institute for Advanced Study

## 1. INTRODUCTION

This talk is a continuation of an introduction to the paper "Supergraphity" (Grisaru & Siegel, 1980) written in collaboration with Marc Grisaru, who presented the first half of the introduction (Grisaru, 1980). He reviewed the superfield approach to supergravity and described the method of introducing the superfield background. I will complete our introduction to the superfield quantization of supergravity by first giving a general description of the various unusual ghosts, with examples from simpler gauge theories, and then applying these methods to find the gauge-fixing and ghost terms in the complete quantum action of superfield supergravity.

## 2. COUNTING GHOSTS

We will first give a simple rule for counting all the ghosts in any gauge theory (Siegel, 1980). In ordinary gauges some of the ghosts may decouple, but in background field gauges all ghosts couple to the background. This coupling of all ghosts can actually be a virtue: e.g., in one-loop calculations in O(4) super-Yang-Mills using O(1) supergraphs, the three chiral ghosts of the background gauge (vs. two in ordinary gauges) exactly cancel the three chiral matter multiplets in all graphs with only vector-multiplet legs (Grisaru, Rocek, & Siegel, 1979).

We begin by deriving the rule for a general component-field gauge theory. To streamline notation, we will drop all indices, and indicate all opposite-statistics fields by primes. The general free Lagrangian for a gauge field can be written as $A\,\Box^n\pi A$, where $\pi$ is a projection operator, and n is an integer (when A is a tensor) or half-integer (when A is a spinor: $\Box^{\frac{1}{2}} \equiv \not\partial$). For physical fields, n = 1 or 1/2. The gauge invariance can be written as $\delta A = \partial\lambda$, with $\pi\partial\lambda = 0$. After gauge fixing, the Lagrangian becomes $A\,\Box^n A$. However, we must cancel the $A = \partial\lambda$ mode, which did not contribute to the original Lagrangian, by introducing a ghost B' through the addition of a similar term to the

Lagrangian with the substitution $A \to A' = \partial B'$. We thus obtain $L = A\Box^n A + (\partial B')\Box^n(\partial B') = A\Box^n A + B'\Box^{n+1}\pi' B'$, where $\pi'$ is a new projection operator. In the simplest cases (e.g., A = the photon field), $\pi' = 1$, and we are through. More generally (e.g., A = an antisymmetric tensor gauge field), B' has a gauge invariance, so we must continue the procedure until no gauge invariance remains. (This is a finite process, since each ghost has one less vector index than its predecessor.) The final Lagrangian is thus $L = A\Box^n A + B'\Box^{n+1}B' + C\Box^{n+2}C + \ldots$ . For tensor fields, a field with kinetic operator $\Box^m$ represents m fields of that type with kinetic operator $\Box$; for spinors, $\Box^m$ represents 2m fields with $\not\partial$. Thus, for physical fields A, the number of successive fields goes as 1,2,3,4,... for tensors (n=1), and as 1,3,5,7, ... for spinors (n=1/2), with alternating statistics. Simple examples where this counting differs from that of naive Faddeev-Popov quantization are the gravitino field, which has 3 ghosts (the third is the Nielsen-Kallosh ghost (Nielsen, 1978; Kallosh, 1978)), and the second-rank antisymmetric tensor, which has only 3 second-generation ghosts (Curtright, 1980; Thierry-Mieg, 1980; Siegel, 1980) (and not the 4 expected from the doubling in naive Faddeev-Popov quantization). When the gauge-fixing terms have no interactions (with either quantum or background fields), all but the original field and its 2 Faddeev-Popov ghosts decouple. (Of course, even the Faddeev-Popov ghosts would then decouple in an Abelian gauge theory.) An exception is the "catalyst" ghosts, which cancel in the counting, but may still couple (see sec. 4). When the only interaction of the gauge-fixing terms is with background fields, the additional ghosts contribute only at 1 loop (with the exception again of catalyst ghosts). The method for modifying the Faddeev-Popov procedure to explicitly determine the interactions will be described in the next section.

Generalization of the counting rule to superfields is straightforward, though each case requires separate consideration, due to the widely varying form of superfield gauge transformations. The simplest example is super-Yang-Mills: The classical free Lagrangian is $V\Box\pi_{1/2}V$, with $\delta V = \Lambda + \bar{\Lambda}$, where $\Lambda$ is chiral ($\bar{D}_\alpha\Lambda=0$) and $\pi_{1/2}$ is the superspin - 1/2 projection operator ($\pi_{1/2} = -D^\alpha\bar{D}^2D_\alpha/8\Box$, $\pi_0 = \{D^2,\bar{D}^2\}/16\Box$). We thus obtain $L = V\Box V + \bar{\phi}'\Box\phi'$, where $\phi'$ is a chiral ghost ($\phi'\Box\phi' + \bar{\phi}'\Box\bar{\phi}'$ gives 0 upon $\int d^4\theta$). The term $\bar{\phi}'\Box\phi'$ is equivalent to 3 of the usual terms $\bar{\phi}'\phi'$, since

$$\delta(\bar{\phi}'\phi')/\delta\begin{pmatrix}\phi'\\ \bar{\phi}'\end{pmatrix} \sim \begin{pmatrix}0 & \bar{D}^2\\ D^2 & 0\end{pmatrix}\begin{pmatrix}\phi'\\ \bar{\phi}'\end{pmatrix} \quad \text{, whereas}$$

$$\delta(\bar{\phi}'\Box\phi')/\delta\begin{pmatrix}\phi'\\ \bar{\phi}'\end{pmatrix} \sim \begin{pmatrix}0 & \bar{D}^2\\ D^2 & 0\end{pmatrix}\Box\begin{pmatrix}\phi'\\ \bar{\phi}'\end{pmatrix} \sim \begin{pmatrix}0 & \bar{D}^2\\ D^2 & 0\end{pmatrix}^3\begin{pmatrix}\phi'\\ \bar{\phi}'\end{pmatrix} \quad . \tag{1}$$

The next simplest example is the antisymmetric-tensor multiplet, described by a chiral gauge spinor $\phi_\alpha$ (Siegel, 1979b). Its classical Lagrangian is $L=(1/8)(D^\alpha\phi_\alpha - \bar{D}^{\dot\alpha}\bar\phi_{\dot\alpha})^2$, and the gauge-fixing term is $(1/8)\alpha(D^\alpha\phi_\alpha + \bar{D}^{\dot\alpha}\bar\phi_{\dot\alpha})^2$. The gauge-fixed Lagrangian is, for the two convenient choices $\alpha = \pm 1$:

$$L_+ = \frac{1}{8}(\phi^\alpha D^2\phi_\alpha + \bar\phi^{\dot\alpha}\bar{D}^2\bar\phi_{\dot\alpha}) \quad (\int d^4\theta L_+ = -\frac{1}{2}\int d^2\theta\phi^\alpha \Box\phi_\alpha + \text{h.c.}),$$

$$L_- = -i\,\bar\phi^{\dot\beta}\partial_{\alpha\dot\beta}\phi^\alpha \qquad . \tag{2a}$$

$L_+$ and $L_-$ can also be obtained directly from one another by field redefinition (Grisaru & Siegel, 1980). The original Lagrangian's gauge invariance $\delta\phi_\alpha = \bar{D}^2 D_\alpha K$, $K = \bar{K}$, thus results in the new Lagrangian $L = L_\pm + V' \Box^2 \pi_{1/2} V'$. (The second term is $\alpha$-independent.) We thus have $\delta V' = \Lambda + \bar\Lambda$, as for super-Yang-Mills, so the final Lagrangian is:

$$L = L_\pm + V' \Box^2 V' + \bar\phi \Box^2 \phi \tag{2b}$$

($\phi$ has normal statistics). The $V'$ term is equivalent to 2 $V'\Box V'$ terms, and the $\phi$ term is equivalent to 5 $\bar\phi\phi$ terms. (Rigorously, we should replace $V' \Box^2 V'$ with $\bar\psi'\Box\,\psi$ to avoid the technicality of $(\Box V')^2 = 0$ due to anticommutativity, but we will consider the statistics of $V'$ merely as a mnemonic to give its loops a factor of $-1$.) An example more analogous to supergravity is the (3/2,1) multiplet, described by a general gauge spinor $\psi_\alpha$ and a compensating chiral scalar $\phi$ (Gates & Siegel, 1980). (The compensating field is necessary for quantization: see sec 5. For convenience in gauge fixing, one may further add a compensating real scalar (Gates & Siegel, 1980), which will cancel with its ghost and so not contribute to the counting.) The Lagrangian is $L= -(1/2)[(\bar{D}^{\dot\beta}\psi^\alpha)(D_\alpha\bar\psi_{\dot\beta})+(1/4)(\bar{D}^{\dot\beta}\psi^\alpha)(\bar{D}_{\dot\beta}\psi_\alpha)+(1/4)(D^\alpha\bar\psi^{\dot\beta})(D_\alpha\bar\psi_{\dot\beta})]$ $-(1/4)\bar\phi\phi+(1/4)(\phi D^\alpha\psi_\alpha + \bar\phi\bar{D}^{\dot\alpha}\bar\psi_{\dot\alpha})$, and the invariance is $\delta\psi_\alpha = \Lambda_\alpha + D_\alpha K$, $\delta\phi = \bar{D}^2 K$ ($\bar{D}_{\dot\beta}\Lambda_\alpha = 0$, $K = \bar{K}$). Although the gauge fixing itself is somewhat complicated in this case (in particular the $\phi\psi_\alpha$ crossterms must be cancelled), the final Lagrangian can be seen to be of the form (due to dimensionality of the fields, and lack of gauge invariances for the first-generation ghosts) $L = i\bar\psi^{\dot\beta}\partial_{\alpha\dot\beta}\psi^\alpha - \frac{1}{8}\bar\phi\phi + i\bar\phi'^{\dot\beta}\partial_{\alpha\dot\beta}\phi'^\alpha + V'\Box V'$. (We have also applied the method of sec. 4 to simplify the $V'$ term.)

We will now consider supergravity itself. The original Lagrangian is of the form (again dropping indices) $L=H\Box\,\pi H + \bar\chi\chi$ ( + $H\chi$ crossterms), where $\pi$ is a sum of projection operators, with the gauge invariance $\delta H = \bar{D}L - D\bar{L}$ (Ferrara & Zumino, 1978), $\delta\chi = \bar{D}^2 DL$ (Siegel, 1979a), in terms of a general spinor parameter $L_\alpha$. After gauge fixing we obtain the Lagrangian $L = H \square H + \bar{\chi}\chi$ (with crossterms cancelled), and adding in the first generation of ghosts (as $H' = \bar{D}\psi' - D\bar{\psi}'$, $\chi' = \bar{D}^2 D\psi'$) gives us $L = H \square H + \bar{\chi}\chi + i\bar{\psi}' \square \not{\partial} \pi' \psi'$ in terms of a new projection operator $\pi'$. Due to the way $L_\alpha$ appears in the transformation laws, $\psi$ has the gauge invariance $\delta\psi'_\alpha = \Lambda_\alpha$, $\bar{D}_{\dot\beta}\Lambda_\alpha = 0$. We thus obtain for the final Lagrangian $L = H \square H + \bar{\chi}\chi + i\bar{\psi}' \square \not{\partial}\psi' + i\bar{\phi} \square \not{\partial}\phi$. (The chiral spinor $\phi_\alpha$ has normal statistics and no new gauge invariance.) The $\psi'$ term represents 3 ghosts appearing as $i\bar{\psi}'\not{\partial}\psi'$, and the $\phi$ term represents 2 ghosts appearing as $i\bar{\phi}\not{\partial}\phi$ or $\phi D^2\phi$ + h.c. (As in eq. ( 2a), $i\bar{\phi} \square \not{\partial}\phi$ is equivalent to $\phi \square D^2\phi$ + h.c. Then $\int d^4\theta\phi \square D^2\phi$ + h.c. $\sim \int d^2\theta\phi \square^2\phi$ + h.c., which is equivalent to taking $\int d^4\theta\phi D^2\phi$ + h.c. $\sim \int d^2\theta\phi \square \phi$ + h.c. twice. Also, one can apply an argument analogous to eq. (1).) We can therefore rewrite our final Lagrangian as

$$L = H^a \square H_a + \bar{\chi}\chi + i \sum_{i=1}^{3} \bar{\psi}'^{\dot\beta}_i \partial_{\alpha\dot\beta}\psi'^{\alpha}_i + \sum_{i=1}^{2} (\phi^{\alpha}_i D^2 \phi_{i\alpha} + \text{h.c.}). \tag{3}$$

We may also find other ghosts which cancel in the counting (i.e., a number of some type of ghosts together with an equal number with opposite statistics, which cancel in 1-loop calculations: see sec. 4). Using the methods described in secs. 3,4, and 5, we will derive eq. (3) with (on-shell) background interactions in sec. 6.

## 3. HIDDEN GHOSTS

The modifications in the naive Faddeev-Popov quantization procedure which result in the ghost counting rules of the previous section are found by a careful analysis of the 't Hooft gauge-averaging trick ('t Hooft, 1971). The first modification (Nielsen, 1978; Kallosh, 1978) is due to the observation that any averaging must be properly normalized: Therefore, a general Gaussian average of a gauge-fixing delta-functional causes the replacement

$$\delta[f(A) - \bar{a}] \rightarrow \int \mathcal{D}a\, \mathcal{D}a' e^{-\int (a\mathcal{O}a + a'\mathcal{O}a')} \delta[f(A) - a]$$

$$= e^{-\int f\mathcal{O}f} \int \mathcal{D}\, a'\, e^{-\int a'\mathcal{O}a'} , \tag{4}$$

where A is the gauge field, $\mathcal{O}$ is an operator which may depend on physical fields, and a' is a "third ghost" (in addition to the first and second of Faddeev and Popov) which normalizes the averaging $\int \mathcal{D}a\mathcal{D}a' e^{-\int (a\mathcal{O}a + a'\mathcal{O}a')} = 1$ due to its having statistics opposite to those of a. In conventional gauge theories and conventional gauges, $\mathcal{O}$ is a constant, and a' decouples. However, when $\mathcal{O}$ has field dependence, as in background-field component supergravity (Nielsen, 1978; Kallosh, 1978), or background-field superfield Yang-Mills (Grisaru et al., 1979), a' must be taken into account.

A more subtle modification must be made when the gauge-fixing function satisfies a constraint (Siegel, 1980). As a simple example, consider a second-rank antisymmetric-tensor gauge field $A_{ab}$ in an external gravitational field with covariant derivative $\nabla_a$ and gauge invariance $\delta A_{ab} = \nabla_{[a}\lambda_{b]}$. The covariant gauge-fixing function $F_a = \nabla^b A_{ab}$ satisfies the constraint $\nabla^a F_a = 0$. The naive Gaussian averaging

$$\int \mathcal{D}f_a \exp(-\int d^4x \sqrt{g}\, f^a f_a)\, \delta(F_a - f_a) = \exp(-\int d^4x \sqrt{g}\, F^a F_a)$$

would give an incorrect result, since the constraint $F_a = f_a$ implies $\nabla^a f_a = 0$, causing $f_a$ to be dependent on the external gravitational field. We would therefore like to average over only the transverse part of $f_a$, as $\int \mathcal{D}f^T \exp(-\int d^4x \sqrt{g}\, f^{T2})\delta(F-f^T)$. Thus we choose $f^{T2} = f^a(\eta_{ab} - \nabla_a \Box^{-1}\nabla_b)f^b$ (where $\Box \equiv \nabla^a\nabla_a$), and we define $\int \mathcal{D}f^T$ as in QED, where also the integrand depends only on the transverse part of the vector field, by gauge-fixing. Choosing the gauge-fixing function $\hat{F} = \nabla^a f_a$, we obtain as our final averaging, including $1^{st}$, $2^{nd}$, and $3^{rd}$ ghosts for $f_a$:

$$\int \mathcal{D}f_a \mathcal{D}\hat{f} \mathcal{D}C_1' \mathcal{D}C_2' \mathcal{D}C_3' \,\delta(F-f^T)\,\delta(\nabla\cdot f - \hat{f}) \cdot$$
$$\cdot \exp[-\int d^4x \sqrt{g}\,(f^{T2} - \hat{f}\,\frac{1}{\Box}\,\hat{f} + C_1'\Box C_1' + C_2'\Box C_2' + C_3'\,\frac{1}{\Box}\,C_3')]$$
$$= \int \mathcal{D}C_1'\mathcal{D}C_2'\mathcal{D}C_3' \exp[-\int d^4x \sqrt{g}\,(F^2 + C_1'\Box C_1' + C_2'\Box C_2' + C_3'\frac{1}{\Box}C_3')]$$
$$= \int \mathcal{D}C' \exp[-\int d^4x \sqrt{g}\,(F^2 + C'\Box C')] \quad . \tag{5}$$

(One could also use $\hat{F} = \Box^{-1}\nabla^a f_a$, and obtain $C_1'C_1' + C_2'C_2' + \bar{C}_3'\Box C_3' \to C'\Box C'$.) Including the 2 vector Faddeev-Popov ghosts $B_a'$ of $A_{ab}$, and their 4 scalar Faddeev-Popov ghosts (due to the invariance $\delta B_a' = \nabla_a\lambda'$), the total quantum Lagrangian is thus of the form

$$A^{ab}\Box A_{ab} + \sum_{i=1}^{2} B_i'^a \Box B_{ia}' + \sum_{i=1}^{3} C_i \Box C_i \ , \tag{6}$$

where we have cancelled one of the commuting ghosts $C_4\Box C_4$ with the anticommuting ghost $C'\Box C'$ of eq. (5). We have thus reproduced the counting of sec. 2. Higher-rank antisymmetric tensors can be treated similarly.

○

The "hidden ghosts" for constrained gauge-fixing functions for superfields can be derived more easily. For example, for the chiral gauge spinor $\phi_\alpha$, the gauge-fixing function $F = \bar{F} = D^\alpha \phi_\alpha + \bar{D}^{\dot{\alpha}} \bar{\phi}_{\dot{\alpha}}$ satisfies the constraint $D^2 F = \bar{D}^2 F = 0$. To prevent the gauge condition $F = f$ from implying the constraint $D^2 f = \bar{D}^2 f = 0$, we can "complete" F to an unconstrained gauge function by adding a chiral term to F, in terms of a chiral scalar $\eta$:

$$F \to \hat{F} = F + (D^2 \Box^{-1} \eta + \bar{D}^2 \Box^{-1} \bar{\eta}) \quad , \tag{7}$$

where $\pi_{1/2} \hat{F} = F$ and $\pi_0 \hat{F} = D^2 \Box^{-1} \eta + \bar{D}^2 \Box^{-1} \bar{\eta}$, so the two terms are orthogonal. We now have $\bar{D}^2 \hat{F} \sim \eta \neq 0$: $\hat{F}$ is unconstrained, so $\int \mathcal{D}f \exp(\int d^4x\, d^4\theta\, f^2)\, \delta(\hat{F} - f)$ is an averaging. The reason for using $D^2 \Box^{-1} \eta + \bar{D}^2 \Box^{-1} \bar{\eta}$, and not simply $\eta + \bar{\eta}$, follows from a component analysis: $\hat{F} = f$ implies $\bar{D}^2 f \sim \eta$, so the components of f with factors of $\theta^2$ or $\bar{\theta}^2$ are set equal to components of $\eta$, without spacetime derivatives (which would not be true for $\hat{F} = F + (\eta + \bar{\eta}) = f \to \bar{D}^2 f = \bar{D}^2 \bar{\eta}$). Therefore, averaging over those components of f does not produce spurious kinetic terms for the corresponding components of $\eta$. The component gauge-fixing functions for $\phi_\alpha$ without spacetime derivatives appear in F at order 1 and $\theta$ (and $\bar{\theta}$). The remaining component of $\hat{F}$ is the $\theta\bar{\theta}$ component, which takes the form $\partial^b A_{ab} + \partial_a \Box^{-1} A$, where A is a component of $\eta$. Averaging of this component gives a term $(\partial^b A_{ab})^2 - A \Box^{-1} A$, with $A \Box^{-1} A$ giving the same contribution as the hidden ghost $C' \Box C'$ of eq. (5). (Therefore, the $\hat{F}_a = F_a + \partial_a \Box^{-1} A$ method can also be applied to the component $A_{ab}$ theory.)

In supergravity, the gauge-fixing function $F_\alpha = \bar{D}^{\dot{\beta}} (H_{\alpha\dot{\beta}} + \ldots)$ also satisfies $\bar{D}^2 F_\alpha = 0$, so we define $\hat{F}_\alpha = F_\alpha + D^2 \Box^{-1} \phi_\alpha$, where $\phi_\alpha$ is chiral. We will discuss supergravity in more detail in sec. 6.

## 4. CATALYST GHOSTS

In order to avoid possible difficulties with infrared divergences (Piguet & Rouet, 1976), it will be necessary to restrict ourselves to propagators with $p^2$ denominators (in contrast to previous quantizations of superfield supergravity (Siegel, 1979a; Namazie & Storey, 1979; Taylor, 1979)). It will also be desirable to do so in a manner which avoids introducing nonlocal terms in the Lagrangian. (Namazie & Storey (1979) have produced gauges which make the linearized action local, but nonlocalities will appear in their interaction terms; Taylor (1979) uses the operator $\Box^{1/2}$ in a ghost kinetic term.) Such unwanted higher-derivative kinetic or nonlocal terms can be eliminated by introducing a new kind of ghost. Since only the numbers and kinds of ghosts derived in sec. 2 will preserve unitarity, these new ghosts must cancel with their own ghosts in the counting. These "catalyst" ghosts will therefore come in pairs with opposite statistics. However, the two members of a pair need not have the same interactions: In particular, we will find that in background-superfield supergravity some of these ghosts will have interactions with quantum fields while their partners do not; their contributions thus cancel at one loop (where only background interactions occur), but not at higher loops. (This should be contrasted with the hidden ghosts of the previous section, which will in our case contribute only at one loop, due to their lack of quantum interactions.)

The basic method is analogous to the Stueckelberg formalism for a massive vector field (Stueckelberg, 1938). Consider such a field coupled to a Dirac spinor, with Lagrangian

$$L = -\frac{1}{4}(\partial_{[a}A_{b]})^2 + \frac{1}{2}m^2A^2 + \bar{\psi}(i\partial\!\!\!/ - e A\!\!\!/)\psi.$$

We now make the redefinitions (of the same form as the gauge transformations of the massless case)

$$A_a \rightarrow A_a + \frac{1}{m}\partial_a B, \quad \psi \rightarrow \exp\left(-i\frac{e}{m}B\right)\psi,$$

introducing the new scalar field B. Our Lagrangian now has the gauge invariance $A' = A + \frac{1}{e}\partial\lambda$, $B' = B - \frac{m}{e}\lambda$, $\psi' = e^{-i\lambda}\psi$. After adding the gauge-fixing term $-\frac{1}{2}(\partial\cdot A - mB)^2$, the final Lagrangian is $L = \frac{1}{2}A\cdot(\Box + m^2)A - \frac{1}{2}B(\Box + m^2)B + \bar{\psi}(i\partial\!\!\!/ - eA\!\!\!/)\psi$. Thus, in this simple example B decouples, the A interactions are unchanged, and the A propagator has changed from $-(\delta_{ab} - p_ap_b/m^2)/(p^2 + m^2)$ to the simpler $-\delta_{ab}/(p^2 + m^2)$, with better ultraviolet behavior. (Also, since the gauge group is Abelian, the Faddeev-Popov ghosts $C_2'(\Box + m^2)C_1'$ decouple.)

Our catalyst ghosts are somewhat different: The interaction terms have no gauge invariance, so the ghosts don't decouple; and there are no mass parameters, so the required redefinitions are nonlocal, and the propagator improvement is with respect to infrared behavior. (In fact, the Stueckelberg formalism is more analogous to gauge fixing with the superconformal-compensating field $\phi$, when coupling to superconformal matter.) For example, consider the linearized Lagrangian $L = A\,\Box[(1-\pi) + \alpha\pi]A$

(where $\pi$ is a projection operator and $\alpha \neq 0,1$). In order to obtain the simpler $L = A\Box A$, the simplest method would be to use the field redefinition $A' = [(1-\pi) + \alpha^{1/2}\pi]A$, but this could produce nonlocal interactions if $\pi$ were nonlocal. The solution is to add a catalyst field B (and its partner of opposite statistics B') of the same type as $\pi A$ (i.e., $\pi B = B$): The Lagrangian now becomes $L = A\Box[(1-\pi) + \alpha\pi]A + \beta B\Box B + B'\Box B'$ ($\beta = \pm 1$). We then make a first shift $B \rightarrow B + a\pi A$, followed by a second shift $A \rightarrow A + bB$. (Note each shift is a transformation on (A,B) with Jacobian = 1.) By choosing a,b, and $\beta$ appropriately ($a = -\beta b = \pm\sqrt{\beta(1-\alpha)}$, $\beta$ = sign of $(1-\alpha)$) we obtain $L \rightarrow A\Box A + \beta\alpha B\Box B + B'\Box B'$ for the linearized Lagrangian. For background interactions, B and B' thus cancel at one loop. Furthermore, local quantum interaction terms are modified as $f(A) \rightarrow f(A + bB)$ and so remain local. This is equivalent to starting with the original Lagrangian, redefining $A \rightarrow A + B$, and fixing the resulting gauge invariance ($\delta A = \lambda$, $\delta B = -\lambda$; $\pi\lambda = \lambda$), as in the Stueckelberg formalism, by a gauge-fixing term

$$\frac{\alpha^2}{1-\alpha}\left(B - \frac{1-\alpha}{\alpha}\pi A\right)\Box\left(B - \frac{1-\alpha}{\alpha}\pi A\right).$$

The first and second (Faddeev-Popov) ghosts can be dropped (they contribute a trivial term of the form $B_1'B_2'$), but the third (Nielsen-Kallosh) ghost contributes a term $B'\Box B'$. By this method we obtain

$$L = A\Box A + \frac{\alpha}{1-\alpha} B\Box B + B'\Box B',$$

and interactions $f(A) \rightarrow f(A + B)$ remain local. (A further redefinition $B \rightarrow bB$ makes L identical to the one obtained by the previous method.)

The above example can be applied to superfields as well as to components: Consider $\pi = \pi_0$, $1 - \pi = \pi_{1/2}$ (the superspin projection operators), with A a general (real) superfield and B chiral ($B = \phi + \bar{\phi}$, $\bar{D}_{\dot{\alpha}}\phi = 0$; the Faddeev-Popov term is $\int d^2\theta\, \phi_1'\phi_2' + \text{h.c.} \rightarrow 0$, not $\int d^4\theta$). The results are identical. This method will be applied frequently when we quantize superfield supergravity in sec. 6.

## 5. UNWANTED GHOSTS

The Faddeev-Popov procedure is a convenient method for performing quantization in a gauge-invariant away. However, unitarity is not obvious, and must be proven directly on the Feynman graphs, or by comparison with canonical quantization in an axial (ghost-free) gauge (see, e.g., Coleman (1975)). In supersymmetric gauge theories, the analog of the axial gauge is the Wess-Zumino gauge (Wess & Zumino, 1974). This is a gauge where all component fields which do not appear in the action are eliminated by nonderivative transformations, and therefore the corresponding Faddeev-Popov ghosts do not propagate. (A full determination of the axial gauge is then made by choosing the usual component-field axial gauges for the remaining derivative transformations.)

Therefore, as a <u>necessary</u> condition for superfield Faddeev-Popov quantization to give the correct unitary result, we must choose a <u>parametrization</u> of the gauge transformations which allows the usual Wess-Zumino gauge to be obtained: i.e., our superfield transformation laws must be written in a form such that the component transformations which follow directly from $\theta$ expansion allow us to gauge away all component fields which do not appear in the action by nonderivative component transformations. For example, if a scalar component field A appears in the gauge superfield but not the action, the superfield gauge transformation law must give a component transformation of the form $\delta A = \lambda$, and <u>not</u> of the form $\delta A = \Box\, \lambda$ or $\delta A = \partial_a \lambda^a$.

As a simple illustration of the importance of choosing a correct parametrization, let's consider the case of nonsupersymmetric Yang-Mills. Instead of the usual transformations $\delta A_a = \nabla_a \lambda$, suppose we choose $\delta A_a = \nabla_a \Box\, \lambda$ ($\Box \equiv \nabla^a \nabla_a$), which is classically equivalent (with appropriate modification of boundary conditions on $\lambda$). However, direct application of the Faddeev-Popov procedure gives a ghost term $C_2' \partial\cdot\nabla \Box\, C_1'$ (instead of just $C_2' \partial\cdot\nabla C_1'$), introducing an extra $\det \Box^2$ into the functional integral, preserving gauge invariance but destroying unitarity. Therefore, the parametrization $\delta A_a = \nabla_a \lambda$ is necessary for the Faddeev-Popov procedure to give the unitary result.

As verified by direct component-field expansions, the parametrizations of sec. 2 are the <u>only</u> correct parametrizations for superfield gauge theories (modulo the trivial type of compensating superfield which can be <u>completely</u> gauged away in a Wess-Zumino gauge: e.g., $\delta A = \lambda$, where A and $\lambda$ are both real superfields). In particular, we note that for the reducible multiplets (namely, $(\frac{3}{2},1)$ and supergravity), certain compensating superfields (in both of these cases, a chiral scalar) are <u>necessary</u> to obtain such a parametrization. Therefore, superfield supergravity can be quantized by the Faddeev-Popov procedure <u>only</u> if one uses the correct set of superfields (a real vector and a chiral scalar, plus perhaps the above-mentioned trivial type of compensating superfields) <u>and</u> the correct gauge superparameter (the general spinor $L_\alpha$; the methods of Namazie & Storey (1979) and Taylor (1979) are lacking with respect to both these points). Elimination of the compensating chiral scalar superfield (by a <u>derivative</u> gauge transformation) thus introduces ghosts, which do not decouple in a background-superfield gauge. In particular, the $\theta^2$ component B of $\phi$ appears in the supergravity Lagrangian with the $\theta^2$

component $B^a$ of $H^a$ as $(\bar{B} + \partial_a \bar{B}^a)(B + \partial_a B^a)$ with gauge invariance $\delta B = \partial_a \lambda^a$, $\delta B^a = -\lambda^a$: $B^a = 0$ is a Wess-Zumino gauge, but $B = 0$ is not. (For nonminimal supergravity, the gauge superparameter is still $L_\alpha$, but the compensating superfield is a linear scalar (Gates & Siegel, 1979), written as $T = \phi + D^\alpha \phi_\alpha$ or $\phi + \bar{D}^{\dot{\alpha}} \bar{\phi}_{\dot{\alpha}}$, $\bar{D}_{\dot{\alpha}} \phi = \bar{D}_{\dot{\alpha}} \phi_\beta = 0$.)

## 6. THE QUANTUM ACTION

We will now combine the methods discussed in the previous sections to derive the complete quantum action for calculations of one-loop S-matrix elements in supergravity. We will work with on-shell background, but we will only use superfields which exist off-shell: All chiral superfields must have only undotted spinor indices, and antichiral superfields only dotted spinor indices, neither having vector indices (Roček (1979), Fischler (1979)). (This is not true of the background-superfield formalism of Namazie & Storey (1979).)

The quantum transformations of the quantum fields are (Grisaru & Siegel, 1980; Grisaru, 1980):

$$\delta H_{\alpha\dot{\beta}} = -\, 8(\bar{\mathcal{D}}_{\dot{\beta}} L_\alpha - \mathcal{D}_\alpha \bar{L}_{\dot{\beta}}) + \mathcal{O}(H),$$

$$\delta\phi = -\,(\bar{\mathcal{D}}^2 + R)L^\alpha \mathcal{D}_\alpha \phi - \tfrac{1}{3}[(\bar{\mathcal{D}}^2 + R)\mathcal{D}^\alpha L_\alpha]\phi \; . \tag{8}$$

(We will use $\mathcal{D}_A$, $R$, $G_a$, and $W_{\alpha\beta\gamma}$ to refer to the covariant derivatives and field strengths which depend only on the background superfields.) It is now convenient to make the redefinition

$$L_\alpha \to \phi^{-3} L_\alpha \tag{9}$$

($= L_\alpha - 3\,\chi\, L_\alpha + \ldots$, where $\phi = 1 + \chi$, so the arguments of the previous section are unaffected). The transformation laws become:

$$\delta H_{\alpha\dot{\beta}} = -\, 8(\bar{\mathcal{D}}_{\dot{\beta}} L_\alpha - \mathcal{D}_\alpha \bar{L}_{\dot{\beta}}) + \mathcal{O}(H) + \mathcal{O}(\chi),$$

$$\delta\chi = -\,\frac{1}{3}\,\bar{\mathcal{D}}^2 \mathcal{D}^\alpha L_\alpha + \mathcal{O}(\chi) \; ,$$

$$\delta\phi^3 = -\,\bar{\mathcal{D}}^2 \mathcal{D}^\alpha L_\alpha \quad . \tag{10}$$

We have included the transformation law of $\phi^3$ because, having made the redefinition of eq. (9), it is independent of quantum fields. This allows us to choose the nonlocal gauge-fixing function

$$F_\alpha = \bar{\mathcal{D}}^{\dot{\beta}} (H_{\alpha\dot{\beta}} + \frac{2i}{5}\, \mathcal{D}_{\alpha\dot{\beta}}\, \Box^{-1}\, \bar{\phi}^3 \;) \tag{11}$$

($\Box \equiv \mathcal{D}^a\mathcal{D}_a$), which will cancel $H\chi$ crossterms in the action without introducing nonlocal interactions for the Faddeev-Popov ghosts. The Faddeev-Popov ghosts will obtain nonlocal kinetic terms, but they can be cancelled by catalyst ghosts. Then, as described in sec. 3, we complete $F_\alpha$ to

$$\hat{F}_\alpha = F_\alpha + \mathcal{D}^2 \Box_+^{-1}\phi_\alpha \text{ , where}$$

$$\Box_+ \equiv \Box + \tfrac{1}{8} W^{\alpha\beta\gamma}\mathcal{D}_\alpha M_{\beta\gamma} \qquad (12)$$

is the chiral form of $\Box$ ($\Box_+\phi_\alpha = \frac{1}{16}\bar{\mathcal{D}}^2\mathcal{D}^2\phi_\alpha$; $M_{\beta\gamma}$ are the Lorentz generators). The 't Hooft gauge averaging is then performed in a manner such that terms quadratic in $\hat{F}$'s and spinor $\mathcal{D}$'s cancel all terms in the action quadratic in H except those of the form H $\Box$ H and $HW\mathcal{D}H$. (The choice of the constant 2/5 in eq. (11) produces the cancellation of $H\chi$ crossterms: $\bar{\phi}^{-3} = 1 + 3\bar{\chi} + \ldots$) No $H\phi_\alpha$ or $\chi\phi_\alpha$ crossterms are produced, and the $\phi_\alpha$ kinetic term is found to be of the trivial form $\int d^2\theta\phi_B^3\,\phi^\alpha\phi_\alpha$ + h.c. (with the background chiral measure $\int d^2\theta\phi_B^3$), after applying an equivalence similar to that of eq. (2a). Thus the field $\phi_\alpha$ may be dropped.

The third ghost $\psi'_{3\alpha}$ obtains a local Lagrangian quadratic in spinor derivatives, which can be cast into the simpler form $\bar{\psi}'^{\dot\beta}_3\, i\,\mathcal{D}_{\alpha\dot\beta}\psi'^\alpha_3$ by use of a nonlocal redefinition, or by the shift

$$\psi'_{3\alpha} \to \psi'_{3\alpha} + \bar{\mathcal{D}}^2\mathcal{D}_\alpha\psi'_3 + \mathcal{D}_\alpha\phi'_3 \quad , \qquad (13)$$

where the complex $\psi'_3$ and chiral $\phi'_3$ are catalysts which cancel with their partners. The first and second ghosts $\psi'_{1\alpha},\psi'_{2\alpha}$ obtain nonlocal kinetic terms, which are made local by the shifts

$$\psi'_{1\alpha} \to \psi'_{1\alpha} + \mathcal{D}_\alpha\psi'_1 \text{ , } \psi'_{2\alpha} \to \psi'_{2\alpha} + \mathcal{D}_\alpha\psi'_2 \quad . \qquad (14)$$

Gauge fixing for the catalysts and for the original invariance $\delta\psi'_{1\alpha} = \Lambda_{1\alpha}$, $\delta\psi'_{2\alpha} = \Lambda_{2\alpha}$ ($\bar{\mathcal{D}}_{\dot\beta}\Lambda_{1\alpha} = \bar{\mathcal{D}}_{\dot\beta}\Lambda_{2\alpha} = 0$; due to $L_\alpha$ occurring only as $\bar{\mathcal{D}}_{\dot\beta}L_\alpha$ in transformation laws) again gives $\psi'_{1\alpha}$ and $\psi'_{2\alpha}$ the desired $\bar{\psi}'^{\dot\beta}\, i\,\mathcal{D}_{\alpha\dot\beta}\psi'^\alpha$ form. $\psi'_1$ and $\psi'_2$ cancel with their partners at one loop, but at higher loops $\psi'_1$ and $\psi'_2$ will contribute through their quantum interactions, which their partners $\psi_1$ and $\psi_2$ lack. For the chiral spinor ghosts arising from fixing the $\Lambda_\alpha$ gauges, the 1st and 3rd ghosts are found to not propagate, while the 2nd ghost contributes $\phi_1^\alpha \mathcal{D}^2 \phi_{1\alpha} + \phi_2^\alpha \mathcal{D}^2 \phi_{2\alpha}$ + h.c. (from $\Lambda_{1\alpha}$ and $\Lambda_{2\alpha}$), after using various catalysts, and performing shifts with respect to $\psi_1$ and $\psi_2$ to eliminate $\psi\phi_\alpha$ crossterms.

The final result for the quantum action in an on-shell background to second order in the quantum fields is:

$$S = \int d^4x\, d^4\theta\, E^{-1}\, [-\tfrac{1}{2} H\cdot \Box H + \tfrac{1}{32} H^{\alpha\dot\beta}(W_\alpha{}^{\gamma\delta}\mathcal{D}_\gamma H_{\delta\dot\beta} + \bar W_{\dot\beta}{}^{\dot\gamma\dot\delta}\bar{\mathcal{D}}_{\dot\gamma} H_{\alpha\dot\delta})$$
$$- \tfrac{18}{5}\bar\chi\chi + \sum_{i=1}^{3} \bar\psi'^{\dot\beta}_i\, i\, \mathcal{D}_{\alpha\dot\beta}\psi'^\alpha_i + \sum_{i=1}^{2} (\phi_i^\alpha \mathcal{D}^2 \phi_{i\alpha} + \text{h.c.})] \,, \qquad (15)$$

where only $\psi'_{i\alpha}$ has abnormal statistics. ($E$ is the determinant of the background supervierbein.) For higher-loop calculations we also include the catalyst terms

$$\sum_{i=1}^{2} (\bar\psi'_i \Box \psi'_i + \bar\psi_i \Box \psi_i) \quad ,$$

which cancel at one loop. Equation (15) is just the covariantization of eq. (3), plus the W terms. A similar situation occurs in (off-shell) background quantization of super-Yang-Mills (Grisaru et al., 1979).

This work was supported by Department of Energy Contract No. DE-AS02-76ER02220. Author's address after September 1, 1980: California Institute of Technology.

REFERENCES

Coleman, S. (1975). Secret symmetry: An introduction to spontaneous symmetry breakdown and gauge fields. In Laws of Hadronic Matter, ed. A. Zichichi, pp. 138-223. Academic Press, New York.

Curtright, T. (1980). Private communication.

Ferrara, S. & Zumino, B. (1978). Structure of linearized supergravity and conformal supergravity. Nucl. Phys. B, 134, pp. 301-326.

Fischler, M. (1979). Globally supersymmetric multiplets without local extensions. Phys. Rev. D, 20, pp. 1842-1845.

Gates, S. J., Jr. & Siegel, W. (1979). Superfield supergravity. Nucl. Phys. B, 147, pp. 77-104.

Gates, S. J., Jr. & Siegel, W. (1980). $(\frac{3}{2},1)$ superfield of O(2) supergravity. Nucl. Phys. B, 164, pp. 484-494.

Grisaru, M. T. (1980). Contribution to this volume.

Grisaru, M. T., Roček, M., & Siegel, W. (1979). Improved methods for supergraphs. Nucl. Phys. B, 159, pp. 429-450.

Grisaru, M. T. & Siegel, W. (1980). Supergraphity. In preparation.

Kallosh, R. E. (1978). Modified Feynman rules in supergravity. Nucl. Phys. B, 141, pp. 141-152.

Namazie, M. A. & Storey, D. (1979). Supersymmetric quantization of linearized supergravity. Nucl. Phys. B, 157, pp. 170-188.

Nielsen, N. K. (1978). Ghost counting in supergravity. Nucl. Phys. B, 140, pp. 499-509.

Piguet, O. & Rouet, A. (1976). Supersymmetric BPHZ renormalization (II). Supersymmetric extension of the pure Yang-Mills model. Nucl. Phys. B, 108, pp. 265-274.

Roček, M. (1979). Private communication.

Siegel, W. (1979a). Supergravity supergraphs. Phys. Lett., 84B, pp. 197-200.

Siegel, W. (1979b). Gauge spinor superfield as scalar multiplet. Phys. Lett., 85B, pp. 333-334.

Siegel, W. (1980). Hidden ghosts. Institute for Advanced Study preprint, to appear in Phys. Lett. B.

Stueckelberg, E. C. G. (1938). Die Wechselwirkungskräfte in der Elektrodynamik und in der Feldtheorie der Kernkräfte. Helv. Phys. Acta, 11, pp. 225-244.

Taylor, J. G. (1979). Quantization of superfield supergravity. King's College preprint.

Thierry-Mieg, J. (1980). BRS structure of antisymmetric tensor gauge fields. Harvard preprint HUTMP-79/B85.

't Hooft, G. (1971). Renormalization of massless Yang-Mills fields. Nucl. Phys. B, 33, pp. 173-199.

Wess, J. & Zumino, B. (1974). Supergauge invariant extension of quantum electrodynamics. Nucl. Phys. B, 78, pp. 1-13.

# SUPERSYMMETRIC DIMENSIONAL REGULARIZATION

W. Siegel

Institute for Advanced Study
Princeton, New Jersey 08540

P.K. Townsend

CERN
Geneva, Switzerland

P. van Nieuwenhuizen

Institute for Theoretical Physics
State University of New York at Stony Brook
Stony Brook, New York 11794

## 1. REGULARIZATION BY DIMENSIONAL REDUCTION

Dimensional regularization ('t Hooft and Veltman, 1972; Bollini and Giambiagi, 1972; Ashmore, 1972; Cicuta and Montaldi, 1972) is the most favored form of regularization for calculational purposes, due to its possession of the following properties: 1) It manifestly preserves almost all symmetries; 2) It requires only one regulating parameter; and 3) It has a simple inherent renormalization procedure (subtraction of poles in the variable D - 4, where D is the dimensionality of spacetime). However, the original form of dimensional regularization does not preserve supersymmetry: If one applies dimensional continuation naively to a supersymmetric component-field action, the Fermi and Bose components will not balance in $D \neq 4$. If one naively continues a superfield action (continuing an N = 1 spinor coordinate $\theta^{\mu}$ in D = 4 to an N = 1 $\theta^{\mu}$ in $D \neq 4$, with a different number of components), ghosts are introduced, so the minimal subtraction scheme of dimensional renormalization would violate unitarity. (For example, the action for a chiral scalar superfield in D = 4 has no gauge invariance because it only contains spin < 1; however, in D > 4 the increase in the number of components of $\theta^\mu$ introduces spins $\geq$ 1, so the lack of gauge invariance introduces ghosts.) Fortunately, there is a simple modification of dimension regularization which does preserve supersymmetry (Siegel, 1979a): dimensional reduction to real D < 4, followed by analytic continuation to complex D. In terms of component fields, this means fixing the ranges of all indices on the fields (and therefore the numbers of Fermi and Bose components). For superfields, it means continuing in the dimensionality of x-space while fixing the dimensionality of $\theta$-space. For example, a vector component field always has 4 components, which can be interpreted as a D-component vector and 4 - D scalars. The same field, appearing as $\theta^\alpha\bar{\theta}^\beta v_{\alpha\beta}$ in the expansion of a real scalar O(1) superfield, always has 2·2 = 4 components.

This regularization procedure allows the simple manipulation of spinor derivatives in supergraph calculations (Grisaru, Roček, and Siegel, 1979). The resulting rules are: 1) First do all algebra exactly as in D = 4; 2) Then do the momentum integrals as in ordinary dimensional regularization.

Unfortunately, this regularization procedure needs extra rules before one can say that it is consistent. In the next section such extra rules needed for superconformal anomalies will be discussed, while in the last section the problems associated with renormalizability and higher order loops are discussed.

The inconsistency present if one does not add extra rules (Siegel, 1980) is closely related to the problem of manifest chiral invariance in ordinary dimensional regularization. There, the D - component Dirac matrices $\gamma_a$ must satisfy $\{\gamma_5,\gamma_a\} = 0$ for all $a$ in order to avoid introducing spurious anomalies when applying the minimal subtraction scheme of dimensional renormalization. However, evaluation of the quantity tr $(\gamma_5 \gamma^e \gamma_e \gamma_a \gamma_b \gamma_c \gamma_d)$ with and without commuting one $\gamma_e$ through all the other $\gamma$'s leads to the identity $(D - 4)$ tr $(\gamma_5 \gamma_a \gamma_b \gamma_c \gamma_d) = 0$. Therefore, a more general definition of the $\gamma$-algebra is needed (Chanowitz, Furman, and Hinchcliffe, 1979). Modifications are also required for supersymmetric dimensional regularization, as will be discussed in the following section. These modifications are expected to be necessary only in graphs where anomalies occur. In fact, in the explicit one-and two-loop examples worked out in section 3 no need for modifications arises, and the scheme works in the sense that the Ward identities considered in global and local supersymmetry are satisfied.

The method can be generalized to local supersymmetry (Siegel, 1979b). Supergravity is described by a real vector superfield $H^m$, appearing only as the operator $H^m\,\partial/\partial x^m$ (Siegel, 1977; Gates and Siegel, 1979). Therefore, it is natural to treat $H^m$ as a D-component vector. If $H^m$ were to be treated as a 4-vector, the multiplet would be reducible in $D<4$ to pure supergravity plus 4-D vector multiplets. This reducibility would introduce independent coupling constants, making the theory (more) nonrenormalizable. (A similar situation occurs when applying dimensional reduction to nonsupersymmetric gauge fields, as will be described in sec.3.) However, there are problems even when using the D-component $H^m$: Although the method seems to make sense in terms of superfields, it does not when expanded in components (except in integral $D \leq 4$). This problem is another consequence of the algebraic inconsistency mentioned above.

## 2. SUPERCONFORMAL ANOMALIES

A check of supersymmetry Ward identities in various models shows no sign of a violation of supersymmetry, nor of an inconsistency of dimensional reduction (Townsend and van Nieuwenhuizen (1979); Sezgin (1980); Capper, Jones, and van Nieuwenhuizen (1980); Majumdar Poggio, and Schnitzer (1980)). It is clear that at least some of the algebraic inconsistencies are related to the existence of superconformal anomalies, and so we are led to the study of anomalies. Let us concentrate on the familiar chiral anomaly. The usual triangle graphs,

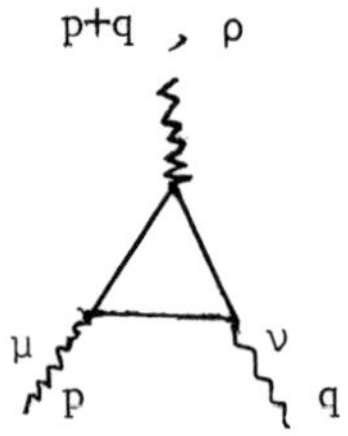


$$+ \left(\begin{matrix}\mu \leftrightarrow \nu \\ p \leftrightarrow q\end{matrix}\right) ,$$

lead to the evaluation of the expression;

$$\int d^4\ell \, \frac{1}{\ell^2 (\ell+q)^2 (\ell+p+q)^2} \, \mathrm{Tr}\{\gamma_5 \gamma_\rho \not{\ell} \gamma_\nu (\not{\ell}+\not{q}) \gamma_\mu (\not{\ell}+\not{p}+\not{q})\} + \left(\begin{matrix}\mu\leftrightarrow\nu \\ p\leftrightarrow q\end{matrix}\right).$$

According to the dimensional reduction method all $\gamma$-matrix algebra is done as in four dimensions. In particular, $\gamma_5$ anticommutes with all $\gamma_\mu$. This prescription leads to an apparent contradiction. Multiplying the above expression by $(p+q)_\rho = (p+q+\ell)_\rho - \ell_\rho$, the single $\ell_\rho$ produces a factor of $\ell^2$ cancelling the $\ell^2$ in the denominator. Similarly, pulling the $(p+q+\ell)_\rho \gamma^\rho$ factor through the $\gamma_5$, and then moving it to the end of the trace by using the cyclic property Tr(ABC) = Tr(BCA) produces a factor of $(\ell+p+q)^2$, which again cancels against this factor in the denominator. We now gather terms with a common denominator, after shifting variables of integration, and then perform the traces by means of the formula $\mathrm{Tr}(\gamma_5\gamma_a\gamma_b\gamma_c\gamma_d) = 4\varepsilon_{abcd}$. In this way we obtain the result $\partial . j^5 = 0$. But we know from the regularization independent Adler-Rosenberg method (Adler 1970) that the general form of the anomaly is

$<\partial.j^5 \, W_\mu(p)W_\nu(q)> \sim Tr(\gamma_5 \not{p}\not{q}\gamma_\mu\gamma_\nu) = 4\varepsilon^{\mu\nu\rho\sigma} p_\rho q_\sigma$ with a non-zero coefficient that can be determined from convergent graphs alone. A vanishing anomaly can only mean that the regularization procedure has forced us to define $Tr(\gamma_5\gamma_a\gamma_b\gamma_c\gamma_d) = 0$ for $D \neq 4$. This is the same problem as occurs in purely algebraic manipulations that allow us to deduce $(D-4)Tr(\gamma_5\gamma_a\gamma_b\gamma_c\gamma_d) = 0$. The use of such a regularization scheme, if not actually inconsistent, would give a different theory at $D \neq 4$ than the one we started with at $D = 4$; i.e. it is nonanalytic. Therefore we reject this option. The only way to make the regularization consistent with the four-dimensional Dirac algebra is to give up one of the assumptions that enters the calculation. One might think of disallowing shifts of integration variables, but this would make the whole regularization prescription unworkable, and in any case does not address the problem of algebraic inconsistencies. The only remaining assumption is the cyclicity of the trace over $\gamma$-matrices. This property is not self-evident as it can fail for infinite dimensional matrices, and it may be that an infinite dimensional representation is required for $D \neq 4$. Fortunately we do not need to assume that $Tr(ABC) = Tr(BCA)$ in order to do the calculation, so we can ask what happens if we drop this assumption (Nicolai and Townsend, (1980a), (1980b)). To proceed with the calculation of $\partial\cdot j^5$ as before we do not pull this factor $(\ell+p+q)_\rho\gamma^\rho$ through the $\gamma_5$ and then to the end of the trace via the cyclicity assumption. Instead, we anticommute $(\ell+p+q)_\rho\gamma^\rho$ through the intervening $\gamma$-matrices to the end of the trace. We then evaluate the traces as before but we do not discard contributions containing a factor of $A_{\sigma\nu\lambda\mu\tau,\rho} \equiv -\varepsilon_{\nu\lambda\mu\tau} g_{\sigma\rho} + \varepsilon_{\sigma\lambda\mu\tau} g_{\nu\rho} - \varepsilon_{\sigma\nu\mu\tau} g_{\lambda\rho} + \varepsilon_{\sigma\nu\lambda\tau} g_{\mu\rho} - \varepsilon_{\sigma\nu\lambda\mu} g_{\tau\rho}$. This tensor is antisymmetric in five indices and must vanish at $D = 4$. But we do not assume that it vanishes for $D \neq 4$. In fact, performing the integrals that multiply this tensor gives rise to a contraction of $A_{\sigma\nu\lambda\mu\tau,\rho}$ with a $D \neq 4$ dimensional $g^{\sigma\rho}$. This gives a factor of $A_{\sigma\nu\mu\tau,\rho}\, g^{\sigma\rho} = (4-D)\,\varepsilon_{\nu\lambda\mu\tau}$. The factors of $(4-D)$ cancel a pole at $D = 4$ coming from this integral, leaving a finite and non-vanishing residue.

Although this result relies upon the assumption of a completeness relation for contractions with $D \neq 4$ dimensional $g_{\mu\nu}$'s, one can also obtain the same non-zero result, with more effort, by direct evaluation of the traces and integrals without any intermediate manipulations.

The difficulties are not yet completely resolved however. Although we can obtain a non-zero anomaly we cannot determine unambiguously whether this anomaly is of the vector or axial vector current. The first method of calculation described above leads to ambiguities proportional to tensors antisymmetric on five indices. The direct evaluation method gives a different result according to whether the traces or the integrals are performed first. In all cases examined this ambiguity amounts to a redistribution of the anomaly between the vector and axial vector currents. There is of course also the ambiguity in where to start the trace in the expression $<j^5_\rho W_\mu W_\nu>$. Therefore, we have succeeded in avoiding an inconsistency in the calculation, only at the expense of an ambiguous result. In the above expression for the triangle graphs one can move the $\gamma_5$ from one vertex to the other quite freely, as it anticommutes with <u>all</u> $\gamma_\mu$. It is then obvious that we cannot hope to prove that the anomaly appears at any particular vertex. In order to put the anomaly into $\partial . j^5$ one must also <u>impose</u> vector current conservation; regularization by dimensional reduction does not automatically preserve it.

We have seen that at least one algebraic inconsistency of dimensional reduction can be rendered innocuous. It is plausible, but not yet established, that all such inconsistencies can be tamed in this way.

## 3. APPLICATIONS

In this section we discuss the renormalizability aspects of the new scheme, and applications to both global and local supersymmetry.

Consider Yang-Mills theory, and decompose in $n < 4$ dimensions the field $W_\mu^a$ into $(W_i^a, W_\sigma^a)$ with $1 \leq i \leq n$ and $n \leq \sigma \leq 4$. All fields depend only on $x^i$ but not on $x^\sigma$. The action splits up into the usual Yang-Mills action but now in n dimensions, plus a minimally coupled Lorentz-scalar iso-vector $W_\sigma^a$ and a Yukawa coupling

$$\mathcal{L} = -\frac{1}{4}(\partial_i W_j^a - \partial_j W_i^a + g\varepsilon^{abc} W_i^b W_j^c)^2 - \tfrac{1}{2}(D_i W_\sigma^a)^2$$

$$-\frac{1}{2} g^2 ((\vec{W}_i \cdot \vec{W}_i)(\vec{W}_j \cdot \vec{W}_j) - (\vec{W}_i \cdot \vec{W}_j)^2)$$

The new local invariance is $\delta\vec{W}_i = D_i\vec{\Lambda}$ and $\delta\vec{W}_\sigma = \vec{W}_\sigma \wedge \vec{\Lambda}$ so that the Yukawa term is separately invariant. Hence the equality of the gauge and the Yukawa coupling constants is accidental, and one expects that quantum divergences will reflect this: probably one can only absorb the divergences by rescaling of the two coupling constants independently. An explicit calculation has confirmed this (Jones and Castellani, 1980). Thus it seems that the new scheme is out for nonsupersymmetric theories. However, it may be possible to fix this up, and the scheme presumably is useful for concrete applications in, for example, QCD when one wants to do complicated gamma matrix algebra in four dimensions.

In higher orders, one should carefully distinguish between 4-dimensional $\delta_{\mu\nu}$ resulting from $W_\mu W_\nu$ contractions (contracting the $W_i^a$ and $W_\sigma^a$ at the same time) and n-dimensional $\delta_{\mu\nu}$ resulting from $p_\mu p_\nu$ momentum integrals. Spinor algebra is always done in 4-dimensions, and this is enormously useful; for example, strings of gamma matrices can always be reduced to 16 Dirac invariants.

Insight into the detailed working of the new scheme is gained by redoing the 2-loop $\beta$-function calculation. To this effect one must calculate the effect of the scalar fields $W_\sigma^a$ on the

2-loop renormalization constants $Z_3(W,W)$, $Z_3(C^*,C)$ and $Z_1(C^*CW)$. The calculation of the contribution of a multiplet of scalars transforming according to an arbitrary representation of the gauge group cannot be applied without modification because, for example, a one-loop contribution of a $W^a_\sigma$ loop to the $W^a_i$ selfenergy is finite (there is an extra factor $\varepsilon = 4 - n$ since there are $\varepsilon$ fields in this loop). The final result is that the β-function up to two-loops is the same (Capper et al. 1980). This is a significant result in that since the leading two terms in $\beta(g)$ should be independent of the regularization/renormalization technique, it shows that the dimensional reduction scheme, while differing non-trivially from the conventional dimensional regularization scheme, leads to the same physical effects. One aspect not tested in this calculation is the different behaviour of the Yukawa coupling constant at the quantum level. For this test one would need a comparison between, for example, $Z_1(W_iW_jW_k)$ and $Z_1(W_iW_\sigma W_\tau)$. A simpler calculation would be to take fermions coupled to Yang-Mills fields at the two-loop level. In $Z_3(\psi\psi)$ as well as in $Z_1(W_i\psi\psi)$ one finds both the one-loop counter term for the $W_\sigma\psi\psi$ vertex and the $W_i\psi\psi$ vertex and it seems likely that again one will find the same β function if one uses this path. This example seems to suggest that one should simply renormalize the various coupling constants independently, and only require that their actual value in the renormalized action is the same.

Let us now turn to supersymmetric theories. In supersymmetric Yang-Mills theory the path-integral is invariant under a change of integration variables given by the global supersymmetry transformations, because the Jacobian is one. Since the spinors and vector bosons have 4-dimensional indices, the classical action is still globally supersymmetric ; the fact that fields depend only on $x^i$ has no influence on the proof of global invariance. In particular, the crucial identity

$$(\lambda^a\gamma_\mu\lambda^c)(\bar{\varepsilon}\gamma^\mu\lambda^b)\varepsilon_{abc} = 0$$

still holds. (This is not true in n-dimensions). One can now derive in the usual way Ward identities, and the following result is obtained(Capper et al., 1980). In the new scheme the Ward identity considered (relating the fermion and boson selfenergies) is valid at the one-loop level for all n, but in ordinary dimensional regularization the identity is only valid at the pole parts but not at the finite parts. The dimensional reduction technique does not violate Ward identities, even when the supersymmetry is explicitly broken by the gauge choice, and this result greatly increases one's confidence in the technique.

Let us now turn to local supersymmetry, i.e., supergravity (Fung et al., 1980) Again one can derive a Ward identity, namely the transversality of the gravitino selfenergy in the gauge $\alpha\gamma^{\mu}\psi_{\mu} + (1-\alpha)\gamma^{a}\delta^{\mu}_{a}\psi_{\mu}$ where $\gamma^{a}$ is constant. (Details about derivations of Ward identities by means of BRST techniques and the functional methods, and relations between these approaches are discussed in detail in the above reference.) If one verifies the transversality of the finite terms at the one-loop level, the calculation becomes sensitive to the regularization method employed. In the ordinary dimensional regularization scheme, the propagators for graviton and gravitino are n-dependent (in a complicated way), but in the new scheme they are n-independent and simple. Direct evaluation shows that only in the new scheme is the local supersymmetry Ward identity satisfied, but not in ordinary dimensional regularization. (Note that in the latter case one replaces $\varepsilon^{\mu\nu\rho\sigma}\gamma_5\gamma_\nu$ by $\{\sigma_{\mu\rho},\gamma_\sigma\}$ which defines the gravitino action in n dimensions).

These explicit examples show that in the cases of interest the scheme works. The problems to be solved are twofold. In non-supersymmetric theories one can perhaps define a consistent renormalization program. In supersymmetric theories one must fix the rules such as to exclude ambiguities (especially in the case of anomalies where ambiguities are necessarily present in this scheme, see section 2). It seems plausible that this can be done, but it should be proved.

## REFERENCES


Adler, S.L. (1970), Lectures at Brandeis Summer Institute, eds. S. Deser, M. Grisaru and H. Pendleton (M.I.T. Press).

Ashmore, J.F. (1972). A Method For Gauge-Invariant Regularization: Lett.Nuo.Cim. 4, pp. 289-290.

Bollini, C.G. and Giambigi, J.J. (1972). Dimensional Renormalization the number of dimensions as a regularizing parameter. Nuo.Cim. 12B, pp. 20-26.

Capper, D.M., Jones, D.R.T. and van Nieuwenhuizen, P., (1980). Regularization by dimensional reduction of supersymmetric and non-supersymmetric gauge theories, Nucl.Phys.B167, pp. 479-499.

Chanowitz, M., Furman, M. and Hinchliffe, I. (1979). The axial current in dimensional regularization. Nucl.Phys. B159, pp. 225-243.

Cicuta, G.M. and Montaldi, E. (1972). Analytic renormalization via continuous space dimension, Lett.Nuo.Cim. 4, pp. 329-332.

Fung, M.K., van Nieuwenhuizen, P. and Jones, D.R.T. (1980). Transversality of the Rarita-Schwinger self-energy Ward identities and dimensional regularization. Phys.Rev.D22.

Gates, S.J., Jr. and Siegel, W. (1979). Superfield supergravity. Nucl.Phys.B147. pp. 77-104.

Grisaru, M.T., Rocek, M. and Siegel, W. (1979). Improved Methods for Supergraphs. Nucl.Phys. B159, pp. 429-450.

Jones, D.R.T. (1980) and Castellani, L. (1980), unpublished.

Majumdar, P., Poggio, E. and Schnitzer, H.J. (1980), Supersymmetry Ward Identity for the Supersymmetric non-Abelian gauge theory, Phys.Rev.D21, pp. 2203-2212.

Nicolai, H. and Townsend, P.K. (1980a). Anomalies and supersymmetric regularization by dimensional reduction. Phys.Lett.93B, pp. 111-115; b "Supersymmetric regularization and superconformal anomalies", Proceedings of the Erice conference on "The unification of the fundamental interactions", S. Ferrara, J. Ellis and P. van Nieuwenhuizen editors.

Sezgin, E. (1980). Dimensional regularization and the massive Wess-Zumino model, Nucl.Phys.B162, pp. 1-11.

Siegel, W. (1977). Supergravity superfields without a supermetric. Harvard preprint HUTP-77/A068.

Siegel, W. (1979a). Supersymmetric dimensional regularization via dimensional reduction, Phys.Lett.84B, pp. 193-196.

Siegel, W. (1979b). Locally supersymmetric dimensional regularization. Phys.Lett.94B, pp. 37-40.

Siegel, W. (1980). Inconsistency of supersymmetric dimensional regularization. Institute for Advanced Study preprint, to appear in Phys.Lett.B.

't Hooft, G. and Veltman, M. (1972). Regularization and renormalization of gauge fields, Nucl.Phys.B44, pp. 189-213.

Townsend, P.K. and van Nieuwenhuizen, P. (1979). Dimensional regularization and supersymmetry at the two loop level, Phys.Rev.D. 20, p. 1832.

# SYMMETRY PRESERVING REGULARIZATION FOR GAUGE AND SUPERGAUGE THEORIES

A. A. Slavnov

Steklov Mathematical Institute
Moscow

Construction of an invariant regularization procedure is an essential part of the renormalizable programme for supergauge theories. This problem was discussed recently (Curtright & Ghandour 1977, Siegel 1979, Townsend & van Nieuwenhuizen 1979, Nicolai & Townsend 1980, 1980a), especially in connection with the supercurrent anomalies. Dimensional regularization which was widely used for ordinary Yang-Mills theories does not work in the supersymmetric case. A modified version of this procedure using a dimensional reduction was proposed by Siegel and was found to be successful for calculations of some particular one and two loop diagrams. However at present no arguments are known for proving the validity of this procedure in the general case.

A manifestly invariant regularization applicable to arbitrary supersymmetric gauge theory (and presumably to local supersymmetry as well) may be constructed as a straightforward generalization of the procedure proposed for the gauge theories some time ago (Slavnov 1977, Krivoshchekov 1978). I shall describe this procedure using the ordinary Yang-Mills theory as a more simple example to make the idea more transparent. The procedure consists of two steps, the first one being the higher invariant derivatives regularization and the second one being a modified version of the Pauli-Villars regularization. To begin with one replaces the Yang-Mills Lagrangian by the higher derivative gauge invariant expression

$$L_{Y.M.} \to \frac{1}{8}\,\mathrm{Tr}(F_{\mu\nu}F_{\mu\nu} + \frac{1}{\Lambda^2}\,\nabla_\alpha F_{\mu\nu}\nabla_\alpha F_{\mu\nu}) \qquad (1)$$

where $\nabla_\alpha$ is the covariant derivative. The free field propagator corresponding to (1) has an asymptotical behaviour $\sim k^{-4}$ therefore improving the convergence of the Feynman diagrams. However to preserve the gauge invariance we were forced to introduce the new vertices which also contain derivatives. As a result the regularization provided by (1) is not complete: one loop diagrams are still divergent (in some cases to make all multi-loop diagrams convergent one needs more than two additional derivatives).

This procedure may be generalized to the supersymmetric case as follows. One replaces the supersymmetric Yang-Mills Lagrangian

$$L(x,\Theta) = -\frac{1}{256}\,\mathrm{Tr}\{(C^{-1})^{\alpha\beta}\left[W^-_\alpha W^-_\beta\bar{\Theta}\Theta^+ + W^+_\alpha W^+_\beta\bar{\Theta}\Theta^-\right]\} \quad (2)$$

where the chiral fields

$$W^\pm_\alpha = -\frac{1}{g}\,\Delta^\pm\{e^{\mp g\Psi}\mathcal{D}^\pm_\alpha e^{\pm g\Psi}\} \quad (3)$$

$$\Delta^\pm = (\bar{\mathcal{D}}\frac{1\pm i\gamma_5}{2}\mathcal{D}) \quad (4)$$

play the role of field strengths, C is the charge conjugation matrix, $\mathcal{D}$ is the super-covariant derivative and

$$\bar{\Theta}\Theta^\pm = \bar{\Theta}\,\frac{1\pm i\gamma_5}{2}\,\Theta,$$

by the higher derivative expression

$$L \to L + L_\Lambda;\; L_\Lambda = \frac{1}{256}\cdot\frac{1}{\Lambda^2}\,\mathrm{Tr}\,\{(C^{-1})^{\alpha\beta}W^+_\alpha\Gamma_+(W^+_\beta)\bar{\Theta}\Theta^- + (+\leftrightarrow -)\} \quad (5)$$

where $\Gamma$ is the supersymmetric analogue of the higher covariant derivative

$$\Gamma_\pm(W^\pm) = \Delta^\mp\{e^{\mp g\Psi}\Delta^\pm(e^{\pm g\Psi}W^\pm_\beta\; e^{\mp g\Psi})e^{\pm g\Psi}\} \quad (6)$$

One can easily verify that under the gauge transformation

$$e^{\pm g\Psi} \to \Omega_\mp\; e^{\pm g\Psi}\Omega_\pm^{-1}$$

$\Gamma$ transforms as follows

$$\Gamma_\pm \to \Omega_\pm\;\Gamma_\pm(W)\Omega_\pm^{-1} \quad (7)$$

providing the invariance of the Lagrangian (5).

The Green function generating functional may be constructed as usual (Slavnov 1975, Krivoshchekov, Slavnov & Faizullaev 1976).

$$Z = \int\exp\{i\int dxd\Theta(L + L_\Lambda + \frac{\mathrm{Tr}}{2\beta}\Psi P\Psi + \kappa\Psi)\}\Pi\Delta(\Psi)d\Psi \quad (8)$$

where $(2\beta)^{-1}\Psi P\Psi$ is a supersymmetric gauge fixing term, and $\Delta(\Psi)$ is the supersymmetric Faddeev-Popov determinant

$$\Delta(\Psi) = \int \exp\{-\tfrac{1}{8}i\int dxd\Theta\ \mathrm{Tr}(\bar{c}Mc)\}d\bar{c}dc \tag{9}$$

$$Mu = (E_+ + E_-)\ \delta\Psi^{\Omega(u)}. \tag{10}$$

$E_\pm$ are the chiral projection operators. It is understood that the gauge fixing term also includes a sufficient number of higher derivatives.

The power counting gives for the maximal index of the diagram

$$\omega_{max} = 2 - 3Eg - 4(L - 1) \tag{11}$$

where $E_g$ is the number of external ghost lines and L is the loop number. Therefore all multiloop diagrams are superficially convergent.

To make all the diagrams convergent the procedure described above must be supplemented by some additional regularization of one loop diagrams. For ordinary Yang-Mills theory the dimensional regularization would do the job, but for the supersymmetric case one needs something different. The procedure which works equally well in both cases is the gauge invariant version of the Pauli-Villars method. Indeed, the one loop diagrams may be always presented as $\det^{-1}A_o$ where $A_o$ is a quadratic form generating one loop diagrams. Pauli-Villars regularization replaces $\det A_o$ by

$$\det{}^{-1}A_o \prod_i (\det A_i)^{-a_i} = \exp(-\mathrm{Tr}\ln A_o - \sum_i a_i \mathrm{Tr}\ln A_i) \tag{12}$$

where the kernels $A_i$ differ from $A_o$ by the mass term

$$A_i = A_o + \frac{m_i^2}{2} \tag{13}$$

and the coefficients $a_i$ satisfy the conditions

$$\Sigma a_i + 1 = 0,\ \Sigma a_i m_i^2 = 0 \tag{14}$$

This procedure clearly makes all the integrals finite, but its straightforward application to the Yang-Mills case would destroy the gauge invariance. The problem is that one loop contributions from the Yang-Mills fields and ghost fields are not invariant separately. Nevertheless as we shall show the divergent part of the one loop diagrams can be written in a manifestly gauge invariant form, hence the regularization (12) will not destroy the symmetry.

We start again with the ordinary Yang-Mills theory. Simply connected one loop diagrams may be presented in the form

$$W(A) = i\ln \int \exp\{ i \int \frac{\partial^2 L}{\partial A_\mu \partial A_\nu} B_\mu B_\nu \, dx \} \prod_x \delta(\partial_\mu B_\mu)\Delta(A)dB \quad (15)$$

Here $L$ is the partially regularized Lagrangian (1) and for definiteness the Lorentz gauge is chosen. The effective action which stands in the exponent is invariant with respect to the transformation

$$A_\mu \to A_\mu + [A_\mu, u] + \partial_\mu u, \; B_\mu \to B_\mu + [B_\mu, u]. \quad (16)$$

However the gauge fixing term and Faddeev-Popov determinant are not invariant. To avoid this problem we pass to the "covariant" gauge

$$\nabla_\mu B_\mu \equiv \partial_\mu B_\mu + g[A_\mu, B_\mu] = 0 \quad (17)$$

It can be done by using the well-known trick. Let us introduce a gauge invariant functional

$$\Delta_c(A,B) \int \delta(\nabla_\mu B_\mu^\Omega)d\Omega = 1 \quad (18)$$

where

$$B_\mu^\Omega = B_\mu + \frac{1}{\varepsilon}(A_\mu^\Omega - A_\mu)$$

$\varepsilon$ is an arbitrary parameter.

Introducing the left hand side of eq.(18) into the integral (15) and changing the variables one has

$$W(A) = i\ln \int \exp\{i \int \frac{\partial^2 L}{\partial A_\mu \partial A_\nu} B_\mu^\Omega B_\nu^\Omega \, dx\} \, \delta(\nabla_\mu B_\mu)\det(\nabla^2)dB \quad (19)$$

where $\Omega$ is defined by the equation

$$\partial_\mu B_\mu^\Omega = 0 \quad (20)$$

One can easily verify that the terms

$$\frac{\partial^2 L}{\partial A_\mu \partial A_\nu} (B_\mu^\Omega - B_\mu)(B_\nu^\Omega - B_\nu) \tag{21}$$

produce only convergent diagrams. Hence being interested only in the divergent part of (15) we can replace $B_\mu^\Omega$ in eq. (19) by $B_\mu$. The resulting expression is manifestly invariant with respect to the transformation (16) and we can apply to it the regularization (12). Namely we replace

$$B_o = \det \nabla^2 = \int \exp \{i \int \nabla_\mu \bar{c} \nabla_\mu c dx\} d\bar{c} dc \tag{22}$$

by $B_o \prod_i B_i^{a_i}$, where

$$B_i = \det (\nabla^2 + m_i^2) \tag{23}$$

and

$$(\det Q_o)^{-\frac{1}{2}} = \int \exp \{i \frac{\partial^2 L}{\partial A_\mu \partial A_\nu} B_\mu B_\nu \, dx\} \, \delta(\nabla_\mu B_\mu) \, dB \tag{24}$$

by $(\det Q_o)^{-\frac{1}{2}} \prod_i (\det Q_i)^{-\frac{1}{2} a_i}$

$$\text{where} \quad Q_i = Q_o + \frac{m_i^2}{2} \tag{25}$$

This replacement makes all one loop diagrams finite preserving the gauge invariance. It is evident that the regularized functional $W(A)$ satisfies generalized Ward identities.

This procedure allows straightforward generalization to the supersymmetric case.

Starting with the one loop diagram generating functional

$$W(\Psi) = i \ln \int d\chi \exp \{i \int dx d\Theta \left( \frac{1}{2} \frac{\partial^2 L}{\partial \Psi^a \partial \Psi^b} \chi^a \chi^b + \frac{1}{2\beta} \mathrm{Tr} \chi P \chi \right)\} \Delta(\Psi) \tag{26}$$

one can transform it into manifestly invariant form along the same lines as before. The divergent part of eq. (26) may be written in the form

$$W(\Psi) = i \ln \{ \Delta_c(\Psi) \det^{-\frac{1}{2}} Q_o(\Psi) \}$$
$$\det^{-\frac{1}{2}} Q_o = \int d\chi \exp\{i \int dx d\Theta (\frac{1}{2} \frac{\partial^2 L}{\partial \Psi^a \partial \Psi^b} \chi^a \chi^b + \frac{1}{2\beta} \mathrm{Tr} T_+(\chi) T_-(\chi))\} \tag{27}$$

where $\Delta_c(\Psi)$ is the covariant Faddeev-Popov determinant

$$\Delta_c(\Psi) = \int \exp\{i \int dx d\Theta \; \mathrm{Tr} \Big( \bar{\eta} e^{g\Psi} \Big[ \Delta^- (e^{-g\Psi} u_- e^{g\Psi}) + \frac{1}{\Lambda^2} \Delta^- \{e^{-g\Psi} \Delta^+ (e^{g\Psi} \Delta^- (e^{-g\Psi} u_- e^{g\Psi}) e^{-g\Psi}) e^{g\Psi}\} \Big] \Big) + (+ \leftrightarrow -, \; \eta \leftrightarrow \xi, \; u_- \leftrightarrow \lambda_+)\} \, d\bar{\eta} du_- d\xi d\lambda_+ \tag{28}$$

and the covariant gauge fixing term is constructed with the help of the operators $T_\pm$

$$T_\pm(\chi) = \pm \frac{1}{g} e^{\pm g\Psi} \{\Delta^{\mp} (e^{\mp g\Psi} \chi^a (\frac{\partial}{\partial \Psi^a} e^{\pm g\Psi}) - \chi^a (\frac{\partial}{\partial \Psi^a} e^{\mp g\Psi}) e^{\pm g\Psi}) + \frac{1}{\Lambda^2} \Delta^{\mp} \Big( e^{\mp g\Psi} \Delta^{\pm} \{ e^{\pm g\Psi} \Delta^{\mp} (e^{\mp g\Psi} \chi^a (\frac{\partial}{\partial \Psi^a} e^{\pm g\Psi}) - \chi^a (\frac{\partial}{\partial \Psi^a} e^{\mp g\Psi}) e^{\pm g\Psi}) e^{\mp g\Psi} \} e^{\pm g\Psi} \Big) \} \tag{29}$$

The eq. (27) is invariant with respect to the transformation

$$\Psi \to \Psi^\Omega; \quad \chi \to \chi^\Omega = \chi \, (\frac{\partial}{\partial \Psi} \Psi^\Omega) \tag{30}$$

To regularize $W(\Psi)$ one should replace

$$\Delta_c \to \Delta_c \prod_i (\Delta_{c_i})^{a_i}, \quad \det^{-\frac{1}{2}} Q_o \to \det^{-\frac{1}{2}} Q_o \prod_i (\det^{-\frac{1}{2}} Q_i)^{b_i} \tag{31}$$

where $Q_i$ and $\Delta_{c_i}$ are obtained from $Q_o$ and $\Delta_o$ by adding to the corresponding quadratic forms the mass terms

$$\mu_i (e^{g\Psi} \bar{\eta} u_- + e^{-g\Psi} \bar{\xi} \lambda_+) \tag{32}$$

$$\frac{1}{2} m_i^2 \; \mathrm{Tr}\{ (\chi^a \frac{\partial}{\partial \Psi^a} e^{g\Psi}) (\chi^b \frac{\partial}{\partial \Psi^b} e^{-g\Psi}) \} \tag{33}$$

These particular terms are chosen to preserve supergauge invariance. Although they contain together with the mass terms additional vertices it does not cause any troubles, because these new vertices enter with the coupling constant proportional to $m^2$ and therefore do not introduce new divergences. It completes the construction of the invariant regularization for supergauge theories.

Although the procedure described above has not been worked out explicitly for the case of local supersymmetry, the general arguments seem to be applicable to this case as well.

The regularized Lagrangian for the supersymmetric Yang-Mills theory looks rather complicated. Nevertheless the very fact of its existence proves that the conserved supercurrent does exist and may be constructed by the usual Noether technique. Of course, this current does not coincide with the classical expression

$$S_\mu = \sigma_{\alpha\beta} F^a_{\alpha\beta} \gamma_\mu \lambda^a \tag{34}$$

and contains also contributions from the higher derivative terms and auxiliary fields. The divergence of the current (34) may be calculated unambiguously giving the known result (Abbot, Grisaru & Schnitzer 1977) for the matrix element $S_\mu \to \Psi + A$

$$\langle p, c; k, b \mid \partial_\mu S_\mu \mid 0\rangle = - \frac{3g^2}{8J_1^2} \varepsilon^*_\nu(k)\bar{u}(p)\{\gamma_\nu(pk) - p_\nu(\gamma k)\} \tag{35}$$

It is worthwhile to emphasize that in local supersymmetry the gauge field would interact with the conserved Noether current and not with the "naive" current (34). That means that the anomaly (35) has nothing to do with the consistency and convergence properties of this theory.

ACKNOWLEDGEMENTS

I would like to thank Professor S.W. Hawking and the Nuffield Foundation for hospitality extended to me at the Supergravity Workshop in Cambridge.

REFERENCES

Abbot, L.F., Grisaru, M.T. & Schnitzer, H.J. (1977) Supercurrent anomaly in a supersymmetric gauge theory. Phys.Rev.D, 16, pp. 2995-3001.

Curtright, T. & Ghandour, G. (1977). Stability and supersymmetry: general formalism and explicit two-loop applications. Annals of Physics, 106, pp. 209-278.

Krivoshchekov, V.K. (1978). Invariant regularization for supersymmetric gauge theories. Theo. & Math.Phys.,36, pp. 745-752.

Krivoshchekov, V.K., Slavnov, A.A. & Faizullaev, B.A. (1976) Superdiagram technique for gauge theories. Theo. & Math.Phys. 26, pp. 95-104.

Nicolai, H. & Townsend, P.K. (1980, 1980a). Supersymmetric regularization and superconformal anomalies. CERN Preprints TH 2819 & TH 2840.

Slavnov, A.A. (1975). Renormalization of supersymmetric gauge theories (II). Non-Abelian case. Nucl.Phys.B, 97 pp. 155-1

Slavnov, A.A. (1977). Pauli-Villars regularization for non-Abelian gauge theories. Theo. & Math.Phys., 33, pp. 977-981.

Siegel, W. (1979). Supersymmetric dimensional regularization via dimensional reduction. Phys.Lett., 84B, pp. 193-196

Townsend, P.K. & van Nieuwenhuizen, P. (1979). Dimensional regularization and supersymmetry at the two loop level. Phys.Rev.D, 20, pp. 1832-1838.

# PROPER TIME APPROACH TO QUANTIZED SUPERGRAVITY

Claudio Teitelboim

The Institute for Advanced Study
Princeton, New Jersey, U.S.A. 08540

ABSTRACT

An approach to the quantization of supergravity based on considering the theory as the square root of general relativity is outlined. The central object is a formal representation for the Feynman propagator of the field as a whole in terms of an integral in which two kinds of proper time -- one commuting, another anticommuting -- appear in a symmetric manner.

A great similarity exists between gauge theories of the Yang-Mills type and generally covariant ones such as Einstein's gravitation theory. It stems from the fact that both types of systems have a geometrical content but are described, for convenience, by means of arbitrarily chosen "coordinates" rather than intrinsically. As a consequence the respective action integrals are invariant under transformations which contain arbitrary functions of space and time (gauge transformations).

There is, however, a fundamental difference between the gauge freedoms available in each case, a difference which is just an expression of the fact that general covariance is not an internal symmetry. It is this: for a Yang-Mills field the action contained in a spacetime region is invariant even when the gauge transformations are not restricted at all on the boundary of that region, whereas invariance under general coordinate transformations exists only when the mapping applies the boundary onto itself. The serious implications that this difference has on the quantization procedure were explored in Ref. 1. Here that discussion is extended to include systems possessing Bose-Fermi general covariance, i.e., local supersymmetry, and specifically to supergravity.

Supergravity may be regarded as the square root of general relativity, much in the same way as the Dirac electron is the square root of a Klein-Gordon particle.[2] In fact, it turns out that most of the features to which this note is devoted already appear in the simpler context of the Dirac particle. Therefore, in order to bring the essential point to light, I will discuss in some detail that simple case and will afterwards briefly indicate the nature of the corresponding generalization to supergravity.

The classical action for a Dirac particle has the form[3]

$$S = \int_{\tau_1}^{\tau_2} d\tau\{\dot{x}p + \frac{i}{2}\,\dot{\theta}\theta - iM\mathcal{S} - N\mathcal{H}\} + \frac{i}{2}\,\theta(1)\theta(2) \qquad (1)$$

where, for a free particle, $\mathcal{S} = \theta_\mu p^\mu + m\theta_5$ and $\mathcal{H} = p^2 + m^2$. Here $x_\mu$, $p^\mu$ and $N$ are commuting variables, whereas the classical analogs $\theta = (\theta_\mu, \theta_5)$ of the Dirac matrices as well as $M$ are anticommuting. The classical equations of motion are obtained by demanding that (1) should not change under small variations of $x$, $p$, $\theta$, $N$ and $M$ under appropriate endpoint conditions.[3] In particular, extremization of the action with respect to the Lagrange multipliers $N$, $M$ yields the (first class) constraint equations

$S \approx 0$, $H \approx 0$. Hence, the connection with quantum mechanics is usually made by observing that if the constraints are imposed as restrictions on physically allowed states, then $S|\psi\rangle = 0$ yields the Dirac equation, whereas $H|\psi\rangle = 0$ reproduces its square ("integrability condition"), the Klein-Gordon equation.

However, the basic object in the relativistic quantum mechanics of a particle in an external field (this will be the analogy of interest for supergravity) is not so much the wave function as it is the Feynman propagator $S_F(2,1)$, which represents the probability amplitude for causal propagation from one space time point to another (and from one spin state to another). This amplitude is more basic than the idea of a wave function because it can simultaneously describe particles going forwards and backwards in time[4] (i.e., particle creation and annihilation), whereas a wave function corresponds to a single time orientation. Indeed, as long as a potential capable of turning particles back in time (producing pairs) is switched on, the system, regarded as a one particle theory without second quantization, possesses no Hamiltonian and no wave function. However, one can still evaluate transition amplitudes by means of the Feynman propagator which replaces for this purpose the ordinary evolution operator.

In the standard treatments $S_F$ is obtained as a causal Green function of the Dirac equation or as the vacuum expectation value of a product of quantized Dirac fields. Those procedures are however not directly extendable to systems of higher dimensionality such as supergravity. I would therefore like to present a method for obtaining $S_F$ directly from the action (1) by means of a path integral.

Summing over histories, the exponential of i times the action (1) is not straightforward because that action is gauge invariant under reparametrizations and supersymmetry transformations. More precisely, if F denotes an arbitrary function of x, p and $\theta$ and one sets in terms of Dirac brackets $\delta F = [F, \varepsilon H + i\eta S]$ as well as $\delta N = \dot{\varepsilon} + i\eta M$ and $\delta M = \dot{\eta}$, then the change in (1) is a boundary term[3] which vanishes <u>if and only if the arbitrary functions $\varepsilon(\tau)$ (commuting) and $\eta(\tau)$ (anticommuting) are set equal to zero at the endpoints $\tau_1, \tau_2$.</u>

Now, as previously stated, the need for demanding that the gauge functions $\varepsilon$, $\eta$ should vanish at the endpoints in order for the action to be invariant is the distinguishing mark of gauge transformations which are not internal. Technically this fact implies that <u>it is not permissible to fix the gauge by imposing a canonical gauge condition</u> (such as $x^0 = \tau$, $\theta_5 = 0$) and <u>one must resort to restrictions on the first time derivatives of the Lagrange multipliers.</u> The simplest permissible conditions turn out to be, in this case,

$$\dot{N} = 0 \; ; \quad \dot{M} = 0 \quad , \tag{2}$$

a pair of equations which will be referred to as defining the <u>proper time gauge</u>.

The unavailability even in principle of a canonical gauge is no obstacle to determining the integration measure for the path integral and one can do so[5] by working directly in the non-canonical gauge (2). The following result is then obtained:

$$S_F(2,1) = \int_0^\infty dT \int d\textcircled{H} \langle 2|\exp\{-iT(H - i0^+) + \textcircled{H} S\}|1\rangle \quad , \qquad (3)$$

Equation (3) is the analog for a Dirac particle of the standard proper time representation of the Klein-Gordon propagator. It yields $S_F$ as an integral over two kinds of proper times, one commuting (T) and another anticommuting ($\textcircled{H}$). The symmetry between T and $\textcircled{H}$ apparent in (3) is quite satisfactory, given the symmetric role played by the commuting and anticommuting variables in the supersymmetric action integral (1). If the integral over $\textcircled{H}$ is performed (it is elementary to do so), one recovers the usual expression giving $S_F$ in terms of the Dirac operator acting on the Klein-Gordon propagator.[6]

The main steps in arriving at (3) are the following: first, one functionally integrates over x, p and the $\theta$'s for fixed N, M restricted by (2). Since $NH + iMS$ plays the role of a Hamiltonian, this integration yields the matrix element between initial and final states of the operator $\exp[-iTH + \textcircled{H} S]$ with the identifications $T = N(\tau_2 - \tau_1)$, $\textcircled{H} = M(\tau_2 - \tau_1)$. Second, one integrates over the Lagrange multipliers N, M (on account of (2) this step involves a single integral in each case instead of one for each value of $\tau$). The corresponding measure turns out to be just $dNdM \equiv dTd\textcircled{H}$ which has the effect of making the arbitrary parameters $\tau_1$, $\tau_2$ drop out from the final result, as should be the case. Lastly, one restricts the integral over T to

$T \geq 0$, thus only including histories which are future oriented in proper time, and hence automatically selecting the correct "outgoing" or "causal" boundary conditions for the propagator. (The negative imaginary part $-i0^+$ added to $H$ in (3) makes the integral convergent for large T and it is not subject to any sign option.) It should be noted that a statement for the Ⓗ integral analogous to $T \geq 0$ is neither necessary (the orientation is fully determined by $T \geq 0$) nor permissible (there is no concept of Berezin integration over a sub-interval).

In the case of supergravity there are two different sets of constraint-generators in the theory. The first set corresponds to those gauge symmetries which must be restricted to be the identity on the boundary in order for the action to be invariant. It contains therefore the analogs of the functions $H$ and $S$ encountered for the spinning particle, namely, the generator of normal deformations $H(x)$ which is quadratic in the gravitational momenta, and its square root,[2] the generator of supersymmetry transformations $S_A(x)$. The second set is formed by the generators $H_i$, $J_{ab}$ of spatial reparametrizations and tetrad rotations, respectively. These transformations are not restricted on the boundary and enter into the theory as if they were internal symmetries of the Yang-Mills type.

Just as for the particle case the Hamiltonian is a linear combination of the constraints and takes the form

$$H = \int d^3x \left( i\, M^A S_A + N\, H + N^i H_i + N^{ab} J_{ab} \right) \qquad (4)$$

where $M^A$, N, $N^i$, $N^{ab}$ are Lagrange multipliers. Since the action is invariant under the various gauge symmetries associated with the respective constraints, it is necessary to fix a gauge and determine the corresponding measure in order to write an expression for the transition amplitude from one field configuration to another.

Due to the fact that normal deformations and supersymmetry transformations have to be the identity on the boundary, it is not permissible to fix the gauge freedom associated to them by means of a canonical gauge condition and one has to resort to a restriction on the first time derivatives of the Lagrange multipliers $N(x,t)$, $M^A(x,t)$. The simplest such condition turns out to be the exact analog of (2), namely, $(\partial N/\partial t) = (\partial M^A/\partial t) = 0$. On the other hand, one would be entitled to use a canonical gauge in order to eliminate the freedom connected with $\mathcal{H}_i$ and $\mathcal{J}_{ab}$. However, it is simpler (and permissible) to impose the non-canonical conditions $N^i = N^{ab} = 0$ plus a canonical condition at the initial time $\tau_1$, the precise form of which need not be of concern here. The gauge $\dot{M}^A = \dot{N} = N^i = N^{ab} = 0$ will be called the <u>proper time gauge in supergravity</u>.

Working in the proper time gauge one is led to an expression analogous to (3) for the Feynman propagation amplitude from one field configuration to another in supergravity. It takes the form

$$S_F(2,1) = \int_{T\geq 0} \prod_{x,A} dT(x)\, d\textcircled{H}^A(x) \times \tag{5}$$

$$\times \langle 2|\exp\{\int d^3x[-iT(x)(\mathcal{H}(x) - i0^+) + \textcircled{H}^A(x)\mathcal{S}_A(x)] - iH^{ghost}\}|1\rangle$$

It is straightforward to determine explicitly the ghost Hamiltonian appearing in (5) from the general relations of Ref. 5 and the algebra of Ref. 2. However, the result that one obtains directly in this way is somewhat complicated and for that reason will not be included here. It is felt that the expression in question may be considerably simplified by appropriate changes of variables analogous to those used to arrive at the compact result reported in Ref. 1 for pure gravity. Indeed one would expect, by analogy with the results of Ref. 1, that up to a minor change in the dTd(H) integration measure, the function $H^{ghost}$ should take the form $i\bar{M}S^{ghost} + NH^{ghost}$, where $S^{ghost}$ and $H^{ghost}$ describe the <u>locally supersymmetric interaction</u> of a set of ghost fields with supergravity. By this I mean that the effective generators $S^{eff} = S + S^{ghost}$, $H^{eff} = H + H^{ghost}$ and corresponding modifications of $H_i$ and $J_{ab}$ should obey a closed graded algebra, and hence could be obtained by writing a locally supersymmetric action principle in Hamiltonian form. This statement remains an unproven conjecture at this point.

Expression (5) takes the place for quantized supergravity of the matrix element of the evolution operator $U(t_2,t_1) = \exp[-iH(t_2 - t_1)]$ in ordinary quantum field theory. Since a canonical gauge is not allowed, there is in fact no Hamiltonian in quantized supergravity, much in the same way as there was no Hamiltonian for the Dirac particle in an external field.[7] In order to evaluate a transition amplitude, one proceeds as if the exponent in (5) were the Hamiltonian but integrates afterwards over all possible separations in both the commuting and anti-commuting proper times at all points of space. That last step has no analog in ordinary quantum field theory.

The ideas described in this **article** may be directly applied to the quantization of the spinning string. That application, as well as a detailed discussion of the procedure for supergravity will be reported elsewhere.

I am grateful to S. Adler, R. Dashen and F. Dyson for much encouragement. This work was supported in part by NSF Grant No. PHY-79-19884. The author holds an Alfred P. Sloan Research Fellowship.

---

REFERENCES

1. Teitelboim, C. (March 1980). Institute for Advanced Study preprint, submitted to Phys. Lett. B.

2. Teitelboim, C. (1977). Phys. Rev. Lett. 38, 1106; Tabensky, R. & Teitelboim, C. (1977). Phys. Lett. 69B, 453. See also Teitelboim, C. (1978). In Current Trends in the Theory of Fields, ed. J.E. Lanutti & P.K. Williams, A.I.P. Conference Proceedings No. 48.

3. There is a vast literature on supersymmetric action principles for Dirac particles. Here we follow the presentation of Galvao, C. & Teitelboim, C. J. Math. Phys., in press.

4. Stückelberg, E.C.G. (1942). Helv. Phys. Acta 15, 23; Feynman, R.P. (1950). Phys. Rev. 80, 440.

5. Batalin, I.A. & Vilkovisky, G.A. (1977). Phys. Lett. 69B, 309; Fradkin, E.S. & Fradkina, T.E. (1978). Phys. Lett. 72B, 343.

6. An anticommuting analog of the proper time for the Dirac particle has also been introduced, independently, by Ravndal, F., Oslo preprint. The starting point in that interesting work is an action principle which is not gauge invariant, a feature which unfortunately prevents direct generalization of the procedure to more complicated systems such as supergravity. See

Footnote 5 of Ref. 1 and related comments therein.

7. See Ref. 1 for a discussion of the analog of "histories proceeding backwards in time" for pure gravity. That analysis, as well as the possibility of closed field loops, the idea of a perturbation theory in the signature $\sigma$ and the implementation of the $-i0^+$ in the exponent in (5) by means of the addition of a negative imaginary part to the cosmological constant, remain unchanged in supergravity.

# III EXTENDED SUPERGRAVITY

# COMPLEX SUPERSPACES AND PREPOTENTIALS FOR N = 2 SUPERGRAVITY

E. Sokatchev

CERN -- Geneva

## 1 INTRODUCTION

Since the appearance of the component (ordinary space) formulation of $N = 2$ supergravity (Fradkin & Vasiliev, 1979 ; de Wit & van Holten, 1979) various attempts have been made to reproduce it in superspace terms. Several versions were first proposed in Breitenlohner & Sohnius, 1980 ; Wess, 1979. Then a more systematic way of derivation was developed in Stelle & West, 1980 ; Gates, 1980a. Finally, the method of gauge completion was applied to obtain superspace results directly from the component ones (Castellani, van Nieuwenhuizen & Gates, 1980). All these approaches have a common feature : they use complicated superfields (vielbeins) as potentials of the theory and then a number of constraints on the torsion tensor is postulated to get rid of the superfluous degrees of freedom. Also, a common problem is how to write down an action for $N = 2$ supergravity.

On the other hand, a more economical approach proved to be very constructive in $N = 1$ supergravity (Ogievetsky & Sokatchev, 1978, 1979a,b ; Siegel & Gates, 1979). There the theory is formulated in terms of unrestricted prepotentials : a vector superfield $H^m$ in the minimal case (Ogievetsky & Sokatchev, 1978, 1979a,b), and a vector ($H^m$) and a spinor ($H^\mu, \overline{H}^{\dot\mu}$) one in the non-minimal case (Siegel & Gates, 1979). Then all quantities like vielbeins, torsion components, etc., are expressed in terms of these prepotentials. The constraints on the torsion follow automatically from the simple basic ideas. The action principle becomes clear and straightforward (Ogievetsky & Sokatchev, 1980c).

The present work is an attempt to generalize the prepotential approach to the $N = 2$ case. We introduce a vector superfield $H^m$ and a spinor one $H^{\mu i}$, $\overline{H}_{\dot\mu i}$ as prepotentials, in close analogy with the non-minimal $N = 1$ case (the similarity between $N = 2$ and non-minimal $N = 1$ supergravities has been pointed out in Gates, 1980a). This allows us to evaluate the torsion components explicitly and thus to derive all the constraints (in the form extracted from the component results and given in Gates, 1980b) except for two of them :

$$T_{\alpha(i,\dot{\beta}j)}{}^{\gamma\dot{\gamma}} = 0$$

$$T_{\alpha\dot{\alpha},}{}^{\alpha(i,\dot{\alpha}j)} + T_{\alpha\dot{\alpha},}{}^{\dot{\alpha}(i,\alpha j)} = 0 \tag{1}$$

(in the second one some non-linear terms are omitted). The latter have to be postulated, which means that our prepotentials are not fully independent. Finally, we consider the invariant volume of the chiral superspaces as a possible action formula for $N = 2$ supergravity.

The content of the paper is as follows. In Section 2, the minimal and non-minimal prepotential formulations of $N = 1$ supergravity are briefly reviewed from a common point of view : the prepotentials are introduced by embedding the real physical superspace in a chiral complex one (minimal case) or a general complex one with invariant chiral subspaces (non-minimal case).

The latter case is generalized in Section 3 to form the basis for the formulation of the $N = 2$ theory. The simplest version of the theory is chosen (the one with a Lorentz structure group), although others are also possible.

In Section 4 the action of the structure group is extracted from the transformation law of the spinor derivatives (in this approach the structure group is induced by the world superspace one). Vielbeins and connections are defined and expressed in terms of the prepotentials.

Section 5 is devoted to the evaluation of the torsion components. To simplify the calculations, the normal gauge technique (Ogievetsky & Sokatchev, 1980a,b) is applied. The results agree with the ones of Gates, 1980b, assuming that the constraints (1) are imposed by hand.

In Section 6 the possibility of using the invariant volume of the chiral superspaces as an action for $N = 2$ supergravity is considered. It has the right dimension to produce equations of motion of the correct order. Using the first of the constraints (1) the chirality of the density factor is checked (a necessary consistency condition).

We use the following notation. The standard places of the spinor and isospinor indices are, e.g.,

$$\theta^{\mu i},\ \bar{\theta}_{\dot{\mu}i},$$

although they can be raised and lowered with the help of $\varepsilon$ symbols the latter are usually absorbed in the notation, e.g.,

$$\theta^{\mu}_{i} = \varepsilon_{ij}\theta^{\mu j},\ \theta^{i}_{\mu} = \varepsilon_{\mu\nu}\theta^{\nu i},$$

$$\theta^{\mu i}\theta_{\mu i} = \varepsilon_{\mu\nu}\varepsilon_{ij}\theta^{\mu i}\theta^{\nu j},\quad \text{etc.}$$

The notation of world vector and isospinor indices can be somewhat confusing (letters $m,n,k,\ell,\ldots$ are used for both) but the latter are usually accompanied by spinor ones. Sometimes a pair of spinor and isospinor indices is denoted as follows

$$A^{\alpha i} = A^{\underline{\alpha}},\ \bar{A}_{\dot{\alpha}i} = \bar{A}_{\underline{\dot{\alpha}}}\ .$$

The symbols like $(ij)$ mean symmetrization :

$$T_{(ij)} = \tfrac{1}{2}(T_{ij}+T_{ji}).$$

## 2 MINIMAL AND NON-MINIMAL $N = 1$ SUPERGRAVITIES

In order to understand better the construction proposed for the $N = 2$ case, we shall first briefly remind ourselves how the prepotentials are introduced in the $N = 1$ theory. There exist two different (although equivalent) formulations : those of Ogievetsky & Sokatchev, 1978, 1979a,b (minimal case) and of Siegel & Gates, 1979 (minimal and non-minimal cases). Here we shall describe both the minimal and non-minimal cases in the framework of Ogievetsky & Sokatchev, 1978, 1979a which is easy to generalize in the $N = 2$ case.

The main idea of the approach is to embed the real physical superspace as a hypersurface in a larger complex superspace, where a general reparametrization group acts. The functions defining the shape of the hypersurface play the rôle of prepotentials in the theory.

### 2.1 Minimal case

Here the simplest possible complex superspace is used : a chiral one $\{(x^m_L,\theta^\mu_L)\}$ together with its conjugate $\{(x^m_R,\bar{\theta}^{\dot{\mu}}_R)\}$. The supergravity gauge group is the general co-ordinate transformation group in these superspaces

$$\delta x^m_L = \lambda^m(x_L,\theta_L),\quad \delta x^m_R = \bar{\lambda}^m(x_R,\bar{\theta}_R)$$
$$\delta\theta^\mu_L = \lambda^\mu(x_L,\theta_L),\quad \delta\bar{\theta}^{\dot{\mu}}_R = \bar{\lambda}^{\dot{\mu}}(x_R,\bar{\theta}_R)\ . \tag{2}$$

The chiral parameters $\lambda^m$, $\lambda^\mu$ (and their conjugates $\bar{\lambda}^m$, $\bar{\lambda}^{\dot{\mu}}$) are arbitrary in the conformal case or restricted by the condition

$$\text{Ber}\left(\frac{\partial(x'_L,\theta'_L)}{\partial(x_L,\theta_L)}\right) = \text{Ber}\left(\frac{\partial(x'_R\,,\,\bar{\theta}'_R)}{\partial(x_R\,,\,\bar{\theta}_R)}\right) = 1 \qquad (3)$$

in the Poincaré case (Ber means the superdeterminant of the trans formations (2)).

The real superspace $\{(x^m,\theta^\mu,\bar{\theta}^{\dot{\mu}})\}$ is introduced as a hyper-surface in the complex one :

$$x^m = \tfrac{1}{2}(x^m_L+x^m_R),\quad \theta^\mu = \theta^\mu_L,\quad \bar{\theta}^{\dot{\mu}} = \bar{\theta}^{\dot{\mu}}_R,$$
$$H^m(x,\theta,\bar{\theta}) = \frac{1}{2i}(x^m_L-x^m_R). \qquad (4)$$

In other words, the imaginary part of the complex four-vector co-ordinate becomes a superfield, i.e., the prepotential for $N = 1$ minimal supergravity.

The components of the parameters $\lambda$ of the supergroup (2) acting on $H^m$ have two functions. Some of them gauge away some components of $H^m$ so that only the relevant fields (dynamical and auxiliary) remain (WZ gauge). The rest become parameters of the physical gauge group like local supersymmetry, local Lorentz group, etc.

On the basis of the supergroup (2) and superfield $H^m$ the familiar differential geometry formalism can be developed (Ogievetsky & Sokatchev, 1979b). In particular, all the torsion constraints postulated in other approaches are obtained automatically here (Ogievetsky & Sokatchev, 1980b).

## 2.2 Non-minimal case

Now we shall use a general complex superspace instead of the chiral one (as J. Gates has informed us, similar ideas had occurred to him too) :

$$\{(x^m_L,\theta^\mu_L,\bar{\theta}^{\dot{\mu}}_L)\},\quad \{(x^m_R,\theta^\mu_R,\bar{\theta}^{\dot{\mu}}_R)\}$$
$$x^m_R = (x^m_L)^*,\quad \bar{\theta}^{\dot{\mu}}_R = \overline{(\theta^\mu_L)},\quad \theta^\mu_R = \overline{(\bar{\theta}^{\dot{\mu}}_L)}. \qquad (5)$$

However, since the concept of chirality proved to be fundamental in the minimal case we shall try to preserve it here. To this end we shall introduce a "projective structure", i.e., a gauge group leaving invariant the chiral subspaces $\{(x^m_L,\theta^\mu_L)\}$, $\{(x^m_R,\bar{\theta}^{\dot{\mu}}_R)\}$ of the general superspace (5)

$$\delta x_L^m = \lambda^m(x_L,\theta_L), \qquad \delta x_R^m = \bar{\lambda}^m(x_R,\bar{\theta}_R),$$

$$\delta\theta_L^\mu = \lambda^\mu(x_L,\theta_L), \qquad \delta\bar{\theta}_R^{\dot{\mu}} = \bar{\lambda}^{\dot{\mu}}(x_R,\bar{\theta}_R), \tag{6}$$

$$\delta\bar{\theta}_L^{\dot{\mu}} = \bar{\rho}^{\dot{\mu}}(x_L,\theta_L,\bar{\theta}_L) \;; \quad \delta\theta_R^\mu = \rho^\mu(x_R,\theta_R,\bar{\theta}_R).$$

The embedding of the real superspace into this structure takes place in a similar way, as in (4)

$$x^m = \tfrac{1}{2}(x_L^m+x_R^m), \; \theta^\mu = \theta_L^\mu, \; \bar{\theta}^{\dot{\mu}} = \bar{\theta}_R^{\dot{\mu}},$$

$$H^m(x,\theta,\bar{\theta}) = \frac{1}{2i}(x_L^m-x_R^m), \tag{7}$$

$$H^\mu(x,\theta,\bar{\theta}) = -\theta_L^\mu + \theta_R^\mu, \quad \overline{H}^{\dot{\mu}}(x,\theta,\bar{\theta}) = \bar{\theta}_L^{\dot{\mu}} - \bar{\theta}_R^{\dot{\mu}}.$$

Comparing (6) and (7), one can easily see that the newly introduced spinor superfield $H^\mu$, $\overline{H}^{\dot{\mu}}$ can be completely gauged away unless the parameters $\rho^\mu$, $\bar{\rho}^{\dot{\mu}}$ are restricted. There is a whole family of such group-covariant restrictions :

$$\left[\mathrm{Ber}\left(\frac{(x_L',\theta_L',\bar{\theta}_L')}{(x_L,\theta_L,\bar{\theta}_L)}\right)\right]^p = \left[\mathrm{Ber}\left(\frac{(x_L',\theta_L')}{(x_L,\theta_L)}\right)\right]^q \tag{8}$$

and the same for the right-handed superspace. Here the powers $p$ and $q$ are numerical parameters (the ratio $p/q$ is related to the parameter $n$ of Siegel & Gates, 1979). Except for the cases $p = q = 0$ (minimal conformal supergravity) and $p = 0$, $q = 1$ (minimal Poincaré supergravity) the gauge freedom is already not enough to gauge away $H^\mu$, $\overline{H}^{\dot{\mu}}$. The description of the theory becomes non-minimal, with prepotentials $H^m$, $H^\mu$, $\overline{H}^{\dot{\mu}}$ instead of just $H^m$ in the minimal case.

## 3 THE N = 2 SUPERGROUP

Trying to generalize the ideas of Section 2 for the $N = 2$ case the simplest, minimal formulation seems to be the one to begin with. However, this attempt fails for reasons which are worth explaining.

As mentioned in Section 2, part of the components of the superparameters $\lambda^m$, $\lambda^\mu$ are used to gauge away certain components of the prepotentials (WZ gauge). The remaining ones have certain geometrical meaning in ordinary space. For instance, in the decomposition of $\lambda^\mu$ in the $N = 1$ case

$$\lambda^{\mu}(x_L,\theta_L) = \varepsilon^{\mu}(x_L) + \theta_L^{\nu}\left[(a(x_L)+ib(x_L))\delta_{\nu}^{\mu}+\Omega_{(\nu}{}^{\mu)}(x_L)\right] + \ldots \quad (9)$$

the parameter $\varepsilon^{\mu}$ clearly generates local supersymmetry transformations. The terms linear in $\theta$ are very important because they determine the linear transformations of the spinor co-ordinates and, respectively, of the component fields in the prepotential superfields. This is just the structure group in $x$ space. In (9) these parameters have a clear interpretation : $a, b, \Omega^{(\mu\nu)}$ generate Weyl, local $\gamma_5$ and local Lorentz transformations.

However, things become more complicated in the $N = 2$ case :

$$\lambda^{\mu i}(x_L,\theta_L) = \varepsilon^{\mu i}(x_L) + \theta_{\nu j}\left[(a(x_L)+ib(x_L))\varepsilon^{\nu\mu}\varepsilon^{ji} + \right.$$
$$\left. + \Omega^{(\nu\mu)}(x_L)\varepsilon^{ji} + G^{(ji)}(x_L)\varepsilon^{\nu\mu} + A^{(\nu\mu)(ji)}(x_L)\right] + \ldots \quad (10)$$

One of the new parameters, $G^{(ji)}(x_L)$, generates <u>complex</u> SU(2) transformations but the other one, $A^{(\nu\mu)(ji)}(x_L)$, is meaningless and must not be allowed to enter the structure group. So, some covariant restriction on the group must be found which eliminates the term $A$ in (10). Unfortunately, this cannot be done without affecting the $x$ sector of the transformations thus restricting the general reparametrization group in $x$ space (the group multiplication law mixes up the parameters $\lambda^m$ and $\lambda^{\mu i}$).

This problem can be solved in the non-minimal framework. Before explaining this in detail let us first write down the extended version of (6) and (7)

$$\delta x_L^m = \lambda^m(x_L,\theta_L), \qquad \delta x_R^m = \bar{\lambda}^m(x_R,\bar{\theta}_R),$$

$$\delta\theta_L^{\mu i} = \lambda^{\mu i}(x_L,\theta_L), \qquad \delta\bar{\theta}^R_{\dot{\mu}i} = \bar{\lambda}_{\dot{\mu}i}(x_R,\bar{\theta}_R), \quad (11)$$

$$\delta\bar{\theta}^L_{\dot{\mu}i} = \bar{\rho}_{\dot{\mu}i}(x_L,\theta_L,\bar{\theta}_L) \; ; \qquad \delta\theta_R^{\mu i} = \rho^{\mu i}(x_R,\theta_R,\bar{\theta}_R).$$

The real superspace and prepotentials are introduced as follows :

$$x^m = \tfrac{1}{2}(x_L^m+x_R^m), \quad \theta^{\mu i} = \theta_L^{\mu i}, \quad \bar{\theta}_{\dot{\mu}i} = \bar{\theta}^R_{\dot{\mu}i}, \quad H^m = \tfrac{1}{2i}(x_L^m-x_R^m),$$
$$H^{\mu i} = -\theta_L^{\mu i} + \theta_R^{\mu i}, \quad \bar{H}_{\dot{\mu}i} = \bar{\theta}^L_{\dot{\mu}i} - \bar{\theta}^R_{\dot{\mu}i} . \quad (12)$$

From (11), (12) the infinitesimal transformation laws of the hypersurface are derived

$$\delta x^m = \tfrac{1}{2}\lambda^m(x+iH,\theta) + \tfrac{1}{2}\bar{\lambda}^m(x-iH,\bar{\theta}),$$
$$\delta\theta^{\mu i} = \lambda^{\mu i}(x+iH,\theta), \quad \delta\bar{\theta}_{\dot{\mu} i} = \bar{\lambda}_{\dot{\mu} i}(x-iH,\bar{\theta}) \;; \tag{13}$$

$$\delta H^m = -\tfrac{i}{2}\lambda^m(x+iH,\theta) + \tfrac{i}{2}\bar{\lambda}^m(x-iH,\bar{\theta}) - \Big[\tfrac{1}{2}(\lambda^n+\bar{\lambda}^n)\partial_n + \lambda^{\underline{\nu}}\partial_{\underline{\nu}} + \bar{\lambda}_{\dot{\underline{\nu}}}\bar{\partial}^{\dot{\underline{\nu}}}\Big] H^m, \tag{14a}$$

$$\delta H^{\mu i} = -\lambda^{\mu i}(x+iH,\theta) + \rho^{\mu i}(x-iH,\theta+H,\bar{\theta}) - \Big[\tfrac{1}{2}(\lambda^n+\bar{\lambda}^n)\partial_n + \lambda^{\underline{\nu}}\partial_{\underline{\nu}} + \bar{\lambda}_{\dot{\underline{\nu}}}\bar{\partial}^{\dot{\underline{\nu}}}\Big] H^{\mu i}, \tag{14b}$$

$$\delta \bar{H}_{\dot{\mu} i} = -\bar{\lambda}_{\dot{\mu} i}(x-iH,\bar{\theta}) + \bar{\rho}_{\dot{\mu} i}(x+iH,\theta,\bar{\theta}+\bar{H}) - \Big[\tfrac{1}{2}(\lambda^n+\bar{\lambda}^n)\partial_n + \lambda^{\underline{\nu}}\partial_{\underline{\nu}} + \bar{\lambda}_{\dot{\underline{\nu}}}\bar{\partial}^{\dot{\underline{\nu}}}\Big] \bar{H}_{\dot{\mu} i} \;. \tag{14c}$$

Now we have a number of new possibilities to restrict the group. The point is that in (11), e.g., the parameter $\rho^{\mu i}$ depends on $\theta_R$ but $\bar{\lambda}^m$ and $\bar{\lambda}_{\dot{\mu} i}$ do not. So, we can restrict this dependence of $\rho^{\mu i}$ in a covariant way without directly affecting $\bar{\lambda}^m$ and $\bar{\lambda}_{\dot{\mu} i}$. However, the question might be asked : how can this solve our problem with constraining $\lambda^{\mu i}$(10) ? The answer is given by the WZ gauge fixing procedure. Suppose that the $\theta_R$ dependence of $\rho^{\mu i}$ has been restricted, i.e., some of the pure $\theta_R$ as well as mixed $\theta_R\bar{\theta}_R$ terms have been fixed. Then, looking at (14b) one sees that the corresponding $\theta$ terms of $\lambda^{\mu i}$ can take over the functions of the fixed pure $\theta_R$ (i.e., $\theta+H$) terms of $\rho^{\mu i}$ in gauging away components of $H^{\mu i}$. Thus, we effectively get rid of some undesired components of $\lambda^{\mu i}$ without worrying about $\lambda^m$ and the group law.

In particular, this enables us to eliminate the term $A^{(\nu\mu)(ji)}$ in (10). To this end we require

$$\partial_R{}^{(\mu}_{(i}\,\rho^{\nu)}_{j)} = 0, \quad \partial_{Ri}^{\;\mu} \equiv \frac{\partial}{\partial\theta_{R\mu}^{\;i}} \;, \tag{15}$$

i.e., eliminate the corresponding term in $\rho^{\mu i}$. One can easily verify that the constraint (15) is compatible with the group multiplication law $(M = m, \underline{\mu}$ or $m, \dot{\underline{\mu}})$

$$\lambda_3^M = \lambda_1^N \partial_N^L \lambda_2^M - \lambda_2^N \partial_N^L \lambda_1^M,$$
$$\rho_3^{\underline{\mu}} = \bar{\lambda}_1^N \partial_N^R \rho_2^{\underline{\mu}} - \bar{\lambda}_2^N \partial_N^R \rho_1^{\underline{\mu}} + \rho_1^{\underline{\nu}} \partial_{\underline{\nu}}^R \rho_2^{\underline{\mu}} - \rho_2^{\underline{\nu}} \partial_{\underline{\nu}}^R \rho_1^{\underline{\mu}}, \quad \text{etc.} \qquad (16)$$

Choosing further restrictions on $\rho^{\mu i}$ (and, effectively, on $\lambda^{\mu i}$) one thus specifies the structure group both in x space and superspace (in the approach discussed, the latter is induced by the former ; see Ogievtsky & Sokatchev, 1979b and also Section 4 for more details). For instance, in the WZ gauge, the requirement

$$\partial^R_{\mu i} \ \rho^{\mu i} = 0 \qquad (17)$$

fixes the Weyl and $\gamma_5$ parameters a and b in (10). Actually, (17) can be replaced by a more general one

$$p\partial^R_{\mu i} \ \rho^{\mu i} = (p-q)(\partial_m \bar{\lambda}^m - \bar{\partial}^{\dot{\mu} i}_R \ \bar{\lambda}_{\dot{\mu} i}), \qquad (18)$$

i.e., a condition of the type (8). Further, the complex SU(2) parameter $G^{(i\ j)}$ in (10) can be made real

$$\partial^{R(i}_{\mu} \ \rho^{\mu j)} = \bar{\partial}_L^{\ \dot{\mu}(i} \ \bar{\rho}^{j)}_{\dot{\mu}} \qquad (19)$$

or completely eliminated

$$\partial^{R(i}_{\mu} \ \rho^{\mu j)} = 0. \qquad (20)$$

Both (19) and (20) are compatible with the group law (16).

So, a variety of N = 2 supergravity theories can be formulated having different (Weyl, $\gamma_5$, SU(2), Lorentz) structure groups in the tangent x and superspaces. Note that we cannot introduce local central charge transformations because there are no suitable parameters. However, according to our final results this is not so essential for describing N = 2 supergravity.

Now we shall restrict ourselves to the simplest case with a Lorentz structure group, i.e. with the conditions (15), (17) (for simplicity we do not consider the whole family (18)), (20). The other possibilities are also worth examining.

Before closing this section we would like to stress an important point. In the N = 1 case and in the WZ gauge the prepotentials contain the relevant fields (dynamical and auxiliary) only. This is not so in the N = 2 case. For instance, $H^m$ contains components with spins 3 and $\frac{5}{2}$. They cannot be gauged away because there are no such gauge parameters. This indicates that some covariant constraint on the prepotentials are needed to obtain an economical set of auxiliary fields. The approach discussed does not explain how to choose them. However, the results of Section 5 show that the number of constraints postulated is considerably smaller than in other superspace approaches.

## 4 DIFFERENTIAL GEOMETRY FORMALISM

The aim of this section is to introduce such concepts and objects as structure group in superspace, vielbeins, and connections, which form the basis of the familiar differential geometry formalism. The construction will closely follow the content of Ogievetsky & Sokatchev, 1979b, therefore a lot of details will be omitted.

First, to show how the world superspace group (11) induces the tangent superspace one, let us introduce a special (left-handed) basis in the real superspace

$$x_L^m = x^m + i H^m, \quad \theta_L^{\mu i} = \theta^{\mu i}, \quad \bar{\theta}^L_{\dot{\mu} i} = \bar{\theta}_{\dot{\mu} i} + \bar{H}_{\dot{\mu} i} . \tag{21}$$

Using (13), (14) it is not hard to show that the right-handed spinor derivative of a scalar superfield $\varphi_L(x_L, \theta_L, \bar{\theta}_L)$ transforms as follows

$$\delta(\bar{\partial}_L^{\dot{\mu}} \varphi_L) = -(\bar{\partial}_L^{\dot{\mu}} \bar{\rho}^L_{\underline{\dot{\nu}}}) \quad \bar{\partial}_L^{\dot{\nu}} \varphi_L - (\lambda_L^m \partial_m^L + \lambda_L^{\nu} \partial_{\underline{\nu}}^L + \bar{\rho}^L_{\underline{\dot{\nu}}} \bar{\partial}_L^{\dot{\nu}}) \, (\bar{\partial}_L^{\dot{\mu}} \varphi_L). \tag{22}$$

Here the index $L$ reminds us that everything is written in the left-handed basis. The second term in (22) is just a translation term. The first one is a linear transformation with a parameter which, with the help of (15), (17), (20) can be rewritten as

$$\bar{\partial}_L^{\dot{\mu} i} \, \bar{\rho}^L_{\dot{\nu} j} = \frac{i}{2} \, \Omega^{ab} \, (\tilde{\sigma}_{ab})^{\dot{\mu}}_{\dot{\nu}} \, \delta^i_j \, . \tag{23}$$

So, this is just a local (in superspace) Lorentz transformation and it will be generalized to define our structure group. Before doing this let us go from the left-handed basis (21) back to the real one $(x^m, \theta^{\mu i}, \bar{\theta}_{\dot{\mu} i})$ :

$$\bar{\partial}_L^{\dot{\mu} i} \, \varphi_L = (\bar{F}^{-1})^{\dot{\mu} i}{}_{\dot{\nu} k} \, \bar{\Delta}^{\dot{\nu} k} \varphi \equiv \bar{\nabla}^{\dot{\mu} i} \, \varphi \tag{24}$$

where

$$\bar{\Delta}^{\dot{\mu} i} = \bar{\partial}^{\dot{\mu} i} - i \bar{\partial}^{\dot{\mu} i} H^n (1 + i H)^{-1 m}_{\ n} \, \partial_m = \bar{\partial}^{\dot{\mu} i} - i \bar{\Delta}^{\dot{\mu} i} H^m \, \partial_m , \tag{25}$$

$$\bar{F}^{\dot{\mu} i}{}_{\dot{\nu} j} = \delta^{\dot{\mu}}_{\dot{\nu}} \, \delta^i_j + \bar{\Delta}^{\dot{\mu} i} \bar{H}_{\dot{\nu} j} , \quad H^m_n = \partial_n H^m . \tag{26}$$

Analogously one finds

$$\partial^R_{\mu i} \, \varphi_R = (F^{-1})_{\mu i}{}^{\nu k} \, \Delta^{\nu k} \equiv \nabla_{\mu i} \varphi , \tag{24'}$$

$$\Delta_{\mu i} = \partial_{\mu i} + i \partial_{\mu i} H^n (1 - i H)^{-1 m}_{\ n} \, \partial_m = \partial_{\mu i} + i \Delta_{\mu i} H^m \, \partial_m , \tag{25'}$$

$$F_{\mu i}{}^{\nu j} = \delta^{\nu}_{\mu}\delta^{j}_{i} + \Delta_{\mu i}\, H^{\nu j} \, . \qquad (26')$$

So, the quantities $\nabla_{\mu i}\varphi$, $\bar{\nabla}^{\dot{\mu} i}\varphi$ transform like Lorentz spinors with the parameter

$$\Omega^{ab} = \frac{i}{8}\left[\nabla^{\mu k}(\sigma^{ab})^{\nu}_{\mu}\, \rho_{\nu k} + \bar{\nabla}_{\dot{\mu} k}(\tilde{\sigma}^{ab})^{\dot{\mu}}_{\dot{\nu}}\, \bar{\rho}^{\dot{\nu} k}\right] . \qquad (27)$$

Generalizing this fact we define the action of our structure group (induced by the world superspace one) on superfields with indices as follows

$$\delta\Phi^{\alpha i} = -\frac{i}{2}\,\Omega^{ab}(\sigma_{ab})^{\alpha}_{\beta}\,\Phi^{\beta i} + \text{t.t.} \, ,$$

$$\delta\Phi_{\dot{\alpha} i} = -\frac{i}{2}\,\Omega^{ab}(\tilde{\sigma}_{ab})^{\dot{\beta}}_{\dot{\alpha}}\,\Phi_{\dot{\beta} i} + \text{t.t.} \, , \qquad (28)$$

$$\delta\Phi_{a} = -2\,\Omega_{a}{}^{b}\,\Phi_{b} + \text{t.t.}, \quad \text{etc.}$$

Here t.t. means translation terms and $\Omega_{ab}$ is defined in (27).

Simultaneously we have introduced spinor covariant derivatives of scalar superfields, namely $\nabla_{\mu i}$, $\bar{\nabla}^{\dot{\mu} i}$ (24'), (24). For superfields with indices, one has to add connections, e.g.,

$$D_{\alpha i}\,\Phi^{\beta j} = \nabla_{\alpha i}\,\Phi^{\beta j} + \omega_{\alpha i}{}^{\beta}{}_{\gamma}\,\Phi^{\gamma j} \, , \qquad (29a)$$

$$D_{\alpha i}\,\Phi_{\dot{\beta} j} = \nabla_{\alpha i}\,\Phi_{\dot{\beta} j} + \omega_{\alpha i \dot{\beta}}{}^{\dot{\gamma}}\,\Phi_{\dot{\gamma} j} \, . \qquad (29b)$$

However, the close examination of the transformation properties of $\nabla_{\alpha i}\,\Phi^{\beta j}$ (using the right-handed basis) shows that there is no need of $\omega_{\alpha i \beta\gamma}$, so

$$\omega_{\alpha i \beta\gamma} = 0. \qquad (30)$$

The following expression

$$\omega_{\alpha i \dot{\beta}\dot{\gamma}} = \frac{1}{6}\left[(\bar{F}^{-1})_{\dot{\beta} i}{}^{\dot{\underline{\mu}}}\,\nabla_{\alpha k}\,\bar{F}_{\dot{\underline{\mu}}\cdot}{}^{k}{}_{\dot{\gamma}} + (\bar{F}^{-1})_{\dot{\beta} k}{}^{\dot{\underline{\mu}}}\,\nabla_{\alpha i}\,\bar{F}_{\dot{\underline{\mu}}\cdot}{}^{k}{}_{\dot{\gamma}}\right] + (\dot{\beta} \leftrightarrow \dot{\gamma}) \qquad (31)$$

has the required transformation properties ; later on we shall show the uniqueness of this definition. Finally, the vector covariant derivative will be defined by the relation

$$D_{a} = -\frac{i}{8}(\sigma_{a})_{\alpha\dot{\alpha}}\,\{D^{\alpha i}, \bar{D}^{\dot{\alpha}}_{i}\} \qquad (32)$$

resembling the flat supersymmetry one.

From the covariant derivatives one can extract expressions for the vielbeins in terms of the prepotentials H (see the Appendix).

So, the basic concepts of differential geometry have been introduced and we can proceed further, evaluating the components of the torsion tensor.

## 5 NORMAL GAUGE, TORSION COMPONENTS, CONSTRAINTS

It has already been demonstrated in the $N=1$ case (Ogievetsky & Sokatchev, 1980a,b) how constructive the normal gauge is when calculating complicated tensor expressions in terms of the derivatives of the prepotentials. Referring the reader to the original papers for full detail, here we shall just remind ourselves that the normal gauge is a gauge in which a number of derivatives of the prepotentials vanish at a given point $z_o$ in superspace. The tensor quantities evaluated in this gauge (and at the point $z_o$) are expressed in terms of the remaining few derivatives, their properties are easily established and then generalized for any gauge.

The investigation of the derivatives of $H^m$, $H^{\mu i}$, $\bar{H}_{\dot{\mu} i}$ and the corresponding tensors will proceed in steps according to the dimensionality of the quantities concerned. Separate formulae for conjugated objects will not be given.

### 5.1 Dimensions $cm^{+1}$, $cm^{+\frac{1}{2}}$

All the respective quantities $(H^m, \partial_{\underline{\mu}} H^m, H^{\underline{\mu}})$ can be made to vanish in normal gauge.

### 5.2 Dimension $cm^o$

Among the dimensionless derivatives

$$\partial_n H^m,\ \partial_{\underline{\mu}}\partial_{\underline{\nu}} H^m,\ \partial_{\underline{\nu}} H^{\underline{\mu}},\ \bar{\partial}_{\underline{\dot{\nu}}} H^{\underline{\mu}},\ \partial_{\underline{\mu}}\bar{\partial}_{\underline{\dot{\nu}}} H^m$$

only the last one survives in normal gauge :

$$(\partial^i_\mu \bar{\partial}_{\dot{\nu} i} H^m)_o = 2(\sigma^m)_{\mu\dot{\nu}} \tag{33a}$$

$$(\partial_{\mu(i}\bar{\partial}_{\dot{\nu}j)} H^m)_o \neq 0. \tag{33b}$$

The nought in (33) means that the value at the fixed point $z_o$ is considered.

The corresponding torsion components are

$$T_{\underline{\alpha},\underline{\beta}}{}^c,\ T_{\underline{\alpha},\underline{\dot{\beta}}}{}^c\ .$$

Using the definition of torsion

$$T_{AB}{}^{C} = -E_A^M(\partial_M E_B^N)E_N^C - \omega_{AB}{}^{C} + (-1)^{AB}(A\leftrightarrow B), \tag{34}$$

the explicit expressions for the vielbeins (see the Appendix), and calculating in normal gauge one easily finds (the results are valid in any gauge)

$$T_{\underline{\alpha},\underline{\beta}}{}^{c} = 0, \quad T^{i}_{\alpha,\dot{\beta}i}{}^{c} = -4i(\sigma^{c})_{\alpha\dot{\beta}} \ . \tag{35}$$

According to (33b) there are no restrictions on the normal gauge value of $T_{\alpha(i,\dot{\beta}j)}{}^{c}$. However, arguments that it should vanish have been found in Stelle & West, 1980 (they are related to the preservation of the central charge representations in the curved case). Other arguments come from the direct comparison with the already known component results (see, e.g., Gates, 1980b). So, for reasons lying outside the framework discussed we have to <u>postulate our first constraint on the prepotentials</u>

$$T_{\alpha(i,\dot{\beta}j)}{}^{c} = iF^{-1}_{\alpha(i}{}^{\underline{\rho}}\,\bar{F}^{-1}_{\dot{\beta}j)}{}^{\underline{\dot{\rho}}}\left[\Delta_{\underline{\rho}}\bar{\Delta}_{\underline{\dot{\rho}}}\right] H^{m}\cdot e_{m}^{c} = 0 \tag{36}$$

(see (34) and the Appendix).

Note that, differentiating (36) covariantly, one obtains relations between the higher order derivatives of the prepotentials. They are actually the essential part of the Bianchi identities (the rest of them follow from the prepotential construction), so no more consistency checks should be performed.

## 5.3 Dimension $cm^{-\frac{1}{2}}$

The only non-vanishing derivatives now are

$$(\partial_{\underline{\nu}}\partial_{\underline{\lambda}}\bar{\partial}_{\underline{\dot{\rho}}}H^{m})_{o}, \quad (\partial_{\underline{\nu}}\bar{\partial}_{\underline{\dot{\lambda}}}H_{\underline{\mu}})_{o} \tag{37a}$$

except

$$(\partial^{i}_{\nu}\bar{\partial}_{\dot{\lambda}i}H_{\mu k})_{o} = 0, \quad (\partial^{i}_{(\nu}\bar{\partial}_{\dot{\nu}j}H_{\mu)i})_{o} = 0 \ . \tag{37b}$$

They are subject to further restrictions obtained from (36)

$$D_{\gamma k}T_{\alpha(i,\dot{\beta}j)}{}^{c} = 0 \rightarrow$$

$$(\partial_{\gamma k}\partial_{\alpha(i}\bar{\partial}_{\dot{\beta}j)}H_{\mu\dot{\mu}})_{o} = 2\varepsilon_{\alpha\mu}(\partial_{\gamma k}\bar{\partial}_{\dot{\beta}(i}\overline{H}_{\dot{\mu}j)})_{o} \tag{38}$$

where

$$H_{\mu\dot{\mu}} = H^{m}(\sigma_{m})_{\mu\dot{\mu}} \ .$$

Splitting this equation and the quantities (37) into irreducible parts (with respect to both the Lorentz and internal indices) it is not hard to show that everything in (37) vanishes except

$$(\bar{\partial}_{\dot{\alpha}i}\partial_{\beta k} H^{\beta k})_o \equiv (\bar{T}_{\dot{\alpha}i})_o,$$

$$(\partial_{\alpha i}\partial_{\beta j}\bar{\partial}_{\dot{\beta}k} H^{\dot{\beta}}_{\gamma})_o = \tfrac{1}{3}\varepsilon_{\alpha\beta}\left[\varepsilon_{ik}(T_{\gamma j})_o + \varepsilon_{jk}(T_{\gamma i})_o\right] - \tag{39}$$

$$-\tfrac{1}{3}\,\varepsilon_{ij}\left[\varepsilon_{\alpha\gamma}(T_{\beta k})_o + \varepsilon_{\beta\gamma}(T_{\alpha k})_o\right] .$$

So, the normal gauge results (37) and the constraint (36) lead to the conclusion that there is only one quantity with dimension $cm^{-\frac{1}{2}}$ : $T_{\alpha}(\bar{T}_{\dot{\alpha}})$. Clearly, the torsion components with the same dimension are expressed in terms of $T_{\alpha}(\bar{T}_{\dot{\alpha}})$ (these expressions are easily obtained from the definition (34) in normal gauge and then generalized to any gauge) :

$$T_{\alpha i,\dot{\beta}j,}{}^{\gamma k} = \tfrac{1}{6}\,\delta^{\gamma}_{\alpha}(\delta^{k}_{i}\,\bar{T}_{\dot{\beta}j} + \delta^{k}_{j}\,\bar{T}_{\dot{\beta}i}), \tag{40a}$$

$$T_{\alpha i,\beta\dot{\beta},}{}^{\gamma\dot{\gamma}} = \tfrac{1}{3}\,\delta^{\dot{\gamma}}_{\dot{\beta}}(\varepsilon_{\alpha\beta}\,T^{\gamma}_{i} + \delta^{\gamma}_{\beta}\,T_{\alpha i}), \tag{40b}$$

$$T_{\alpha i,\beta j,}{}^{\dot{\gamma}k} = 0, \tag{40c}$$

$$T_{\alpha i,\beta j,}{}^{\gamma k} = 0. \tag{40d}$$

On the other hand, (40a,b) can be regarded as a definition of the tensor quantity $T_{\alpha}(\bar{T}_{\dot{\alpha}})$ in terms of prepotentials (the left-hand sides of Eqs. (40a,b) can be calculated explicitly using (34)).

Notice that the connection $\omega_{\alpha i\dot{\beta}\dot{\alpha}}$ (31) is defined uniquely. Indeed, there are no tensors constructed out of $T_{\alpha}(\bar{T}_{\dot{\alpha}})$ with the symmetry properties of $\omega_{\alpha i\dot{\beta}\dot{\alpha}}$ which could have modified it. However, the vanishing connection $\omega_{\alpha i\beta\gamma}$ (30) can be replaced by the tensor

$$\varepsilon_{\alpha\beta}\,T_{\gamma i} + \varepsilon_{\alpha\gamma}\,T_{\beta i}$$

which would lead to a rearrangement of the torsion components (40). In such cases of ambiguity in the choice of connections we prefer to apply the criterion that the connections should vanish in normal gauge (as, for instance, the Christoffel symbols vanish in what is the analogue of normal gauge in general relativity). Both definitions (30) and (31) obey this rule.

Finally, we stress that the results (40) agree with those of Gates, 1980b. So, no new constraints are needed at this dimensional level.

## 5.4 Dimension $cm^{-1}$

The non-vanishing derivatives are arranged in two groups according to the odd or even number of their dotted and undotted spinor indices :

$$
\begin{aligned}
&(\partial_{\mu(i}\bar{\partial}_{\dot{\nu}j)}\partial_n H^m)_o, \quad (\partial_{\underline{\nu}}\partial_{\underline{\lambda}}\bar{\partial}_{\underline{\dot{\rho}}}\bar{\partial}_{\underline{\dot{\kappa}}} H^m)_o, \\
&(\partial_{\underline{\nu}}\partial_{\underline{\lambda}}\bar{\partial}_{\underline{\dot{\rho}}} H^{\underline{\mu}})_o ;
\end{aligned}
\qquad (41a)
$$

$$
\begin{aligned}
&(\partial_{\underline{\nu}}\partial_{\underline{\lambda}}\partial_{\underline{\rho}}\bar{\partial}_{\underline{\dot{\kappa}}} H^m)_o, \\
&(\bar{\partial}_{\underline{\dot{\lambda}}}\bar{\partial}_{\underline{\dot{\rho}}}\partial_{\underline{\nu}} H^{\underline{\nu}})_o, \quad (\bar{\partial}_{\underline{\dot{\lambda}}}\bar{\partial}_{\underline{\dot{\rho}}}\partial_{\nu(i} H^{\mu}_{j)})_o .
\end{aligned}
\qquad (41b)
$$

The restrictions on the derivatives (41a) following from the constraint (36) are given by the equation

$$
D_{\underline{\alpha}}\bar{D}_{\underline{\dot{\beta}}}T_{\gamma(i,\dot{\delta}j)}{}^{\varepsilon\dot{\varepsilon}} = 0 \qquad (42)
$$

and its conjugate. Written out in normal gauge these equations, besides the derivatives (41a), contain also terms bi-linear in the dimension $cm^{-\frac{1}{2}}$ quantity $T_{\underline{\alpha}}(\bar{T}_{\underline{\dot{\alpha}}})$ (39). To simplify the calculations we have not considered these terms, although they can always be restored. So, after a careful analysis of the irreducible parts of (41a), (42) one finds that the derivatives (41a) are expressed in terms of the normal gauge values of two independent quantities : a complex vector $A^{\alpha\dot{\alpha}}(\bar{A}^{\alpha\dot{\alpha}})$ and a vector-isovector

$$
B^{\alpha\dot{\alpha}}_{ij} = B^{\alpha\dot{\alpha}}_{ji} = \bar{B}^{\alpha\dot{\alpha}}_{ij}
$$

(we omit the lengthy formulae here). Respectively, the dimension $cm^{-1}$ torsion components with odd numbers of dotted and undotted spinor indices are also expressed in terms of $A$ and $B$, in normal as well as in any gauge (thus providing definitions for the tensors $A$, $B$) :

$$
\begin{aligned}
T_{\alpha\dot{\alpha},\beta j,\gamma k} = -\frac{i}{4}\{&\tfrac{1}{4}\varepsilon_{\alpha\gamma}\left[3\varepsilon_{jk}(A-\bar{A})_{\beta\dot{\alpha}} - 2B_{\beta\dot{\alpha}jk}\right] + \\
&+ \varepsilon_{jk}\left[\varepsilon_{\beta\alpha}(A+\bar{A})_{\gamma\dot{\alpha}} + (\alpha\leftrightarrow\gamma)\right] + 2\left[\varepsilon_{\beta\alpha}B_{\gamma\dot{\alpha}jk} + (\alpha\leftrightarrow\gamma)\right]\} .
\end{aligned}
\qquad (43)
$$

The value of $T_{ab}{}^{c}$ depends on the choice of the vector connection $\omega_{ab}{}^{c}$. Indeed, in normal gauge one finds from (34)

$$(T_{ab}{}^{c})_{o} = -(\omega_{ab}{}^{c})_{o} + (\omega_{ba}{}^{c})_{o}.$$

Given some expression for $\omega_{abc}$ (e.g., the one following from (32)), and having calculated $T_{ab}{}^{c}$ with it we can always redefine $\omega_{abc}$

$$\omega'_{abc} = \omega_{abc} + \tfrac{1}{2}(T_{abc} - T_{bca} + T_{cab})$$

so that our normal gauge criterion

$$(\omega'_{abc})_{o} = 0$$

will be fulfilled. Thus we obtain

$$T'_{abc} = 0. \qquad (44)$$

Finally, here is the expression for $\overline{D}_{\dot{\alpha}i}\, T_{\beta k}$ in terms of A and B

$$\overline{D}_{\dot{\alpha}i}\, T_{\beta k} = \tfrac{3}{2}\,\varepsilon_{ik}(A_{\beta\dot{\alpha}} - 3\bar{A}_{\beta\dot{\alpha}}) - 3B_{\beta\dot{\alpha}ik}\ . \qquad (45)$$

Now we consider the "even" derivatives (41b). The counterpart of (42) in this case is

$$D_{\underline{\alpha}} D_{\underline{\beta}}\, T_{\gamma(i,\dot{\delta}j)}{}^{\mu\dot{\mu}} = 0. \qquad (46)$$

Omitting once again the terms bi-linear in $T_{\underline{\alpha}}$ one finds that the derivatives (41b) are reduced to a complex scalar $C(\bar{C})$, two antisymmetric tensors,

$$D_{ab} = (\sigma_{ab})^{\mu\nu}\, D_{\mu\nu} + (\tilde{\sigma}_{ab})^{\dot{\mu}\dot{\nu}}\,\overline{D}_{\dot{\mu}\dot{\nu}}\ ,$$

$$E_{ab} = (\sigma_{ab})^{\mu\nu}\, E_{\mu\nu} + (\tilde{\sigma}_{ab})^{\dot{\mu}\dot{\nu}}\,\bar{E}_{\dot{\mu}\dot{\nu}}$$

and a complex scalar-isovector

$$F_{ij} = F_{ji},\ \bar{F}_{ij} = \bar{F}_{ji}\ .$$

Further, the "even" tensors are expressed in terms of the quantities C, D, E, F :

$$T_{\alpha\dot{\alpha},\beta i,\dot{\gamma}j} = -\tfrac{i}{4}\big[\varepsilon_{\dot{\alpha}\dot{\gamma}}(\varepsilon_{\beta\alpha}\, F_{ij} + \varepsilon_{ij}\, E_{\beta\alpha}) + \varepsilon_{\alpha\beta}\varepsilon_{ij}(\varepsilon_{\dot{\alpha}\dot{\gamma}}\,\bar{C} - 2\,\overline{D}_{\dot{\alpha}\dot{\gamma}})\big]\ , \qquad (47a)$$

$$D_{\alpha i}T_{\beta k} = 3(\varepsilon_{\alpha\beta}F_{ik} + \varepsilon_{ik}E_{\alpha\beta}) - \varepsilon_{ik}\varepsilon_{\alpha\beta}\bar{C}. \tag{47b}$$

Note that (47a) contains only one quantity $(\bar{D})$ independent of $D_{\underline{\alpha}}T_{\underline{\beta}}$.

Comparing (47) with the results of Gates, 1980b, we find a single difference : the isovector $F_{ij}$ is real in Gates, 1980b. This means that we have to impose a second constraint

$$T_{\alpha\dot{\alpha},}{}^{\alpha(i,\dot{\alpha}j)} = -T_{\alpha\dot{\alpha},}{}^{\dot{\alpha}(i,\alpha j)} \tag{48a}$$

or, equivalently

$$D^{\alpha}_{(i}\,T_{\alpha j)} = \bar{D}_{\dot{\alpha}(i}\bar{T}^{\dot{\alpha}}_{j)}\ . \tag{48b}$$

Strictly speaking, this constraint must include bi-linear terms

$$T^{\alpha}_{i}T_{\alpha j},\quad \bar{T}_{\dot{\alpha}i}\,\bar{T}^{\dot{\alpha}}_{j}$$

which we have omitted.

The examination of the tensor quantities with dimensions higher than $cm^{-1}$ does not lead to any new (imposed by hand) constraints.

Concluding this section we stress once again that in the framework proposed, the majority of superspace constraints ((35), (40), (43), (44), (47)) appear in a natural way, following from the basic geometric construction. Only two of them, ((36), (48)) have to be postulated.

## 6 A POSSIBLE ACTION FORMULA FOR N = 2 SUPERGRAVITY

The problem of finding a superspace action formula for $N = 2$ supergravity has not yet been solved. As it is well known, in the $N = 1$ case the action is

$$S = \frac{1}{\kappa^2}\int d^4x\, d^2\,\theta\, d^2\,\bar{\theta}\ \mathrm{Ber}(E^A_M) \tag{49}$$

where Ber is the superdeterminant of the vielbeins and $\kappa$ is the coupling constant, $[\kappa] = cm$. In other words, (49) is the invariant volume of the superspace. Simple dimensional arguments show that (49) gives rise to an equation of motion of the right order for the gravitational field. However, in the $N = 2$ case (49) is replaced by

$$S = \int d^4\,x\, d^4\,\theta\, d^4\,\bar{\theta}\ \mathrm{Ber}(E^A_M) \tag{50}$$

and the same arguments show that one gets a fourth order equation for the graviton. The cause of the trouble is, obviously, the increased number of $\theta$'s in (50).

Then one can try another formula with <u>the right dimension of the Lagrangian</u>

$$S_L = \frac{1}{\kappa^2} \int d^4 x_L \, d^4 \theta_L \, \mathrm{Ber}(\ell_M^A) \,. \tag{51}$$

This is the invariant volume of the left-handed chiral superspace (actually, the real sum $S_L + S_R$ should be considered). The objects $\ell_M^A$, the so-called "left-handed vielbeins" (this concept was introduced in the $N = 1$ case in Ogievetsky & Sokatchev, 1979b) will be now defined.

First, consider a left-handed chiral superfield

$$\bar{D}_{\underline{\dot{\alpha}}} \varphi = 0 \tag{52}$$

in the left-handed basis (21). There (52) takes the form

$$\bar{\partial}^L_{\underline{\dot{\alpha}}} \varphi_L = 0. \tag{53}$$

Now, using (53), rewrite the derivatives $D_A \varphi$ $(A = \alpha, a)$ in this basis

$$\begin{aligned} D_A \varphi = E_A^M \partial_M \varphi &= (E_A^m + i E_A^N \partial_N H^m) \partial_m^L \varphi_L + E_A^{\underline{\mu}} \partial^L_{\underline{\mu}} \varphi_L \equiv \\ &\equiv \ell_A^M \partial_M^L \varphi_L \end{aligned} \tag{54}$$

where, by definition, $M$ in $\ell_A^M$ is $m$ or $\underline{\mu}$. It is not hard to show that $\ell_A^M$ transforms under the group (11), (15), (17), (20) as follows

$$\delta \ell_A^M = L_A{}^B \ell_B{}^M + \ell_A{}^N \partial_N{}^L \lambda_L{}^M - \lambda_L{}^N \partial_N{}^L \ell_A{}^M \,. \tag{55}$$

Here $L_A{}^B$ denotes Lorentz transformations of the type (28) (just like the transformations of $D_A \varphi$ in (54)). Using (55) one finds

$$\begin{aligned} \delta \left[\mathrm{Ber}(\ell_M^A)\right] \equiv \delta\left[\mathrm{Ber}^{-1}(\ell_A^M)\right] &= -\mathrm{Ber}(\ell_M^A) \cdot \ell_N^B \, \delta \ell_B^N (-1)^N = \\ &= -\mathrm{Ber}(\ell_M^A) \cdot \partial_N^L \, \lambda_L^N (-1)^N - \lambda_L^N \partial_N^L \left[\mathrm{Ber}(\ell_M^A)\right] \,, \end{aligned} \tag{56}$$

i.e., this is exactly the transformation law of a scalar density for the left-handed superspace.

However, we cannot put $\mathrm{Ber}(\ell_M^A)$ into (51) unless it is chiral, i.e., depends on $x_L^m$, $\theta_L^{\underline{\mu}}$ only. Whether this is true or not should be checked explicitly since $\ell_A^M$ in (54) are not chiral themselves.

So, using the definition (54) and the Appendix we obtain

$$\mathrm{Ber}(\ell_M^A) = \mathrm{Ber}^{-1}(\ell_A^M) = \det{}^{-1}(\hat{\ell}_a^m)\,\det{}^{-1}(\delta_{\underline{\alpha}}^{\underline{\mu}} + \Delta_{\underline{\alpha}} H^{\underline{\mu}}) \tag{57}$$

where

$$\hat{\ell}_a^m = \tfrac{1}{4}(\sigma_a)_{\alpha\dot{\alpha}}(1 + \Delta H)^{-1\alpha i \underline{\beta}}(1 + \overline{\Delta}\,\overline{H})^{-1\alpha}{}_{\dot{i}}{}^{\underline{\dot{\beta}}}\,\bar{\Delta}_{\underline{\dot{\beta}}}\,\Delta_{\underline{\beta}}\,H^m .$$

We have to prove that

$$\bar{D}_{\underline{\dot{\alpha}}}\,\mathrm{Ber}(\ell_M^A) = 0. \tag{58}$$

Note that this equation is covariant because the weight factor

$$\partial_N^L\,\lambda_L^N\,(-1)^N$$

in (56) is chiral, i.e., it goes through the operator $\bar{D}_{\underline{\dot{\alpha}}}$ in (58). For the same reason we have the right to carry out the check in normal gauge. So, using (33), (36), (37), (39), we find

$$(\bar{D}_{\underline{\dot{\alpha}}}\,\mathrm{Ber}(\ell_M^A))_o = -\left[\mathrm{Ber}(\ell_M^A)\,(\bar{\partial}_{\underline{\dot{\alpha}}}\,\hat{\ell}_a^m\cdot\hat{\ell}_m^a + \bar{\partial}_{\underline{\dot{\alpha}}}\,\partial_{\underline{\beta}}\,H^{\underline{\beta}})\right]_o =$$

$$= -\tfrac{1}{4}(\sigma_a)_{\rho\dot{\rho}}\left[\bar{\partial}_{\underline{\dot{\alpha}}}((1+\Delta H)^{-1\rho k\beta j}\,\bar{\Delta}_k^{\dot{\rho}}\,\Delta_{\beta j}\,H^a)\right]_o - (\bar{\partial}_{\underline{\dot{\alpha}}}\,\partial_{\underline{\beta}}\,H^{\underline{\beta}})_o =$$

$$= \tfrac{1}{4}(\bar{\partial}_{\underline{\dot{\alpha}}}\,\bar{\partial}_k^{\dot{\rho}}\,\partial^{\rho k}\,H_{\rho\dot{\rho}})_o = 0 \quad .$$

Note that the chirality of $\mathrm{Ber}(\ell_M^A)$ (and thus the consistency of (51)) follows from the constraint (36). So, (51) is a non-trivial invariant. Of course, to prove the conjecture that it is an action for $N = 2$ supergravity one should vary the action and obtain the correct equations of motion. This is not a straightforward procedure because the prepotentials here are constrained. However, due to the small number of constraints in the present approach the task seems relatively easy. This problem is under investigation.

Finally, we would like to point out that another density has been proposed in Gates, 1980b

$$\mathrm{Ber}(E_M^A)\,R^{-2} \tag{59}$$

where $R$ is a dimension $\mathrm{cm}^{-1}$ chiral superfield (corresponding to $C$ in (47)). Recalling the real basis form of the left-handed volume element in $N = 1$ supergravity (Siegel & Gates, 1979)

$$d^4 x \, d^2 \theta \, d^2 \bar{\theta} \, \mathrm{Ber}(E_M^A) \cdot R^{-1}$$

one can guess that (51) and (59) are just different forms of the same invariant.

## 7 CONCLUSIONS

In the present paper we have generalized the ideas of Ogievetsky & Sokatchev, 1978, 1979a,b and Siegel & Gates, 1979 for the $N = 2$ case. We have constructed a prepotential formulation of $N = 2$ supergravity in close analogy with the non-minimal $N = 1$ case. However, our prepotentials are actually semi-prepotentials, i.e., they are constrained (although by a small number of constraints). So, the question of finding true (independent) prepotentials is still open. A possible solution would be to express $H^m$, $H^{\underline{\mu}}$, $\bar{H}^{\underline{\dot{\mu}}}$ as derivatives of another superfield, although it is hard to imagine how the world index $m$ will occur.

The approach proposed can easily be applied to the cases with $N > 2$. However, one point is not quite clear. The component results indicate that for $N > 2$ physical spinor fields are needed. Their existence seems to spoil the constraint (40c) which is a necessary condition for the preservation of the chiral representation, a basic concept in our approach. On the other hand, a deeper understanding of the rôle of the central charges can alter this conclusion and lead to a better formulation of extended supergravities.

## ACKNOWLEDGEMENTS

The author would like to cordially thank Professor V. Ogievetsky and A. Galperin for useful discussions at the early stages of this work. I am very grateful to Professor S. Hawking and M. Roček for their hospitality and for the excellently organized Nuffield Supergravity Workshop in Cambridge, 1980. It is a pleasure to thank J. Gates, W. Siegel, K. Stelle, P. West, M. Sohnius and many other participants of the Workshop for stimulating discussions.

## APPENDIX : VIELBEINS IN TERMS OF PREPOTENTIALS

From the definition of covariant derivatives

$$D_A = E_A^M \partial_M + \omega_A$$

and equations (24)-(26'), (30)-(32) one can find the following expressions for $E_A^M$ :

$$E_{\underline{\alpha}}^{\underline{\mu}} = F^{-1}{}_{\underline{\alpha}}^{\underline{\mu}}, \quad E_{\underline{\alpha}}^{\dot{\underline{\mu}}} = 0, \quad E_{\underline{\alpha}}^{m} = i \nabla_{\underline{\alpha}} H^m,$$

$$E_a^{\underline{\mu}} = -\tfrac{i}{8}(\sigma_a)_{\alpha\dot{\alpha}}(\bar{\nabla}_i^{\dot{\alpha}} F^{-1\alpha i\underline{\mu}} - \bar{\omega}^{\dot{\alpha}i\alpha\gamma} F^{-1}{}_{\gamma i}^{\underline{\mu}}),$$

$$E_a^m = e_a^m + iE_a^{\underline{\mu}} \Delta_{\underline{\mu}} H^m - iE_a^{\dot{\underline{\mu}}} \bar{\Delta}_{\dot{\underline{\mu}}} H^m,$$

$$e_a^m = -\tfrac{1}{8}(\sigma_a)_{\alpha\dot{\alpha}} F^{-1\alpha i\underline{\beta}} \bar{F}^{-1\dot{\alpha}\dot{\underline{\beta}}}_i \left[\Delta_{\underline{\beta}} \bar{\Delta}_{\dot{\underline{\beta}}}\right] H^m$$

and their conjugates for the remaining ones.

Inverting $E_A^M$, one gets expressions for $E_M^A$

$$E_{\underline{\mu}}^{\underline{\alpha}} = F_{\underline{\mu}}^{\underline{\alpha}} + i\Delta_{\underline{\mu}} H^m \cdot e_m^a E_a^{\underline{\nu}} F_{\underline{\nu}}^{\underline{\alpha}},$$

$$E_{\underline{\mu}}^{\dot{\underline{\alpha}}} = i\Delta_{\underline{\mu}} H^m \cdot e_m^a E_a^{\dot{\underline{\nu}}} \bar{F}_{\dot{\underline{\nu}}}^{\dot{\underline{\alpha}}},$$

$$E_{\underline{\mu}}^{a} = -i\Delta_{\underline{\mu}} H^m \cdot e_m^a ,$$

$$E_m^{\underline{\alpha}} = -e_m^a E_a^{\underline{\mu}} F_{\underline{\mu}}^{\underline{\alpha}}, \quad E_m^a = e_m^a$$

and conjugates.

## REFERENCES

Breitenlohner P. & Sohnius, M. (1980). Superfields, auxiliary fields and tensor calculus for N = 2 extended supergravity. Nucl. Phys., B165, pp. 483-510.

Castellani L., van Nieuwenhuizen P. & Gates, J. (1980). The constraints for N = 2 superspace from extended supergravity in ordinary space. Stony Brook Preprint, ITP-5B-80-9.

Fradkin, E. & Vasiliev, M. (1979). Minimal set of auxiliary fields and S matrix for extended supergravity. Lett. Nuovo Cimento, 25, pp. 79-87 ; Minimal set of auxiliary fields in SO(2) extended supergravity. Phys.Letters, 85B, pp. 47-51.

Gates, J. (1980a). Supercovariant derivatives, super-Weyl groups, and N = 2 supergravity. Harvard Preprint HUTP-80/A011.

Gates, J. (1980b). Another solution for N = 2 superspace Bianchi identities. Harvard Preprint HUTP-80/A042.

Ogievetsky, V. & Sokatchev, E. (1978). Structure of the supergravity group. Phys.Letters, 79B, pp. 222-224.

Ogievetsky, V. & Sokatchev, E. (1979a). The simplest group of Einstein supergravity. Dubna Preprint E2-12469.

Ogievetsky, V. & Sokatchev, E. (1979b). The gravitational axial superfield and the formalism of differential geometry. Dubna Preprint E2-12511.

Ogievetsky, V. & Sokatchev, E. (1980a). Normal gauge in supergravity. Dubna Preprint E2-80-127.

Ogievetsky, V. & Sokatchev, E. (1980b). Torsion and curvature in terms of the axial gravitational superfield. Dubna Preprint E2-80-138.

Ogievetsky, V. & Sokatchev, E. (1980c). Equation of motion for the axial gravitational superfield. Dubna Preprint E2-80-139.

Siegel, W. & Gates, J. (1979). Superfield supergravity. Nucl.Phys., B147, pp. 77-104.

Stelle, K. & West, P. (1980). Algebraic derivation of N = 2 supergravity constraints. Imperial College Preprint ICTP/79-80/9.

Wess, J. (1979). Supergravity. Lectures given at the Boulder Summer School, June 1979. Karlsruhe Preprint.

de Wit, B. & van Holten, J. (1979). Multiplets of linearized SO(2) supergravity. Nucl.Phys., B155, pp. 530-542.

# TOWARD AN UNEXTENDED SUPERFIELD FORMULATION OF N=2 SUPERGRAVITY*

S. James Gates, Jr.**†

Lyman Laboratory of Physics
Harvard University
Cambridge, MA 02138

## INTRODUCTION

There are at least two reasons why we believe an N=1 superfield formulation of the N=2 supergravity theory is desirable. First, the coupling of the N=1 (2,3/2) supermultiplet to all lower spin N=1 supermultiplets is known except for the (3/2,1) supermultiplet. For all other supermultiplets a simple minimal coupling procedure is sufficient. In the case of gauge multiplets this means defining properly their field strengths. The N=2 theory must contain the solution to this problem. Once an understanding of this is achieved, several (3/2,1) multiplets may be coupled to N=1 supergravity. This might be desirable for even higher extended theories. Secondly, once an unextended formulation of the N=2 theory exists, then N=1 superfield perturbation methods can be used to study in a systematic way the extended theory.

In this talk we shall discuss the "geometrical" formulation of this problem and describe a procedure for its resolution. The first part of the talk is a discussion of the N=1 linearized theory. We emphasize the role played by compensating superfields and describe how these arise. The purpose of this discussion is to show why a "geometrical" derivation of the coupled (2,3/2) - (3/2,1) system is necessary. Otherwise, we might miss some important compensating superfield. The second part concerns the main problem at hand.

## LINEARIZED N=1 SUPERGRAVITY

In a previous work (Gates & Siegel, 1979) the complete nonlinear solution to the kinematic constraints for supergravity with minimal and nonminimal auxiliary fields was presented. Ogievetsky and Sokatchev (1977) were the first to develop a description of linearized supergravity. This theory has also been discussed by Ferrara and Zumino (1978) and Howe and Tucker (1979) for the case of minimal auxiliary fields. In all of these works, however, the existence of the compensating, chiral scalar superfield (Siegel, 1978) was ignored. This superfield plays an important role. Neglecting this quantity leads to the component field description given first by Stelle and West (1978). The formulation of Ferrara and van Nieuwenhuizen (1978) requires the compensating superfield. Similarly, any superspace description of the nonminimal theory also demands the use of a compensating superfield. This is important since the nonminimal N=1 theory will play a prominent role in extended supergravity. Furthermore a compensating superfield is necessary for the introduction of a true cosmological term in supergravity. Otherwise complications arise at the boundary of the action integral. Finally, Namazie and Storey (1979) found that the absence of the compensating superfield leads to the need for nonlocal gauge-fixing terms to quantize superfield supergravity.

We begin by introducing the anholonomized coordinate derivative $E_A \equiv E_A{}^M \partial_M$. We define the object of anholonomity, Lorentz covariant derivative, torsion and curvature tensors as usual

$$[E_A, E_B\} = C_{AB}{}^D E_D, \mathcal{D}_A = E_A + \frac{1}{2} \phi_{Ac}{}^d M_d{}^c,$$
$$[\mathcal{D}_A, \mathcal{D}_B\} = T_{AB}{}^D \mathcal{D}_D + \frac{1}{2} R_{ABc}{}^d M_d{}^c, \qquad (1)$$

where $M_d{}^c$ is the Lie algebra generator of Lorentz transformations. The curvature and torsion can be expressed in terms of the object of anholonomity and the spin-connection superfield.

Flat superspace is the result which arises in the limit where

$$E_\alpha = \partial_\alpha + i\sigma^c{}_{\alpha\dot\beta} \bar\theta^{\dot\beta} \partial_c \equiv D_\alpha,$$
$$E_{\dot\alpha} = -(\overline{E_\alpha}) = \bar\partial_{\dot\alpha} + i\sigma^c{}_{\beta\dot\alpha} \theta^\beta \partial_c \equiv \bar D_{\dot\alpha},$$
$$E_b = i\frac{1}{4} \sigma_b{}^{\alpha\dot\beta} [E_\alpha, E_{\dot\beta}\} = \partial_b, \ \phi_{Abc} = 0. \qquad (2)$$

In order to introduce a supergravity disturbance on the flat background, we introduce superfields $H_A{}^B$ which describe linearized supergravity. We define

$$E_\alpha = D_\alpha + H_\alpha{}^\beta D_\beta + H_\alpha{}^{\dot\beta} \bar D_{\dot\beta} + H_\alpha{}^b \partial_b,$$
$$E_a = \partial_a + H_a{}^\beta D_\beta + \bar H_a{}^{\dot\beta} \bar D_{\dot\beta} + H_a{}^b \partial_b. \qquad (3)$$

The local symmetries are, of course, super coordinate invariance and Lorentz invariance. The infinitesimal transformations are generated as

$$\delta\mathcal{D}_A = [iK, \mathcal{D}_A], \ iK = K^M \partial_M + \frac{1}{2} K_a{}^b M_b{}^a, \qquad (4)$$

in analogy with a Yang-Mills theory.

We now begin to place constraints on the superspace geometry.

Demanding that $C_{\alpha\beta}{}^{d}$ and $C_{\alpha\beta}{}^{\dot\gamma}$ maintain their flat space values implies that $E_\alpha$ is given by

$$E_\alpha = D_\alpha + (H_\alpha{}^\gamma + D_\alpha H^\gamma)D_\gamma + (D_\alpha \bar H^{\dot\gamma})\bar D_{\dot\gamma} + (D_\alpha H^d - i2\sigma^d{}_{\alpha\dot\gamma}\bar H^{\dot\gamma})\partial_d \tag{5}$$

from which it follows,

$$[E_\alpha, E_\beta\} = (D_\alpha H_\beta{}^\gamma + D_\beta H_\alpha{}^\gamma)D_\gamma \equiv C_{\alpha\beta}{}^\gamma\, E_\gamma. \tag{6}$$

Thus, at this point we have the "prepotentials" $H^d$ and $H^\gamma$ and the potential $H_\alpha{}^\gamma$. As first noted in a previous work (Gates, 1980a), the vanishing of $C_{\alpha\beta}{}^{d}$ and $C_{\alpha\beta}{}^{\dot\gamma}$ is crucial in the existence of chiral substances. The solution in (5) "senses" the presence of the chiral subspaces by possessing an invariance group. It remains invariant under the transformation,

$$\delta H^\gamma = -2\,\bar D^2 L^\gamma,\quad \delta H_\alpha{}^\gamma = +2\,\underline{D_\alpha \bar D^2}\,L^\gamma$$
$$\delta H^b = i8\sigma^b{}_{\beta\dot\gamma}\, D^\beta \bar L^{\dot\gamma},\quad (\bar L^{\dot\gamma} = \overline{L^\gamma}) \tag{7}$$

where $L^\gamma$ is a totally unconstrained spinor superfield. These transformations are associated with the presence of the chiral subspaces. This phenomenon is already familiar from supersymmetric Yang-Mills theories (Gates & Shapiro, 1978). We can further constrain the linearized supergravity superfields by imposing the constraint $T_{\alpha\dot\beta}{}^c = i2\,\sigma^c{}_{\alpha\dot\beta}$ which has the solution,

$$E_a = \partial_a + i\frac{1}{4}\sigma_a{}^{\beta\dot\gamma}[i2\,\partial_{\beta\dot\gamma}H^\delta + \bar D_{\dot\gamma}H_\beta{}^\delta - i\sigma^b{}_{\beta\dot\gamma}H_b{}^\delta]D_\delta$$
$$+ i\frac{1}{4}\sigma_a{}^{\beta\dot\gamma}[i2\,\partial_{\beta\dot\gamma}\bar H^{\dot\delta} + D_\beta \bar H_{\dot\gamma}{}^{\dot\delta} - i\sigma^b{}_{\beta\dot\gamma}\bar H_b{}^{\dot\delta}]\bar D_{\dot\delta}$$
$$+ i\frac{1}{4}\sigma_a{}^{\beta\dot\gamma}[\bar D_{\dot\gamma}D_\beta H^b + D_\beta \bar D_{\dot\gamma}\bar H^b + i2(\sigma^b{}_{\delta\dot\gamma}H_\beta{}^\delta + \sigma^b{}_{\beta\dot\delta}\bar H_{\dot\gamma}{}^{\dot\delta})]\partial_b. \tag{8}$$

With this solution we find

$$[E_\alpha, E_{\dot\beta}\} = i2^c{}_{\alpha\dot\beta}E_c + i\sigma^b{}_{\alpha\dot\beta}(H_b{}^\gamma E_\gamma + \bar H_b{}^{\dot\gamma}E_{\dot\gamma}),$$
$$\rightarrow T_{\alpha\dot\beta} = i2\sigma^c{}_{\alpha\dot\beta},\quad C_{\alpha\dot\beta}{}^{\dot\gamma} = i\sigma^b{}_{\alpha\dot\beta}\bar H_b{}^{\dot\gamma}. \tag{9}$$

So the constraint on $T_{\alpha\dot\beta}{}^c$ implies that at this point $E_a$ contains one more independent potential $H_a{}^\gamma$. However, by imposing additional constraints we can eliminate this potential as an independent variable. In particular, we impose the constraints

$$C_{\alpha(bc)} - \frac{1}{2}\eta_{bc}C_{\alpha d}{}^{d} = 0, \quad C_{\alpha\dot{\gamma}}{}^{\dot{\gamma}} - \frac{1}{3}C_{\alpha d}{}^{d} = 0 \tag{10}$$

which have the solution given by,

$$\begin{aligned}\bar{H}_a{}^{\dot{\gamma}} &= i\,\frac{1}{8}\,\sigma_a{}^{\alpha\dot{\gamma}}[i\,\frac{1}{8}\,D^2\bar{D}^{\dot{\beta}}(H_{\alpha\dot{\beta}}-\bar{H}_{\alpha\dot{\beta}}) + D_\alpha(2H_\beta{}^\beta+\bar{H}_{\dot{\beta}}{}^{\dot{\beta}})] \\ &\quad + \frac{1}{4}\,\sigma_a{}^{\alpha\dot{\beta}}[\frac{1}{8}\,D^2\bar{D}_{(\dot{\beta}}(H_\alpha{}^{\dot{\gamma})}-\bar{H}_\alpha{}^{\dot{\gamma})}) + i\,D_\alpha\bar{H}_{(\dot{\beta}}{}^{\dot{\gamma})}]. \end{aligned} \tag{11}$$

($H_{\alpha\dot{\beta}} \equiv \sigma^b{}_{\alpha\dot{\beta}}H_b$) so that $\bar{H}_a{}^{\dot{\gamma}}$ is no longer an independent potential. In computing (11) we have made use of the fact that

$$C_{\alpha d}{}^{d} = \frac{3}{2}\,[i\,\frac{1}{8}\,D^2\bar{D}^{\dot{\gamma}}(H_{\alpha\dot{\gamma}}-\bar{H}_{\alpha\dot{\gamma}}) + D_\alpha(2H_\beta{}^\beta+\bar{H}_{\dot{\beta}}{}^{\dot{\beta}})] \equiv \frac{3}{2}\,T_\alpha \tag{12}$$

which also follows from (10).

Thus far we have not made use of any of the gauge freedoms of the theory. These can be exploited to simplify the expressions (5), (8), and (11). From equation (4) we note that $K^a$ may be used to set the real part of $H^a$ equal to zero. Similarly $K^\alpha$ can be used to set $H^\alpha$ equal to zero. Finally $K^{ab}$ can be used to set the symmetric part of $H_\alpha{}^\beta$ equal to zero. Thus, there exists a W-Z gauge wherein

$$H^\alpha \to 0, \; H^d \to i\,U^d \quad (U^d = \bar{U}^d), \; H_\alpha{}^\beta \to \delta_\alpha^\beta\,\eta \tag{13}$$

for some superfields $U^d$ and $\eta$. In terms of these we have,

$$E_\alpha = D_\alpha + \eta\,D_\alpha + i(D_\alpha U^d)\partial_d, \tag{14}$$

$$E_a = \partial_a + \Psi_a{}^\delta D_\delta + \bar{\Psi}_a{}^{\dot{\delta}}\,\bar{D}_{\dot{\delta}} + h_a{}^b\partial_b, \tag{15}$$

where the superfields $\Psi_a{}^\delta$ and $h_{ab}$ are defined as,

$$\Psi_a{}^\delta = i\,\frac{1}{16}\,\sigma_a{}^{\beta\dot{\gamma}}[\bar{D}^2D^\delta U_{\beta\dot{\gamma}} - \frac{3}{4}\,\delta_\beta^\delta\bar{D}^2D^\alpha U_{\alpha\dot{\gamma}} - 2\delta_\beta^\delta\bar{D}_{\dot{\gamma}}(2\bar{\eta}-\eta)], \tag{16a}$$

$$h_{ab} = -\,\frac{1}{4}\,\sigma_a{}^{\beta\dot{\gamma}}[[\bar{D}_{\dot{\gamma}},D_\beta]U_b + 2\sigma_{b\beta\dot{\gamma}}(\eta+\bar{\eta})]. \tag{16b}$$

We have chosen the symbols $\Psi_a{}^\delta$ and $h_{ab}$ for these superfields because at $\theta=0$ these correspond precisely to the gravitino and linearized graviton. In this special gauge, the superfield $T_\alpha$ defined in (12) is given by

$$T_\alpha = [-\,\frac{1}{4}\,D^2\bar{D}^{\dot{\gamma}}U_{\alpha\dot{\gamma}} + 2D_\alpha(2\eta+\bar{\eta})]. \tag{17}$$

There are two other supertensors of importance. We define one additional superfield (denoted by $G_a$) by the equation,

$$G_a = \frac{1}{2}[i(C_{a\beta}{}^{\beta} - C_{a\dot\beta}{}^{\dot\beta}) + \frac{1}{72}\sigma_a{}^{\alpha\dot\beta}(D_\alpha C_{\dot\beta d}{}^{d} - \bar D_{\dot\beta} C_{\alpha d}{}^{d})]. \qquad (18)$$

Once again working in our W-Z gauge we find in terms of $U_{\alpha\dot\beta}$ and $\eta$

$$G_a = \frac{11}{16}[D^\alpha \bar D^2 D_\alpha + \bar D^{\dot\alpha} D^2 \bar D_{\dot\alpha}]U_a - \frac{1}{48}\sigma_a{}^{\alpha\dot\beta}[D_\alpha \bar D^2 D^\gamma U_{\gamma\dot\beta} + \bar D_{\dot\beta} D^2 \bar D^{\dot\gamma} U_{\alpha\dot\gamma}]$$

$$+ i\frac{2}{3}\partial_a(\eta - \bar\eta)$$

$$= \frac{11}{8} D^\alpha \bar D^2 D_\alpha U_a - \frac{1}{24}\sigma_a{}^{\alpha\dot\beta} D_\alpha \bar D^2 D^\gamma U_{\gamma\dot\beta} + i\frac{2}{3}\partial_a(\eta - \bar\eta). \qquad (19)$$

The last superfield (denoted by $\bar R$) is defined by

$$\bar R = -i\frac{1}{2}\sigma^{a\alpha}{}_{\dot\beta}\, C_{a\alpha}{}^{\dot\beta} = -\frac{1}{4}D^2(2\eta - \bar\eta) \qquad (20)$$

where the latter form is valid only in the linearized, W-Z gauge theory of equation (13).

At this point we can note the distinction between the minimal and nonminimal theories. In the minimal theory, the additional constraint is $T_\alpha = 0$. From equation (17) the most general solution to the resulting differential equation is given by

$$2\eta + \bar\eta = -\frac{1}{4}D^\beta \bar D^{\dot\gamma} U_{\beta\dot\gamma} + c\bar\phi, \quad D_\alpha \bar\phi = 0, \qquad (21)$$

where c is an arbitrary constant of proportionality which we may set equal to -3. This is the difference between the Stelle-West (1978) and Ferrara-van Nieuwenhuizen (1978) formulations of the minimal theory. The S and P component auxiliary fields "live" locally in $\phi$ and nonlocally in $U_{\alpha\dot\beta}$.

It is also interesting to note that certain functions are independent of $\eta$. It can be seen that the following equations are satisfied.

$$\frac{\delta}{\delta\eta}(h_{ab} - \frac{1}{4}\eta_{ab} h_c{}^c) = \frac{\delta}{\delta\eta}(\Psi_a{}^\beta + \frac{1}{4}\sigma_a{}^{\beta\dot\gamma}\sigma^b{}_{\alpha\dot\gamma}\Psi_b{}^\alpha) = 0$$

$$\frac{\delta}{\delta\bar\eta}(\Psi_a{}^\beta + \frac{1}{4}\sigma_a{}^{\beta\dot\gamma}\sigma^b{}_{\alpha\dot\gamma}\Psi_b{}^\alpha) = \frac{\delta}{\delta\eta}(G_a - \Box^{-1}\partial_a\partial^b G_b) = 0. \qquad (22)$$

Indeed, the last of these is the determining factor for the ratio of the two sets of terms which appear in equation (18). These equations are simply consequences of the fact that the "transverse" parts of h, $\Psi$, and G are superconformally invariant.

Returning now to the nonminimal theory, we still must place some constraint on $\eta$ to obtain an irreducible representation. For the nonminimal theory there are many ways parametrized by a real number n (Gates & Siegel, 1979) to choose this final constraint.

However, the component formulations of N=2 supergravity (Fradkin & Vasiliev, 1979; de Wit & van Holten, 1979) possess the "geometric" n = -1 theory as the pure N=1 supergravity subset. The final constraint for the n = -1 theory is very simple.

$$D^2\eta = 0. \tag{23}$$

This differential equation states that $\eta$ is a linear superfield thus,

$$\eta = D_\alpha \Phi^\alpha \equiv T. \tag{24}$$

Since the spinor superfield $\phi_\alpha$ only enters the theory through $\eta$, we may subject $\phi_\alpha$ to the variation

$$\delta\phi^\alpha = D_\beta f^{(\alpha\beta)} \tag{25}$$

without changing the value of $\eta$. We note that $\phi_\alpha$ is the truly unconstrained auxiliary variable in nonminimal supergravity. The analog of this superfield exists in the full nonlinear theory also. A superfield $\hat{T}$ exists in the full theory (Gates & Siegel, 1979) which can be utilized to define a world spinor $\bar{\phi}^{\dot\mu}$ via the equation $\hat{T} = \bar{\partial}_{\dot\mu}\bar{\phi}^{\dot\mu}$ where $\bar{\phi}^{\dot\mu}$ transforms as

$$\delta\bar{\phi}^{\dot\mu} = -\ (\Lambda^m\partial_m + \Lambda^\mu\partial_\mu)\bar{\phi}^{\dot\mu} + \Lambda^{\dot\mu}(\bar{\partial}_{\dot\nu}\bar{\phi}^{\dot\nu})$$
$$-\left(\frac{n+1}{3n+1}\right)(\partial_m\Lambda^m - \partial_\mu\Lambda^\mu)\bar{\phi}^{\dot\mu} + \bar{\partial}_{\dot\nu}\bar{\Lambda}^{(\dot\mu\dot\nu)}. \tag{26}$$

The first three terms are the transformation law under the chiral general coordinate group, while the last term is the generalization of (25) to curved superspace.

We can now explain why our parameter $L^\gamma$ in (7) is unconstrained whereas in other treatments of the linearized theory this is not true. Note from (7) that

$$\delta\eta = D_\alpha \bar{D}^2 L^\alpha. \tag{27}$$

For the different cases of compensating superfields (nonminimal/minimal) this implies (from (21) and (24))

$$\delta\phi_\alpha = \bar{D}^2 L_\alpha \quad \text{or} \quad \delta\phi = \bar{D}^2 D^\alpha L_\alpha. \tag{28}$$

Thus in the minimal theory if we set $\phi$ equal to a constant, its variation must vanish. This implies a constraint on the gauge parameter $L_\alpha$. Similarly in the nonminimal theory $L_\alpha$ can be used to gauge away the chiral part of $\phi_\alpha$. In order to maintain this condition the gauge parameter must be constrained. Thus we see why compensating superfields are important.

## UNEXTENDED SUPERFIELD FORMULATION OF N=2 SUPERGRAVITY

We earlier stated that constraints should be imposed on torsion and/or curvature supertensors. Yet in equation (23) we seem to have departed from this. We must show that this constraint is equivalent to a covariant procedure. In superfield supergravity, any appropriately transforming function of the object of anholonomity can serve as the Lorentz connection superfield. We choose (Gates, 1980b)

$$\phi_{\alpha bc} = -\frac{1}{2}\,[C_{\alpha[bc]} - i\,\frac{1}{6}\,(\sigma_{bc})_{\alpha}{}^{\beta}\,C_{\beta d}{}^{d}], \tag{29}$$

which in terms of $U_{\alpha\dot\beta}$ and $\eta$ (see appendix) becomes,

$$\phi_{\alpha bc} = \frac{1}{4}\,[\sigma_{[b\alpha\dot\beta}D^{2}\bar{D}^{\dot\beta}U_{c]} - i\,\frac{1}{2}\,(\sigma_{bc})_{\alpha}{}^{\gamma}(D^{2}\bar{D}^{\dot\delta}U_{\gamma\dot\delta}-8D_{\gamma}\eta)]. \tag{30}$$

The linearized curvature $R_{\alpha\beta cd}$ is thus given by

$$R_{\alpha\beta cd} = D_{(\alpha}\phi_{\beta)cd} = i(\sigma_{cd})_{\alpha\beta}\,D^{2}\eta \tag{31}$$

and it can be seen that the constraint in (23) is indeed a covariant one. The remaining Lorentz connection superfield can be chosen so as to satisfy

$$R_{\alpha\dot\beta cd} = i\,\frac{1}{8}\,\sigma_{a}{}^{\gamma\dot\delta}\,[\bar{D}_{\dot\delta}T_{\gamma}-D_{\gamma}\bar{T}_{\dot\delta}+T_{\gamma}\bar{T}_{\dot\delta}]\varepsilon^{ab}{}_{cd}\,\sigma_{b\alpha\dot\beta}, \tag{32}$$

which has the linearized solution

$$\phi_{abc} = i\,\frac{1}{4}\,\sigma_{\alpha}{}^{\alpha\dot\beta}D_{\alpha}(\delta^{d}_{a}\phi_{\dot\beta bc}+i\,\frac{1}{4}\,\varepsilon_{abc}{}^{d}\bar{T}_{\dot\beta}) + \text{h.c.} \tag{33}$$

We have made this presentation in such a way so that our results are valid for either the minimal or nonminimal theories. We have identified all of the component fields as the $\theta=0$ value of certain functions of the "prepotentials" $U_{\alpha\dot\gamma}$ and $\eta$. For the minimal theory (Wess & Zumino, 1978), we have

$$(h_{ab},\Psi_{a}{}^{\alpha},G_{a},\bar{R}) \tag{34}$$

while in the nonminimal theory (Gates & Siegel, 1979) these are

$$(h_{ab},\Psi_{a}{}^{\alpha},G_{a},T_{\alpha}). \tag{35}$$

The remaining "extra" component fields are generated from spinorial differentiation of $T_{\alpha}$ (Brown & Gates, 1980). The action for both forms of the theory is the same (Siegel, 1977) and is given by,

$$\frac{1}{\kappa^{2}}\int d^{4}x\,d^{4}\theta E^{-1} \cong \frac{1}{2\kappa^{2}}\int d^{4}x\,\frac{1}{16}\,D^{2}\bar{D}^{2}\{(-)^{A}H_{A}{}^{B}H_{B}{}^{A} + [(-)^{A}H_{A}{}^{A}]^{2}\} \tag{36}$$

where $H_{A}{}^{B}$ is defined by (3), (14), and (15) in terms of $U_{\alpha\dot\beta}$ and $\eta$.

By operating on the expression in the curly brackets with the spinorial derivatives we obtain the familiar action in terms of the component fields. Variation of the action with respect to $U_{\alpha\dot{\beta}}$ yields the equation of motion,

$$G_{\alpha\dot{\beta}} = 0. \qquad (37)$$

Variation with respect to $\eta$ also gives an equation of motion, but this equation depends on whether we consider the minimal or nonminimal theory. For the minimal theory the variation with respect to $\Phi$ yields,

$$R = 0. \qquad (38)$$

For the nonminimal theory the variation with respect to $\Phi^{\alpha}$ yields the spinorial equation of motion,

$$T_{\alpha} = 0. \qquad (39)$$

These last two results show the need for some type of compensating superfield in superfield supergravity. Otherwise, we cannot produce the full set of equations of motion (Wess & Zumino, 1978) even in the minimal theory. In the presence of matter, source terms appear on the rhs of these equations.

THE N=2 SUPERGRAVITY AND N=1 SUPERFIELDS

The kinematic constraints for N=2 superspace are, by now, well known. These have been obtained by a number of authors (Wess, 1977; Breitenlohner & Sohnius, 1980; Castellani, van Nieuwenhuizen, and Gates, 1980) in various ways. Although these constraints are not identical, they are equivalent and related to each other by superspace contorsion transformations. One form of the constraints (Gates, 1980c) is given by

$$F_{\alpha i,\beta j}{}^{kl} - C_{\alpha\beta}\delta_i{}^{[k}\delta_j{}^{l]} = F_{\alpha i,\dot{\beta} j}{}^{kl} = T_{\alpha i,\beta j}{}^{d} = 0,$$

$$T_{\alpha i,\dot{\beta} j}{}^{d} - i2\sigma^{d}{}_{\alpha\dot{\beta}}\delta_{ij} = T_{\alpha i,\beta j}{}^{\dot{\delta} k} = 0,$$

$$T_{\alpha i,\beta j}{}^{\gamma k} = \delta^{ij} R_{\alpha i,\dot{\beta} jcd} = 0,$$

$$T_{\alpha i,\dot{\beta} j}{}^{\dot{\gamma} k} = \frac{1}{4}\,\delta_{\dot{\beta}}^{\dot{\gamma}}\,(\sigma^{\Gamma})_{ij}(\sigma_{\Gamma})^{kl}T_{\alpha l}. \qquad (40)$$

The superspace contorsion can be used to "shuttle" $T_{\alpha i}$ among $T_{\alpha i,\dot{\beta} j}{}^{\delta k}$, $T_{\alpha i,b}{}^{d}$, and $T_{\alpha i,\beta j}{}^{\delta k}$. However, $T_{\alpha i}$ must appear in at least two of these three tensors. We also note that superspace contorsion is a more general concept than its ordinary space analog (Gates & Siegel, 1980a). It can be used to redefine $\phi_{\alpha bc}$, $\phi_{abc}$, and $E_a{}^{\mu}$. The first of these in terms of components corresponds to a redefinition of the supersymmetry generator by Lorentz transformations which are field dependent. Redefinition of $\phi_{abc}$ implies changes in the component Lorentz connection and corresponds to ordinary space contorsion. Finally redefinition of

$E_a{}^{\mu i}$ implies a redefinition of the component gravitino.

The constraints above are defined for a supercovariant derivative of the form,

$$\mathcal{D}_A = E_A{}^M \partial_M + \frac{1}{2} \phi_{Ab}{}^c M_c{}^b + \frac{1}{2} \phi_A{}^{ij} \partial_{ji} \tag{41}$$

where $E_A{}^M$, $\phi_{Abc}$, and $\phi_A{}^{ij}$ are the extended vielbein, Lorentz connection, and central charge connection superfields, respectively. We denote the coordinate derivative, Lorentz generator, and central charge generator by $\partial_M$, $M_{bc}$, and $\partial_{ij}$, respectively. The constraints in (40) lead to an N=2 component theory which may be interpreted as the direct sum of the nonminimal N=1 (2,3/2) multiplet (Breitenlohner, 1977; de Wit & Grisaru, 1978) with an N=1 (3/2,1) multiplet. The latter multiplet has been discussed within the context of global N=1 supersymmetry (Ogievetsky & Sokatchev, 1976; Gates & Siegel, 1980b) and is described by, in one case, a set of prepotentials ($\Psi_\alpha$,V,$\phi'$) much akin to the prepotentials (($U_{\alpha\dot\beta}$,$\phi$) or ($U_{\alpha\dot\beta}$,$\phi^\alpha$)) of the supergravity multiplet. The component fields of the N=1 (3/2,1) multiplet are generated from a set of supertensors analogous to (33). These are

$$B_a = -i(D_\alpha \bar\Psi_{\dot\beta} + \bar D_{\dot\beta}\Psi_\alpha + i[D_\alpha, \bar D_{\dot\beta}]V),$$

$$\Psi'_{a\delta} = -i\,\frac{1}{4}\,\sigma_a{}^{\beta\dot\gamma}([\bar D_{\dot\gamma}, D_\beta]\Psi_\delta + C_{\beta\delta}\bar D_{\dot\gamma}\bar\phi'),$$

$$\lambda_\alpha = -\frac{1}{4}\,(\bar D^2\Psi_\alpha - 2\bar D_{\dot\beta} D_\alpha \bar\Psi^{\dot\beta} + D_\alpha\phi' - i2\,\bar D^2 D_\alpha V),$$

$$J = -\frac{1}{8}\,(D^\alpha \bar D^2 \Psi_\alpha - i\,4\partial_{\alpha\dot\beta}\bar D^{\dot\beta}\Psi^\alpha - \bar D^2\bar\phi'),$$

$$P = i\,\frac{1}{8}\,[D^\alpha \bar D^2\Psi_\alpha - \bar D^{\dot\alpha} D^2 \bar\Psi_{\dot\alpha} - i\,(D^\alpha \bar D^2 D_\alpha + \bar D^{\dot\alpha} D^2\,\bar D_{\dot\alpha})V]. \tag{42}$$

At $\theta$=0 these correspond to certain component fields. The remaining fields may be obtained from spinorial differentiation. It is clear that by taking variations of these we can derive differential equations for the variation of the (3/2,1) prepotentials in terms of the variations of the "component" superfields (ie. those on the lhs). The variations of the "component" superfields are known from the N=2 transformation laws and can via equations (16), (17), (18), (20), and (21) be expressed in terms of the prepotentials $U_{\alpha\dot\beta}$ and $\phi^\alpha$. The same can be done to express the variations of the supergravity prepotentials in terms of $\Psi_\alpha$, V, and $\phi'$. Thus we might try to construct the operator in (41) in terms of $U_{\alpha\dot\beta}$, $\phi^\alpha$, $\Psi_\alpha$, V, and $\phi'$ where we identify $\theta^{\alpha 1}$ with the $\theta$'s in these superfields. However, even if we were successful in this, we have no guarantee that we have not missed some important compensating superfield. (This would be analogous to attempting to describe the N=1 theory in terms of $U_{\alpha\dot\beta}$ alone!) So we need to follow a more "geometrical" approach. We can use a notation where $\theta^{\alpha 1} \equiv \theta^\alpha$, $\theta^{\alpha 2} \equiv \xi^\alpha$, $\partial_\alpha \equiv \partial/\partial\theta^\alpha$, and $\delta_\alpha \equiv \partial/\partial\xi^\alpha$. In this notation the spinorial supercovariant derivative can be written as

$$\mathcal{D}_{\alpha i} = e^1{}_i[\nabla_\alpha + \xi^\beta\nabla_{\alpha\beta}{}^{(1)} + \bar{\xi}^{\dot\beta}\nabla_{\alpha\dot\beta}{}^{(2)} + \xi^2\nabla_\alpha{}^{(3)} + \bar{\xi}^2\nabla_\alpha{}^{(4)}$$
$$+ \xi^\beta\bar{\xi}^{\dot\gamma}\nabla_{\alpha\beta\dot\gamma}{}^{(5)} + \xi^2\bar{\xi}^{\dot\beta}\nabla_{\alpha\dot\beta}{}^{(6)} + \bar{\xi}^2\xi^\beta\nabla_{\alpha\beta}{}^{(7)} + \bar{\xi}^2\xi^2\nabla_\alpha{}^{(8)}]$$
$$+ e^2{}_i[\delta_\alpha + \Delta_\alpha{}^{(0)} + \xi^\beta\Delta_{\alpha\beta}{}^{(1)} + \bar{\xi}^{\dot\beta}\Delta_{\alpha\dot\beta}{}^{(2)} + \xi^2\Delta_\alpha{}^{(3)} + \bar{\xi}^2\Delta_\alpha{}^{(4)}$$
$$+ \xi^\beta\bar{\xi}^{\dot\gamma}\Delta_{\alpha\beta\dot\gamma}{}^{(5)} + \xi^2\bar{\xi}^{\dot\beta}\Delta_{\alpha\dot\beta}{}^{(6)} + \bar{\xi}^2\xi^\beta\Delta_{\alpha\beta}{}^{(7)} + \bar{\xi}^2\xi^2\Delta_\alpha{}^{(8)}], \tag{43}$$

where each operator $\nabla^{(i)}$ and $\Delta^{(i)}$ (suppressing lower indices) is an N=1 superfield operator. For instance we have,

$$\nabla_\alpha = \mathcal{D}_\alpha + \Psi_\alpha{}^\varepsilon\delta_\varepsilon + \Psi_\alpha{}^{\dot\varepsilon}\bar{\delta}_{\dot\varepsilon} + \phi_\alpha\partial,\ \partial = \frac{1}{2}J^{ij}\partial_{ji},$$
$$\nabla^{(i)}_{\alpha\beta} = P^{(i)B}_{\alpha\beta}\mathcal{D}_B + \frac{1}{2}X^{(i)c}_{\alpha\beta b}M_c{}^b + \Psi^{(i)\varepsilon}_{\alpha\beta}\delta_{\dot\varepsilon} + \phi^{(i)}_{\alpha\beta}\partial \quad (\text{for } i \geq 1),$$
$$\Delta_\alpha = J_\alpha{}^B\mathcal{D}_B + \frac{1}{2}L_{\alpha b}{}^cM_c{}^b + \chi_\alpha{}^\varepsilon\delta_\varepsilon + \chi_\alpha{}^{\dot\varepsilon}\bar{\delta}_{\dot\varepsilon} + \chi_\alpha\partial,$$
$$\mathcal{D}_\alpha = E_\alpha{}^M\hat{\partial}_M + \frac{1}{2}\phi_{\alpha b}{}^cM_c{}^b,\ \hat{\partial}_M \equiv (\partial_\mu, \bar{\partial}_{\dot\mu}, \partial_m). \tag{44}$$

Thus it is simple to express the N=2 spinorial supercovariant derivative in terms of N=1 superfields. To simplify (43) we can choose a W-Z gauge. The parameter of gauge transformations, iK, is expanded over the same algebra as is $\mathcal{D}_{\alpha i}$.

$$iK = K^M\partial_M + \frac{1}{2}K_c{}^dM_d{}^c + \frac{1}{2}K^{ij}\partial_{ji}. \tag{45}$$

Furthermore this operator is an N=2 superfield and consequently can be expanded as

$$iK = K^{(0)} + \xi^\beta K_\beta{}^{(1)} + \bar{\xi}^{\dot\beta}\bar{K}_{\dot\beta}^{(1)} + \xi^2K^{(2)} + \bar{\xi}^2\bar{K}^{(2)}$$
$$+ \xi^\beta\bar{\xi}^{\dot\gamma}\,K_{\beta\dot\gamma}{}^{(3)} + \bar{\xi}^2\xi^\beta\,K_\beta{}^{(4)} + \xi^2\bar{\xi}^{\dot\beta}\,\bar{K}_{\dot\beta}{}^{(4)} + \xi^2\bar{\xi}^2\,K^{(5)},$$
$$K^{(0)} = \bar{K}^{(0)},\ \overline{K_{\beta\dot\gamma}{}^{(3)}} = K_{\gamma\dot\beta}{}^{(3)},\ K^{(5)} = \bar{K}^{(5)}. \tag{46}$$

where each operator $K^{(i)}$ is of the same form as the $\nabla^{(i)}$ and $\Delta^{(i)}$ operators. We can "use up" all of the gauge degrees of freedom in $K^{(i)}$ for $i \geq 1$ to simplify the $\Delta_\alpha$ operator. In such a gauge,

$$\mathcal{D}_{\alpha i} = e^1_i\hat{\nabla}_\alpha + e^2{}_i[\delta_\alpha + \xi^\beta\Delta_{(\alpha\beta)} + i\,\bar{\xi}^{\dot\beta}\Delta_{\alpha\dot\beta} + \xi^2\Delta'_\alpha + \xi^\beta\bar{\xi}^{\dot\gamma}\Delta_{\alpha\beta\dot\gamma}$$
$$+ \xi^2\bar{\xi}^{\dot\beta}\Delta'_{\alpha\dot\beta} + \bar{\xi}^2\xi^\beta(iC_{\alpha\beta}\Delta + \Delta'_{(\alpha\beta)}) + \bar{\xi}^2\xi^2\Delta_\alpha{}''],$$
$$\overline{\Delta_{\alpha\dot\beta}} = \Delta_{\beta\dot\alpha},\ \bar{\Delta} = \Delta. \tag{47}$$

In order to maintain this gauge, the $K^{(i)}$ for $i \geq 1$ are now determined functions of $K^{(0)}$ and the remaining $\Delta$ operators in (47). This implies that the remaining gauge freedom is parametrized by $K^{(0)}$ (i.e. in terms of N=1 superfields). For instance in order to perform a supersymmetry transformation with respect to $\xi$, iK is given by

$$iK = K^{\varepsilon}\delta_{\varepsilon} + \bar{K}^{\dot{\varepsilon}}\bar{\delta}_{\dot{\varepsilon}} + i(\xi^{\alpha}\bar{K}^{\dot{\beta}} + \bar{\xi}^{\dot{\beta}}K^{\alpha})\Delta_{\alpha\dot{\beta}}$$
$$+ \xi^{\alpha}K^{\beta}\Delta_{(\alpha\beta)} + \bar{\xi}^{\dot{\alpha}}\bar{K}^{\dot{\beta}}\bar{\Delta}_{(\dot{\alpha}\dot{\beta})} + O(\xi^2) \quad (48)$$

where $K^{\alpha}$ is an N=1 superfield. All $\xi$ dependent terms in iK are completely determined in order to maintain the gauge of $\hat{\Delta}_{\alpha}$. Using this gauge function we see that

$$\delta\nabla_A = -(\nabla_A K^{\varepsilon})\delta_{\varepsilon} - (\nabla_A\bar{K}^{\dot{\varepsilon}})\bar{\delta}_{\dot{\varepsilon}} + i(\Psi_A{}^{\gamma}\bar{K}^{\dot{\beta}} + \Psi_A{}^{\dot{\beta}}K^{\gamma})\Delta_{\gamma\dot{\beta}}$$
$$+ \Psi_A{}^{\beta}K^{\gamma}\,\Delta_{(\beta\gamma)} + \Psi_A{}^{\dot{\beta}}\bar{K}^{\dot{\gamma}}\Delta_{(\dot{\beta}\dot{\gamma})}$$
$$+ K^{\beta}\nabla^{(1)}_{A\beta} + \bar{K}^{\dot{\beta}}\nabla^{(2)}_{A\dot{\beta}}, \quad (49)$$

$$\rightarrow\delta E_A{}^M = i(\Psi_A{}^{\gamma}\bar{K}^{\dot{\beta}} + \Psi_A{}^{\dot{\beta}}K^{\gamma})J_{\gamma\dot{\beta}}{}^D E_D{}^M$$
$$+ [\Psi_A{}^{\beta}K^{\gamma}J_{(\beta\gamma)}{}^D + \Psi_A{}^{\dot{\beta}}\bar{K}^{\dot{\gamma}}J_{(\dot{\beta}\dot{\gamma})}{}^D]E_D{}^M$$
$$+ [K^{\beta}P^{(1)D}_{A\beta} + \bar{K}^{\dot{\beta}}P^{(2)D}_{A\dot{\beta}}]E_D{}^M \equiv \Pi_A{}^D E_D{}^M,$$

$$\rightarrow E^{-1} = -E^{-1}(-)^A\,\Pi_A{}^A. \quad (50)$$

Thus we find the result that under the second supersymmetry transformation the action for Poincaré supergravity undergoes a rescaling. The constraints in (40) will determine most of the quantities in terms of $E_{\alpha}{}^M$, $\Psi_{\alpha}{}^{\beta}$, $\Psi_{\alpha}{}^{\dot{\beta}}$, and $\phi_{\alpha}$. These quantities are then related to a set of prepotentials. The expression in (47) can be used to "translate" the N=2 kinematic constraints of (40) into a set of N=1 kinematic constraints. The derivation and solution of these constraints will then lead to an unconstrained N=1 formulation of N=2 supergravity. This method obviously generalizes to higher N theories. At this point, however, we will simply discuss some features of the linearization of this formulation. We hope to discuss the full nonlinear problem in the future.

The flat superspace limit of the N=2 supercovariant derivative is given by

$$\mathcal{D}_{\alpha i} = e_i^1 D'_{\alpha} + e_i^2\tilde{D}_{\alpha},\; D'_{\alpha} = D_{\alpha} + \xi_{\alpha}\partial,\; \tilde{D}_{\alpha} = \delta_{\alpha} + i\bar{\xi}^{\dot{\gamma}}\partial_{\alpha\dot{\nu}}, \quad (51)$$

$$\mathcal{D}_a = i\,\frac{1}{8}\,\sigma_a{}^{\alpha\dot{\beta}}\delta^{ij}[\mathcal{D}_{\alpha i},\bar{\mathcal{D}}_{\dot{\beta}j}\} = \partial_a. \quad (52)$$

We now linearize (47) about this configuration which means we write

$$\mathcal{D}_{\alpha i} = e_i^1[D'_{\alpha} + \hat{\nabla}_{\alpha}] + e_i^2\,[\tilde{D}_{\alpha} + (\hat{\Delta}_{\alpha}-\delta_{\alpha})]. \quad (53)$$

In this expression $\hat{\nabla}_{\alpha}$ and $\hat{\Delta}_{\alpha}$ are the operators in (47), but are now considered as infinitesimals. The constraints in (40) imply

$$[\mathcal{D}_{\alpha i},\mathcal{D}_{\beta j}\} = - C_{\alpha\beta}[\partial_{ij} - 4\,\bar{W}_{\dot{\gamma}\dot{\delta}ij}\bar{M}^{\dot{\gamma}\dot{\delta}}]. \tag{54}$$

We will work to order $\xi^2$ in $\mathcal{D}_{\alpha i}$ in the following. This implies that we compute the spinor-spinor curvatures and torsions to order $\xi$. Since we use (52) as our definition of the vectorial derivative, we only have this operator to order $\xi$ also. So we only will be able to compute spinor-vector torsions and curvatures and vector-vector torsions and curvatures to order $\xi^0$. This will simplify our discussion. By choosing i = j = 2 (54) implies,

$$\xi^0:\ \Delta_{(\alpha\beta)} = 0,$$

$$\xi:\ 2C_{\delta(\alpha}\Delta'_{\beta)} - i\,\frac{1}{4}\,[L_{(\beta\delta)\alpha}{}^{\varepsilon} + L_{(\alpha\delta)\beta}{}^{\varepsilon}]\delta_{\varepsilon} + i[\chi_{(\beta\delta)}{}^{\dot{\gamma}}\partial_{\alpha\dot{\gamma}} + \chi_{(\alpha\delta)}{}^{\dot{\gamma}}\partial_{\beta\dot{\gamma}}] = 0,$$

$$\bar{\xi}:\ \Delta_{(\alpha\beta)\dot{\delta}} - \frac{1}{4}\,L_{(\alpha\dot{\delta}\beta)}{}^{\varepsilon}\delta_{\varepsilon} - \chi_{(\alpha\dot{\delta}}{}^{\dot{\varepsilon}}\partial_{\beta)\dot{\varepsilon}} = 0. \tag{55}$$

The first two equations simply imply the vanishing of $\Delta_{(\alpha\beta)}$ and $\Delta'_{\beta}$. (Note that $L_{(\beta\delta)\alpha}{}^{\varepsilon}$ and $\chi_{(\beta\delta)\alpha}{}^{\dot{\gamma}}$ are the coefficient superfunctions of $M_{dc}$ and $\delta_{\dot{\gamma}}$ in $\Delta_{(\beta\delta)}$ by equation (44). $\Delta_{(\alpha\beta)} = 0 \rightarrow L_{(\beta\delta)\alpha}{}^{\varepsilon} = \chi_{(\beta\delta)\alpha}{}^{\dot{\gamma}} = 0$). The third equation implies

$$\Delta_{\alpha\beta\dot{\delta}} = C_{\alpha\beta}\Delta'''_{\dot{\delta}} + \frac{1}{8}\,L_{(\alpha\dot{\delta}\beta)}{}^{\varepsilon}\delta_{\varepsilon} + \frac{1}{2}\,\chi_{(\alpha\dot{\delta}}{}^{\dot{\varepsilon}}\partial_{\beta)\dot{\varepsilon}} \tag{56}$$

for some operator $\Delta'''_{\dot{\delta}}$.

By choosing i = j = 1 (54) implies,

$$\xi^0:\ D_{(\alpha}\nabla_{\beta)} + i2\,H_{(\alpha}{}^{\dot{\varepsilon}}\partial_{\beta)\dot{\varepsilon}} + i\,\frac{1}{4}\,\phi_{(\beta\alpha)}{}^{\delta}D_{\delta} + \Psi_{(\alpha\beta)}\partial = 0,$$

$$\xi:\ D_{(\alpha}\nabla^{(1)}_{\beta)\gamma} - i2\,H^{(1)\dot{\varepsilon}}_{(\alpha}\partial_{\beta)\dot{\varepsilon}} - i\,\frac{1}{4}\,[\phi_{(\alpha\beta)\gamma}\partial + \phi^{(1)}_{(\alpha\gamma\beta)}{}^{\delta}D_{\delta}] - \Psi^{(1)}_{(\alpha\gamma\beta)}\partial = 0,$$

$$\bar{\xi}:\ D_{(\alpha}\nabla^{(2)}_{\beta)\dot{\gamma}} - i2\,H^{(2)\dot{\varepsilon}}_{(\alpha\dot{\gamma}}\partial_{\beta)\dot{\varepsilon}} - i\,\frac{1}{4}\,\phi^{(2)}_{(\alpha\dot{\gamma}\beta)}{}^{\delta}D_{\delta} + \Psi^{(2)}_{(\alpha\dot{\gamma}\beta)}\partial = 0. \tag{57}$$

We next set i = 1, j = 2 to find,

$$\xi^0:\ i\,\frac{1}{4}\,\phi_{\alpha\beta}{}^{\varepsilon}\delta_{\varepsilon} + i\,\Psi_{\alpha}{}^{\dot{\varepsilon}}\partial_{\beta\dot{\varepsilon}} + \nabla^{(1)}_{\alpha\beta} = 4C_{\alpha\beta}(\bar{W}_{\dot{\gamma}\dot{\delta}})_0\,\bar{M}^{\dot{\gamma}\dot{\delta}}$$

$$\xi:\ -D_{\alpha}\Delta_{(\beta\delta)} + i\,\frac{1}{4}\,L_{(\beta\delta)\alpha}{}^{\gamma}D_{\gamma} + \chi_{(\beta\delta)\alpha}\partial + i2\,J_{(\beta\delta)}{}^{\dot{\varepsilon}}\partial_{\alpha\dot{\varepsilon}} + i\,\frac{1}{4}\,(P^{(1)B}_{\alpha\delta}\phi_{B\beta}{}^{\varepsilon} + X^{(1)\varepsilon}_{\alpha\delta\beta})\delta_{\varepsilon} + i\Psi^{(1)\dot{\varepsilon}}_{\alpha\delta}\partial_{\beta\dot{\varepsilon}} + 2C_{\delta\beta}\nabla^{(3)}_{\alpha} = 4C_{\alpha\beta}(\delta_{\delta}\bar{W}_{\dot{\gamma}\dot{\delta}})_0\,\bar{M}^{\dot{\gamma}\dot{\delta}}$$

$$\xi:\ -D_{\alpha}\Delta_{\beta\dot{\delta}} + i\,\frac{1}{4}\,L_{\beta\dot{\delta}\alpha}{}^{\gamma}D_{\gamma} + i2\,J_{\beta\dot{\delta}}{}^{\dot{\varepsilon}}\partial_{\alpha\dot{\varepsilon}} + \chi_{\beta\dot{\delta}\alpha}\partial - i\,\frac{1}{4}\,\phi_{\alpha\beta}{}^{\varepsilon}\partial_{\varepsilon\dot{\delta}} + i\,\frac{1}{4}\,(P^{(2)B}_{\alpha\dot{\delta}}\phi_{B\beta}{}^{\varepsilon} + X^{(2)\varepsilon}_{\alpha\dot{\delta}\beta})\delta_{\varepsilon} + i\Psi^{(2)\dot{\varepsilon}}_{\alpha\dot{\delta}}\partial_{\beta\dot{\varepsilon}}$$

$$= 4\, C_{\alpha\beta} (\bar{\delta}_{\dot{\delta}} \bar{W}_{\dot{\gamma}\dot{\delta}})_0\, \bar{M}^{\dot{\gamma}\dot{\delta}} \tag{58}$$

where the notation $(\ )_0$ denotes the $\xi = 0$ value of the N=2 superfield. Thus we find a set of N=1 superfield equations which follow from the N=2 extended formulation. Of course, equation (54) is not the full set of constraints for N=2 supergravity. The remaining constraints of (40) generate more N=1 equations. At this point, we end our discussion of this aspect of the problem. We will defer a discussion of the solutions of these equations until the future.

We note that the gauge choice of $\mathfrak{D}_{\alpha i}$ suggests that an exponential solution similar to the N=1 theory (Siegel, 1978) should play a role in the N=2 theory. For the N=2 theory it is simple to see that a semicovariant vielbein may be defined as

$$\hat{E}_{\alpha i} = \exp[-F^M \partial_M][e_i^1 \partial_\alpha + e_i^2 \delta_\alpha] \exp[F^M \partial_M], \tag{59}$$

where $F^M \partial_M$ is an N=2 superfield operator. If we expand $F^M$ to all orders in $\theta$ and x but to zero order in $\xi$ we find

$$E_{\alpha i} = \exp[-\Psi] \exp[-W][e_i^1 \partial_\alpha + e_i^2 \delta_\alpha] \exp[W] \exp[\Psi]$$
$$W = W^M \hat{\partial}_M, \quad \Psi = \Psi^\varepsilon \delta_{\dot{\varepsilon}} + \bar{\Psi}^\varepsilon \delta_{\dot{\varepsilon}}. \tag{60}$$

Thus the N=1 semicovariant derivative, $\hat{E}_\alpha$, appears gauge covariantized with respect to the second supersymmetry by $\Psi$. The expression in (59) obviously leads to a fully covariant derivative of the form of (47). Restoring the $\xi$ terms, but choosing a Wess-Zumino gauge with iK and exploiting the $\xi$-chiral invariance of the $\delta_\alpha$ term implies that $E_{\alpha i}$ can be written as,

$$E_{\alpha i} = e_i^1[e^{-F} e^{-G} e^{-\Psi} \hat{E}_\alpha e^{\Psi} e^{G} e^{F}] + e_i^2[e^{-F} \delta_\alpha e^{F}]$$

$$F = i\xi^\beta \bar{\xi}^{\dot{\gamma}} f_{\beta\dot{\gamma}} + \xi^2 \bar{\xi}^{\dot{\beta}} \bar{f}_{\dot{\beta}} + i\xi^2 \bar{\xi}^2 f, \quad G = \bar{\xi}^{\dot{\beta}} \bar{g}_{\dot{\beta}} + \bar{\xi}^2 g \tag{61}$$

where $f_{\beta\dot{\gamma}}$, $\bar{f}_{\dot{\beta}}$, f, $\bar{g}_{\dot{\beta}}$, and g are all N=1 superfields expanded over $\partial_M$. (It is probably convenient in some applications to expand these superfields over $\hat{E}_\alpha$, $\hat{E}_{\dot{\alpha}}$, $\hat{E}_a$, $\delta_\varepsilon$, and $\bar{\delta}_{\dot{\varepsilon}}$ instead of $\partial_M$.)

The semicovariant vielbein superfield can be utilized to define a fully covariant one

$$E_{\alpha i} = N_{\alpha i}{}^{\beta j} \hat{E}_{\beta j}, \quad \det(N) = 1. \tag{62}$$

Thus as an alternative to utilizing (47) to generate a set of N=1 constraints, it is possible to solve some of the constraints in (40) in terms of N=2 superfields first and then solve others in terms of N=1 superfields. This requires a constraint analysis which has been described previously (Gates & Siegel, 1980a). We note that some constraints simply can be solved algebraically to determine a superfield. For instance

$$T_{\alpha i[bc]} = i\,\frac{1}{6}\,(\sigma_{bc})_\alpha{}^\delta\, T_{\delta i d}{}^d \quad \&\quad \delta^{ij}\, R_{\alpha i, \dot{\beta} j c d} = 0 \tag{63}$$

which define $\phi_{Abc}$ follows from (40). Similarly the constraints

$$\delta^{ij} C_{\alpha i,\dot\beta j}{}^{d} - i4\sigma^{d}{}_{\alpha\dot\beta} = C_{\alpha i,\dot\beta j}{}^{\dot\beta j} - C_{\alpha i,b}{}^{b} = 0,$$

$$C_{\alpha i(\dot\beta j,\dot\gamma)}{}^{j} - i(\bar\sigma^{bd})_{\dot\beta\dot\gamma}\, C_{\alpha ibd} = 0, \tag{64}$$

define the vielbein superfield $E_a{}^M$. Similarly the equation

$$\delta^{ij} F_{\alpha i,\beta j}{}^{kl} = 0, \tag{65}$$

defines $\phi_{aij}$. Other constraints give rise to the N=2 prepotentials. These constraints are

$$C_{\alpha i,\beta j}{}^{\dot\delta k} = C_{\alpha i,\beta j}{}^{d} = F_{(\alpha i,\beta)j}{}^{kl} = 0. \tag{66}$$

All of the remaining constraints

$$(\sigma^{\Gamma})^{ij} C_{\alpha i,\dot\beta j}{}^{d} = (\sigma^{\Gamma})_{k}{}^{j} C_{\alpha i(\dot\beta j,\dot\gamma)}{}^{k} = 0$$

$$(\sigma^{\Gamma})_{k}{}^{j} C_{\alpha i,\dot\beta j}{}^{\dot\beta k} + \frac{1}{3}(\sigma^{\Gamma})_{i}{}^{j} C_{\alpha jd}{}^{d} = 0,$$

$$(\sigma^{\Gamma})^{ij} F_{\alpha i,\dot\beta j}{}^{kl} = F_{\alpha i,\ j}{}^{\alpha\ kl} - \delta_{i}{}^{[k}\delta_{j}{}^{l]} = 0. \tag{67}$$

are actual constraints on the N=2 prepotentials $(F^M, \phi^{kl})$ and must be solved in terms of N=1 superfields.

## APPENDIX A: LINEARIZED N=1 OBJECT OF ANHOLONOMITY

Here we present expressions for the linearized N=1 superspace object of anholonomity. These expressions are given only in terms of the superfields $(U_{\alpha\dot\beta}, \eta)$ and are only valid in the gauge of equation (13).

$$C_{\alpha\beta}{}^{d} = 0,\ C_{\alpha\beta}{}^{\dot\delta} = 0,\ C_{\alpha\dot\beta}{}^{d} = i2\sigma^{d}{}_{\alpha\dot\beta},$$

$$C_{\alpha\dot\beta}{}^{\dot\delta} = \frac{1}{4}[D^2\bar D^{\dot\delta} U_{\alpha\dot\beta} - \frac{3}{4}\delta_{\dot\beta}^{\dot\delta} D^2\bar D^{\dot\gamma} U_{\alpha\dot\gamma} + 2\,\delta_{\dot\beta}^{\dot\delta} D_{\alpha}(2\eta+\bar\eta)],$$

$$C_{\alpha\beta}{}^{\delta} = [D_{(\alpha}\eta]\delta_{\beta)}^{\delta},$$

$$C_{\alpha b}{}^{d} = -\frac{1}{4}\sigma_{[b\alpha\dot\beta} D^2\bar D^{\dot\beta} U^{d]} + \frac{3}{32}\sigma^{d}{}_{\alpha\dot\beta}\sigma_{b}{}^{\gamma\dot\beta} D^2\bar D^{\dot\delta} U_{\gamma\dot\delta}$$
$$+ \frac{3}{4}\delta_{b}^{d} D_{\alpha}(2\eta+\bar\eta) - i\,\frac{1}{4}(\sigma_{b}{}^{d})_{\alpha}{}^{\beta} D_{\beta}(2\eta-\bar\eta),$$

$$C_{a\beta}{}^{\delta} = -i\,\frac{1}{16}\sigma_{a}{}^{\kappa\dot\gamma}[D_{\beta}\bar D^2 D^{\delta} U_{\kappa\dot\gamma} - \frac{3}{4}\delta_{\kappa}^{\delta} D_{\beta}\bar D^2 D^{\varepsilon} U_{\varepsilon\dot\gamma}]$$
$$+ i\,\frac{1}{8}\sigma_{a}{}^{\delta\dot\gamma} D_{\beta}\bar D_{\dot\gamma}(2\bar\eta-\eta) + (\partial_{a}\eta)\delta_{\beta}^{\delta},$$

$$C_{a\beta}{}^{\dot\delta} = i\,\frac{1}{16}\sigma_{a\beta}{}^{\dot\delta} D^2(2\eta-\bar\eta)$$

$$C_{ab}{}^{\delta} = \partial_{[a}\Psi_{b]}{}^{\delta},\quad C_{ab}{}^{d} = \partial_{[a}h_{b]}{}^{d}.$$

The components of the object not given here can be constructed by superspace conjugation. We emphasize also that these are valid <u>subject</u> to the constraints of (10). Since we have given these expressions in terms of $\eta$ they are equally valid for the minimal or nonminimal theory. We need only substitute the appropriate expression for $\eta$.

Minimal:

$$\eta = (\phi - 2\bar{\phi}) - \frac{1}{12}\left[2D^{\beta}\bar{D}^{\dot{\gamma}}U_{\beta\dot{\gamma}} + \bar{D}^{\dot{\gamma}}D^{\beta}U_{\beta\dot{\gamma}}\right], \quad \bar{D}_{\dot{\beta}}\phi = 0.$$

Nonminimal:

$$\eta = D_{\alpha}\phi^{\alpha}.$$

REFERENCE LIST

Breitenlohner, P. (1977). A geometric interpretation of local supersymmetry. Phys. Lett. 67B, pp. 49-51; ibid. Some invariant Lagrangians for local supersymmetry. Nucl. Phys. B124, pp. 500-510.

Breitenlohner, P. & Sohnius, M. (1980). Superfields, auxiliary fields and tensor calculus for N=2 extended supergravity. Nucl. Phys.B165, pp. 483-510.

Brown, M. & Gates, S. (1980). Superspace geometry and N=1 Nonminimal supergravity. Nucl. Phys. B165, pp. 445-461.

Castellani, L., van Nieuwenhuizen, P. & Gates, S. (1980). The constraints for N=2 superspace from extended supergravity in ordinary space. Stony Brook preprint ITP-SB-80-9 (unpublished).

de Wit, B. & Grisaru, M. (1978). Covariant quantization of supergravity. Phys. Lett. 74B, pp. 57-60.

de Wit. B. & van Holten, J. (1979). Multiplets of linearized SO(2) supergravity. Nucl. Phys. B155, pp. 530-542.

Ferrara, S. & van Nieuwenhuizen, P. (1978). The auxiliary fields of supergravity. Phys. Lett. 74B, pp. 333-335.

Ferrara, S. & Zumino, B. (1978). Structure of linearized supergravity and conformal supergravity. Nucl. Phys. B134, pp. 301-306.

Fradkin, E. & Vasiliev, M. (1979). Minimal set of auxiliary fields and S-matrix for extended supergravity. Lett. Al. Nuovo Cimento, No. 25, pp. 79-87; ibid. Minimal set of auxiliary fields in SO(2)-Extended supergravity. Phys. Lett. 85B, pp. 47-51.

Gates, S. (1980a). Superconformal transformations and six dimensional spacetime. Nucl. Phys. B162, pp. 79-92.

Gates, S. (1980b). Supercovariant derivatives, super-Weyl groups, and N=2 supergravity. Harvard preprint HUTP-80/A011 (unpublished).

Gates, S. (1980c). Another solution for N=2 superspace Bianchi identities. Harvard preprint HUTP-80/A042 (unpublished).

Gates, S. & Shapiro, J. (1978). Local supersymmetry in superspace. Phys. Rev. D 18, pp. 2768-2775.

Gates, S. & Siegel. W. (1979). Superfield supergravity. Nucl. Phys. B147, pp. 77-104.

Gates, S. & Siegel, W. (1980a). Understanding constraints in superspace formulations of supergravity. Nucl. Phys. B163, pp. 519-545.

Gates, S. & Siegel, W. (1980b). (3/2,1) superfield of O(2) supergravity. Nucl. Phys. B164, pp. 484-494.

Howe, P. & Tucker, R. (1979). A note on the linearization of supergravity. Journ. of Phys. A12, pp. L21-24.

Namazie, M. & Storey, D. (1979). Supersymmetric quantization of linearized supergravity. Nucl. Phys. B157, pp. 170-188.

Ogievetsky, V. & Sokatchev, E. (1976). Gauge spinor superfield. (Zh. Eksp. Teor. Fiz. Pis'ma Red. 23, pp. 66-68.) JETP Lett. 23, pp. 58-59.

Ogievetsky, V. & Sokatchev. E. (1977). On a vector superfield generated by the supercurrent. Nucl. Phys. B124, pp. 309-316.

Siegel. W. (1977). The superfield supergravity action, Harvard preprint HUTP-77/A080 (unpublished).

Siegel, W. (1978). Solution to constraints in Wess-Zumino supergravity formalism. Nucl. Phys. B142, pp. 301-305.

Stelle, K. & West, P. (1978). Minimal auxiliary fields for supergravity. Phys. Lett. 74B, pp. 333-335.

Wess, J. (1977). Contribution to the EPS Conference, Geneva. (unpublished).

Wess, J. & Zumino, B. (1978). Superfield Lagrangian for supergravity. Phys. Lett. 74B, pp. 51-53.

*Research supported in part by the National Science Foundation under Grant No. PHY77-22864.

** Junior Fellow, Harvard Society of Fellows.

† Address after September 1, 1980: California Institute of Technology, Pasadena, CA 91125.

# EXTENDED CONFORMAL SUPERGRAVITY AND ITS APPLICATIONS

E. Bergshoeff, M. de Roo, J.W. van Holten
Instituut-Lorentz, University of Leiden

B. de Wit
NIKHEF-H, Amsterdam

A. Van Proeyen
Instituut voor Theoretische Natuurkunde, K.U. Leuven

## 1. INTRODUCTION

There exist different kinds of supergravity theories which can be viewed as extensions to local supersymmetry of Weyl, Poincaré and de Sitter gravity. In the first part of this talk we discuss the status of the extended superconformal theories, i.e. Weyl supergravity with N independent supersymmetries ($N \leq 4$). This study of conformal supergravity is not motivated by the hope that these theories have direct physical applicability. In fact our initial motivation for using superconformal concepts was to gain insight in the complicated structure of N=2 Poincaré supergravity. This strategy was chosen because conformal supergravity is considerably simpler owing to its higher degree of invariance. The extra symmetry then poses strong restrictions on possible deviations from rigid-supersymmetry results.

However, the use of superconformal concepts goes beyond merely providing a clarification of the de Sitter or Poincaré supergravity results. In the second part of this talk we demonstrate for N=2 supergravity how to construct Poincaré or de Sitter supergravity directly from the conformal theory by means of appropriately chosen gauge conditions and compensating fields. The crucial point is that the superconformal calculus suffices to obtain a large variety of invariant actions for all supergravity theories, so that there is no need to develop a selfcontained calculus for Poincaré supergravity.

In section 2 we discuss the multiplets corresponding to N=2 and N=4 conformal supergravity. From the N=2 Weyl multiplet and a superconformal vector-multiplet we construct in section 3 the so-called minimal field representation, which is the starting point for the reduction to Poincaré and de Sitter supergravity theories. The incorporation of central charges in the superconformal framework is discussed in section 4. In section 5 we construct Poincaré and de Sitter supergravity, by adding suitable multiplets to the minimal field representation and by imposing corresponding gauge conditions. We present our conclusions in section 6.

## 2. WEYL MULTIPLETS

The starting point for the superconformal theories is the graded conformal algebra SU(2,2|N) (Ferrara, Kaku, Townsend & van Nieuwenhuizen, 1977), whose generators are listed in Table 1. The corresponding gauge fields are those of the conformal group SU(2,2), the supersymmetry transformations, and the chiral group U(N) or SU(N). (For N=4 the graded algebra only implies SU(4) ). The corresponding curvatures are denoted by $R^{a}_{\mu\nu}(P)$, $R^{ab}_{\mu\nu}(M)$, etc. The transformation rules of gauge fields and curvatures are directly determined by the structure constants of the SU(2,2|N) algebra. In addition these quantities transform under general coordinate transformations, which at this stage are still completely separated from the superconformal symmetries. In particular the commutator of two Q-supersymmetry transformations yields a local translation (P), which is unrelated to a general coordinate transformation. To convert the local translations into general coordinate transformations, which will bring us back to the conventional supersymmetry algebras, requires some of the gauge fields to be expressed in terms of the remaining ones. This is done by imposing constraints on the curvatures:

$$\begin{aligned} R^{a}_{\mu\nu}(P) + \dots &= 0 \,, \\ R^{ab}_{\mu\nu}(M)e^{\nu}{}_{b} + \dots &= 0 \,, \\ \gamma^{\mu}R^{i}_{\mu\nu}(Q) + \dots &= 0 \,. \end{aligned} \tag{2.1}$$

We have only exhibited the characteristic curvature components of these constraints and we refer to the literature for explicit examples (Kaku, Townsend & van Nieuwenhuizen, 1978; de Wit, 1979; de Wit, van Holten & van Proeyen, 1980a). The constraints (2.1) eliminate $\omega_{\mu}{}^{ab}$, $f_{\mu}{}^{a}$, and $\phi_{\mu}{}^{i}$ as independent gauge fields, and thus reduce the number of bosonic degrees of freedom by 24 + 16, and the fermionic ones by 16N.

The implementation of (2.1) has two immediate consequences. First the transformation rules of $\omega_{\mu}^{ab}$, $f_{\mu}{}^{a}$, and $\phi_{\mu}{}^{i}$ are no longer fully determined by the superconformal algebra. The reason is that the constraints are not invariant under superconformal transformations. Therefore additional variations for $\omega_{\mu}^{ab}$, $f_{\mu}{}^{a}$, and $\phi_{\mu}{}^{i}$, must compensate this lack of invariance. This leads to the second consequence, which is that the algebra is no longer guaranteed to close. To see that this is indeed the case one may count the bosonic and fermionic degrees of freedom. We find

$45 - 24 - 16 = 5$ degrees of freedom for P,M,D,K gauge fields,

$3N^2$ or $3(N^2 - 1)$ degrees of freedom for C and A gauge fields,

$24N - 16N = 8N$ degrees of freedom for Q and S gauge fields.

Table 1: Generators and corresponding gauge fields of the superconformal group. The numbers in the right column denote the degrees of freedom represented by the gauge fields. The U(1) symmetry is absent in the case of N=4.

| superconformal gauge symmetries | | | gauge fields | |
|---|---|---|---|---|
| 4 | translations | P | $e_\mu{}^a$ | 45 |
| 6 | Lorentz rotations | M | $\omega_\mu{}^{ab}$ | |
| 1 | dilatation | D | $b_\mu$ | |
| 4 | conformal boosts | K | $f_\mu{}^a$ | |
| 4N | supersymmetries | Q | $\psi_\mu{}^i$ | 24N |
| 4N | special supersymmetries | S | $\phi_\mu{}^i$ | |
| $N^2-1$ | chiral SU(N) | C | $V_\mu{}^i{}_j$ | $3(N^2-1)$ |
| 1 | chiral U(1) | A | $A_\mu$ | 3 |

Since the number of bosonic degrees of freedom exceeds the number of fermionic ones for $N > 1$ the algebra will no longer close. This signals the need for matter fields in the basic field representation of conformal supergravity. In fact, this was already known for N=2, where we found 7 bosonic, and 8 fermionic degrees of freedom which were not represented by gauge fields (de Wit & van Holten, 1979). For N=4 the corresponding numbers are 78 and 96, as we shall discuss later.

The field representations of conformal supergravity, which we call the Weyl multiplets, can be represented in terms of their gauge covariant components as a chiral superfield. For N=2 this superfield is an antidual Lorentz tensor $W^{ij}_{ab}$, antisymmetric in SU(2) indices i,j. We have the following decomposition

$$\begin{aligned} W^{ij}_{ab}(z^\mu,\theta^k) &= T^{-ij}_{ab} + 4\bar\theta^{[i}\mathcal{R}_{ab}{}^{j]}(Q) + \bar\theta^k\theta^{[i}\mathcal{R}_{ab}{}^{j]}{}_k(\mathcal{V}) \\ &- 2\bar\theta^i\sigma_{cd}\theta^j\mathcal{R}^{cd}_{ab}(M) + \frac{2}{3}\varepsilon_{k\ell}\,\bar\theta^k\sigma_{cd}\theta^\ell\bar\theta^m\sigma_{cd}\,\mathcal{R}_{abm}(S)\varepsilon^{ij} \\ &- \frac{1}{12}(\varepsilon_{mn}\bar\theta^m\sigma_{cd}\theta^n)^2\,\varepsilon^{ij}\varepsilon^{k\ell}(\partial_{[a}\partial_c T^+_{cb]k\ell} - \varepsilon_{abde}\partial_d\partial_c T^+_{cek\ell}). \end{aligned} \tag{2.2}$$

This field is subject to a constraint, which reads in flat superspace as

$$(\varepsilon_{ij}\bar{D}^{i}\sigma_{cd}D^{j})^2(\partial_a W^{k\ell}_{ab}\varepsilon_{k\ell})^* = -96\,\Box\,\partial_a W^{k\ell}_{ab}\varepsilon_{k\ell}. \qquad (2.3)$$

The restriction (2.3) is precisely equivalent to the previously mentioned curvature constraints (2.1) combined with the Bianchi identities of the superconformal gauge theory. The various components of $W^{ij}_{ab}$ consist of modified curvatures, which are superconformally covariant and contain the spinor matter field $\chi_C{}^i$ (related to $\sigma^{ab}\mathcal{R}^{i}_{ab}(Q)$), and the Lorentz scalar $D_C$ (related to $\mathcal{R}^{ab}_{ab}(M)$). The Lorentz tensor $T^{-ij}_{ab}$ occurs explicitly in (2.2).

The full structure of $W^{ij}_{ab}$ is known with the corresponding transformation properties and constraints. Some of these results were found from a study of N=2 Poincaré supergravity (de Wit et al., 1980a), and more complete results will be published elsewhere (Bergshoeff, de Roo & de Wit, 1980). The linearized results can also be obtained from the N=2 multiplet of currents (Sohnius, 1979).

The superfield

$$W^2 = W^{ij}_{ab}W^{k\ell}_{ab}\varepsilon_{ij}\varepsilon_{k\ell} \qquad (2.4)$$

is a chiral superfield with Weyl weight w=2. Therefore its highest component has Weyl weight w=4, and can be used for the construction of a superconformally invariant action (see de Roo, van Holten, de Wit & Van Proeyen (1980), where a general formula is given). This yields the N=2 superconformal theory.

In section 3 we give a method to construct invariant densities from chiral superfields with arbitrary Weyl weight w. It can be applied to the superfield $W^{ij}_{ab}$ to construct counterterms for N=2 Poincaré supergravity. In fact, on using the Poincaré field equations this yields all nonvanishing counterterms. To investigate these counterterms, we must rewrite the components of $W^{ij}_{ab}$ in terms of Poincaré fields. Upon using the field equations we obtain

$$T^{-ij}_{ab} = -\sqrt{2}F(B)^{ij}_{ab} \qquad \text{-dual,}$$

$$\mathcal{R}^{i}_{ab}(Q) = \partial_{[a}\psi^{i}_{b]} \qquad \text{-dual,} \qquad (2.5)$$

$$\mathcal{R}^{cd}_{ab}(M) = \left(C(\omega)^{cd}_{ab} + \frac{\kappa^2}{16}\,[T^{-ij}_{ab}T^{+}_{cdij} + \text{h.c}]\right) \qquad \text{-dual,}$$

where $C^{cd}_{ab}$ represents the Weyl tensor.

The one-loop counterterm is equal to the highest component of $W^2$, and is related to the square of the Weyl tensor. For pure gravity the square of the Weyl tensor can be written as a total derivative because of the Gauss-Bonnet theorem. By supersymmetry we expect this to be true for the full counterterm as well. To check this we have verified that the $T^4$ contribution from $\mathcal{R}(M)^2$ cancels against $T^4$ contributions generated by other terms in the density formula (de Roo et al., 1980). A candidate for the two-loop counterterm, constructed out of $W^3$, vanishes because of the antisymmetry of the Lorentz-indices of $W^{ij}_{ab}$. At the three-loop level it is possible to construct a nonvanishing counterterm from the superfield $W^2\mathbb{T}(W^2)$ (where $\mathbb{T}(W^2)$ is the kinetic multiplet corresponding to $W^2$) or from $W^2(W^*)^2$. This counterterm is related to the square of the Bel-Robinson tensor (Deser, Kay & Stelle, 1977). This situation is the same as in the N=1 case (Ferrara & van Nieuwenhuizen, 1978).

These results can be extended to N=4. The gauge-covariant components of conformal supergravity can again be represented by a chiral superfield, which in this case is a Lorentz scalar. The expansion in $\theta^i$ is as follows:

$$\begin{aligned} W(z^\mu,\theta^i) = {} & C + \bar\theta^i\Lambda_i + \tfrac12\bar\theta^i\theta^j E_{ij} + \tfrac12\bar\theta^i\sigma_{ab}\theta^j\varepsilon_{ijk\ell}T^{k\ell}_{ab} \\ & + \frac13\,\bar\theta^i\sigma_{ab}\theta^j\varepsilon_{ijk\ell}\{\bar\theta^m\sigma_{ab}\chi^{k\ell}_m + \bar\theta^k\mathcal{R}^\ell_{ab}(Q)\} \\ & + \frac{1}{12}\,\bar\theta^i\sigma_{ab}\theta^j\varepsilon_{ijk\ell}\{\tfrac12\bar\theta^k\theta^m\mathcal{R}^{\ \ell}_{abm}(\mathcal{V}) + \bar\theta^m\sigma_{ab}\theta^n D^{k\ell}_{mn} \\ & - \bar\theta^k\sigma_{cd}\theta^\ell\mathcal{R}^{cd}_{ab}(M)\} + \dots \text{ higher } \theta \text{ terms}\ . \end{aligned} \tag{2.6}$$

This multiplet has also been discussed by Howe & Lindström (1980) and Siegel (1980), and is related to a multiplet presented by Taylor (1980). It contains scalar fields C, $E_{ij}$ and $D^{k\ell}_{mn}$. C and E are a complex singlet and a complex symmetric tensor, respectively. $D^{k\ell}_{mn}$ is antisymmetric in both pairs of indices, traceless, and corresponds to the real 20-dimensional representation. The spinors $\Lambda_i$ and $\chi^{k\ell}_m$ are left-handed, and transform as the complex 4- and 20-dimensional representations, respectively. The Lorentz tensors $T^{ij}_{ab}$, $\mathcal{R}^{\ \ \ell}_{ab\ m}(\mathcal{V})$, $\mathcal{R}^{cd}_{ab}(M)$ and the left-handed tensor-spinor $\mathcal{R}^\ell_{ab}(Q)$ are all antiselfdual in their Lorentz-indices, and are the analogues of the components of the N=2 superfield $W^{ij}_{ab}$ in (2.2). The algebraic structure of (2.6) implies for the modified curvatures of N=4

$$\begin{aligned} & \sigma\cdot\mathcal{R}^i(Q) = 0 \\ & \mathcal{R}^{cd}_{ab}(M) = \mathcal{R}(M)^{ab}_{cd} \\ & \mathcal{R}^{ab}_{ab}(M) = 0\ . \end{aligned} \tag{2.7}$$

The superfield W has Weyl weight w=0 and is, as in N=2, subject to a constraint similar to (2.3). These restrictions show, that the tensor fields $\mathcal{R}_{ab}{}^{\ell}{}_{m}(\mathcal{V})$, $\mathcal{R}^{cd}_{ab}(M)$ and $\mathcal{R}^{\ell}_{ab}(Q)$ correspond to field strengths, and satisfy the curvature constraints and Bianchi identities of the superconformal algebra SU(2,2|4). The remaining components C, $\Lambda_i$, $E_{ij}$, $T^{k\ell}_{ab}$, $\chi^{k\ell}_{m}$ and $D^{k\ell}_{mn}$ correspond to the matter fields of the N=4 superconformal multiplet. They represent 78 bosonic and 96 fermionic degrees of freedom, as we anticipated in the beginning of this section. The transformation rules of the components of W can be easily determined from (2.6).

Another way to obtain the linearized N=4 superconformal theory is to construct the multiplet of currents of the superconformal symmetries (Ferrara & Zumino, 1975). We have applied this method to the N=4 supersymmetric Yang-Mills theory (Brink, Scherk & Schwarz, 1977; Gliozzi, Olive & Scherk, 1977). For this construction knowledge of the on-shell theory suffices. We have obtained in this manner the Q-supersymmetry transformation rules of the superconformal multiplet, expressed in terms of the gauge potentials. Also the S-supersymmetry transformations have been obtained from the commutator of a Q and a K transformation. These results are in complete agreement with those obtained from the superfield (2.6). We conclude therefore, that the components of (2.6) indeed represent the curvatures and matter fields of N=4 conformal supergravity.

This conclusion is confirmed by off-shell counting of massive states (de Wit & Ferrara, 1979). If one considers the linear combination of the superconformal and Poincaré Lagrangians

$$\mathcal{L}_C + m^2\mathcal{L}_P \tag{2.8}$$

the multiplets of states should correspond to massive and massless representations of the N=4 supersymmetry algebra. Since these representations are known, one obtains in this way information about the content of $\mathcal{L}_P$ and $\mathcal{L}_C$. We shall now indicate how the massive multiplets arise. The terms $C\Box^2C$, $(\partial E)^2$ and $(D^{ij}_{k\ell})^2$ in the conformal action must combine with terms in the Poincaré action to give 42 massive spin-0 states. For C and E this implies that $\mathcal{L}_P$ must contain terms $C\Box C$ and $E^2$, indicating that C is a physical field of the Poincaré theory. We expect that D, as in N=2, is the divergence of a vector field: $D^{ij}_{k\ell} = \partial\cdot V^{ij}_{k\ell}$. This is indeed consistent with the transformation character of D under supersymmetry: $\delta D = \partial_\mu(\bar{\varepsilon}\gamma^\mu\chi)$. The spin-½ fields give rise to contributions $\bar{\Lambda}\not\partial^3\Lambda$ and $\bar{\chi}\not\partial\chi$ in $\mathcal{L}_C$. The field $\Lambda_i$ will generate 8 massive spin-½ states, if the Poincaré theory contains a term $\bar{\Lambda}\not\partial\Lambda$. Therefore we identify $\Lambda$ with the physical spin-½ field of N=4 Poincaré supergravity. For the field $\chi$ we can obtain the correct number of massive states by using a similar mechanism as in N=2 (de Wit & van Holten, 1979), i.e. $\bar{\chi}\not\partial\chi$ combines with terms of the form $\bar{\chi}\lambda$ and $\bar{\lambda}\not{\partial}\lambda$ in $\mathcal{L}_P$ to give 40 massive spin-$\frac{1}{2}$ states. The field $\lambda$ will be an auxiliary field of N=4 Poincaré supergravity. The field $T^{ij}_{ab}$ and the SU(4) gauge field $\mathcal{V}_\mu{}^i{}_j$ contribute 12 and 15 massive spin-1 states, respectively, giving the 27 spin-1 states of the massive N=4 representation. The spin-3/2 and spin-2 contributions are obtained in the same way as for N=1 and N=2. Altogether, this leads to 128 + 128 massive states.

We conclude that N=4 is rather analogous to N=2. However, the full commutator algebra is more complicated. It contains extra field-dependent transformations which are absent in N=2. In particular, this is the case for the commutator of an S and a Q-supersymmetry transformation, which now contains a $\Lambda$-dependent S transformation. Also, the commutator of two Q transformations has a more complicated structure: for instance it contains an E-dependent SU(4), a $\Lambda$-dependent Q, and a $\chi$- and $\Lambda$-dependent S transformation. Full details of $N \leq 4$ superconformal theories will be discussed elsewhere (Bergshoeff et al., 1980).

This concludes our discussion of the Weyl multiplets. In the next sections we return to the N=2 conformal theory. We shall show how conformal supergravity allows reductions to Poincaré and de Sitter supergravity theories.

## 3. THE N=2 MINIMAL FIELD REPRESENTATION AND A COMPENSATING MULTIPLET

To the 24 + 24 components of the N=2 Weyl multiplet we add a superconformal vector multiplet. This multiplet consists of a complex Lorentz scalar $a$, a Majorana spinor doublet $\xi_i$ of positive chirality, a Lorentz scalar $S_{ij}$, symmetric in SU(2) indices and subject to an SU(2)-invariant constraint

$$S^{ij} \equiv (S_{ij})^* = \varepsilon^{ik}\varepsilon^{j\ell}S_{k\ell}, \tag{3.1}$$

and a gauge field $B^{ij}_\mu$, antisymmetric in SU(2) indices. The gauge field has a modified field strength whose selfdual components are denoted by $t^+_{ab}{}^{ij}$ and $t^-_{abij}$, which is defined by

$$t^-_{abij}\varepsilon^{ij} + t^+_{ab}{}^{ij}\varepsilon_{ij} = 2\sqrt{2}\,\hat{F}^{ij}_{ab}(B)\varepsilon_{ij} + aT^+_{abij}\varepsilon^{ij} + a^*T^-_{ab}{}^{ij}\varepsilon_{ij}\,, \tag{3.2}$$

with $\hat{F}^{ij}_{ab}(B)$ the conventional supercovariant field strength of $B^{ij}$. For future purposes we give the Q-and S-supersymmetry transformations of $a$ and $\xi_i$, and refer to the literature for more complete results (de Wit, van Holten & Van Proeyen, 1980b,c).

$$\delta a = \bar{\varepsilon}^{i}\xi_{i},$$
$$\delta\xi_{i} = 2\not{D}a\varepsilon_{i} + 2S_{ij}\varepsilon^{j} - \tfrac{1}{2}\sigma\cdot t^{-}_{ij}\varepsilon^{j} + 2a\eta_{i} \; . \tag{3.3}$$

The reducible set of 32 + 32 degrees of freedom consisting of the Weyl multiplet and this vector multiplet will be referred to as the <u>minimal field representation</u>. This name is chosen since it is the smallest representation that allows the incorporation of multiplets with a central charge. Moreover it is a minimum set for passing to the Poincaré supersymmetry algebra by means of a choice of gauge. Before discussing this in any detail we first introduce the superconformal algebra based on the minimal field representation. This algebra coincides with the original graded algebra $SU(2,2|N)$, except for the commutator of two Q-supersymmetry transformations. The latter now yields a general coordinate transformation together with further field-dependent symmetry transformations. The precise form is

$$\begin{aligned}[\delta_{Q}(\varepsilon_{1}), \delta_{Q}(\varepsilon_{2})] &= 2(\bar{\varepsilon}^{i}_{2}\gamma^{a}\varepsilon_{1i} + \text{h.c.})D_{a} + \delta_{M}(\bar{\varepsilon}^{i}_{1}T^{+}_{ij}{}^{ab}\varepsilon^{j}_{2} + \text{h.c.}) \\ &+ \delta_{K}\left(-\tfrac{1}{2}(2+a)D_{C}\bar{\varepsilon}^{i}_{2}\gamma^{a}\varepsilon_{1i} + \bar{\varepsilon}^{i}_{1}D_{b}T^{+}_{baij}\,\varepsilon^{j}_{2} + \text{h.c.}\right) + \delta_{S}(3\bar{\varepsilon}^{[i}_{1}\varepsilon^{j]}_{2}\chi_{Cj}) \\ &+ \delta_{Z}(2a^{*}\varepsilon_{ij}\bar{\varepsilon}^{i}_{2}\varepsilon^{j}_{1} + \text{h.c.}) \; , \end{aligned} \tag{3.4}$$

where the supercovariant derivative $D_a$ indicates a covariant translation. The field-dependent Lorentz, special conformal, and S-supersymmetry transformations depend on the matter components of the Weyl multiplet, whereas the central charge transformation originates from the vector multiplet. At this point the central charge only acts on the gauge field $B^{ij}_{\mu}$ by means of a gauge transformation:

$$\delta_{Z}B^{ij}_{\mu} = -\sqrt{2}\,\varepsilon^{ij}\partial_{\mu}z \; . \tag{3.5}$$

We now discuss those aspects of the minimal field representation related to Poincaré supersymmetry and to the use of the vector multiplet as a compensating field. How to incorporate nontrivial central charges will be the subject of the next section. To pass to the Poincaré theory we make a choice of gauge which breaks some of the superconformal symmetries that are not contained in the Poincaré supersymmetry algebra. For the minimal field representation the obvious gauge choice is

$$a = 1, \quad \xi_{i} = 0, \quad b_{\mu} = 0. \tag{3.6}$$

These gauge conditions break the invariance under local dilatations, chiral U(1), S supersymmetry and conformal boosts.

Only a field-dependent linear combination of Q, S, and K transformations is left as a fermionic symmetry. This is expressed by a uniform decomposition rule, which defines the Poincaré supersymmetry transformations in terms of superconformal transformations

$$\delta(\varepsilon_i) = \delta_Q(\varepsilon_i) + \delta_S\big((S_{ij} + \tfrac{1}{4}\sigma\cdot t^-_{ij})\varepsilon^j - i\not{A}\varepsilon_i\big)$$

$$+ \delta_K\big(\tfrac{1}{2}\bar{\varepsilon}^i\gamma_a\chi_{Ci} + \tfrac{1}{2}\bar{\varepsilon}^i(\hat{R}_{ai} - \tfrac{1}{3}\gamma_a\gamma\cdot\hat{R}_i) + \text{h.c.}\big) \quad . \qquad (3.7)$$

The field-dependent S component is required to maintain the second condition (3.6), the K component by the need to keep the third condition (3.6) unaffected. This demonstrates how the transition to Poincaré supersymmetry can be made by means of a gauge choice. The procedure is precisely analogous to the one used in N=1 supergravity (Kaku & Townsend, 1978; Das, Kaku & Townsend, 1978).

It is an obvious necessity in Poincaré supergravity to construct actions from multiplets of arbitrary Weyl weight. This can be done in the superconformal context by making use of the vector submultiplet of the minimal field representation as a compensating field. For example, in the calculus of chiral superfields one introduces a chiral superfield $F_\xi(z^\mu,\theta)$ which represents the gauge covariant components of the vector multiplet ($F_\xi$ plays a similar role as the chiral superfield $W^{ij}_{ab}$ for the Weyl multiplet (see section 2) ). In increasing order of $\theta$ the components of $F_\xi$ are given by

$$F_\xi : a\,,\ \xi_i\ S_{ij},\ t^-_{abij},\quad \not{D}\xi^i,\quad 2\Box^C a^* + \ldots\ , \qquad (3.8)$$

where the dots indicate further modifications. Since $F_\xi$ constitutes a chiral superfield with Weyl weight w=1, we can change the Weyl weight of any chiral superfield by multiplying with an appropriate power of $F_\xi$. Because of the Poincaré gauge condition (3.6) the first component of $F_\xi$ ultimately becomes equal to 1. Therefore, multiplying by powers of $F_\xi$ amounts to merely adding nonlinear modifications to the original components. Using $F_\xi$ as a compensating field we can now insist on using the conformal calculus. For instance, to construct an invariant density for a chiral superfield with arbitrary Weyl weight we first multiply by $(F_\xi)^{2-w}$ to obtain a superfield with weight 2. For the latter a superconformally invariant density exists. It is straightforward to find further applications for the compensating superfield (de Wit et al., 1980c).

Hence we will adopt the following strategy. In the construction of supersymmetric invariants we consistently use the superconformal calculus. To compensate for possible discrepancies in Weyl weights appropriate powers of the vector submultiplet are to be inserted. At that point conformally invariant densities are obtained. The last step is to specify whether the supersymmetry will be restricted to the symmetry of Poincaré or the Sitter supergravity. This is achieved by imposing the appropriate gauge conditions, from which one obtains the restricted invariants. The transformation rules are then determined by decomposition rules such as (3.7).

## 4. CENTRAL CHARGES

The fact that the superconformal algebra (3.4) exhibits a field-dependent central charge can be used to incorporate multiplets with nontrivial central charges in the superconformal framework. This is achieved by imposing this algebra on a known representation of rigid supersymmetry. In this section we will describe this procedure for the scalar multiplet (Fayet, 1976; Sohnius, 1978). A scalar multiplet consists of scalar fields $A^{\alpha}_{\;i}$ (with i the local chiral SU(2) index, and $\alpha$ the index of a second rigid chiral SU(2) group) and a Majorana spinor doublet $\zeta^{\alpha}$ taken with positive chirality. The scalar multiplet has a central charge, which generates a series of equivalent multiplets (Breitenlohner & Sohnius, 1980):

$$(A^{\alpha}_{\;i},\zeta^{\alpha}) \xrightarrow{Z} (A^{(z)\alpha}_{\;i},\, \zeta^{(z)\alpha}) \xrightarrow{Z} (A^{(zz)\alpha}_{\;i},\, \zeta^{(zz)\alpha}) \longrightarrow \text{etc.} \tag{4.1}$$

The transformations under Q and S supersymmetry are rather simple.

$$\begin{aligned} \delta A^{\alpha}_{\;i} &= 2\bar{\zeta}^{\alpha}\varepsilon_{i} + 2\varepsilon^{\alpha\beta}\varepsilon_{ij}\bar{\zeta}_{\beta}\varepsilon^{j} \quad , \\ \delta\zeta^{\alpha} &= \not{D}A^{\alpha}_{\;i}\varepsilon^{i} + a A^{(z)\alpha}_{\;i}\varepsilon^{ij}\varepsilon_{j} + A^{\alpha}_{\;i}\eta^{i} \ . \end{aligned} \tag{4.2}$$

This result was found by imposing the algebra (3.4) on $A^{\alpha}_{\;i}$ and $\zeta^{\alpha}$, using the Q-supersymmetry variation of $A^{\alpha}_{\;i}$ as input. Note that derivatives are fully covariant with respect to the superconformal symmetries <u>and</u> the central charge transformations. The gauge field of the central charge is $B^{ij}_{\mu}$ (see (3.5)). When imposing the supersymmetry algebra, one finds that closure is achieved if and only if the following conditions hold

$$\begin{aligned} &\varepsilon^{\alpha\beta}\not{D}\zeta_{\beta} + \tfrac{1}{2}A^{(z)\alpha}_{\;i}\xi^{i} + a^{*}\zeta^{(z)\alpha} + \frac{1}{8}\,\sigma\cdot T^{+}_{ij}\varepsilon^{ij}\zeta^{\alpha} - \frac{3}{2}\,\varepsilon^{ij}\chi_{Ci}A^{\alpha}_{\;j} = 0 , \\ &(\Box^{C} + \frac{3}{2}\,D_{C})A^{\alpha}_{\;i} + |a|^{2}A^{(zz)\alpha}_{\;i} - A^{(z)\alpha}_{\;j}S^{jk}\varepsilon_{ik} \\ &+ \varepsilon^{\alpha\beta}\bar{\xi}_{i}\zeta^{(z)}_{\beta} - \varepsilon_{ij}\bar{\xi}^{j}\zeta^{(z)\alpha} = 0 \ . \end{aligned} \tag{4.3}$$

These equations show that the scalar multiplet has $A_i^{\alpha}$, $\zeta^{\alpha}$, and $A_i^{(z)\alpha}$ as independent components. All further components that are generated by applying the central charge can be re-expressed in terms of these 8 + 8 components by virtue of (4.3), or by similar conditions obtained by repeated use of the central charge on (4.3).

It is possible to form linear multiplets (with a central charge) from products of two scalar multiplets. For a linear multiplet there exists a superconformal density, in which the minimal vector submultiplet is again required as a compensating field. We refer to the literature for details (de Wit et al., 1980b), and give the most pertinent terms of two superconformally invariant Lagrangians quadratic in two scalar multiplets denoted by $(A_i^{\alpha}, \zeta^{\alpha})$ and $(A_i'^{\alpha}, \zeta'^{\alpha})$.

$$
\begin{aligned}
e^{-1}\mathcal{L} = \Big\{ & 4a^{*}\bar{\zeta}^{\alpha}\zeta'^{\beta} + |a|^2 A_i^{\alpha} A_j'^{(z)\beta}\varepsilon^{ij} - |a|^2 A_i^{(z)\alpha} A_j'^{\beta}\varepsilon^{ij} \\
& + A_i^{\alpha} A_j'^{\beta}\left(S^{ij} - \tfrac{1}{2}\bar{\psi}^{(i}\cdot\gamma\xi^{j)} - 2a^{*}\bar{\psi}_{\mu}^{i}\sigma^{\mu\nu}\psi_{\nu}^{j}\right) \\
& + \tfrac{1}{4}\sqrt{2}\,\varepsilon_{ij} B_a^{ij} A_k^{\alpha} \overleftrightarrow{D}_a A_\ell'^{\beta}\varepsilon^{k\ell} + \ldots \Big\}\eta_{\alpha\beta} + \text{h.c.},
\end{aligned}
\tag{4.4}
$$

where the multiplication tensor $\eta_{\alpha\beta}$ equals either $\varepsilon_{\alpha\beta}$ or $\delta_{\alpha\beta}$.

Eq. (4.4) has two important applications. The first one is obtained by choosing

$$
\begin{aligned}
& (A_i'^{\alpha}, \zeta'^{\alpha}) = (A_i^{(z)\alpha}, \zeta^{(z)\alpha}) , \\
& \eta_{\alpha\beta} = \varepsilon_{\alpha\beta} .
\end{aligned}
\tag{4.5}
$$

In the invariant action (4.4) we then have terms proportional to $a^{*}\zeta^{(z)\alpha}$ and $|a|^2 A_i^{(zz)\alpha}$, for which we can use the constraints (4.3). The result is a superconformal Lagrangian which contains the kinetic terms for the scalar multiplet components $A_i^{\alpha}$ and $\zeta^{\alpha}$.

$$
\begin{aligned}
e^{-1}\mathcal{L}_{kin} = {} & A_{\alpha}^{i}(\hat{\Box} + \tfrac{3}{2} D_C)A_i^{\alpha} + \left(|a|^2 + \tfrac{1}{8}(\varepsilon_{ij}B_a^{ij})^2\right)A_{\alpha}^{(z)k} A_k^{(z)\alpha} \\
& -4\left(\bar{\zeta}^{\alpha} - \tfrac{1}{2}A_k^{\alpha}\bar{\psi}^{k}\cdot\gamma\right)\left(\hat{\not{D}}\zeta_{\alpha} + \tfrac{3}{2}A_{\alpha}^{i}\chi_{Ci} - \tfrac{1}{8}\sigma\cdot T_{ij}^{+}\varepsilon^{ij}\varepsilon_{\alpha\beta}\zeta^{\beta}\right) \\
& + \text{h.c.} .
\end{aligned}
\tag{4.6}
$$

In (4.6) the d'Alembertian $\hat{\Box}$ and the derivative $\hat{D}_a$ are no longer covariant with respect to central charge transformations and contain no terms that depend on $A_i^{(z)\alpha}$. The second application follows from substituting into (4.4)

$$(A'^{\alpha}_{i}, \zeta'^{\alpha}) = (A^{\alpha}_{i}, \zeta^{\alpha}) \,, \qquad \eta_{\alpha\beta} = \delta_{\alpha\beta} \,. \tag{4.7}$$

This leads to a superconformal Lagrangian that contains mass-like terms for the scalar multiplet

$$\begin{aligned} e^{-1}\mathcal{L}_m = \Big\{ & 4a^*\bar{\zeta}^{\alpha}\zeta^{\alpha} + 2\big(|a|^2 + \tfrac{1}{8}(\varepsilon_{ij}B^{ij}_a)^2\big)A^{\alpha}_k A^{(z)\alpha}_{\ell}\varepsilon^{k\ell} \\ & + A^{\alpha}_i A^{\alpha}_j(S^{ij} - \bar{\psi}^i\cdot\gamma\xi^j - 2a^*\bar{\psi}^i_{\mu}\sigma^{\mu\nu}\psi^j_{\nu}) \\ & + \tfrac{1}{4}\sqrt{2}\,\varepsilon_{ij}B^{ij}_a A^{\alpha}_k\hat{D}_a A^{\alpha}_{\ell}\varepsilon^{k\ell} + \ldots\Big\} + \text{h.c.} \,, \end{aligned} \tag{4.8}$$

where we have only given the most characteristic terms.

We now impose the gauge condition (3.6). In that case the field-dependent central charge in the algebra (3.4) becomes of the standard field-independent type. In addition the Lagrangian (4.8) now contains the mass terms for the scalar multiplet. Hence, the sum of (4.6) and (4.8) subject to the conditions (3.6) defines the Lagrangian for the massive scalar multiplet in Poincaré supergravity. If we add this result to the Poincaré supergravity Lagrangian and eliminate the auxiliary fields, the result coincides with the invariant action constructed by Zachos (1978). We refer to the literature for further details (de Wit et al., 1980c).

This application illustrates the strategy outlined in the previous section. Using the superconformal calculus we have constructed invariant actions. Upon a suitable gauge choice this leads to actions invariant under the restricted transformations of Poincaré supersymmetry, including those features that have an intrinsic dimension, such as central charges or dimensionful couplings in a Lagrangian (note that we have consistently chosen the gravitational coupling constant $\kappa=1$).

## 5. POINCARE AND DE SITTER SUPERGRAVITY THEORIES

In the previous sections we have discussed the superconformal transformation rules and invariant actions of matter multiplets. Furthermore we have demonstrated how to restrict the superconformal symmetry to Poincaré supersymmetry by means of a gauge choice. This was all done in the context of the minimal field representation, which is the reducible sum of the superconformal N=2 Weyl multiplet and a vector multiplet. The latter often plays the role of a compensating field, and is essential in incorporating central charges and in making the transition to the

Fig. 1: Choices of gauge leading to different formulations of N=2 supergravity theories.

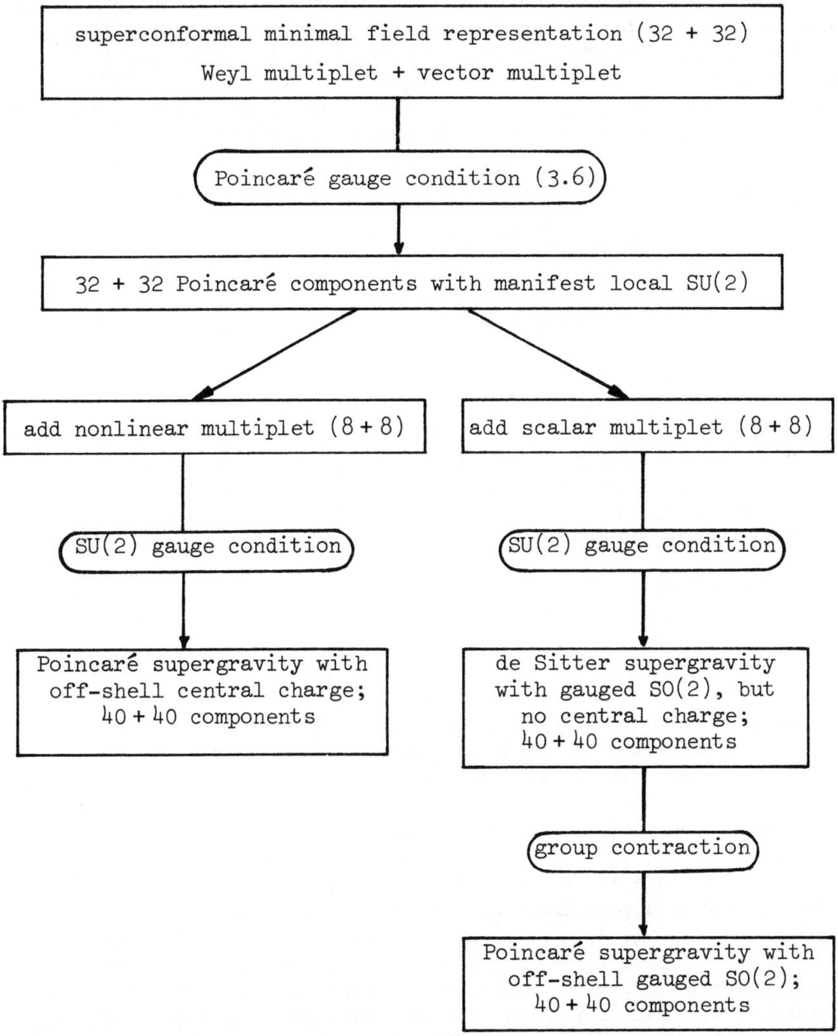


Poincaré supersymmetry algebra.

Within this framework we can consider various supermultiplets, such as chiral, vector and linear multiplets. We take the minimal field representation subject to the Poincaré gauge (3.6), which still allows a local SU(2) invariance. For such a theory we will now discuss the possibility of imposing further (inequivalent) gauge choices. In this way we will derive two different theories as we have indicated schematically in Fig. 1. In addition we will construct the corresponding actions.

One might attempt the construction of an invariant action for Poincaré supergravity based on the 32 + 32 components of the minimal field representation. The obvious candidate for this is the action of the vector submultiplet (de Roo et al., 1980), which upon imposing the gauge condition (3.6) generates many of the terms that are known to occur in the supergravity action.

$$\begin{aligned} e^{-1}\mathcal{L}_{min} &= |\partial_\mu a|^2 + (D_C - \tfrac{1}{6} R)|a|^2 - \tfrac{1}{2}|S_{ij}|^2 + |a|^2 A_a^2 \\ &+ \tfrac{1}{2}\bar{\xi}^i \not{\partial} \xi_i + a^* \bar{\chi}_{Ci} \xi^i + a \bar{\chi}_C^i \xi_i \\ &- \frac{1}{64}\left\{\tfrac{1}{2}(a T^+_{abij} \varepsilon^{ij} - t^{+\,ij}_{ab} \varepsilon_{ij})^2 - (t^{+\,ij}_{ab} \varepsilon_{ij})^2 + \text{h.c.}\right\} \\ &+ \ldots\ldots , \end{aligned} \tag{5.1}$$

where we have only given the most pertinent terms with a suitable overall normalization. Indeed when subject to the conditions (3.6) some of the terms in (5.1) coincide with those of the Poincaré supergravity Lagrangian. We note the presence of a Riemann scalar curvature R, of the squares of the auxiliary fields $A_a$, $t_{ab}^{+ij}$, $S_{ij}$, and of the field strength of $B_\mu{}^{ij}$ represented by a linear combination of $t_{ab}^{+ij}$ and $T_{ab\,ij}^{+}$ (see eq. (3.2)). However, the relative coefficients of these terms are not those appearing in the Poincaré supergravity Lagrangian. In fact (5.1) does not constitute a meaningful action at all, since the corresponding field equations are not consistent (Breitenlohner, 1979; Breitenlohner & Sohnius, 1980; de Wit, 1980). This forms an indication that more than 32 + 32 field components are required in order that a Poincaré action will exist.

It is a priori not known how many extra fields are required, and whether the addition is unique. We have found two different ways to complete the minimal field representation to a multiplet that allows a supersymmetric Poincaré action, which we have indicated in Fig. 1. The first possibility is to add a "non-linear" multiplet to be introduced below, for which one may choose a second gauge condition to break the local SU(2). This leads to the familiar form of SO(2) supergravity with auxiliary fields (Fradkin & Vasiliev, 1979a,b; de Wit & van Holten, 1979; de Wit et al., 1980a; Breitenlohner & Sohnius, 1980). It turns out that this formulation allows an extension with local SO(2) invariance (de Wit, van Holten & Van Proeyen, 1980c). An alternative formulation is obtained by choosing the second option to extend the minimal field representation. In that case one adds a scalar multiplet and breaks the SU(2) invariance by a corresponding gauge condition. The result is a different formulation of Poincaré supergravity, which has no central charge. It is based on an entirely different field representation, which seems to be related to the one advocated by MacDowell (1980). For this representation there exists a second invariant density which yields a cosmological term upon use of the field equations for the auxiliary fields. This is comparable to the case of N=1 where a cosmological term can be obtained in a similar way but without a gauging (Ferrara, Grisaru & van Nieuwenhuizen, 1978). The theory has no central charge transformations, but a complicated abelian invariance which contains a field-dependent SO(2) transformation. On-shell this transformation reduces to an ordinary local SO(2) transformation, which becomes a central charge after a Wigner-Inönü group contraction. In the remaining part of this chapter we will briefly discuss some further details of these two formulations.

We first turn to the formulation of Poincaré supergravity which is obtained by adding a nonlinear multiplet $\{\Phi^i_\alpha, \lambda^i, M^{ij}, V_a\}$ to the minimal field representation (see Fig. 1). In this multiplet $\Phi^i_\alpha$ denotes a set of Lorentz scalars which form an element of SU(2). It transforms under two different SU(2) groups: the local SU(2) acting on the index i from the left, and a new rigid SU(2) acting on the index $\alpha$ from the right. The field $\lambda^i$ represents a spinor doublet with negative chirality, while $M^{ij}$ is an antisymmetric scalar and $V_a$ a Lorentz vector. Because $\Phi$ is an element of SU(2) the supersymmetry variations of the multiplet are nonlinear in terms of its components. We do not present all the nonlinear variations under Q and S supersymmetry here; complete results will be given elsewhere (de Wit et al., 1980c).

$$
\begin{aligned}
\delta\Phi^i{}_\alpha &= (2\bar{\varepsilon}^i\lambda_j - \delta^i{}_j\bar{\varepsilon}^k\lambda_k - \text{h.c.})\Phi^j{}_\alpha ,\\
\delta\lambda^i &= -\tfrac{1}{2}\not{V}\varepsilon^i - \tfrac{1}{2}M^{ij}\varepsilon_j + \Phi^i{}_\alpha \not{D}\Phi^\alpha{}_j\varepsilon^j + \text{nonlinear terms} + \eta^i , \qquad (5.2)\\
\delta M^{ij} &= 4\bar{\varepsilon}^{[i}\not{D}\lambda^{j]} + 6\bar{\varepsilon}^{[i}\chi_C{}^{j]} + \text{nonlinear terms} ,\\
\delta V_a &= 4\bar{\varepsilon}^i\sigma_{ab}D_b\lambda_i + \frac{3}{2}\bar{\varepsilon}^i\gamma_a\chi_{Ci} + \text{nonlinear terms} + \bar{\eta}^i\gamma_a\lambda_i + \text{h.c.}.
\end{aligned}
$$

The superconformal algebra closes on this field representation if the following supersymmetric constraint holds:

$$
\begin{aligned}
&D\cdot V - 3D_C - \tfrac{1}{2}V_a^2 - \tfrac{1}{4}|M^{ij}|^2 - |\Phi^i{}_\alpha D_a\Phi^\alpha{}_j|^2 + 2\bar{\lambda}^i\overleftrightarrow{\not{D}}\lambda_i\\
&\quad - \tfrac{1}{2}(\bar{\lambda}^i\sigma\cdot T^+_{ij}\lambda^j + \text{h.c.}) + 3(\bar{\chi}_C^i\lambda_i + \text{h.c.}) = 0 . \qquad (5.3)
\end{aligned}
$$

This constraint allows one to extend the action (5.1) to a meaningful Lagrangian for Poincaré supergravity. Namely we add the supersymmetric expression (5.3), multiplied by $e|a|^2$ to maintain conformal invariance, to (5.1):

$$\mathcal{L}_{\text{Poincaré}} = \mathcal{L}_{\min} + e|a|^2(D\cdot V - 3D_C - \tfrac{1}{2}V_a^2 + \dots). \qquad (5.4)$$

On imposing the gauge condition (3.6) the terms linear in $D_C$ will cancel. To see this, one expresses the divergence of $V_a$ according to

$$D\cdot V = D^P\cdot V - 2f_\mu{}^\mu + \dots = D^P\cdot V + 2D_C - \frac{1}{3}R + \dots , \qquad (5.5)$$

where $D^P_a$ denotes a derivative covariant with respect to Poincaré supergravity. Modulo a total derivative (5.4) is now precisely equal to a locally SU(2)-invariant formulation of N=2 supergravity (de Wit, 1980; a similar formulation has been reported by Breitenlohner (1980)).

The above procedure leaves local SU(2) symmetry unaffected. One may break this symmetry by an additional gauge choice:

$$\Phi^i{}_\alpha = \delta^i{}_\alpha , \qquad (5.6)$$

upon which one regains previous off-shell formulations of N=2 Poincaré supergravity (Fradkin & Vasiliev, 1979a,b; de Wit & van Holten, 1979; Breitenlohner & Sohnius, 1980; de Wit et al., 1980a). Precisely as in those formulations the central charge is only realized on the vector field $B_\mu^{ij}$ by means of the gauge transformation (3.5). This is the result of the fact that the non-linear multiplet (5.2) vanishes under the central charge.

The latter forms precisely the crucial difference with the second extension of the minimal field representation (see Fig. 1). Namely we add a scalar multiplet $\{A^\alpha{}_i, \zeta^\alpha\}$, which does transform under central charge transformations. For the scalar multiplet we have two invariants at our disposal, given in (4.4) and (4.8). If we add these to the Lagrangian (5.1), and choose the gauge conditions (3.6) and

$$A^\alpha{}_i = e^u \delta^\alpha{}_i \qquad (5.7)$$

to break SU(2)-invariance, we are left with (the gauge (5.7) is equivalent to the gauge that was actually presented during the workshop):

$$\begin{aligned} e^{-1}\mathcal{L}_{\text{de Sitter}} &= e^{-1}\mathcal{L}_{\min} - \tfrac{1}{2}e^{-1}\mathcal{L}_{\text{kin}} + \tfrac{1}{2}g e^{-1}\mathcal{L}_m \\ &= -\frac{1}{6}(1 + 2e^{2u})(R + \bar\psi_\mu{}^i R^\mu{}_i + \bar\psi_{\mu i}R^{\mu i}) + D_C(1 - e^{2u}) \\ &+ A_a{}^2 - \frac{1}{64}\left(\tfrac{1}{2}(T^+_{abij}\varepsilon^{ij} - t_{ab}{}^{+ij}\varepsilon_{ij})^2 - (t_{ab}{}^{+ij}\varepsilon_{ij})^2 + \text{h.c.}\right) \\ &- \tfrac{1}{2}|S^{ij}|^2 + g e^{2u}S^{ii} + \tfrac{1}{4}e^{2u}|V_a{}^i{}_j|^2 + \tfrac{1}{2}g\sqrt{2}\, e^{2u}B_a{}^{ij}V_a{}^i{}_j \\ &- e^{2u}\left(1 + 8(\varepsilon_{k\ell}B_a{}^{k\ell})^2\right)\left(|G^i{}_j|^2 - 2g\varepsilon^{ij}G^i{}_j\right) + \dots \end{aligned} \qquad (5.8)$$

Here we have introduced an arbitrary parameter g, the Rarita-Schwinger field equation $R^i_\mu$, and the quantities

$$\begin{aligned} \lambda^\alpha &= e^{-u}\zeta^\alpha , \\ G^\alpha{}_i &= e^{-u}A^{(z)\alpha}{}_i . \end{aligned} \tag{5.9}$$

The Lagrangian (5.8) implies the following field equations (among others):

$$\begin{aligned} u &= 0 , \\ S_{ij} &= g\delta_{ij} , \\ G^i{}_j &= g\varepsilon_{ij} , \\ V_a{}^i{}_j &= g\sqrt{2}\, B_a{}^{ij} . \end{aligned} \tag{5.10}$$

When reinserted into (5.8) the first three equations (5.10) generate a cosmological term in the Lagrangian with positive coefficient $3g^2$. The fourth equation is responsible for a minimal coupling of $B^{ij}_\mu$ to SO(2) covariant quantities. This indicates that when all auxiliary fields are eliminated by corresponding field equations, we will find the Lagrangian for N=2 de Sitter supergravity with local SO(2) invariance in an on-shell formulation (Das & Freedman, 1977; Fradkin & Vasiliev, 1976). It is straightforward to show that this is indeed the case.

The off-shell version of the theory we have derived here, does not have a simple local SO(2) invariance. The gauge condition (5.7) leaves only a field-dependent linear combination of central charge and SU(2) transformations as an invariance. This is expressed by a decomposition rule

$$\delta_{\text{de Sitter}}(z) = \delta_Z(z) + \delta_{SU(2)}\left(z(G^i{}_j - \tfrac{1}{2}\delta^i{}_j G^k{}_k)\right) . \tag{5.11}$$

The symmetry (5.11) will only reduce to SO(2) uniformly when the auxiliary fields satisfy their field equations.

The limit $g \to 0$ gives the contraction of de Sitter supergravity to Poincaré supergravity. However, we obtain a formulation of the latter that differs from the off-shell formulation discussed previously, cf. (5.4). This difference is manifest in the presence of the extra scalar fields $G^i{}_j$, instead of the vector field $V_a$, and in the invariance (5.11) which occurs in the Poincaré supersymmetry algebra, and which reduces to a central charge in the on-shell formulation. One may also consider the N=1 reduction of this theory. The result is a nonminimal formulation of N=1 Poincaré supergravity, which differs from the one proposed by Breitenlohner (1977a,b) by the exchange of a vector by four scalar auxiliary fields. Full details and further results concerning these N=2 supergravity theories will be published in a forthcoming paper (de Wit et al., 1980c).

## 6. CONCLUSIONS

We have demonstrated how to restrict the symmetries of N=2 extended superconformal gravity to obtain a variety of Poincaré and de Sitter supergravity theories. This is achieved by making appropriate use of gauge conditions and compensating fields. This elegant approach has two advantages. First, it is now sufficient to develop a superconformal multiplet calculus in order to construct all supergravity theories. Secondly, this procedure is well suited to explore different off-shell formulations of Poincaré supergravity that are not equivalent. An example of this was discussed in section 5, where we presented two such formulations for N=2 supergravity.

The general applicability of the superconformal theories further motivates an independent study of their properties for general N. We have done this in section 2 where we have reviewed their status and presented some new results for $N \leq 4$.

B. de Wit and J.W. van Holten acknowledge many invaluable discussions with the participants of the Nuffield Workshop on Supergravity, Cambridge. We thank M. Roček for pointing out the existence of a singular limit for the scalar multiplet coupled to supergravity, which gives rise to supergravity with local SO(2) invariance. This led us to study the variety of Poincaré gauge conditions described in section 5.

This work was presented at the Workshop by B. de Wit.

REFERENCES

Bergshoeff, E., de Roo, M. & de Wit, B. (1980). Extended conformal supergravity (in preparation).

Breitenlohner, P. (1977a). A geometric interpretation of local supersymmetry. Phys.Lett. 67B, pp. 49-51; (1977b). Some invariant Lagrangians for local supersymmetry. Nucl.Phys.B 124, pp. 500-510; (1979). In Supergravity, eds. van Nieuwenhuizen & Freedman, North Holland, Amsterdam, pp. 123-132; (1980). Talk presented at the Europhysics study conference on "Unification of the fundamental interactions", Erice.

Breitenlohner, P. & Sohnius, M. (1980). Superfields, auxiliary fields and tensor calculus for N=2 extended supergravity. Nucl.Phys. B 165, pp. 483-510.

Brink, L., Scherk, J. & Schwarz, J.H. (1977). Supersymmetric Yang-Mills theories. Nucl.Phys. B 121, pp. 77-92.

Das, A. & Freedman, D.Z. (1977). Gauge internal symmetry in extended supergravity. Nucl.Phys. B 120, pp. 221-230.

Das, A., Kaku, M. & Townsend, P.K. (1978). Unified approach to matter coupling in Weyl and Einstein supergravity. Phys.Rev. Lett. 40, pp. 1215-1218.

de Roo, M., van Holten, J.W., de Wit, B. & Van Proeyen, A. (1980). Chiral superfields in N=2 supergravity. Nucl.Phys. B, to be published.

Deser, S., Kay, J.H. & Stelle, K.S. (1977). Renormalizability properties of supergravity. Phys.Rev.Lett. 38, pp. 527-530.

de Wit, B. (1979). In Supergravity, eds. van Nieuwenhuizen & Freedman, North Holland, Amsterdam, pp. 113-122; (1980). Talk presented at the Europhysics study conference on "Unification of the fundamental interactions", Erice.

de Wit, B. & Ferrara, S. (1979). On higher-order invariants in extended supergravity. Phys.Lett. 81B, pp. 317-20.

de Wit, B. & van Holten, J.W. (1979). Multiplets of linearized SO(2) supergravity. Nucl.Phys. B 155, pp. 530-542.

de Wit, B., van Holten, J.W. & Van Proeyen, A. (1980a). Transformation rules of N=2 supergravity multiplets. Nucl.Phys. B167, pp. 186-204, B172, pp. 543-44; (1980b). Central charges and conformal supergravity. Phys.Lett. 95B, pp. 51-55; (1980c). Structure of N=2 supergravity (in preparation).

Fayet, P. (1976). Fermi-Bose hypersymmetry. Nucl.Phys. B 113, pp. 135-55.

Ferrara, S., Grisaru, M. & van Nieuwenhuizen, P. (1978). Poincaré and conformal supergravity models with closed algebras. Nucl.Phys. B 138, pp. 430-444.

Ferrara, S., Kaku, M., Townsend, P.K. & van Nieuwenhuizen, P. (1977). Unified field theories with U(N) internal symmetries: gauging the superconformal group. Nucl.Phys. B 129, pp. 125-134.

Ferrara, S. & van Nieuwenhuizen, P. (1978). Structure of supergravity. Phys.Lett. 78B, pp. 573-6.

Ferrara, S. & Zumino, B. (1975). Transformation properties of the supercurrent. Nucl.Phys. B 87, pp. 207-20.

Fradkin, E.S. & Vasiliev, M.A. (1976). Lebedev preprint; (1979a). Minimal set of auxiliary fields and S-matrix for extended supergravity. Lett. Nuovo Cim. 25, pp. 79; (1979b). Minimal set of auxiliary fields in SO(2)-extended supergravity. Phys.Lett. 85B, pp. 47-51.

Gliozzi, F., Olive, D. & Scherk, J. (1977). Supersymmetry, supergravity and the dual spinor model, Nucl.Phys. B 122, pp.253-290.

Howe, P.S. & Lindström, U. (1980). Counterterms for extended supergravity. Talk presented at this meeting.

Kaku, M. & Townsend, P.K. (1978). Poincaré supergravity as broken superconformal gravity. Phys.Lett. 76B, pp. 54-58.

Kaku, M., Townsend, P.K. & van Nieuwenhuizen, P. (1978). Properties of conformal supergravity. Phys.Rev. D 17, pp. 3179-3187.

MacDowell, S. (1980). Talk presented at this conference.

Siegel, W. (1980). On-shell O(N) supergravity in superspace. Preprint, Institute for Advanced Study, Princeton.

Sohnius, M. (1978). Supersymmetry and central charges. Nucl.Phys. B 138, pp. 109-21; (1979). The multiplet of currents for N=2 extended supergravity. Phys.Lett. 81B, pp. 8-10.

Taylor, J.G. (1980). Talk presented at this meeting.

Zachos, C. (1978). N=2 supergravity with a gauged central charge. Phys.Lett. 76B, pp. 329-332.

# CONSTRUCTION OF EXTENDED SUPERGRAVITY WITH AUXILIARY FIELDS

S. W. MacDowell

Yale University, U.S.A.

## INTRODUCTION

The path integral quantization of Supergravity requires that a Lagrangian be found, invariant under an algebra of transformations that closes under commutation.

We have developed a method for constructing $SO_N$-supergravity[5] as a realization of the supersymmetry algebra generated by a set of vector fields $\mathcal{D}_A$, which is a covariant basis for the tangent space of a bundle of frames. This bundle is a fibration of the supergroup $G=OSp(N|4)$ with respect to the Lie subgroup $H=S\ell(2C)\otimes SO_N$.[7]

Our construction consists in imposing a set of constraints on components of the torsion, which can be interpreted physically as conditions for minimal coupling. Taking these constraints into the Bianchi identities one finds that all the torsion and curvature components are determined in terms of a set of independent and auxiliary fields and the gauge potentials.[5,8,9]

In order to obtain a specific model one has to express the auxiliary fields in terms of the independent ones in a manner consistent with the Bianchi identities. They will then provide a complete set of transformation laws and equations of motion for the independent fields and the gauge potentials. From these equations an action can be constructed, which reproduces them as the Euler-Lagrange equations for its extremum. By construction, the set of equations of motion is invariant under an algebra of infinitesimal transformations which closes under commutation, but the action itself is not.

On the other hand, if the auxiliary fields are not constrained to be functions of the independent fields, all the equations of motion actually turn out to be transformation laws for the auxiliary fields. We thus obtain a realization of the algebra on a space of functions not restricted by constraints (such as equations of motion). One can then construct an action, invariant under the algebra of supersymmetry transformations, from which the equations of motion for the original fields and the constraints for the auxiliary fields can be obtained by a variational principle. For simple supergravity our set of auxiliary fields reduces to a minimal set found by van Nieuwenhuizen and Ferrara[3].

In this report I shall first prove a general theorem showing that a necessary and sufficient condition for an algebra of local transformations to close is that the Bianchi identities for its generators be satisfied. This is analogous to Lie's theorem, which requires the structure constants to satisfy the Jacobi identities.

Then we give preliminary results on the construction of an unconstrained realization of the algebra for $SO_N$-supergravity, and outline the procedure to obtain the invariant action.

## 1. THE ALGEBRA OF GENERALIZED GAUGE TRANSFORMATIONS

We consider a locally supersymmetric gauge theory in four dimensions and denote by $h_\mu^A$ the gauge potentials, where $\mu$ is a coordinate index and A is a group index.

In general an infinitesimal gauge transformation with parameters $\varepsilon^A(x)$ is of the form:

$$\delta h_\mu^A = \partial_\mu \varepsilon^A + h_\mu^B \varepsilon^C F_{CB}{}^A \tag{1}$$

where $F_{CB}{}^A$ are as yet unspecified functions of the coordinates.

We shall make the following assumptions:

i) The functions $F_{CB}{}^A$, which are not gauge fields, transform by:

$$\delta F_{CB}{}^A = \varepsilon^D \mathcal{D}_D F_{CB}{}^A \tag{2}$$

where the operators $\mathcal{D}_D$ will be called the generators of the infinitesimal transformations.

ii) The generators $\partial_\mu$ of translations are linear combinations of the generators $\mathcal{D}_D$, so that general coordinate transformations are obtained as transformations of the gauge group. On account of general covariance they are of the form

$$\partial_\mu = h_\mu^D \mathcal{D}_D \quad . \tag{3}$$

iii) The operators $\partial_\mu$ are left invariant under the gauge transformations (1,2).

Now let us require that the commutator algebra of these

infinitesimal transformations closes:

$$[\delta_1,\delta_2] = \delta_3 h_\mu^A = \partial_\mu \varepsilon_3^A + h_\mu^B \varepsilon_3^C F_{CB}^{\;\;A} \tag{4}$$

This is a necessary and sufficient condition for these transformations to form a local group of transformations. Taking into account the postulates i, ii, iii, we have:

$$\begin{aligned}
\delta_3 h_\mu^A &= \partial_\mu(\delta_2\varepsilon_1^A - \delta_1\varepsilon_2^A) + [(\delta_2 h_\mu^B \varepsilon_1^C - \delta_1 h_\mu^B \varepsilon_2^C) + h_\mu^B(\delta_2\varepsilon_1^C - \delta_1\varepsilon_2^C)]F_{CB}^{\;\;A} \\
&\quad + h_\mu^B(\varepsilon_1^C \delta_2 F_{CB}^{\;\;A} - \varepsilon_2^C \delta_1 F_{CB}^{\;\;A}) \\
&= \partial_\mu(\delta_2\varepsilon_1^A - \delta_1\varepsilon_2^A + \varepsilon_2^B\varepsilon_1^C F_{CB}^{\;\;A}) - \varepsilon_2^B\varepsilon_1^C h_\mu^D \mathcal{D}_D F_{CB}^{\;\;A} \\
&\quad + h_\mu^B(\varepsilon_1^C\varepsilon_2^D - \varepsilon_2^C\varepsilon_1^D)\mathcal{D}_D F_{CB}^{\;\;A} + h_\mu^B(\delta_2\varepsilon_1^C - \delta_1\varepsilon_2^C + \varepsilon_2^D\varepsilon_1^E F_{ED}^{\;\;C})F_{CB}^{\;\;A} \\
&\quad - h_\mu^B \varepsilon_2^D \varepsilon_1^E F_{ED}^{\;\;C} F_{CB}^{\;\;A} + h_\mu^E(\varepsilon_2^D F_{DE}^{\;\;B}\varepsilon_1^C - \varepsilon_1^D F_{DE}^{\;\;B}\varepsilon_2^C)F_{CB}^{\;\;A} \\
&= \partial_\mu \varepsilon_3^A + h_\mu^B \varepsilon_3^C F_{CB}^{\;\;A} \\
&\quad - (-1)^{\sigma_B\sigma_C}\varepsilon_2^B h_\mu^D \varepsilon_1^C \sum_{(BCD)} (-1)^{\sigma_B\sigma_D}(\mathcal{D}_B F_{CD}^{\;\;A} + F_{BC}^{\;\;E} F_{ED}^{\;\;A})
\end{aligned} \tag{5}$$

where:

$$\varepsilon_3^A = \delta_2\varepsilon_1^A - \delta_1\varepsilon_2^A + \varepsilon_2^B\varepsilon_1^C F_{CB}^{\;\;A} \tag{6}$$

Therefore it follows that for the transformation $\delta_3 h_\mu^A$ to be of the form (1) the F's must satisfy the Bianchi identities:

$$(BCD)^A = \sum_{(BCD)} (-1)^{\sigma_B\sigma_D}(\mathcal{D}_B F_{CD}^{\;\;A} + F_{BC}^{\;\;E} F_{ED}^{\;\;A}) = 0 \tag{7}$$

From (6) it follows that the parameters $\varepsilon$ are in general functions of the fields $F_{CB}^{\;\;A}$. Therefore, $\delta_2\varepsilon_1^A = \varepsilon_2^B \mathcal{D}_B \varepsilon_1^A$.

Notice that, had we required closure only for the commutator of two spinor supersymmetry transformations, one would find that only the Bianchi identities with at least two lower spinor indices need be satisfied. This however would not be a sufficient condition for the whole algebra to close. The closure of the commutator algebra applied to the functions $F_{BC}^{\;\;A}$ gives the same composition law and the commutator:

$$[\mathcal{D}_B, \mathcal{D}_C\} = -F_{BC}^{\;\;A}\, \mathcal{D}_A \tag{8}$$

Taking the indices A in one-to-one correspondence with the generators of the superalgebra $L$ for the supergroup $G = OSp(N|4)$, and assuming that the infinitesimal transformations corresponding to the sub-algebra $L_0$ of $H = S\ell(2C)\otimes SO_N$ are linear, we recover the structure of a bundle space as introduced in ref. 5.

The F's are the components in the covariant basis $\{\mathcal{D}_A\}$ of a two form[6]:

$$F = d\phi \tag{9}$$

taking values in $L$; the one-form $\phi$ is the sum of the connection form $\omega$ taking values in $L_0$ and the solder form h taking values in the vector space $L_1$ which is the complement of $L_0$ in $L$. The Bianchi identities (7) are the components in the covariant basis of the equation $dF = 0$. The relation between the components of $F$ with vector indices in a coordinate basis and the covariant components is:

$$F^A(\mathcal{D}_\mu,\mathcal{D}_\nu) = F_{\mu\nu}{}^A = (-1)^{\sigma_{B_1}\sigma_{C_1}} h_\mu^{B_1} h_\nu^{C_1} F_{B_1C_1}{}^A \tag{10}$$

where:

$$\mathcal{D}_\mu = h_\mu^{B_1}\mathcal{D}_{B_1} = \partial_\mu - h_\mu^{A_0}\mathcal{D}_{A_0} \tag{11}$$

$h_\mu^{B_1}$ are the "vielbein" and $h_\mu^{A_0}$ the connection coefficients. We denote by $A_0$ and $A_1$ indices of $\mathcal{D}_A$ in $L_0$ and $L_1$ respectively. Then $F_{A_0B}{}^C = f_{A_0B}{}^C$ is a structure constant for $L$, $F_{B_1C_1}{}^{A_0} = R_{B_1C_1}{}^{A_0}$ is a curvature component and $F_{B_1C_1}{}^{A_1} = T_{B_1C_1}{}^{A_1}$ is a torsion component.

We shall be using the following notation:

(i,j,k...) to denote indices for the vector representation $(\frac{1}{2},\frac{1}{2})$ of $S\ell(2C)$.

($\mathfrak{a},\mathfrak{b},\mathfrak{c}$...) to denote indices for the spinor representation $[(\frac{1}{2},0) + (0,\frac{1}{2})]$ of $S\ell(2C)$.

(a,b,c...) to denote indices for the vector representation of $SO_N$.

($\mu,\nu,\lambda$...) to denote space-time coordinate indices.

## 2. REALIZATION OF THE SUPERSYMMETRY ALGEBRA FOR OSp(N|4).

The starting point for the algebraic construction of $SO_N$-supergravity is the set of constraints:

$$T_{\mathfrak{a}a,\mathfrak{b}b}{}^i = f_{\mathfrak{a}\mathfrak{b}}{}^i\eta_{ab}, \qquad T_{\mathfrak{a}ai}{}^j = 0 \tag{12}$$

The physical implication of these constraints is the absence of derivative coupling involving the spin - $\frac{3}{2}$ field $h_\mu^{\alpha a}$.

Replacing these constraints into the Bianchi identities one finds that there exists an irreducible realization of the algebra of transformations (8) on a set of independent fields which includes $h_\mu^i$, $h_\mu^{\alpha a}$, $h_\mu^{[ab]}$ and a subset of components of $F_{BC}{}^A$, which is completely specified by the constraints (12) and the following additional ones.

$$T_{\alpha a,\beta b}{}^{\gamma c}\left\{C^{\alpha\beta}C_{\gamma\delta} - \gamma_5^{\alpha\beta}\gamma_{5\gamma\delta} + \tfrac{1}{2}(\gamma_5 f_i)^{\alpha\beta}(\gamma_5 f^i)_{\alpha\beta}\right\} = 0 \quad (13)$$

$$T_{\alpha a,\beta b}{}^{\gamma c}\left|(N-3)(\delta^a_c\delta^b_d + \delta^b_c\delta^a_d) - 5\eta^{ab}\eta_{cd}\right| = 0 \quad (14)$$

$$T_{j\alpha a}{}^{\beta b}(2\delta^j_i + f^j f_i)_\beta{}^\alpha\ f_{[cd]b}{}^a = 0 \quad (15)$$

The first constraint (13) may be taken as an equation for the spinor component of the internal symmetry connection; the other two eliminate independent supermultiplets starting off with a singlet axial-vector field and a vector field antisymmetric in two internal symmetry indices. It should be noticed that these constraints may not be unique, but in order to obtain an irreducible realization of the algebra, the set of torsion components in (12-15) must be given in terms of the independent fields. Our choice was made so as to give the conventional model[1] for N = 4.

All of the torsion and curvature components with a lower spinor index are then determined in terms of $F_{ij}{}^A$ and the following set of representations of $S\ell(2C)\otimes SO_N$:

A) Bosons: $T^5_{i(ab)}$, $T^0_{(ab)}$, $T^5_{(ab)}$, $T'^5_{i(ab)}$, $T^{[k\ell]}_{[ab]}$, $T'^{[k\ell]}_{[ab]}$, $T_{i[ab]}$,

$R^0_{[abcd]}$, $R^5_{[abcd]}$, $R^0_{[abc]d}$, $R^5_{[abc]d}$ .

B) Fermions: $T^\delta_{[abc]}$, $T^\delta_{[ab]c}$, $T^\delta_{(abc)}$, $T^\delta_a$, $R^\delta_{[ab]c}$ .

These representations transform as irreducible tensors under $SO_N$ except for the boson fields which decompose into their traces and traceless parts; the traces of $T^5_{i(ab)}$ and $T'^5_{i(ab)}$ are related by (21). The indices 0,5 stand for even or odd parity. For N = 8, $R^{0,5}_{[abcd]}$ are not irreducible. An irreducible realization of the supersymmetry algebra is obtained by imposing the duality constraints:

$$R^{0,5}_{[abcd]} = (\begin{smallmatrix}-&+\\+&-\end{smallmatrix})\ R^{0,5}_{[a'b'c'd']}\varepsilon_{abcd}{}^{a'b'c'd'} \quad (16)$$

A different set of auxiliary fields[2,4] has been obtained for N = 2. The constraint (12) is modified to $T_{\alpha ai}{}^j = \tfrac{1}{4}T'_{\alpha a}\delta^j_i$, and the set includes new fields $T'_{\alpha a}$, $T_i$, $T^0_{[ab]}$, $T^5_{[ab]}$; however, in the sets A) and B), $T^5_{i(ab)}$ and $T^{[k\ell]}_{[ab]}$ are related respectively to $T'^5_{i(ab)}$ and $T'^{[k\ell]}_{[ab]}$, only the trace of $T^0_{(ab)}$ and the traceless part of $T^5_{(ab)}$ remain and $T^{d}_{(abc)} = 0$. The boson curvatures $R^{0,5}_{[abcd]}$, $R^{0,5}_{[abc]d}$ and the fermions $T^{d}_{[ab]c}$, $R^{d}_{[ab]c}$ vanish identically for N = 2.

We outline here the information obtained from each Bianchi identity as one imposes the constraints (12-15). Complete results will be given in a forthcoming publication. The identities with upper vector index:

$$1.\quad (\alpha a, \beta b, \gamma c)^k = 0, \quad 2.\quad (\alpha a, b, j)^k = 0, \quad 3.\quad (\alpha a, i, j)^k = 0$$

determine the torsion and curvature components with a lower spinor index. In particular the torsion components are given by:

$$T_{\alpha a,\beta b}{}^{\gamma c} = \{T^{d}_{[abd]}(C_{\alpha\beta}C_{\delta}{}^{\gamma} + \gamma_{5\alpha\beta}\gamma_{5\delta}{}^{\gamma}) + (T_{\alpha a(bd)}\gamma_{5\beta}{}^{\gamma} + T_{\beta b(ad)}\gamma_{5\alpha}{}^{\gamma})\}\eta^{cd} \tag{17}$$

where:

$$T_{\alpha a(bc)} = T_{\alpha(bcd)} + \tfrac{1}{2}(T_{\alpha[ab]c} + T_{\alpha[ac]b}) + [(N-4)(T_{\alpha b}\eta_{ac} + T_{\alpha c}\eta_{ab}) - (N-7)T_{\alpha a}\eta_{bc}] \tag{18}$$

and

$$T_{i\alpha a}{}^{\beta b} = \{T_{j[ac]}f_i f^j + T^{[jk]}_{[ac]}f_i f_{[jk]} + T'^{[jk]}_{[ac]}f_{[jk]}f_i + T^5_{i(ac)}\gamma_5 + T^0_{(ac)}f_i + T^5_{(ac)}\gamma_5 f_i + T^{[jk]}_{i(ac)}f_{[jk]}\}_{\alpha}{}^{\beta}\eta^{bc} \tag{19}$$

where:

$$T^{[\ell n]}_{i(ac)}f_{[\ell n]j}{}^{k} = \varepsilon_{ij}{}^{k\ell}T'^5_{\ell(ac)} + T'^{k}_{ij}\eta_{ac} \tag{20}$$

$$(3T'^5_{i(ac)} + (N-4)T^5_{i(ac)})\eta^{ac} = \tfrac{1}{2}T_{[abc]}\gamma_5 f_i T^{[abc]} \tag{21}$$

$$T'^{k}_{ij} - T'^{k}_{ji} = T_{ij}{}^{k} \tag{22}$$

Notice that equation (19) follows from (15) and the condition:

$$T_{i\alpha a}{}^{\gamma c}f_{\gamma\beta}{}^{i}\eta_{cb} + T_{i\beta b}{}^{\gamma c}f_{\gamma\alpha}{}^{i}\eta_{ca} = 0 \tag{23}$$

The independent curvatures $R^{0,5}_{[abcd]}$ and $R^{0,5}_{[abc]d}$ are defined by:

$$R^{0,5}_{[abcd]} = \frac{1}{6} \sum_{(abc)} (R_{\alpha a,\beta b[cd]} + R_{\alpha c,\beta d[ab]})(C^{\alpha\beta}, \gamma_5^{\alpha\beta}) \tag{24}$$

$$R^{0,5}_{[abc]d} = \frac{1}{2} \sum_{(abc)} (R_{\alpha a,\beta b[cd]} - R_{\alpha c,\beta d[ab]})(C^{\alpha\beta}, \gamma_5^{\alpha\beta}) \tag{25}$$

where $R_{\alpha a,\beta b[cd]} = R_{\alpha a,\beta b}{}^{[gh]} f_{[gh]c}{}^{e} \eta_{de}$.

The identities with upper spinor index,

4. $(\alpha a, \beta b, \gamma c)^{\delta d} = 0$ , 5. $(\alpha a, \beta b, j)^{\delta d} = 0$

provide transformation laws for the fermionic and bosonic independent torsions and determine $R_{i\alpha a[bc]} = R_{i\alpha a}{}^{[gh]} f_{[gh]b}{}^{e} \eta_{ce}$ in terms of torsion components and $R_{\alpha a[bc]}$ defined by:

$$R_{\alpha a[bc]} = R_{i\gamma a'[b'c']} f_{\beta\alpha}{}^{i} C^{\beta\gamma} \{\tfrac{1}{3}(2\delta_b^{b'}\delta_c^{c'}\delta_a^{a'} + \delta_b^{b'}\delta_a^{c'}\delta_c^{a'} + \delta_a^{b'}\delta_c^{c'}\delta_b^{a'}) - \frac{1}{N+1}(\eta^{a'b'}\eta_{ab}\delta_c^{c'} - \eta^{a'c'}\eta_{ac}\delta_b^{b'})\} \tag{26}$$

The identities

6. $(\alpha a, \beta b, \gamma c)^{[gh]} = 0$ , 7. $(\alpha a, \beta b, j)^{[gh]} = 0$

provide transformation laws for the fermionic and bosonic curvatures.

The identity:

8. $(\alpha a, i, j)^{\delta d} = 0$

determines the torsion $T_{ij}{}^{k}$.

The remaining identities do not give new information. In fact, first, identities with upper index [ij] are satisfied once the Lorentz connection $h_\mu^{[ij]}$ is determined by the equation for the torsion $T_{ij}{}^{k}$. Secondly, the identity

$$[\alpha a, i, j]^{[gh]} = 0$$

is satisfied by virtue of the relation:

$$\mathcal{D}_{\alpha a}(\beta b, i, j)^{\delta d} + \mathcal{D}_{\beta b}(\alpha a, i, j)^{\delta d} = \sum_A (-1)^{\sigma_A}\{(\beta b, i, j)^A F_{A\alpha a}{}^{\delta d} + (\alpha a, i, j)^A F_{A\beta b}{}^{\delta d}\} \tag{27}$$

Finally, the following relation,

$$\sum_{A_1B_1C_1} h_\mu^{A_1} h_\nu^{B_1} h_\lambda^{C_1} (-1)^{(\sigma_{A_1}\sigma_{B_1}+\sigma_{B_1}\sigma_{C_1}+\sigma_{C_1}\sigma_{A_1})} (A_1,B_1,C_1)^D = 0 \quad (28)$$

which is identically satisfied, implies that

$$(i,j,k)^D = 0$$

will obtain as a consequence of the set of identities with the same upper index D.

When all the torsion and curvature components with a lower spinor index are expressed in terms of the independent fields and of the vector components $T_{ij}^{A_1}$, $R_{ij}^{A_0}$, these vector components are in turn determined by means of equation (10). The transformation laws for the fields $h_\mu^A$ become then completely specified.

The Bianchi identities do not completely determine the supersymmetry transformations of all the fields in the sets A) and B). One has to use further the commutator (8). As one proceeds, new auxiliary fields may emerge. However, one is assured that the algebra closes after a finite number of steps, because a product of more than 4N generators $\mathcal{D}_{a_a}$ can always be reduced to terms with fewer factors. Therefore, following this procedure, a realization of the algebra is obtained.

Knowing the algebra of invariant transformations, the equations of motion for the dynamical variables, and the equations of constraint for the auxiliary fields, one can find an action from which these equations are obtained. This action will be invariant under the transformations (1,2). We remark that, although the action can be written in the first order formalism (giving the equation for $T_{ij}^{k}$) as well as in second order, only in the latter case the supersymmetry algebra closes.

REFERENCES

1. E. Cremmer, J. Scherk and S. Ferrara, Phys. Lett. 74B, 409 (1978); E. Cremmer and J. Scherk, Nucl. Phys. B127, 259 (1978).

2. B. de Wit and J. W. van Holten, Nucl. Phys. B155, 530 (1979); B. de Wit, J. W. van Holten and A. Van Proeyen, Nucl. Phys. B167, 186 (1980).

3. S. Ferrara and P. van Nieuwenhuizen, Phys. Lett. 74B, 333 (1978)

4. E. S. Fradkin and M. A. Vasiliev, Lett. Nuovo Cim. 25, 79 (1979) Phys. Lett. 85B, 47 (1979).

5. S. W. MacDowell, Phys. Lett. 80B, 212 (1979).

6. S. W. MacDowell, Supergravity (p. 77), North Holland Publishing Company, Amsterdam (1979), Eds. P. van Nieuwenhuizen and D. Z. Freedman.

7. Y. Ne'eman and T. Regge, Phys. Lett. 74B, 31 (1978); Rivista del Nuovo Cimento, Vol. 1, 5, 1 (1978).

8. J. Wess and B. Zumino, Phys. Letters 66B, 361 (1977); Phys. Letters 79B, 394 (1978);
J. Wess, R. Grim and B. Zumino, Phys. Letters 73B, 415 (1978); Nuclear Phys. B152, 255 (1979);
B. Zumino, Proc. 19th Int. Conf. High Energy Physics Tokyo, (1978) p 555.

9. L. Brink, M. Gell-Mann, P. Ramond and J. Schwarz, Phys. Lett. 74B, 336 (1978); Phys. Lett. 76B, 417 (1978).

# SUPERGRAVITIES IN 5 DIMENSIONS

E. Cremmer

Laboratoire de Physique Théorique
de l'Ecole Normale Supérieure, Paris

Dedicated to Joël Scherk

## 1 WHY 5 DIMENSIONS ?

In supersymmetry the consideration of theories in dimensions $D > 4$ has been very fruitful. In particular the supersymmetric N=4 Yang-Mills theory has been derived from the N=1 supersymmetric Yang-Mills theory in 10 dimensions (Gliozzi, Olive & Scherk, 1977 ; Brink, Scherk & Schwarz, 1977). More recently, starting from the N=1 supergravity in 11 dimensions (Cremmer, Julia & Scherk, 1978) the N=8 supergravity in 4 dimensions has been derived with its unexpected symmetries $E_7$ global x SU(8) local (Cremmer & Julia, 1978 and 1979).

We would like today to concentrate on supergravities in 5 dimensions for essentially four reasons :

(i) For extended supergravities (especially N=8) the structure is simpler in 5 dimensions than in 4 dimensions because all the invariances are invariances of the Lagrangian instead of invariances of equations of motion. (This is related to the duality transformations on vector fields for the theories in 4 dimensions). This could therefore lead to a better understanding of these extended supergravities.

(ii) From the theories in 5 dimensions we can obtain spontaneously broken supersymmetric theories in 4 dimensions by a generalized dimensional reduction (Scherk & Schwarz, 1979). In particular, spontaneously broken N=8 supergravity with 4 mass parameters has been constructed in this way (Cremmer, Scherk & Schwarz, 1979).

(iii) The knowledge of the theory on-shell in 5 dimensions allows one to have an off-shell formulation in 4 dimensions modulo some differential constraints on the fields using the dimensional reduction by Legendre transformation (Sohnius, Stelle & West, 1980). In particular, an off-shell formulation of extended N=8 supersymmetry has been derived (Cremmer, Ferrara, Stelle & West, 1980).

(iv) From the "Lagrangian builder" point of view it shows how the conjecture of the bosonic symmetries allows one to construct the complete theory up to a few coefficients.

The plan of my talk will be the following :

(1) I shall give some notation and definitions of symplectic spinors in 5 dimensions.

(2) I shall give the particle contents of all supergravities in 5 dimensions.

(3) I shall briefly recall some facts about global and local symmetries in supergravity.

(4) The main part of the talk will be devoted to the description and construction of the N=8 supergravity in 5 dimensions.

(5) Finally, I shall give the consistent sets of truncation which lead to the N=6, 4 and 2 supergravities in 5 dimensions.

## 2 SYMPLECTIC SPINORS

The metric of the 5-dimensional spacetime is

$$\eta_{rs} = (+, -, -, -, -)$$

The $\gamma$ matrices are defined by their anticommutation relations

$$\{\gamma_r , \gamma_s\} = 2\,\eta_{rs}$$

$\gamma_0 , \gamma_1 , \gamma_2 , \gamma_3$ are the same as in 4 dimensions and are pure imaginary. Since $(\gamma_5)^2 = +1$ we must define

$\gamma_4 = i\,\gamma_5$ which is real.
This shows that there are no Majorana spinors in 5 dimensions.
$\gamma_0$ and $\gamma_5$ are antisymmetric and $\gamma_1 , \gamma_2 , \gamma_3$ are symmetric.
The five $\gamma$ matrices are related by

$$\gamma_{rstuv} = \varepsilon_{rstuv}$$

where $\gamma_{rstuv}$ is the totally antisymmetric product of $\gamma_r\ \gamma_s\ \gamma_t\ \gamma_u\ \gamma_v$ and $\varepsilon_{rstuv}$ is the usual Levi-Civita symbol with 5 indices ( $\varepsilon_{01234} = +1$).

In 5 dimensions the N extended supersymmetry algebra can only be defined for even N and it has a natural isomorphism which is the USp(N) symplectic symmetry (compact)

$$\{\bar{Q}^a_\alpha , Q^b_\beta\} = \Omega^{ab}(\gamma^\mu)_{\beta\alpha}\, P_\mu$$

( $\mu = 0 \ldots 4$ ; $\alpha = 1 \ldots 4$ ; $a = 1 \ldots N$)

The charges $Q^a_\alpha$ (and consequently the spinor fields) satisfy a generalized Majorana condition

$$Q^a_\alpha = C_5 \, \bar{Q}^{ta}_\alpha$$

where $C_5$ satisfies $\quad C_5 \gamma_\mu C_5^{-1} = \gamma^t_\mu$

$$\bar{Q}^a = (Q_a^*)^t \gamma_0$$

$\Omega^{ab}$ is the real symplectic metric and is used to raise or lower indices

$$Q_a = \Omega_{ab} Q^b$$

from which we deduce $\quad \bar{Q}_a = \Omega_{ab} \bar{Q}^b = -(Q^a)^{*t} \gamma_0$

We can choose $C_5 = \gamma_0 \gamma_5$. In this case the symplectic spinors are defined by

$$Q^a = \gamma_5 (Q_a)^*$$

From these definitions we deduce the important property of bilinear expressions in Fermi fields :

$$\bar{\psi}^a \gamma_{\mu_1} \ldots \gamma_{\mu_M} \chi^b = \bar{\chi}^b \gamma_{\mu_M} \ldots \gamma_{\mu_1} \psi^a \qquad , \quad \forall M$$

Finally let us give the Fierz transformation in 5 dimensions :

$$\bar{\varepsilon}_1 \varepsilon_2 \bar{\varepsilon}_3 \varepsilon_4 = -\frac{1}{4} \{ \bar{\varepsilon}_1 \varepsilon_4 \bar{\varepsilon}_3 \varepsilon_2 + \bar{\varepsilon}_1 \gamma_r \varepsilon_4 \bar{\varepsilon}_3 \gamma^r \varepsilon_2 - \frac{1}{2} \bar{\varepsilon}_1 \gamma_{rs} \varepsilon_4 \bar{\varepsilon}_3 \gamma^{rs} \varepsilon_2 \}$$

## 3 SUPERGRAVITIES IN 5 DIMENSIONS

The physical states of 2N extended supersymmetric massless multiplets in 5 dimensions are classified by USp(2N) (compact) as the states of the massive multiplet with central charge in 4 dimensions. It is the 5th dimension which is related to the central charge in 4 dimensions.

In the simplest multiplets, the representations of USp(2N) which appear are the antisymmetric and traceless tensors $R^{abc\ldots m}$ with

$$\Omega_{ab} R^{abc\ldots m} = 0$$

with $m \leqslant N$ . For $N < m \leqslant 2N$ an antisymmetric traceless tensor is automatically zero because the Levi-Civita tensor with 2N indices can be written in terms of $\Omega_{ab}$

$$\varepsilon_{a_1 a_2 \ldots a_{2N-1} a_{2N}} \sim \Omega_{[a_1 a_2} \Omega_{a_3 a_4} \ldots \Omega_{a_{2N-1} a_{2N}]}$$

The fields also satisfy the same kind of generalized reality condition as the spinor charges

$$A_\mu^{\ ab} = (A_{\mu\ ab})^*$$
$$\chi^{abc} = \gamma_5 (\chi_{abc})^*$$

The lowest spin supermultiplet for 2N supersymmetry has states from spin s up to s=N (SU(2) is the little group of Lorentz group in 5 dimensions) and has the following content

| s=N | s=N−1/2 | s=N−1 | ... | s=0 |
|---|---|---|---|---|
| $\phi$ | $\phi^a$ | $\phi^{ab}$ | ... | $\phi^{a_1 \ldots a_N}$ |

where all $\phi^{ab\ldots}$ are antisymmetric and traceless.
Other multiplets can be obtained by combining this multiplet with states of angular momentum J and an arbitrary representation of USp(2N) (Ferrara & Zumino, 1979).

This allows a simple construction of the representations of extended supergravities in 5 dimensions. They are given, in the following table, as well as the lowest spin supermultiplet.

| s | 2 | 3/2 | 1 | 1/2 | 0 | group |
|---|---|---|---|---|---|---|
| N=8 | 1 | 8 | 27 | 48 | 42 | USp(8) |
| N=6 | 1 | 6 | 14+1 | 14'+6 | 14 | USp(6) |
| | (J=$\frac{1}{2}$) | ⊗[1 | 6 | 14 | 14'] | |
| N=4 | 1 | 4 | 5+1 | 4 | 1 | USp(4) |
| | (J=1) | ⊗ | [1 | 4 | 5] | |
| N=2 | 1 | 2 | 1 | | | USp(2) |
| | | | (J=$\frac{3}{2}$) | ⊗ [1 , | 2] | |

As in 4 dimensions the N=8 supergravity multiplet is also the lowest spin supermultiplet.

## 4 GLOBAL AND LOCAL SYMMETRIES IN SUPERGRAVITY

The dimensional reduction shows that for maximal extended supergravities in D dimensions obtained from N=1 supergravity in 11 dimensions (Cremmer & Julia, 1979 ; Cremmer, 1980), the theory is invariant under the product of a non-compact global group and a compact local group

$$SL(11 - D,R)_{\text{global}} \times SO(11 - D)_{\text{local}}$$

SO(11 - D)local acts on Fermi fields and scalar fields.
SL(11 - D,R)global acts on tensor fields and scalar fields.
The scalar fields which come from the metric in 11 dimensions are described by the coset GL(11 - D,R)/SO(11 - D) (after a Weyl rescaling). We expect that all scalar fields can be described in this geometric way by a coset G/H (i.e. a matrix of G defined up to a local transformation of H) as the vielbein $e_\mu^r$ is described by the coset GL(D,R)/SO(D - 1, 1). If G is non compact, there is no problem of positivity if H is the maximal compact subgroup of G. The symmetries G and H can be conjectured by simple counting arguments if we remember that H is the maximal group linearly realised on all fields. (This H is the diagonal subgroup of $G_{\text{global}} \times H_{\text{local}} \supset H_{\text{global}} \times H_{\text{global}} \supset H$).

We shall give below the content and the symmetries of maximal supergravities in D=9...3 after duality transformations which convert a p tensor field into a (D - 2 - p) tensor field

D=9 $\quad GL(2, R)_{\text{global}} \otimes SO(2)_{\text{local}}$

$1\, e_\mu^r$ , $2\, \Psi_\mu$ , $1A_{\mu\nu\rho}$ , $2A_{\mu\nu}$ , $3A_\mu$ , $4\, \chi$ , 3 scalars

D=8 $\quad E_{3(+3)} = SL(3,R) \times SL(2,R)_{\text{global}} \otimes [SO(3) \times SO(2)]_{\text{local}}$

$1\, e_\mu^r$ , $2\, \Psi_\mu$ , $1\underset{\sim}{A_{\mu\nu\rho}}$ , $3A_{\mu\nu}$ , $6A_\mu$ , $6\, \chi$ , 7 scalars

D=7 $\quad E_{4(+4)} = SL(5,R)_{\text{global}} \otimes SO(5)_{\text{local}}$

$1\, e_\mu^r$ , $4\, \Psi_\mu$ , $5A_{\mu\nu}$ , $10A_\mu$ , $16\, \chi$ , 14 scalars

D=6 $\quad E_{5(+5)} = SO(5, 5)_{\text{global}} \otimes SO(5) \times SO(5)_{\text{local}}$

$1\, e_\mu^r$ , $4\, \Psi_\mu$ , $5\underset{\sim}{A_{\mu\nu}}$ , $16A_\mu$ , $20\, \chi$ , 25 scalars

D=5 $\quad E_{6(+6)\text{global}} \otimes USp(8)_{\text{local}}$

$1\, e_\mu^r$ , $8\, \Psi_\mu$ , $27A_\mu$ , $48\, \chi$ , 42 scalars

D=4 $\quad E_{7(+7)\text{global}} \otimes SU(8)_{\text{local}}$

$1\, e_\mu^r$ , $8\, \Psi_\mu$ , $\underset{\sim}{28A_\mu}$ , $56\, \chi$ , 70 scalars

D=3 $\quad E_{8(+8)\text{global}} \otimes SO(16)_{\text{local}}$

$1\, e_\mu^r$ , $16\Psi_\mu$ , $128\, \chi$ , 128 scalars

Let us note that in 3 dimensions there is no degree of freedom for the graviton and the gravitinos. The underlined tensor fields need duality transformations to form a representation of the global group. The global symmetry will not be a symmetry of the Lagrangian but only of the equations of motion : the symmetry will exchange the Bianchi identity for the field strength of the tensor with its equation of motion.

It has been seen that in 4 dimensions, for all extended supergravities, the scalar fields are described by a coset, the local symmetry being U(N). In the same way, we can conjecture that all extended supergravities in 5 dimensions have a global symmetry G and a local symmetry USp(2N), the scalar fields being described by G/USp(2N). This gives the following table

| | |
|---|---|
| N=8 | $E_{6(+6)}$ global ⊗ USp(8) local |
| N=6 | $SU^*(6)$ global ⊗ USp(6) local |
| N=4 | USp(4) x R global ⊗ USp(4) local |
| N=2 | USp(2) global ⊗ USp(2) local |

## 5 N=8 SUPERGRAVITY IN 5 DIMENSIONS

As we have seen, the free particle spectrum is described by the fields $h_{\mu\nu}$ , $\Psi_\mu{}^a$, $A_\mu{}^{ab}$, $\chi^{abc}$ and $\phi^{abcd}$ where $a$ =1...8 and these fields are pseudoreal in the sense previously defined, completely antisymmetric and traceless in the internal USp(8) indices.

We have seen that we expect the theory to have a global symmetry $E_6$ and a local symmetry USp(8). Let us first briefly describe $E_6$. It has 78 generators and the fundamental representation has dimension 27. We are interested in the non-compact form which has 42 non-compact generators and 36 compact ones which generate the maximal subgroup USp(8). The 27 representation acts in the vector space spanned by $Z^{\alpha\beta}$ ( $\alpha$ , $\beta$ = 1...8) such that

$$Z^{\alpha\beta} = - Z^{\beta\alpha} = (Z_{\alpha\beta})^*$$

$$\Omega_{\alpha\beta} Z^{\alpha\beta} = 0 \quad , \quad Z_{\alpha\beta} = \Omega_{\alpha\gamma} \Omega_{\beta\delta} Z^{\gamma\delta}$$

and the infinitesimal transformations of $E_6$ are given by

$$\delta Z^{\alpha\beta} = \Lambda^\alpha{}_\gamma Z^{\gamma\beta} + \Lambda^\beta{}_\gamma Z^{\alpha\gamma} + \Sigma^{\alpha\beta\gamma\delta} Z_{\gamma\delta}$$

where $\Lambda^\alpha{}_\gamma$ is an antihermitian matrix such that $\Lambda_{\alpha\gamma}$ is symmetric and $\Sigma^{\alpha\beta\gamma\delta}$ is totally antisymmetric, traceless and pseudoreal

$$\Sigma^{\alpha\beta\gamma\delta} = (\Sigma_{\alpha\beta\gamma\delta})^*$$

There is no quadratic invariant for $E_6$ : $27 \times 27 \not\ni 1$ in particular $Z^{\alpha\beta} Z_{\alpha\beta}$ is not invariant for $E_6$. We can form an invariant from $27 \times \overline{27}$ where $\overline{27}$ is spanned by $\tilde{Z}^{\alpha\beta}$

$$\tilde{Z}^{\alpha\beta} = - \tilde{Z}^{\alpha\beta} = - (Z_{\alpha\beta})^* \quad ; \quad \Omega_{\alpha\beta} \tilde{Z}^{\alpha\beta} = 0$$

which transforms under $E_6$ by

$$\delta\tilde{Z}^{\alpha\beta} = \Lambda^{\alpha}{}_{\gamma} \tilde{Z}^{\gamma\beta} + \Lambda^{\beta}{}_{\gamma} \tilde{Z}^{\alpha\gamma} - \Sigma^{\alpha\beta\gamma\delta} \tilde{Z}_{\gamma\delta}$$

$Z_{\alpha\beta} \tilde{Z}^{\alpha\beta}$ is an invariant under $E_6$.
Both $Z_{\alpha\beta} Z^{\alpha\beta}$ and $Z_{\alpha\beta} \tilde{Z}^{\alpha\beta}$ are invariant under the subgroup USp(8). There exists a trilinear invariant for $E_6$ : $27 \times 27 \times 27 = 1+$

$$J = Z^{\alpha\beta} \Omega_{\beta\gamma} Z^{\gamma\delta} \Omega_{\delta\varepsilon} Z^{\varepsilon\lambda} \Omega_{\lambda\alpha}$$

These properties of $E_6$ are all we need to obtain the general structure of the theory.

The fields of the N=8 supergravity are :

- the graviton $e_\mu^r$ , an element of GL(5,R)/SO(4, 1)
- the 8 gravitinos $\Psi_\mu^a$ which are in the representation 8 of USp(8) and singlets for $E_6$
- the 27 vector fields $A_\mu^{\alpha\beta}$ which are singlets for USp(8) and in the 27 representation of $E_6$
- the 48 spin 1/2 fields $\chi^{abc}$ which are in the representation 48 of USp(8) and singlets for $E_6$
- the 42 scalar fields will be described by an element $\mathcal{V}_{\alpha\beta}{}^{ab}$ of the coset $E_{6(+6)}$/USp(8) (78 - 36 = 42). It transforms as $\overline{27}$ under $E_6$ and 27 under USp(8). The indices $\alpha$, $\beta$ =1...8 are the 'curved' indices of $E_6$ and a , b =1...8 are the flat indices of USp(8) and $\mathcal{V}_{\alpha\beta}{}^{ab}$ is a 27-bein connecting these two groups.

The self-interaction of the scalar fields is described by a non linear $\sigma$-model associated to the coset $E_6$/USp(8) and therefore by the Lagrangian

$$\mathcal{L}_s \sim D_\mu \mathcal{V}_{\alpha\beta}{}^{ab} D^\mu \tilde{\mathcal{V}}_{ab}{}^{\alpha\beta} = - \mathrm{Tr}\, (\tilde{\mathcal{V}} D_\mu \mathcal{V})^2$$

where $\tilde{\mathcal{V}}$ is the inverse of $\mathcal{V}$

$$\tilde{\mathcal{V}}_{ab}{}^{\alpha\beta} \mathcal{V}_{\alpha\beta}{}^{cd} = \frac{1}{2} (\delta_a{}^c \delta_b^d - \delta_a{}^d \delta_b^c) + \frac{1}{8} \Omega_{ab} \Omega^{cd}$$

$D_\mu$ is the covariant derivative with respect to USp(8) using the associated connexion $\Omega_{\mu\, a}{}^b$ . Since there is no kinetic term for $\Omega_{\mu\, a}{}^b$ we can solve its equations of motion. Since $\mathcal{V}$ is an element of $E_6$ we have the following decomposition for $\mathcal{V}^{-1} \partial_\mu \mathcal{V}$ which is in the Lie algebra of $E_6$

$$\tilde{\mathcal{V}}_{cd}{}^{\alpha\beta} \partial_\mu \mathcal{V}_{\alpha\beta}{}^{ab} = 2\, Q_{\mu[c}{}^{[a} \delta_{d]}^{b]} + P_\mu{}^{ab}{}_{cd}$$

where $Q_{\mu\ c}^{\ a}$ belongs to the Lie algebra of USp(8) and $P_{\mu\ abcd}$ is in the orthogonal part to USp(8) (with respect to the Killing metric). For $\Omega_\mu$ we get

$$\Omega_{\mu\ a}^{\ b} = Q_{\mu\ a}^{\ b}$$

The Lagrangian then becomes

$$\mathcal{L}_S \sim |P_{\mu\ abcd}|^2$$

$Q_{\mu a}^{\ b}$ and $P_{\mu abcd}$ are obviously invariant under $E_6$. If we restrict ourselves to the scalar fields, as in the case of general relativity, we can describe them by a metric $G_{\alpha\beta,\gamma\delta}$ instead of the 27-bein $\mathcal{V}_{\alpha\beta}^{\ ab}$ (to be compared to $g_{\mu\nu}$ and $e_\mu^{\ r}$) The metric is invariant under the local group USp(8) and covariant under $E_6$. It is given by

$$G_{\alpha\beta,\gamma\delta} = \mathcal{V}_{\alpha\beta}^{\ ab}\ \Omega_{ac}\ \Omega_{bd}\ \mathcal{V}_{\gamma\delta}^{\ cd}$$

and is characterized by the property :

$$G_{\alpha\beta,\gamma\delta} = G_{\gamma\delta,\alpha\beta}$$

The Lagrangian is then written as

$$\mathcal{L}_S \sim \partial_\mu\ G_{\alpha\beta,\gamma\delta}\ \partial^\mu\ (G^{-1})^{\alpha\beta,\gamma\delta}$$

This metric must also be used to describe the interaction of the vector fields since there is no quadratic invariant for $E_7$. The generalized "kinetic" term for the vectors is then given by

$$\mathcal{L}_{V^2} \sim G_{\alpha\beta,\gamma\delta}\ F_{\mu\nu}^{\ \alpha\beta}\ F_{\rho\sigma}^{\ \gamma\delta}\ g^{\mu\rho}\ g^{\nu\sigma}$$

As in 11 dimensions there also exists a trilinear gauge invariant coupling (up to a total derivative) of the vectors which is required by supersymmetry. Since there is a trilinear $E_6$ invariant J, we do not need the scalar metric (nor the metric tensor $g_{\mu\nu}$)

$$\mathcal{L}_{V^3} \sim \varepsilon^{\mu\nu\rho\sigma\lambda}\ \Omega_{\alpha\beta}\ F_{\mu\nu}^{\ \beta\gamma}\ \Omega_{\gamma\delta}\ F_{\rho\sigma}^{\ \delta\varepsilon}\ \Omega_{\varepsilon\eta}\ A_\lambda^{\eta\alpha}$$

The couplings to the fermions can no longer be described by the metric, but require the 27-bein $\mathcal{V}$. The "kinetic" terms for the fermionic fields $\Psi_\mu^{\ a}$ and $\chi^{abc}$ will be covariant with respect to the local Lorentz group SO(4, 1) with the connexion $\omega_{\mu\ rs}$ and the local group USp(8) with the connexion $Q_{\mu\ a}^{\ b}$

$$D_\mu\ \Psi_\rho^{\ a} = (\partial_\mu\ \delta_b^a - Q_{\mu\ b}^{\ a} + \frac{1}{4}\ \omega_{\mu rs}\ \gamma^{rs}\ \delta_b^a)\ \Psi_\rho^{\ b}$$

$$D_\mu\ \chi^{abc} = (\partial_\mu\ \delta^{[a}{}_d - 3Q_\mu{}^{[a}{}_d + \frac{1}{4}\ \omega_{\mu rs}\ \gamma^{rs}\ \delta^{[a}{}_d)\ \chi^{bc]\ d}$$

As usual there exists a Noether-type coupling required by supersymmetry

$$P_\rho{}^{abcd}\,\bar{\Psi}_{\mu\, a}\,\gamma^\rho\,\gamma^\mu\,\chi_{bcd}$$

The coupling of fermions to $F_{\mu\nu}{}^{\alpha\beta}$ must occur only through the $E_6$ invariant (and scalar under the general coordinate transformation)

$$F_{rs}{}^{ab} = \mathcal{V}_{\alpha\beta}{}^{ab}\,F_{\mu\nu}{}^{\alpha\beta}\,e^\mu{}_r\,e^\nu{}_s$$

Let us note that $\mathcal{L}_{v^2}$ can be written as

$$\mathcal{L}_{v^2} \sim (F_{rs}{}^{ab})^2$$

but $F_{rs}{}^{ab}$ in no longer a curl.

The supersymmetry transformation laws $\delta\phi$ are conjectured to be covariant with respect to USp(8) and $E_6$. Therefore $\mathcal{L}$ and $\delta\phi$ are now defined up to numerical coefficients, quartic fermionic terms for $\mathcal{L}$ and trilinear fermionic terms for $\delta\,\Psi_{\mu\, a}$ and $\delta\,\chi_{abc}$. In particular all the non-polynomial structure in the scalar fields is fixed. Supersymmetry is used to get rid of the remaining arbitrariness.

(i) Numerical coefficients (and Lorentz structure) in $\delta\,\Psi$ and $\delta\,\chi$ are determined by checking the supersymmetry invariance of $\mathcal{L}$ in the terms of the type $\bar{\varepsilon}\,\Psi$, $\bar{\varepsilon}\,\chi$

(ii) Quartic terms in $\mathcal{L}$ and trilinear terms in $\delta\,\Psi$ and $\delta\,\chi$ are determined in two independent ways :

- we require supercovariant equations of motion for fermionic fields
- we require the closure of the supersymmetry algebra on the bosonic fields

$$[\delta_{\varepsilon_2},\,\delta_{\varepsilon_1}] = \delta_G + \delta_{\varepsilon'} + \delta_L + \delta_{USp(8)} + \delta_{U(1)}$$

where $\delta_G$ is the general coordinate transformation, $\delta_{\varepsilon'}$ a new supersymmetry transformation, $\delta_L$ a local Lorentz transformation, $\delta_{USp(8)}$ a local USp(8) transformation and $\delta_{U(1)}$ an Abelian gauge transformation on the vector fields. At this stage only the $\chi^4$ terms in $\mathcal{L}$ are still undetermined. They are determined by checking $\delta\mathcal{L}$ in the terms of the type $\varepsilon\,\chi^3$ or by looking at the closure on fermionic fields which requires the fermionic equations of motion.

The Lagrangian is then written, (we have put K=1)

$$e^{-1}\mathcal{L} = -\frac{1}{4}\,R(\omega) - \frac{i}{2}\,\bar{\Psi}_\mu{}^a\,\gamma^{\mu\nu\rho}\,D_\nu\,\Psi_{\rho\, a} - \frac{1}{8}\,g^{\mu\rho}g^{\nu\sigma}G_{\alpha\beta,\gamma\delta}F_{\mu\nu}{}^{\alpha\beta}F_{\rho\sigma}{}^{\gamma\delta}$$

$$+ \frac{i}{12}\, \bar{\chi}^{abc}\, \gamma^{\mu}\, D_{\mu}\, \chi_{abc} + \frac{1}{24}\, g^{\mu\nu}\, P_{\mu\, abcd}\, P_{\nu}{}^{abcd}$$

$$- \frac{e^{-1}}{12}\, \varepsilon^{\mu\nu\rho\sigma\lambda} (F_{\mu\nu})^{\alpha}{}_{\beta} (F_{\rho\sigma})^{\beta}{}_{\gamma}\, (A_{\lambda})^{\gamma}{}_{\alpha} + \frac{i}{3\sqrt{2}}\, P_{\rho\, abcd}\, \bar{\Psi}_{\mu}{}^{a}\, \gamma^{\rho}\, \gamma^{\mu}\, \chi^{bcd}$$

$$+ \frac{i}{4}\, \mathcal{V}^{ab}_{\alpha\beta}\, F^{\alpha\beta}_{\mu\nu} \{\bar{\Psi}^{\rho}_{a}\, \gamma_{[\rho}\, \gamma^{\mu\nu}\, \gamma_{\sigma]}\, \Psi^{\sigma}_{b} + \frac{1}{\sqrt{2}}\, \bar{\Psi}^{c}_{\rho}\, \gamma^{\mu\nu}\, \gamma^{\rho}\, \chi_{abc} + \frac{1}{2}\, \bar{\chi}_{acd}\, \gamma^{\mu\nu}\, \chi_{b}{}^{cd}\}$$

$$+ e^{-1} \mathcal{L}_{4}$$

$\mathcal{L}_4$ represents the quartic fermionic terms. Except for the $\Psi^4$ terms it is not enough to replace $\omega_{\mu\, rs}$, $P_{\rho\, abcd}$ and $F_{ab}$ by $(\omega + \hat{\omega})/2$, $(P + \hat{P})/2$, $(F + \hat{F})/2$ (^ means supercovariant extension) to reabsorb all the quartic terms. For completeness we give $\mathcal{L}_4$ below :

$$e^{-1}\mathcal{L}_{4} = - \frac{1}{16} [\bar{\Psi}^{\rho a} \gamma^{rs} \Psi_{\rho a} \bar{\Psi}^{\sigma b} \gamma_{rs} \Psi_{\sigma b} - \bar{\Psi}^{\rho a} \gamma^{rs} \Psi^{\sigma}{}_{a} \bar{\Psi}_{\rho}{}^{b} \gamma_{rs} \Psi_{\sigma b}]$$

$$+ \frac{e^{-1}}{8}\, \varepsilon^{\mu\nu\rho\sigma\lambda}\, \bar{\Psi}_{\rho}{}^{a}\, \Psi_{\sigma a} \bar{\Psi}_{\lambda}{}^{b} \gamma_{\nu} \Psi_{\mu b} - \frac{1}{4}\, \bar{\Psi}^{\rho a} \Psi^{\sigma}{}_{a} \bar{\Psi}^{b}{}_{\rho} \Psi_{\sigma b}$$

$$+ \frac{1}{8} [\bar{\Psi}^{\rho a} \gamma^{\lambda} \Psi_{\sigma a} \bar{\Psi}^{b}{}_{\lambda} \gamma^{\sigma} \Psi_{\rho b} - 2\, \bar{\Psi}^{\rho a} \gamma^{\sigma} \Psi_{\sigma a} \bar{\Psi}^{\lambda b} \gamma_{\lambda} \Psi_{\rho b}]$$

$$+ \frac{1}{2\sqrt{2}}\, \bar{\chi}^{abc}\, \gamma^{\mu}\, \gamma^{\rho\sigma}\, \Psi_{\mu c}\, \bar{\Psi}_{\rho a}\, \Psi_{\sigma b}$$

$$+ \frac{1}{96} [\bar{\chi}^{abc}\, \gamma^{\rho\sigma}\, \chi_{abc}\, \bar{\Psi}^{\mu d}\, \gamma_{\rho\sigma}\, \Psi_{\mu d}\ \ -\bar{\chi}^{abc}\, \gamma^{\rho\sigma}\, \chi_{abc}\, \bar{\Psi}_{\rho}{}^{d}\, \Psi_{\sigma d}$$

$$+ \bar{\chi}^{abc}\, \gamma^{\mu\rho\sigma}\, \chi_{abc}\, \bar{\Psi}_{\rho}{}^{d}\, \gamma_{\mu}\, \Psi_{\sigma d}]$$

$$+ \frac{1}{32} [\bar{\chi}^{abc} (\gamma^{\rho\sigma} - 3\, g^{\rho\sigma})\, \chi^{d}{}_{ab}\, \bar{\Psi}_{\rho c}\, \Psi_{\sigma d}$$

$$+ \bar{\chi}^{abc} (-\gamma^{\rho\sigma\lambda} + \gamma^{\sigma} g^{\rho\lambda} + \gamma^{\rho} g^{\sigma\lambda} + 3\gamma^{\lambda} g^{\rho\sigma})\, \chi^{d}{}_{ab}\, \bar{\Psi}_{\rho c}\, \gamma_{\lambda}\, \Psi_{\sigma d}$$

$$+ \frac{1}{2}\, \bar{\chi}^{abc} (\gamma^{\rho\sigma\mu\nu} - 3\, g^{\rho\sigma} \gamma^{\mu\nu} - 2\, g^{\rho\mu} \gamma^{\sigma\nu} - 2\, g^{\sigma\mu} \gamma^{\rho\nu} - 2\, g^{\sigma\mu} g^{\rho\nu}) \chi^{d}{}_{ab} \bar{\Psi}_{\rho c} \gamma_{\mu\nu} \Psi_{\sigma d}]$$

$$- \frac{1}{8\sqrt{2}}\, \bar{\chi}^{abc}\, \gamma_{\rho\sigma}\, \chi^{d}{}_{bc}\, \bar{\chi}_{ade} (g^{\mu\rho}\, \gamma^{\sigma} + \frac{1}{6}\, \gamma^{\mu\rho\sigma})\, \Psi_{\mu}{}^{e}$$

$$+ \frac{1}{80} [\bar{\chi}_{bcd}\, \gamma^{r}\, \chi^{bc}{}_{e}\, \bar{\chi}^{d}{}_{fg}\, \gamma_{r}\, \chi^{efg}$$

$$- \bar{\chi}_{bcd}\, \gamma^{rs}\, \chi^{bc}{}_{e}\, \bar{\chi}^{d}{}_{fg}\, \gamma_{rs}\, \chi^{efg}$$

$$+ \frac{7}{48}\, \bar{\chi}_{abc}\, \gamma^{rs}\, \chi^{abc}\, \bar{\chi}_{efg}\, \gamma_{rs}\, \chi^{efg}]$$

The supersymmetry transformation laws are given by

$$\delta e_\mu{}^r = -i\,\bar{\varepsilon}^a \gamma^r \Psi_{\mu a}$$

$$\tilde{\mathcal{V}}^{\alpha\beta}{}_{cd}\,\delta\mathcal{V}_{\alpha\beta,ab} = -2i\sqrt{2}(\bar{\varepsilon}_{[a}\chi_{bcd]} + \tfrac{3}{4}\Omega_{[ab}\bar{\varepsilon}_e\chi^e{}_{cd]})$$

$$\delta A_\mu{}^{\alpha\beta} = 2i\,\tilde{\mathcal{V}}^{\alpha\beta}{}_{ab}(\bar{\varepsilon}^a\Psi_\mu{}^b + \tfrac{1}{2\sqrt{2}}\bar{\varepsilon}_c\gamma_\mu\chi^{abc})$$

$$\delta\Psi_{\mu a} = (D_\mu(\hat{\omega})\delta_a{}^b + Q_{\mu a}{}^b)\varepsilon_b - \tfrac{1}{6}\hat{F}_{rsab}(\gamma^{rs}\gamma_\mu + 2\gamma^r e_\mu{}^s)\varepsilon^b$$

$$+\frac{i\sqrt{2}}{4}(3\,\bar{\varepsilon}^b\Psi_\mu{}^c\chi_{abc} - \bar{\varepsilon}^b\gamma^r\Psi_\mu{}^c\gamma_r\chi_{abc})$$

$$-\frac{i}{12}(\gamma_{\rho\mu}+2g_{\rho\mu})\varepsilon_d\,\bar{\chi}_{abc}\gamma^\rho\chi^{bcd} - \frac{i}{12}(\gamma_{\rho\sigma\mu}+\gamma_\rho g_{\sigma\mu})\varepsilon_d\,\bar{\chi}_{abc}\gamma^{\rho\sigma}\chi^{bcd}$$

$$\delta\chi_{abc} = \sqrt{2}\,\hat{P}_{\mu\ abcd}\varepsilon^d - \frac{3}{2\sqrt{2}}\gamma^{rs}(\hat{F}_{rs[ab}\varepsilon_{c]} + \tfrac{1}{3}\Omega_{[ab}\hat{F}_{rsc]d}\varepsilon^d)$$

$$+\frac{3i\sqrt{2}}{8}[3\,\varepsilon_g\bar{\chi}^{gf}{}_{[a}\chi_{bc]f} - \varepsilon_g\bar{\chi}^{gfd}\chi_{fd[a}\Omega_{bc]}$$

$$-\gamma_r\varepsilon_g\bar{\chi}^{gf}{}_{[a}\gamma^r\chi_{bc]f} + \frac{\gamma_r}{3}\varepsilon_g\bar{\chi}^{gfd}\gamma^r\chi_{fd[a}\Omega_{bc]}$$

$$+\frac{1}{2}\gamma_{rs}\varepsilon_g\bar{\chi}^{gf}{}_{[a}\gamma^{rs}\chi_{bc]f} - \frac{\gamma_{rs}}{6}\varepsilon_g\bar{\chi}^{gfd}\gamma^{rs}\chi_{fd[a}\Omega_{bc]}]$$

with :

$$\hat{\omega}_{\mu rs} = \omega^\circ_{\mu rs}(e) + \tfrac{i}{2}[\bar{\Psi}_\mu{}^a\gamma_s\Psi_{ra} - \bar{\Psi}^a{}_r\gamma_\mu\Psi_{sa} + \bar{\Psi}^a{}_s\gamma_r\Psi_{\mu a}] - \frac{i}{24}\bar{\chi}^{abc}\gamma_{\mu rs}\chi_{abc}$$

$$\hat{P}_{\rho\ abcd} = P_{\rho\ abcd} + 2i\sqrt{2}(\bar{\Psi}_{\rho[a}\chi_{bcd]} + \tfrac{3}{4}\Omega_{[ab}\bar{\Psi}_{\rho e}\chi^e{}_{cd]})$$

$$\hat{F}_{\rho\sigma ab} = F_{\rho\sigma ab} - 2i[\bar{\Psi}_{\rho[a}\Psi_{\sigma b]} + \tfrac{1}{8}\Omega_{ab}\bar{\Psi}_\rho{}^c\Psi_{\sigma c} - \tfrac{1}{\sqrt{2}}\bar{\Psi}_{[\rho}{}^c\gamma_{\sigma]}\chi_{abc}]$$

Let us note that since $\mathcal{V}^{-1}\delta\mathcal{V}$ has no component in USp(8), $Q_{\mu\ a}{}^b$ is supercovariant by itself but $P_{\mu\ abcd}$ is not.

The fermionic equations of motion are :

$$-i\,\gamma^{\mu\nu\rho}\hat{\mathcal{D}}_\mu\Psi_{\nu a} + \frac{i}{3\sqrt{2}}\hat{P}_{\mu\ abcd}\gamma^\mu\gamma^\rho\chi^{bcd} + \frac{i}{4\sqrt{2}}\gamma^{\mu\nu}\gamma^\rho\chi_{abc}\hat{F}_{\mu\nu}{}^{bc}$$

$$-\frac{1}{8\sqrt{2}}(\tfrac{1}{6}\gamma^{\rho\mu\nu} - g^{\rho\mu}\gamma^\nu)\chi_{abc}\bar{\chi}^{bde}\gamma_{\mu\nu}\chi^c{}_{de} = 0$$

$$\frac{i}{6}\gamma^{\mu}\hat{\mathscr{D}}_{\mu}\chi_{abc} + \frac{i}{4}\gamma^{\mu\nu}(\chi^{d}{}_{[ab}\hat{F}_{\mu\nu\, c]d} - \text{Trace abc})$$
$$-\frac{1}{20}[\gamma^{\mu}\chi_{e[ab}\bar{\chi}_{c]fg}\gamma_{\mu}\chi^{efg} - \gamma^{\rho\sigma}\chi_{e[ab}\bar{\chi}_{c]fg}\gamma_{\rho\sigma}\chi^{efg}$$
$$+\frac{1}{24}\gamma^{\rho\sigma}\chi_{abc}\bar{\chi}_{efg}\gamma_{\rho\sigma}\chi^{efg} - \text{Trace abc}] = 0$$

$\hat{\mathscr{D}}_{[\mu}\Psi_{\nu]a}$ and $\hat{\mathscr{D}}_{\mu}\chi_{abc}$ are supercovariant extensions of $D_{[\mu}\Psi_{\nu]a}$ and $D_{\mu}\chi_{abc}$ defined such that their variation by supersymmetry has no derivative of $\varepsilon$.

The algebra of supersymmetry is given by

$$[\delta_{\varepsilon_2},\delta_{\varepsilon_1}] = \delta_G(\xi_\mu) + \delta_S(\varepsilon') + \delta_L(\Sigma^{rs}) + \delta_{USp(8)}(\Lambda_a{}^b) + \delta_{U(1)}(U^{\alpha\beta})$$

with

$$\xi_\mu = -i\,\bar{\varepsilon}_1{}^c\gamma_\mu\varepsilon_{2c}$$
$$\bar{\varepsilon}'^{a} = -\xi^t\bar{\Psi}^a{}_t + \frac{i\sqrt{2}}{4}(3\,\bar{\varepsilon}_{1b}\varepsilon_{2c}\bar{\chi}^{abc} - \bar{\varepsilon}_{1b}\gamma^\rho\varepsilon_{2c}\bar{\chi}^{abc}\gamma_\rho)$$
$$\Sigma_{rs} = \xi^t\hat{\omega}_{t,rs} + \frac{i}{3}\hat{F}_{uv,ab}\,\bar{\varepsilon}_1^a(\gamma_{rs}{}^{uv} + 4\delta^u_r\delta^v_s)\varepsilon_2^b$$
$$-\frac{1}{6}\bar{\varepsilon}_{1a}\gamma_{rst}\varepsilon_{2b}\bar{\chi}^a{}_{cd}\gamma^t\chi^{bcd}$$
$$+\frac{1}{6}\bar{\varepsilon}_{1a}(\gamma_{rstu} + \eta_{rt}\eta_{su})\varepsilon_{2b}\bar{\chi}^a{}_{cd}\gamma^{tu}\chi^{bcd}$$
$$\Lambda_a{}^b = \xi^t Q_{ta}{}^b - \frac{8}{3}[(\bar{\varepsilon}_2{}^{[b}\chi^{cde]} - \frac{3}{4}\Omega^{[bc}\bar{\varepsilon}_{2f}\chi^{de]f})$$
$$\times(\bar{\varepsilon}_{1[a}\chi_{cde]} + \frac{3}{4}\Omega_{[ac}\bar{\varepsilon}_{1f}\chi_{de]}{}^f) - \varepsilon_1\leftrightarrow\varepsilon_2]$$
$$U^{\alpha\beta} = -\xi^\rho A_\rho{}^{\alpha\beta} + 2i\,\tilde{\mathscr{V}}^{\alpha\beta}{}_{ab}\bar{\varepsilon}_1^a\varepsilon_2^b$$

We have seen that the conjectured $E_6$ global $\otimes$ USp(8) local was the clue to construct the N=8 supergravity in 5 dimensions. There still remain some complications in the quartic fermionic terms. This could be a sign that there is still some structure to be discovered. We can hope that it would be easier to discover it in 5 than in 4 dimensions. Another problem could also be more easily solved in 5 dimensions : the construction of the multiplet containing the connexion of USp(8), $Q_{\mu\,a}{}^{b}$. This is of crucial importance in 4 dimensions where we conjectured that the local SU(8) could become dynamical at the quantum level (Cremmer & Julia, 1979, 1980). Some conjectures have been made on this multiplet which could lead to a grand unified model based on SU(5) with 3 families (Ellis, Gaillard, Maiani & Zumino, 1980 ; Zumino, 1980).

## 6 SUPERGRAVITIES N=6, 4, 2

In order to derive supergravities N=6, 4, 2 from N=8 by consistent truncations, it is useful to choose a particular representation for $\Omega_{ab}$ namely

$$\Omega_{ab} = \begin{pmatrix} 0 & 1 & & & & & & \\ -1 & 0 & & & & & & \\ & & 0 & 1 & & & & \\ & & -1 & 0 & & & & \\ & & & & 0 & 1 & & \\ & & & & -1 & 0 & & \\ & & & & & & 0 & 1 \\ & & & & & & -1 & 0 \end{pmatrix}$$

We shall describe the consistent truncations for N=6, 4, 2 and give the complete results only for N=2.

### 6.1 N=6 Supergravity

The invariance of the theory is SU*(6)global x USp(6). We note by $a$ =1...6 the indices of USp(6) and by $\alpha$ =1...6 the indices of SU*(6). We keep the fields $\Psi_{\mu a}$, $\chi_{abc}$ $\chi_{a\,78}$, $A_\mu{}^{\alpha\beta}$, $A_\mu{}^{78}$, $\mathcal{V}_{\alpha\beta}{}^{ab}$, $\mathcal{V}^{ab}{}_{78}$, $\mathcal{V}_{\alpha\beta}{}^{78}$ and $\mathcal{V}_{78}{}^{78}$. (For $\mathcal{V}$ we have also to use the local USp(8) invariance before making the truncation). We must take the traceless condition for N=8 into account

$$2\,\Omega_{78}\,A_\mu{}^{78} + \Omega_{\alpha\beta}\,A_\mu{}^{\alpha\beta} = 0$$

this implies that for $\mathcal{V}^{-1}\,\partial_\mu\,\mathcal{V}$ (N=8) only the following components $Q$ and $P$ remain

$$Q_{\mu\,a}{}^{b},\ P_{\mu\,abcd},\ P_{\mu\,ab78}\ (P_{\mu\,78ab})$$

with the trace conditions

$$\Omega^{ab}\,\Omega^{cd}\,P_{\mu\ abcd} = 0\ ,\quad \Omega^{ab}\,P_{\mu\ ab78} = 0$$

$$\Omega^{ab}\,P_{\mu\ abcd} + 2\,\Omega^{78}\,P_{\mu\ 78cd} = 0$$

and $P_{\mu\ abcd}$ totally antisymmetric : this is equivalent to an antisymmetric $P_{\mu\ ab}$ traceless which can be identified as $P_{\mu\ ab\ 78}$ corresponding to the 14 scalar fields ($P_{\mu\ abcd} \sim \varepsilon_{abcdef}\, P_{\mu}^{ef}$). The theory is invariant under the N=6 supersymmetry obtained by restricting $\varepsilon$ to $\varepsilon^{a}{}_{r}$ ($\varepsilon^7=\varepsilon^8=0$). The content of the theory is as expected 1 graviton $e_{\mu}$, 6 gravitinos $\Psi_{\mu}{}^{a}$, 14+1 vector fields $A_{\mu}{}^{\alpha\beta}$, 14+6 spinor fields and 14 scalar fields.

## 6.2 N=4 Supergravity

The symmetries are (USp(4) x R) global ⊗ USp(4) local. We keep $\Psi_{\mu}{}^{a}$ (a=1...4); $A_{\mu}{}^{\alpha\beta}$ ($\alpha$ =1...4) $A_{\mu}{}^{56}$ with the condition

$$\Omega_{\alpha\beta}\,A_{\mu}{}^{\alpha\beta} + 4\,\Omega_{56}\,A_{\mu}{}^{56} = 0$$

as well as $A_{\mu}{}^{78}$ defined for N=6 in terms of $A_{\mu}{}^{\alpha\beta}$ and $A_{\mu}{}^{56}$. In the same way we keep $\chi_{abc}$ and $\chi_{56c}$ ($\chi_{a\ 78}$ being a function of the previous ones) with the condition

$$\Omega^{ab}\,\chi_{abc} + 4\,\Omega^{56}\,\chi_{56\ c} = 0$$

This corresponds to 4 spin 1/2 since $\chi_{abc} \sim \varepsilon_{abcd}\,\chi^{d}$ from which we deduce $\chi_{56c} \sim \chi_{c}$ . On $\mathcal{V}^{\alpha\beta}{}_{ab}$ we make the same truncation as on $A_{\mu}{}^{\alpha\beta}$ . This implies, for $P_{\mu\ abcd}$ and $P_{\mu\ 56\ cd}$, the relation

$$\Omega^{ab}\,P_{\mu\ abcd} + 4\,\Omega^{56}\,P_{\mu\ 56\ cd} = 0$$

This corresponds to 1 scalar field : $P_{\mu\ abcd} \sim \varepsilon_{abcd}\,\phi$

$P_{\mu\ ab\ 56} \sim \Omega_{ab}\,\phi$

The remaining $Q_{\mu a}{}^{b}$ has the form $\mathcal{V}^{-1}\,\partial_{\mu}\,\mathcal{V}$ where $\mathcal{V}$ is an element of USp(4) and therefore being a pure gauge it can be reabsorbed by redefining the fermionic field with the USp(4) transformation . The theory is invariant under N=4 supersymmetry with parameter $\varepsilon^{a}$ . The content is 1 graviton, 4 gravitinos, 5+1 vector fields, 4 spin 1/2 fields and 1 scalar field.

## 6.3 N=2 Supergravity

From N=4 we keep $\Psi_{\mu\ a}$ ( a =1, 2) $A_{\mu\ ab}$ with the relation

$$\Omega^{ab}\,A_{\mu\ ab} + 6\,\Omega^{34}\,A_{\mu\ 34} = 0$$

There is no $X_{abc}$ left and we can replace $\mathcal{V}^{ab}{}_{\alpha\beta}$ by "1". The truncation can also be directly made from N=8, keeping $\Psi_{\mu a}$, $A_{\mu\,12}$, $A_{\mu\,34}$, $A_{\mu\,56}$ and $A_{\mu\,78}$ with the relations

$$A_{\mu\,78} = A_{\mu\,56} = A_{\mu\,34} = -\frac{1}{3} A_{\mu\,12}$$

After renormalization of the vector fields we get the Lagrangian for N=2 supergravity with field $e_\mu{}^r$, $\Psi_\mu{}^a$ (a =1, 2) and $A_\mu$.

$$e^{-1}\mathscr{L} = -\frac{1}{4}R(\omega) - \frac{i}{2}\bar{\Psi}_\mu{}^a \gamma^{\mu\nu\rho} D_\nu\left(\frac{\omega+\hat{\omega}}{2}\right)\Psi_{\rho a} - \frac{1}{4}F_{\mu\nu}F_{\rho\sigma}\, g^{\mu\rho} g^{\nu\sigma}$$

$$+ \frac{e^{-1}}{6\sqrt{3}}\varepsilon^{\mu\nu\rho\sigma\lambda} F_{\mu\nu}F_{\rho\sigma}A_\lambda - \frac{i\sqrt{3}}{16}(F_{\mu\nu}+\hat{F}_{\mu\nu})\bar{\Psi}^{\rho c}\gamma_{[\rho}\,\gamma^{\mu\nu}\,\gamma_{\sigma]}\,\Psi^\sigma{}_c$$

with

$$\hat{F}_{\mu\nu} = F_{\mu\nu} + \frac{\sqrt{3}}{4}\bar{\Psi}_\mu{}^c\,\Psi_{\nu c}$$

and $\omega$ is given by the 1st order formalism if $\hat{\omega}$ is defined as

$$\hat{\omega}_{\mu rs} = \omega_{\mu rs} + \frac{i}{4}\bar{\Psi}^{\rho a}\,\gamma_{\mu\,rs\,\rho\sigma}\,\Psi^\sigma{}_a$$

After solving the equations of motion for $\omega$ we get, as usual

$$\hat{\omega}_{\mu rs} = \omega^0{}_{\mu rs}(e) + \frac{i}{2}(\bar{\Psi}_\mu{}^a\,\gamma_s\,\Psi_{ra} - \bar{\Psi}^a{}_r\,\gamma_\mu\,\Psi_{sa} + \bar{\Psi}_s{}^a\,\gamma_r\,\Psi_{\mu a})$$

is invariant under the following N=2 supersymmetry transformation

$$\delta e_\mu{}^r = -i\,\bar{\varepsilon}^a\,\gamma^r\,\Psi_{\mu a}$$

$$\delta\Psi_{\mu a} = [D_\mu(\hat{\omega}) + \frac{1}{4\sqrt{3}}\hat{F}_{\rho\sigma}(\gamma^{\rho\sigma}\gamma_\mu + 2\gamma^\rho\,\delta_\mu{}^\sigma)]\,\varepsilon_a$$

$$\delta A_\mu = -\frac{\sqrt{3}}{4}\bar{\varepsilon}^a\,\Psi_{\mu a}$$

All quartic terms are contained in the replacement of $\omega$ and $F$ by $(\omega+\hat{\omega})/2$ and $(F+\hat{F})/2$ in the bilinear fermionic terms. We note that N=2 supergravity in 5 dimensions has exactly the same structure as the N=1 supergravity in 11 dimensions where everything comes from. This should be compared with the partial purely geometric results obtained by D'Auria & Fré, 1980 ; D'Auria, Fré & Regge, 1980 for this N=2 supergravity in 5 dimensions.

REFERENCES

Brink, L., Scherk, J. & Schwarz, J.H. (1977). Supersymmetric Yang-Mills Theories. Nucl. Phys. B, 121, pp.77-92.

Cremmer, E. (1980). N = 8 Supergravity. In Proc. Europhysics Study Conf. "Unification of the Fundamental Interactions", Erice.

Cremmer, E., Ferrara, S., Stelle, K. & West, P.C. (1980). Off-Shell N = 8 Supersymmetry with Central Charges. Phys. Lett. 94B, pp.349-354.

Cremmer, E. & Julia, B. (1978). The N = 8 Supergravity Theory I The Lagrangian. Phys. Lett., 80B, pp.48-51.
(1979). The SO(8) Supergravity. Nucl. Phys. B, 159, pp.141-212.

Cremmer, E., Julia, B. & Scherk, J. (1978). Supergravity Theory in 11 Dimensions. Phys. Lett., 76B, pp.409-412.

Cremmer, E., Scherk, J. & Schwarz, J.H. (1979). Spontaneously broken N = 8 supergravity. Phys. Lett., 84B, pp.83-86.

D'Auria, R. & Fré, P. (1980). About Bosonic Rheonomic Symmetry and the Generation of a Spin 1 Field in D = 5 Supergravity. Torino preprint IFTT 323. (To be published in Nucl. Phys.B).

D'Auria, R., Fré, P. & Regge, T. (1980). Supergravity and Cohomology Theory : Progress and Problems in d = 5. In Proc. Europhysics Study Conf. "Unification of the Fundamental Interactions", Erice.

Ellis, J., Gaillard, M.K., Maiani, L. & Zumino, B. (1980). Attempts at Superunification. In Proc. Europhysics Study Conf. "Unification of the Fundamental Interactions", Erice.

Ferrara, S. & Zumino, B. (1979). The Mass Matrix of N = 8 Supergravities. Phys. Lett., 86B, pp.279-282.

Gliozzi, F., Olive, D. & Scherk, J. (1977). Supersymmetry, Supergravity Theories and the Dual Spinor Model. Nucl. Phys. B, 122 pp.253-290.

Scherk, J. & Schwarz, J.H. (1979). Spontaneous Breaking of Supersymmetry through Dimensional Reduction. Phys. Lett., 82B, pp.60-64.
(1979). How to Get Masses from Extra-Dimensions. Nucl. Phys. B, 153, pp.61-88.

Sohnius, M., Stelle, K. & West, P.C. (1980). Supersymmetric Yang-Mills Theories. In Proc. Europhysics Study Conf. "Unification of the Fundamental Interactions", Erice.

Zumino, B. (1980). Contribution to this conference.

# REPRESENTATIONS OF EXTENDED SUPERSYMMETRY

Martin F. Sohnius and Kellogg S. Stelle
The Blackett Laboratory, Imperial College, London SW7

and

Peter C. West

Department of Mathematics, King's College, London WC2

## 1. INTRODUCTION

In this article we present some recent results of the quest for off-shell formulations of supersymmetric theories.

In such formulations the Lagrangians are constructed from sets of fields which form representations of the sypersymmetry algebra. It is well known that such representations almost without exception contain fields which do not correspond to particle states at the classical level; such fields are called auxiliary fields. The necessary auxiliary fields are automatically contained in superfield formulations of these theories. Unconstrained superfields, however, do not correspond to irreducible representations of supersymmetry, and superspace formulations therefore require constraints. The knowledge of these constraints is then equivalent to the knowledge of the auxiliary field structure of the theory when formulated in ordinary space.

The technique of dimensional reduction by Legendre transformation (Sohnius, Stelle and West, 1980a ) which we will describe in section four, achieves the goal of constructing off-shell formulations of all supersymmetric theories in four space-time dimensions: transformation laws on a set of fields can be deduced which close without use of the equations of motion, which represent supersymmetry, and which leave a Lagrangian invariant. This Lagrangian can be obtained as the "Hamiltonian" from a higher dimensional theory. In the more interesting cases, however, namely for gauge theories, the Lagrangian does not immediately allow an interpretation of the physics described by it: one or more of the fields have to satisfy constraints on their x-space dependence, and these constraints do not follow from the Lagrangian itself, but rather have to be implemented "from outside". In the case of abelian gauge groups, this can be done by solving the constraints and replacing the constrained fields by derivatives of unconstrained ones. No such solutions are known for the non-abelian case, i.e. for an interacting theory. In this case, the constraints give rise to severe difficulties when one attempts to calculate the physical content of these theories.

On our way to eventually solving these problems for the maximally extended theories, we return here to the more familiar territory of N=2 Yang-Mills theories. After discussing some of the more useful representations of N=2 supersymmetry in section three we, in section five, apply the general programme (Gates, Stelle and West, 1980 ) of finding superspace constraints to discover the necessary set of constraints on the N=2 Yang-Mills field strengths. Also in section five, we evaluate the further constraints which are implied by the Yang-Mills Bianchi-identities.

Section six is devoted to showing the consequences of the above constraints upon the component field content of the N=2 Yang-Mills theory in ordinary space. The above constraints are weaker but contained by either of the sets of superspace constraints required to specify the two known N=2 supersymmetric formulations. These two formulations have the same physical spectrum but differ in the way their physical degrees of freedom are described by fields. The formulation with the particles described by the usual field assignments was found by Grimm, Sohnius and Wess (1978) and the one which has a field with constrained x-space dependence was found by Sohnius, Stelle and West (1980a,b). In section seven we discuss the additional constraints required to yield these two formulations.

The remainder of this article is devoted to implementing the constraints that arise in the technique of dimensional reduction by Legendre transform in a supersymmetric manner using the method of Lagrange multipliers. Two simple examples of the supersymmetric implementation of constraints are given. We obtain the Wess-Zumino model in section eight, and a theory which has the same content as the N=2 hypermultiplet in section nine. The more interesting case of the N=2 abelian gauge theory is given in section ten. This last formulation includes both of the known N=2 abelian formulations. A procedure to deduce this last formulation by dimensional reduction is given in section eleven.

In the final section we discuss the implications of the preceding discussion for the maximally extended supersymmetric Yang-Mills theory and conjecture on a new formulation of this theory.

## 2. RELEVANCE OF OFF-SHELL FORMULATIONS

We divide considerations on the question whether off-shell formulations are desirable or even necessary into two separate issues: a) are they necessary in principle for gaining correct answers, or b) do they only lead more quickly to the same answers as other formulations do (i.e. on-shell supersymmetry without auxiliary fields)?

Taking up the second point first, we note that we can obtain a manifestly supersymmetric formulation only from an off-shell version of the theory. Calculations for manifestly supersymmetric theories are much quicker than for formulations without auxiliary fields. This is true for the construction of invariants at the classical level as well as for the calculation of quantum corrections with the help of superspace Feynman rules.

The first point, that of principle, can be asked in the following form: given a classically supersymmetric theory in its on-shell form, i.e. constructed from fields which do not carry a representation of supersymmetry and which do not contain auxiliary fields, does such a theory

(i) have an extension to an off-shell version?

(ii) lead to a supersymmetric quantum theory?

(iii) give the correct impression of the symmetries of the supersymmetric quantum theory?

By question (i) we mean to ask whether we can add to the theory a set of auxiliary fields which lead to a closing algebra. We know from ghost counting arguments which are reviewed in Sohnius, Stelle and West (1980c) that the structure of the auxiliary fields and the structure of those fields which correspond to physical states are related. For example, in the only known off-shell formulation of the N=4 gauge theory we have scalar auxiliary fields only. It is then essential that one of the spin-zero particles be represented by an antisymmetric tensor. It is an open question whether a particle of given spin in a supersymmetric theory can be represented by any field representation appropriate to that spin, and whether such a theory possesses an off-shell extension. In this context we note an interesting observation by W. Siegel in an unpublished communication: if in N=8 supergravity one keeps the fields in the representation provided by the dimensional reduction from eleven space-time dimensions, then one finds the theory to be one-loop finite. This is not the case for the theory with redefined fields as given by Cremmer and Julia (1979). Such an observation leads one to wonder whether the N=8 theory in the version of Julia and Cremmer has an off-shell extension at all.

We are told that the answer to question (ii) seems to be a cautious yes (de Wit and van Holten, 1978).

The point at issue in question (iii) is illustrated by the case of the N=4 Yang-Mills theory. The on-shell formulation of Gliozzi, Scherk and Olive (1978) and Brink, Schwarz and Scherk (1977) has a global SU(4) symmetry whereas the off-shell formulation (Sohnius, Stelle and West, 1980a,b) has an SP(4), and only its classical field equations have the SU(4) symmetry. It is unknown whether there exists an alternative off-shell formulation with the larger SU(4) symmetry, whether the SP(4) invariant formulation maintains its hidden SU(4) symmetry even for the quantum correction, or whether SP(4) is actually the maximal good symmetry of the theory, in spite of the fact that the possible conformal invariance of the theory suggests SU(4) to be the true symmetry. The reason for the differences in these symmetries is the occurrence of an off-shell central charge in the only known off-shell formulation, an off-shell central charge being one which is zero on-shell, but non-zero off-shell. In the presence of a central charge the commutator of two supersymmetry charges is given by

$$\{Q_{\alpha i}, Q_{\beta j}\} = 2 \varepsilon_{\alpha\beta}\Omega_{ij} Z \quad ,$$

where $\Omega_{ij} = - \Omega_{ji}$.

The maximal possible symmetry off-shell is then the automorphism group of the supersymmetry algebra with a central charge, namely SP(N). On shell, where the central charge is zero, it is SU(N). Consequently, if there were to exist a formulation of the N=4 Yang-Mills theory with a manifest SU(4) symmetry it would not possess a central charge. It is tempting to contemplate whether any of the maximally extended theories do exist without a central charge and if off-shell central charges have any physical significance. We will return to this point in the final section.

In any case, the presence of central charges makes an enormous difference for the type of field representations that can occur for a given theory. This is most easily seen by application of the ghost counting arguments: the representations in which the ghosts can occur are much smaller if a central charge, even an off-shell one, is present. It is our inexperience in dealing with central charges, or, equivalently, with extra bosonic dimensions, which partly explains the lack of results about extended supersymmetries. Certainly, central charges and the constructions they facilitate form a vital part of the discussions in this article, and hopefully we may have learnt more about constructing four-dimensional theories whose fields are defined over a five-dimensional space.

## 3. MULTIPLETS OF N=2 SUPERSYMMETRY

In the study of extended supersymmetry theories, the first task is to obtain a general understanding of the representations of supersymmetry on multiplets of fields. At a later stage it becomes clear which of these may be necessary to formulate physically interesting theories. At this initial algebraic stage of the discussion it is necessary to consider all the generic features of the supersymmetry algebra, and not be too prejudicial about which are physically "relevant". Thus, for N-extended supersymmetry we consider the superalgebra with anti-commutation relations

$$\{Q_{\alpha i},\bar{Q}_{\dot\beta}{}^{j}\} = 2\delta_i{}^j(\sigma^\mu)_{\alpha\dot\beta}P_\mu$$

$$\{Q_{\alpha i},Q_{\beta j}\} = 2\varepsilon_{\alpha\beta}(a^e)_{ij}Z_e \quad ; \{\bar{Q}_{\dot\alpha}{}^{i},\bar{Q}_{\dot\beta}{}^{j}\} = 2\varepsilon_{\dot\alpha\dot\beta}(a^{e*})^{ij}Z_e$$

where $(a^e)_{ij} = -(a^e)_{ji}$, $i=1...N$ and we use two component notation for the spinors. A generic feature that we have retained without prejudice of its physical interpretation is the presence of the central charges $Z_e$.

The presence of central charges in the algebra affects the internal symmetry group according to which fields with the indices i,j transform. In our discussion it will be sufficient to consider the case where a single central charge Z is active. Then the superalgebra has an automorphism algebra that is the direct product of the Lorentz algebra and the algebra of the symplectic group SP(N) (for N even). In this case, we can combine the two-component Weyl spinors $Q_{\alpha i}$, $\bar{Q}_{\dot\beta}^{j}$ into SP(N) - Majorana spinors satisfying

$$Q_{\alpha i} = i(\gamma_5 C)_{\alpha\beta}\, \Omega_{ij} \bar{Q}^{\beta j}, \qquad (3.2)$$

where $\Omega_{ij}$ is the SP(N) metric.

In the case of N=2 supersymmetry with one central charge, we may let the central charge be a pseudoscalar and rewrite the anticommutation relations as

$$\{Q_{\alpha i}, \bar{Q}^{\beta j}\} = 2\delta_i^{\ j}\left((\gamma^\mu)_\alpha^{\ \beta} P_\mu - (\gamma_5)_\alpha^{\ \beta} P_5\right) \qquad (3.3)$$

where $P_5 = Z$ is the central charge.

The action of the central charge on fields may be realized in two ways. The first is to perform an abelian phase transformation on complex fields. Since the central charge generator has the dimensions of mass, the eigenvalue of a particular multiplet of complex fields must also have dimensions of mass. In other words, central charges can be realized as phase rotations only if a non-zero mass parameter is present. Moreover, the particular value of the mass parameter is specified by the field equations, so this way of realizing central charge transformations is essentially restricted to "on-shell" supersymmetry.

If we wish to realize the central charge action generally off-shell, and also in the case where no mass parameters are present, the way is suggested by the anticommutation relations (3.3). Since $Z=P_5$ occurs on the same footing as $P_\mu$, we may realize the transformations which Z generates by defining our fields to depend upon a fifth coordinate $x^5$, and then performing shifts in $x^5$. Note that although it is convenient to use the notation $x^5$ for the extra coordinate, we do not wish to imply that five-dimensional Lorentz invariance must always be maintained. Indeed, we shall see that this is a requirement that we definitely do not wish to make.

Fields defined over the superspace $(x^\mu, x^5; \theta^{\alpha i}, \bar{\theta}^{\dot{\alpha}}_{i})$ constitute the representations of supersymmetry with one central charge. However, an essential feature of superspace representations is that unconstrained superfields do not give irreducible representations of the supersymmetry. This is an extension of the situation which prevails in N=1 supersymmetry, where a general scalar superfield is reducible. The origin of this feature is that the coordinates of superspace form a reducible representation under the 4-dimensional Lorentz group. This permits the setting to zero of some fermi or bose derivatives on a superfield without breaking supersymmetry. Thus, we must consider superfields that satisfy differential constraints in order to obtain irreducible representations.

To make this clear, we consider the first few representations of N=2 supersymmetry. The general scalar superfield $\phi(x^\mu, x; \theta^{\alpha i}, \bar{\theta}^{\dot{\alpha}}_{i})$ defines a highly reducible representation. In fact, the smallest representation in terms of fields defined over four-space is the scalar hypermultiplet (Fayet, 1976). This is a superfield with an SU(2)= SP(2) isospinor index $\phi_i(x,\theta)$ satisfying the constraints (Sohnius, 1978):

$$\sum_{(ij)} D_{\alpha i}\phi_j \equiv \frac{1}{2}(D_{\alpha i}\phi_j + D_{\alpha j}\phi_i) = 0. \tag{3.4a}$$

$$\sum_{(ij)} \bar{D}_{\dot{\alpha} i}\phi_j = 0 \tag{3.4b}$$

The covariant derivatives in (3.4) that preserve superfield transformation type are given in two component spinor notation by

$$\begin{aligned} D_{\alpha i} &= \frac{\partial}{\partial\theta^{\alpha i}} + i\,(\not{\partial}\bar{\theta})_{\alpha i} + \varepsilon_{\alpha\beta}\,\varepsilon_{ij}\,\theta^{\beta j}\frac{\partial}{\partial z} \\ \bar{D}_{\dot{\alpha}}{}^{i} &= -\frac{\partial}{\partial\bar{\theta}^{\dot{\alpha}}{}_{i}} - i(\theta\not{\partial})_{\dot{\alpha}}{}^{i} - \varepsilon_{\dot{\alpha}\dot{\beta}}\,\varepsilon^{ij}\,\bar{\theta}^{\dot{\beta}}{}_{j}\frac{\partial}{\partial z} \end{aligned} \tag{3.5}$$

with anticommutation relations

$$\begin{aligned} \{D_{\alpha i}, \bar{D}_{\dot{\beta}}{}^{j}\} &= -2i\delta_i{}^j\sigma^\mu{}_{\alpha\dot{\beta}}\partial_\mu \equiv -2i\,\delta_i{}^j\,\not{\partial}_{\alpha\dot{\beta}} \\ \{D_{\alpha i}, D_{\beta j}\} &= 2\,\varepsilon_{\alpha\beta}\,\varepsilon_{ij}\,\frac{\partial}{\partial z} \\ \{\bar{D}_{\dot{\alpha}}{}^{i}, \bar{D}_{\dot{\beta}}{}^{j}\} &= 2\,\varepsilon_{\dot{\alpha}\dot{\beta}}\,\varepsilon^{ij}\,\frac{\partial}{\partial z}\,. \end{aligned} \tag{3.6}$$

In the usual fashion, we define the members of the supermultiplet of fields over x corresponding to $\phi_i$ by the θ- and z-independent parts of the superfields resulting from various covariant differential operations acting on $\phi_i$. Thus, $(D_\alpha{}^i\phi_i, \bar{D}_{\dot\alpha}{}^i\phi_i)$ constitutes a Dirac spinor which we may rename ψ. By use of the constraints (3.4) together with the algebra (3.6) one can show that

$$\not\partial^{(v)} \psi \equiv (\gamma^\mu\partial_\mu - \gamma_5\partial_5)\,\psi = 0 \tag{3.7a}$$

and

$$\pentagon\, \phi_i \equiv \left( \Box\phi_i - \frac{\partial^2}{\partial z^2}\,\phi_i \right) = 0 \tag{3.7b}$$

The relations (3.7) are nothing other than the free field equations for a field defined over five-dimensional space-time. These relations have the effect of reducing the multiplet to a finite number of fields defined over four-space, essentially just the Cauchy data for the equations (3.7).

The component fields of the scalar hypermultiplet are a doublet of isospinor scalars, an isosinglet Dirac spinor and a further doublet of isospinor scalars, of dimension 2 if the first doublet is of dimension 1.

The next representation of N=2 supersymmetry with which we shall be concerned is the linear multiplet (Breitenlohner and Sohnius, 1980). This is obtained from a superfield with an isovector index, or equivalently a symmetric pair of isospinor indices $K_{ij}$, satisfying the constraints

$$\sum_{(ijk)} D_{\alpha i}\,K_{jk} = 0 \tag{3.8a}$$

$$\sum_{(ijk)} \bar{D}_{\dot\alpha i}\,K_{jk} = 0 \quad , \tag{3.8b}$$

where now the brackets indicate total symmetrization in (ijk).

In the case of the linear multiplet the $x^5$-derivatives are not automatically related to the $x^\mu$-derivatives as in the case of the scalar hypermultiplet. In fact, this means that the linear multiplet may be further reduced by requiring, e.g., that $\partial_5 K_{ij} = 0$. We shall find it useful to consider the general linear multiplet, where this condition of vanishing central charge is not imposed.

The linear multiplet is of great importance because it is the basis of an action formula that involves only two fermi derivatives. The integral

$$\int d^4x(D^iD^j - \bar{D}^i\bar{D}^j)K_{ij} \qquad (3.9)$$

is a supersymmetric and a central charge invariant, as may be checked from (3.6),(3.8). Note that it is not necessary to restrict the $x^5$ dependence of all component fields to get a central charge invariant. The scalar component of the linear multiplet, which is the integrand in (3.9), has the property that its Z-variation is a total $x^\mu$-derivative.

An example of the use of the linear multiplet action formula is the kinetic term and mass term for the scalar hypermultiplet. Due to the hypermultiplet constraints (3.4), the quadratic expressions (Sohnius 1978)

$$K^{kin}_{ij} = i\sum_{(ij)} \phi^*_i \overleftrightarrow{\frac{d}{dx^5}} \phi_j \qquad (3.10)$$

and

$$K^{mass}_{ij} = \sum_{(ij)} \phi^*_i\phi_j \qquad (3.11)$$

are linear multiplets and may be used to form supersymmetric kinetic and mass terms respectively. Note that in order to obtain sensible dynamics in the four-dimensional theory, it has been necessary to include an explicit $x^5$-derivative in (3.10). Five-dimensional Lorentz invariance is not in general required or even desirable for the integrands.

The linear multiplet may be considered to represent a supersymmetric matter theory in its own right (compare also section 10). In this case, we require that there be only a finite number of fields over four dimensions. In order to achieve this, we require that

$$\frac{d}{dx^5} K_{ij} = 0 \qquad (3.12)$$

In this case, we have the following four-dimensional fields: a scalar isotriplet $K_{(ij)}$, a spinor doublet $\varphi_i$, a vector $V_\mu$ subject to the constraint $\partial^\mu V_\mu = 0$ which follows from imposing simultaneously (3.12) and (3.8), plus a scalar and a pseudoscalar, of dimension 2 if the isotriplet is of dimension 1.

In order to write down a supersymmetric action for the linear multiplet, we need an action formula that can accommodate a quadratic product of linear multiplets. Repeating the pattern that took us from the scalar hypermultiplet to the linear multiplet, this brings us to the next multiplet in our discussion. We consider a superfield with four totally symmetric isospin indices $T_{ijk\ell}$ , subject to the constraints

$$\sum_{(rijk\ell)} D_{\alpha(r} T_{ijk\ell)} = 0 \tag{3.13}$$

$$\sum_{(rijk\ell)} \bar{D}_{\dot{\alpha}(r} T_{ijk\ell)} = 0 \tag{3.14}$$

In this case, the brackets signify total symmetrization in the five indices $(rijk\ell)$. As before, the constraints restrict the isospin content of component fields to be no higher than that of the external indices on the whole superfield, which in this case is pure isospin two. The isospin two multiplet also gives rise to an action formula:

$$\int d^4x\, D^i D^j \bar{D}^k \bar{D}^\ell\, T_{ijk\ell} \tag{3.15}$$

In order for (3.15) to be a supersymmetric invariant, it is necessary that $T_{ijk\ell}$ have zero central charge :

$$\frac{d}{dx^5} T_{ijk\ell} = 0. \tag{3.16}$$

The kinetic action for the linear multiplet with zero central charge is then

$$\int d^4x\, D^i D^j \bar{D}^k \bar{D}^\ell\, K_{(ij}K_{k\ell)} \;. \tag{3.17}$$

A classification of representations of extended representations using Casimir operators can be found in Taylor (1980).

## 4. DIMENSIONAL REDUCTION BY LEGENDRE TRANSFORMATION

The action formulas that we have seen above involve integration over four spatial dimensions only. How can we achieve invariance of such actions under shifts in the fifth dimension as well? If there is a field in a multiplet that varies into a total $\partial_\mu$ derivative under central charge transformations, then we may use it. We have seen this in the action formula (3.9). The means by which the required transformation property is achieved in a quadratic or higher nonlinear expression such as (3.10) needs to be clarified. One answer to this problem can be lifted from any mechanics text.

In order to obtain an action that is invariant under shifts in the unintegrated coordinate $x^5$, we must form a first integral of the five-dimensional equations of motion. Treating the fifth dimension for this purpose as a "time" (although its signature must be space-like in order to have the correct sign for physical energies), we can construct the "Hamiltonian" of the five-dimensional theory. Then, subject to the imposition of the five-dimensional field equations as constraints, we have an action for four dimensions that is central charge invariant without the necessity of integrating over the fifth dimension. The crucial point is that the five-dimensional equations of motion do not imply the four-dimensional equations :

$\hat{\Box}\,\phi = 0$

does not imply $\Box\phi = 0$.

The "Hamiltonian" can in principle be constructed simply by finding the "energy" first integral of the five-dimensional field equations. In practice, it is easier to start from the "on-shell Lagrangian" in five dimensions, the functional of physical (not auxiliary) fields whose variation yields the five dimensional field equations. From the on-shell Lagrangian the Hamiltonian may be constructed by a Legendre transformation (Sohnius, Stelle and West, 1980a). In this construction, we have to introduce momenta canonically conjugate to the physical fields. The introduction of these momenta removes all explicit $\frac{d}{dx^5}$ derivatives from the "Hamiltonian". In fact, these new fields are nothing but the auxiliary fields of the theory. Thus from the Lagrangian density $L_{(5)}$, we form the Hamiltonian density

$$\mathcal{H} = \sum_{\phi} \frac{d\phi}{dx^5} \frac{\partial L_{(5)}(\partial_\mu\phi, \partial_5\phi, \phi)}{\partial\left(\frac{d\phi}{dx^5}\right)} - L_{(5)} \tag{4.1}$$

The four-dimensional action is then minus the Hamiltonian

$$I_{(4)} = -H = -\int d^4x\, \mathcal{H}\,. \tag{4.2}$$

The action formula for the scalar hypermultiplet previously given in (3.10) may easily be derived by this technique. Another important example is the N=2 Abelian supersymmetric gauge theory. Working directly in components, we start from the five-dimensional on-shell Lagrangian for a five-vector $A_a$, a Dirac spinor $\lambda$ and a scalar $\phi$:

$$L_{(5)} = -\frac{1}{4} F_{ab}F^{ab} - \frac{i}{4} \bar{\lambda}\gamma^a \overleftrightarrow{\partial}_a \lambda + \frac{1}{2} \partial^a\phi\partial_a\phi \tag{4.3}$$

where $a = 0\ldots3;5$.

The five-dimensional supersymmetry transformations are

$$\delta A_a = \frac{i}{2}(\bar{\zeta}\gamma_a\lambda - \bar{\lambda}\gamma_a\zeta)$$
$$\delta\phi = \frac{i}{2}(\bar{\zeta}\lambda - \bar{\lambda}\zeta) \tag{4.4}$$
$$\delta\lambda = \frac{i}{2}\sigma_{ab}\zeta F^{ab} + \gamma^a\zeta\partial_a\phi$$

Varying the Lagrangian with respect to the $x^5$-derivatives of the physical fields, we have the canonical momenta

$$\pi(A^\mu) = F_{5\mu} \equiv V_\mu \tag{4.5a}$$
$$\pi(\phi) = -\partial_5\phi \equiv H \tag{4.5b}$$
$$\pi(\lambda) = -\frac{i}{4}\bar{\lambda}\gamma_5\ ,\ \pi(\bar{\lambda}) = \frac{i}{4}\gamma_5\lambda \tag{4.5c}$$
$$\pi(A^5) = 0 \tag{4.5d}$$

Constructing the "Hamiltonian density", we get the four-dimensional Lagrangian density :

$$L_{(4)} = -\mathcal{H}_{(4)} = -\frac{1}{4}F_{\mu\nu}F^{\mu\nu} - \frac{i}{4}\bar{\lambda}\gamma^\mu\overleftrightarrow{\partial_\mu}\lambda + \frac{1}{2}\partial_\mu\phi\partial^\mu\phi$$
$$-\frac{1}{2}V_\mu V^\mu + \frac{1}{2}H^2. \tag{4.6}$$

In order for the above expression to give a supersymmetric action, we must impose the five-dimensional field equations as constraints In addition to the defining relations for the momenta, (4.5), we have

$$0 = \partial^\mu V_\mu \tag{4.7a}$$
$$\partial_5 V_\mu = -\partial^\nu F_{\mu\nu} \tag{4.7b}$$
$$\partial_5 H = -\Box\phi \tag{4.7c}$$
$$\partial_5\lambda = \gamma^\mu\gamma_5\partial_\mu\lambda\ . \tag{4.7d}$$

The equations (4.7b-d) together with the relations (4.5) just give the central charge transformations of the various fields except for $A^5$. This reduces the multiplet to a finite number of fields defined over four-space. The central charge variation of $A^5$ is undetermined, for it has no conjugate momentum. In fact, in the original Lagrangian (4.3), $A^5$ has the role of a Lagrange multiplier enforcing the condition (4.7a), which involves only four-space derivatives. This is of course familiar from the canonical analysis of any gauge theory, where there must be a 1↔1 correspondence between constraints and gauge invariances. In

Maxwell's theory, we have the Lagrange multiplier $A^o$ for the constraint $\vec{\nabla}\cdot\vec{E}=0$. As in Maxwell theory, once the constraint is enforced, the variation of $A^5$ is irrelevant since it drops out of the action.

In the Abelian theory, it is a simple matter to solve the constraint by setting

$$V_\mu = \partial^\nu A^*_{\mu\nu} , \tag{4.8}$$

in which case the term $V_\mu V^\mu$ is the action for a spin zero field realized as an antisymmetric tensor, and the formalism becomes one which was first obtained by a consideration of ghost states in higher-derivative extensions of the theory (Sohnius, Stelle, West 1980b). In the non-Abelian case, the constraint becomes

$$0 = \mathcal{D}^\mu V_\mu + \frac{1}{2}\{\bar{\lambda},\gamma_5\lambda\} - i[H,\phi] . \tag{4.9}$$

The technique of dimensional reduction by Legendre transformation is applicable to all extended supersymmetric theories, and by truncation of the extended theories, to the unextended N=1 theories as well. We give another illustration of the method in the case of the supersymmetric theory which seems to have the greatest chance of a genuine physical application, the maximally extended N=8 supergravity theory (Cremmer et al. 1980), but we consider the linearized level only.

The linearized on-shell Lagrangian for N=8 supergravity in five dimensions is

$$\begin{aligned} L_{(5)} = {} & \frac{1}{4}(2\partial^a h_{ab}\partial^b h_c{}^c + \partial_c h_{ab}\partial^c h^{ab} \\ & -2\partial^a h_{ab}\partial^c h_c{}^b - \partial^a h_b{}^b\partial_a h_c{}^c) \\ & - \frac{i}{2}\bar{\psi}^i_a \gamma^{abc}\partial_b \psi_{ci} - \frac{1}{8} F_{ab}{}^{ij}F_{ij}{}^{ab} \\ & + \frac{i}{12}\bar{\lambda}^{ijk}\gamma^a\partial_a \chi_{ijk} + \frac{1}{48}\partial^a\phi^{ijk\ell}\partial_a\phi_{ijk\ell} \end{aligned} \tag{4.10}$$

where i,j= 1...8; a,b,c = 0....3;5. $\gamma^{abc}$ is the totally antisymmetrized product of three $\gamma$'s and the fields carry SP8 indices. The graviton field $h_{ab}$ is an SP8 singlet, the gravitino field $\psi_a{}^i$ is in the SP8-Majorana representation with a multiplicity equal to 8 ordinary Majorana spinors, the vectors $A_a^{ij}$ are in the $\underline{27}$, the spinors $\chi_{ijk}$ in the $\underline{48}$ and the scalars $\phi_{ijk\ell}$ in the $\underline{42}$ dimensional representations of SP(8).

Varying the Lagrangian (4.10) to obtain the canonical momenta, we find

$$\pi(h_{\mu\nu}) = -\frac{1}{2}\left(2\eta^{\mu\nu}\partial^{\lambda} h_{\lambda 5} - \partial^{\mu}h^{\nu}_{5} - \partial^{\nu}h^{\mu}_{5} + \partial_5(h^{\mu\nu} - \eta^{\mu\nu}h_{\lambda}{}^{\lambda})\right) \equiv \pi^{\mu\nu} \tag{4.11a}$$

$$\pi(\psi_{\mu i}) = -\frac{1}{2}\bar{\psi}_{\nu}{}^{i}\gamma^{\nu\mu}\gamma_5 \tag{4.11b}$$

$$\pi(A^{\mu}_{ij}) = \frac{1}{2}F_{5\mu}{}^{ij} \equiv \pi_{\mu}{}^{ij} \tag{4.11c}$$

$$\pi(\chi_{ijk}) = -\frac{1}{12}\bar{\chi}^{ijk}\gamma_5 \tag{4.11d}$$

$$\pi(\phi_{ijk\ell}) = -\frac{1}{24}\partial_5\phi^{ijk\ell} \equiv H^{ijk\ell} \tag{4.11e}$$

$$\pi(h_{5\mu}) = \pi(h_{55}) = \pi(\psi_5{}^{i}) = \pi(A_5{}^{ij}) = 0. \tag{4.11f}$$

After performing the Legendre transformation, the four-dimensional Lagrangian is :

$$\begin{aligned} L_{(4)} = & \frac{1}{4}(2\partial^{\mu}h_{\mu\nu}\partial^{\nu}h_{\lambda}{}^{\lambda} + \partial_{\mu}h_{\nu\rho}\partial^{\mu}h^{\nu\rho} \\ & - 2\partial^{\mu}h_{\mu\nu}\partial^{\rho}h_{\rho}{}^{\nu} - \partial^{\mu}h_{\rho}{}^{\rho}\partial_{\mu}h_{\nu}{}^{\nu}) \\ & + \pi_{\mu\nu}\pi^{\mu\nu} - \frac{1}{3}(\pi_{\nu}{}^{\nu})^2 - \frac{i}{2}\bar{\psi}_{\mu}{}^{i}\gamma^{\mu\nu\rho}\partial_{\nu}\psi_{\rho i} - \frac{1}{8}F^{ij}_{\mu\nu}F^{\mu\nu}_{ij} \\ & + \frac{i}{12}\bar{\chi}^{ijk}\gamma^{\mu}\partial_{\mu}\chi_{ijk} + \frac{1}{48}\partial_{\mu}\phi^{ijk\ell}\partial^{\mu}\phi_{ijk\ell} \\ & + \frac{1}{48}H^{ijk\ell}H_{ijk\ell} - \pi^{\mu ij}\pi_{\mu ij} \; . \end{aligned} \tag{4.12}$$

As always, the five-dimensional field equations must be imposed to get invariance of the action. Due to the gauge invariances on the fields,

$$\delta h_{ab} = \partial_a\eta_b + \partial_b\eta_a \tag{4.13a}$$

$$\delta\psi_a{}^{i} = \partial_a\zeta^{i} \tag{4.13b}$$

$$\delta A_a{}^{ij} = \partial_a\Lambda^{ij} \tag{4.13c}$$

the five-dimensional field equations contain the constraints:

$$\Box h_\mu{}^\mu - \partial^\mu\partial^\nu h_{\mu\nu} = 0 \tag{4.14a}$$

$$\partial^\mu \pi_{\mu\nu} = 0 \tag{4.14b}$$

$$\gamma^{\mu\nu}\partial_\mu \psi_\nu{}^i = 0 \tag{4.14c}$$

$$\partial^\mu \pi_\mu{}^{ij} = 0 \tag{4.14d}$$

The last of these constraints can easily be solved as before

$$\pi_\mu{}^{ij} = \frac{1}{4}\partial^\nu \varepsilon_{\mu\nu\rho\lambda} A^{\rho\lambda ij} \quad , \tag{4.15}$$

so 27 spin zero degrees of freedom are represented by anti-symmetric tensors. Unfortunately, the other constraints cannot be solved by a similar single derivative construction, even in the linearized theory. Consequently, the only way we can extract the physical consequences of the constraints at this stage is to give up manifest supersymmetry and put the constraints back into the Lagrangian with Lagrange multipliers. Treating the 27 spin zero degrees of freedom mentioned above in the same fashion for completeness, we have the addition

$$L'_{(4)} = A^\mu \partial^\nu \pi_{\mu\nu} + \frac{\phi}{\sqrt{3}} \partial^\mu (\partial_\mu h_\rho{}^\rho - \partial^\nu h_{\mu\nu}) + \frac{1}{2}\phi^{ij}\partial^\mu \pi_{\mu ij} + \sqrt{\frac{2}{3}}\, i\, \bar{\chi}^i \gamma^{\mu\nu}\partial_\mu \psi_{\nu i} \quad . \tag{4.16}$$

The field $\pi_{\mu\nu}$ can be eliminated using its algebraic field equation to give a Maxwell Lagrangian in terms of $A_\mu$: $-\frac{1}{4} F_{\mu\nu} F^{\mu\nu}$. Similarly, elimination of $\pi_{\mu ij}$ gives scalar kinetic terms for the 27 fields $\phi^{ij}$: $\frac{1}{4}\partial_\mu \phi^{ij}\partial^\mu \phi_{ij}$. For the other constraints, we make the field redefinitions $\tilde{h}_{\mu\nu} = h_{\mu\nu} + 1/\sqrt{3}\,\eta_{\mu\nu}$ ; $\tilde{\psi}_\mu^i = \psi_\mu^i - 1/\sqrt{6}\,\gamma_\mu \chi^i$. This kills the off-diagonal terms in the Lagrangian and gives the new terms

$$\frac{1}{2}\partial^\mu \phi\, \partial_\mu \phi + \frac{i}{2}\bar{\chi}^i \gamma^\mu \partial_\mu \chi_i \quad .$$

To summarize, we have the following degrees of freedom from the original fields and from the constraints:

| Field: | Original degrees of freedom | Additional degrees of freedom from constraints |
|---|---|---|
| $h_{\mu\nu}$ | 1 spin-2 | 1 spin-0 |
| $\pi_{\mu\nu}$ | | 1 spin-1 |
| $\psi_\mu{}^i$ | 8 spin-$\frac{3}{2}$ | 8 spin-$\frac{1}{2}$ |
| $A_\mu{}^{ij}$ | 27 spin-1 | |
| $\pi_\mu{}^{ij}$ | | 27 spin-0 |
| $\chi_{ijk}$ | 48 spin-$\frac{1}{2}$ | |
| $\phi_{ijk\ell}$ | 42 spin-0 | |

This gives a total of one spin-2, eight spin $\frac{3}{2}$, 28 spin-1, 56 spin-$\frac{1}{2}$ and 70 spin-0 degrees of freedom.

## 5. PRESERVATION OF REPRESENTATIONS AND THE STRUCTURE OF N=2 GAUGE FIELD MULTIPLETS

The incorporation of gauge invariances into supersymmetric theories poses some difficulties that are not encountered in ordinary gauge theories. At first sight the way to approach their formulation would seem to be a straightforward extension of the differential geometric point of view for ordinary space-time gauge theories into superspace. However, this leads immediately to problems which are once again related to the reducibility of the superspace coordinates' Lorentz representation. And, once again, we must cure the problems by imposing constraints upon the gauge multiplet. These constraints have a dual role: they eliminate unwanted reducible parts of the gauge multiplet and serve as integrability conditions for the differential constraints on gauge-covariant irreducible matter multiplets, thus preserving these irreducible representations in the face of gauge coupling.

We introduce a set of Lie algebra valued connections $\mathcal{A}_A$ defined on the N=2 superspace, where A runs over all the superspace indices including the central charge index 5. Then the gauge-and super-covariantized derivatives

$$\mathcal{D}_A = D_A + i\mathcal{A}_A \qquad (5.1)$$

satisfy the Ricci identity

$$[\mathcal{D}_A, \mathcal{D}_B\} = -2i\sigma_{AB}{}^C{}_C + iF_{AB} \qquad (5.2)$$

where, for N=2 supersymmetry,

$$\sigma^5_{\alpha i\beta j} = i\varepsilon_{\alpha\beta}\varepsilon_{ij} \; ; \; \sigma^{5i}_{\dot\alpha}{}^j_{\dot\beta} = i\varepsilon_{\dot\alpha\dot\beta}\varepsilon^{ij} \; ; \; \sigma^\mu_{\alpha i\dot\beta}{}^j = \delta_i{}^j \sigma^\mu_{\alpha\dot\beta}$$

The field strengths $F_{AB}$ thus have the form

$$F_{AB} = D_A \mathcal{A}_B \pm D_B \mathcal{A}_A + i[\mathcal{A}_A, \mathcal{A}_B\} + 2i\sigma_{AB}{}^C \mathcal{A}_C \qquad (5.3)$$

(plus sign for fermi-fermi components).

The presence of the last term in (5.3) is due to the torsion in the flat superspace algebra. It has the consequence that expressions exist which are constructed from the fermionic connections $\mathcal{A}_{\alpha i}$ and $\bar{\mathcal{A}}_{\dot\alpha}{}^i$ only, but transform like connections with bosonic indices μ and 5. This circumstance is analogous to that in general relativity where the Christoffel symbol can be constructed from derivatives of the metric, making it unnecessary to introduce an independent connection. Thus we can adopt the conventional constraints

$$F_{\alpha i\dot\beta}{}^i = 0 \qquad (5.4a)$$

and

$$F_{\alpha i}{}^{\alpha i} - F^{\dot\alpha}{}_{i\dot\alpha}{}^i = 0 \qquad (5.4b)$$

which can be solved to find, respectively

$$\mathcal{A}_\mu = \frac{i}{8} (\bar\sigma_\mu)^{\dot\beta\alpha} \left( D_{\alpha i} \bar{\mathcal{A}}^i_{\dot\beta} + \bar D^i_{\dot\beta} \mathcal{A}_{\alpha i} + i\{\mathcal{A}_{\alpha i}, \bar{\mathcal{A}}^i_{\dot\beta}\} \right)$$

(5.5a,b)

$$\mathcal{A}_5 = \frac{1}{16} \left( 2 D_{\alpha i} \mathcal{A}^{\alpha i} + i\{\mathcal{A}_{\alpha i}, \mathcal{A}^{\alpha i}\} - 2 \bar D^{\dot\alpha}_i \bar{\mathcal{A}}^i_{\dot\alpha} - i\{\bar{\mathcal{A}}^{\dot\alpha}_i, \bar{\mathcal{A}}^i_{\dot\alpha}\} \right)$$

The most important constraints that must be imposed upon the gauge connections serve as integrability conditions for covariant defining conditions that generalize the defining conditions given in section 3 for irreducible supersymmetric matter representations. The simplest example of this kind of integrability condition is that which permits the definition of a Yang-Mills covariant chiral superfield. If we wish to solve the differential equation

$$\mathcal{D}_{\alpha i}\phi = 0 \qquad (5.6)$$

for some complex scalar superfield $\phi$ that carries a non-trivial Yang-Mills representation, we immediately require

$$\{\mathcal{D}_{\alpha i}, \mathcal{D}_{\beta j}\} \phi = 0 \qquad (5.7a)$$

ie.

$$2\varepsilon_{\alpha\beta}\, \varepsilon_{ij}\, \mathcal{D}_5 \phi + i\, F_{\alpha i \beta j}\, \phi = 0 \qquad (5.7b)$$

Equation (5.7b) implies two conditions. The first is a differential equation on $\phi$ governing its central charge dependence

$$\mathcal{D}_5 \phi + 8i\, F^{\alpha i}{}_{\alpha i}\, \phi = 0. \qquad (5.8)$$

The analogous condition in the absence of gauge coupling required that the chiral field have no central charge dependence. In this case it requires that the central charge transformation of $\phi$ be given by a field-dependent gauge transformation (with dependence on $A_5$ and $F^{\alpha i}{}_{\alpha i}$).

Substituting (5.8) back into (5.7b) yields an equation in which $\phi$ appears purely algebraically, without any derivatives. This cannot be solved non-trivially by imposing any further condition on $\phi$ but rather requires an integrability condition that constrains the gauge connections themselves :

$$F_{\alpha i \beta j} + F_{\alpha j \beta i} = 0 \ . \qquad (5.9a)$$

The integrability condition for chiral fields of the opposite chirality is the complex conjugate of (5.9),

$$F_{\dot\alpha}{}^{i}{}_{\dot\beta}{}^{j} + F_{\dot\alpha}{}^{j}{}_{\dot\beta}{}^{i} = 0 \ . \qquad (5.9b)$$

The chiral multiplet determined by (5.6) is not yet irreducible. One way to define an irreducible multiplet is to impose a further condition of second order in fermionic derivatives (Wess, 1975) that requires the scalar isovector component field to be real, i.e. $(\bar{\mathcal{D}} \vec{\tau} \bar{\mathcal{D}} + \mathcal{D} \vec{\tau} \mathcal{D})\, \phi = 0$ . Such second order defining conditions do not require any further integrability conditions on the Yang-Mills supermultiplet.

A further integrability condition is needed, however, in order to permit the Yang-Mills covariant generalization of the scalar hypermultiplet discussed in section 3. The covariantized defining conditions rule out covariantly transforming isospin-1 fields in the multiplet:

$$\mathcal{D}_{\alpha i}\, \phi_j + \mathcal{D}_{\alpha j}\, \phi_i = 0 \ ; \ \bar{\mathcal{D}}_{\dot\alpha i}\, \phi_j + \bar{\mathcal{D}}_{\dot\alpha j}\, \phi_i = 0 \ . \qquad (5.10)$$

Equations (5.10) require

$$\sum_{(ijk)} \{\mathcal{D}_{\alpha i}, \bar{\mathcal{D}}_{\dot\beta j}\}\, \phi_k = 0, \text{ i.e. } \sum_{(ijk)} F_{\alpha i \dot\beta j}\, \phi_k = 0 \ , \qquad (5.11)$$

which requires

$$F_{\alpha i\dot{\beta}j} + F_{\alpha j\dot{\beta}i} = 0 \ . \tag{5.12}$$

Taking together the constraints (5.4), (5.5), (5.9) and (5.12), the only non-zero spinor-spinor field strengths are

$$F_{\alpha i\beta j} = 2i\ \varepsilon_{\alpha\beta}\varepsilon_{ij}\ W \ ; \quad F_{\dot{\alpha}\ \dot{\beta}}^{\ i\ j} = -2i\ \varepsilon_{\dot{\alpha}\dot{\beta}}\varepsilon^{ij}\ W \ , \tag{5.13}$$

where the superfield W is real. Equations (5.13) are only the lowest dimensional consequence of the constraints, however. The field strengths $F_{AB}$ are constructed from the potentials $\mathcal{A}_A$, and consequently must satisfy the Bianchi identities for N=2 supersymmetric gauge theories. The technique of analysing the higher dimensional consequences of the constraints through the Bianchi identities is well known. Here we are content to list the results, with the relevant identity specified by $I_{ABC}$, corresponding to the Bianchi identity derived from the Jacobi identity on the appropriately symmetrized sum of triple derivatives $\mathcal{D}_A \mathcal{D}_B \mathcal{D}_C$. For compactness, we go over to the notation with SU(2)-Majorana spinors and get:

| Identity: | Results: | |
|---|---|---|
| $I_{\alpha i\beta j\gamma k}$ | $F_{5\alpha i} = -(\gamma_5 \mathcal{D})_{\alpha i}\ W$ and $F_{\mu\alpha i} = -(\gamma_\mu \mathcal{D})_{\alpha i}\ W$ | (5.14) |
| $I_{5\alpha i\beta j}$ | $\bar{\mathcal{D}}\gamma_\mu\gamma_5\mathcal{D}\,W = 8i\ F_{5\mu}$ and $\bar{\mathcal{D}}\gamma_\mu\vec{\tau}\mathcal{D}\,W = 0$ | (5.15) |
| $I_{\mu\alpha i\beta j}$ | $\bar{\mathcal{D}}\sigma_{\mu\nu}\mathcal{D}\ W = 8\ F_{\mu\nu}$ and $\bar{\mathcal{D}}\gamma_5\vec{\tau}\mathcal{D}\,W = 0$ | (5.16) |

The other independent superfield of the same dimension as $F_{5\mu}$ is the real superfield $\bar{\mathcal{D}}\vec{\tau}\mathcal{D}\,W$ , which we shall henceforth designate by $\vec{X}$. It is convenient to give $F_{5\alpha}$ and $F_{5\mu}$ new names as well, and we get the following multiplet of gauge-covariant superfields:

$$W$$
$$\lambda_{\alpha i} \equiv (\gamma_5)_\alpha^{\ \beta}\ F_{5\beta i}$$
$$V_\mu \equiv F_{5\mu}$$
$$\vec{X} \equiv \frac{1}{8}\bar{\mathcal{D}}\ \vec{\tau}\mathcal{D}\ W$$
$$F_{\mu\nu} = \partial_\mu \mathcal{A}_\nu - \partial_\nu \mathcal{A}_\mu + i\,[\mathcal{A}_\mu , \mathcal{A}_\nu] \qquad ,$$

and central charge variations $W^{(1)} \equiv \mathcal{D}_5\ W$ , $\lambda^{(1)} \equiv \mathcal{D}_5\ \lambda$ , etc. Due to one of the Bianchi-identities, we have in particular

$$F^{(1)}_{\mu\nu} = \mathcal{D}_\mu\ V_\nu - \mathcal{D}_\nu\ V_\mu$$

## 6. THE FIELD CONTENT OF THE GAUGE MULTIPLET

We have seen in the previous section how the application of the general program of finding superspace constraints (Gates, Stelle and West, 1980) has led, in the case of the N=2 Yang-Mills theory, to the demand that all the dimension-1 field strengths ($F_{\alpha i\beta j}$) be zero except for one real scalar component:

$$F_{\alpha i\beta j} = 2i\, \mathbb{1}_{\alpha i\beta j} W \quad ; \quad W = W^{+} \tag{6.1}$$

The symbol $\mathbb{1}$ stands for the SU(2)-covariant "spinor metric"

$$\mathbb{1}_{\alpha i\beta j} = \varepsilon_{ij}(i\gamma_5 C)_{\alpha\beta} \, .$$

Since it is the gauge potentials $\mathcal{A}_{\alpha i}$, and not the field strengths which are the fundamental objects, we must use the Bianchi-identities to find out which additional constraints are implied by (6.1). We have found the only additional constraints to be

$$0 = \bar{\mathcal{D}}^{i}\gamma_{\mu}\, \vec{\tau}_{i}{}^{j}\, \mathcal{D}_{j} W = \bar{\mathcal{D}}^{i}\gamma_{5}\, \vec{\tau}_{i}{}^{j}\mathcal{D}_{j} W. \tag{6.2}$$

The component field content of a real superfield subject to equations (6.2) (called a gauge superfield, or equivalently a gauge multiplet) can in principle be found by expanding $\mathcal{A}_{\alpha i}(x,z,\theta_{\alpha i})$ in a Taylor series in $\theta$ and $z$, and then applying the constraint (6.2) on the field strength to restrict the coefficients in this expansion. The supersymmetry transformations of the component fields are induced by the supersymmetry transformations on superspace. However, this approach is loaded with the necessity to calculate many non-gauge-covariant terms, instead of only the covariant components of W.

A better method which extracts only the covariant components of W is to first consider all superfields formed by the action of the covariant derivatives on W, namely the superfields

$$W,\ \mathcal{D}_{\alpha i}W,\ \mathcal{D}_{\mu}W,\ \mathcal{D}_{z}W,\ \mathcal{D}_{\alpha i}\mathcal{D}_{\beta j}W, \text{ etc.} \tag{6.3}$$

Application of the rules for covariant derivatives in conjunction with (6.2) leads to the discovery that some of these expressions are zero, and others may be expressed in terms of spatial derivatives and products of some of the lower dimensional superfields. The field components of W may then be obtained by taking the $\theta$=z=0 part of those of the above superfields which are truly independent. The supersymmetry transformations with parameter $\varepsilon_{\alpha i}$ of these field components are given by operating with $i\epsilon^{-\alpha i}\mathcal{D}_{\alpha i}$ on the corresponding superfields and then setting $\theta=z=0$. This is the correct procedure since at $\theta=z=0$ the covariant derivative and the supersymmetry generator differ at most by a field dependent gauge transformation (and not at all in the Wess-Zumino gauge $\mathcal{A}_{\alpha i}|_{\theta=z=0} = 0$). This is also true for $\mathcal{D}_z$ and the central-charge generator, and we may derive the action of a central charge transformation with parameter $\omega$ on the fields by operating with $\omega\mathcal{D}_z$ on the associated superfield and setting $\theta=z=0$. It should be noted that the field components obtained by this method may differ from those of a Taylor expansion, but are related to them by field redefinitions.

Using the described method we find the following infinite set of field components for W:

$$\begin{array}{llll} W, & W^{(1)}, & W^{(2)}, \ldots & \\ \lambda_{\alpha i}, & \lambda^{(1)}_{\alpha i}, & \ldots & \\ V_\mu, & V^{(1)}_\mu, & \ldots & \\ f_{\mu\nu}, & f^{(1)}_{\mu\nu}, & \ldots & \\ \vec{X}, & \vec{X}^{(1)}, & \ldots & \text{etc.} \end{array} \tag{6.4}$$

subject to

$$\mathcal{D}^{\nu}\,{}^{*}f_{\mu\nu} = 0 \quad \text{i.e.} \quad f_{\mu\nu} = \partial_\mu \mathcal{A}_\nu - \partial_\nu \mathcal{A}_\mu + i[\mathcal{A}_\mu, \mathcal{A}_\nu]$$

$$f^{(1)}_{\mu\nu} = \mathcal{D}_\mu V_\nu - \mathcal{D}_\nu V_\mu$$

$$f^{(2)}_{\mu\nu} = \mathcal{D}_\mu V^{(1)}_\nu - \mathcal{D}_\nu V^{(1)}_\mu$$

etc.

In this context, we have used W to denote $W(x,\theta,z)|_{\theta=z=0}$. The infinite set of component fields arise because of the presence of the central charge and its associated bosonic coordinate z. The action of central charge transformations on the fields is given by

$$\delta_\omega W = \omega\mathcal{D}_z W = \omega W^{(1)}; \; \delta_\omega \lambda_{\alpha i} = \omega\mathcal{D}_z \lambda_{\alpha i} = \omega\lambda^{(1)}_{\alpha i},$$
$$\delta_\omega V_\mu = \omega\mathcal{D}_z V_\mu = \omega V^{(1)}_\mu; \ldots; \; \delta_\omega W^{(1)} = \omega\mathcal{D}_z W^{(2)} = \omega W^{(2)}, \text{etc.} \tag{6.5}$$

In the abelian limit the fields in each column of equation (6.4) correspond to the coefficient of a given $z^n/n!$ in the Taylor expansion of W around z=0, viz.

$$W(x,z,\theta) = W(x,\theta) + z\, W^{(1)}(x,\theta) + \frac{z^2}{2!} W^{(2)}(x,\theta) + \ldots \qquad (6.6)$$

One may wonder if W(z) is analytic at z=0 and thus can be expanded into the above series. We will return to this point later.

The supersymmetry and central charge transformations of the component fields in $W(x,\theta)$ are given by

$$\delta W = i\bar{\varepsilon}\lambda + \omega W^{(1)}$$

$$\delta\lambda = \frac{i}{2}\sigma^{\mu\nu}\varepsilon f_{\mu\nu} + \gamma^\mu \varepsilon \mathcal{D}_\mu W - \gamma_5 \varepsilon W^{(1)}$$

$$- \gamma^\mu \gamma_5 \varepsilon V_\mu + i\vec{\tau}\varepsilon \vec{X} + \omega \lambda^{(1)} \qquad (6.7)$$

$$\delta f_{\mu\nu} = -i\bar{\varepsilon}(\gamma_\mu \mathcal{D}_\nu - \gamma_\nu \mathcal{D}_\mu)\lambda + \omega(\mathcal{D}_\mu V_\nu - \mathcal{D}_\nu V_\mu)$$

$$\delta V_\mu = -i\bar{\varepsilon}\gamma_5 \mathcal{D}_\mu \lambda + i\bar{\varepsilon}\gamma_\mu \lambda^{(1)} + \omega V_\mu^{(1)}$$

$$\delta\vec{X} = \bar{\varepsilon}\vec{\tau}\gamma^\mu \mathcal{D}_\mu \lambda - \bar{\varepsilon}\vec{\tau}\gamma_5 \lambda^{(1)} + i[\bar{\varepsilon}\vec{\tau}\lambda, W] + \omega \vec{X}^{(1)}$$

The transformations for the remaining fields are obtained by taking the central charge variations of (6.7) and using the "Ricci" identities

$$[\mathcal{D}_z, \mathcal{D}_{\alpha i}]\, W = -i\,[(\gamma_5 \lambda_{\alpha i}), W]$$

$$[\mathcal{D}_z, \mathcal{D}_\mu]\, W = i\,[V_\mu, W]$$

which gives, e.g. for the components of $W^{(1)}(x,\theta)$:

$$\delta W^{(1)} = i\bar{\varepsilon}\lambda^{(1)} - [\bar{\varepsilon}\gamma_5 \lambda, W] + \omega W^{(2)}$$

$$\delta\lambda^{(1)} = i\,\sigma^{\mu\nu}\varepsilon \mathcal{D}_\mu V_\nu + \gamma^\mu \varepsilon \mathcal{D}_\mu W^{(1)} + i\gamma^\mu \varepsilon\,[V_\mu, W]$$

$$- \gamma_5 \varepsilon W^{(2)} - \gamma^\mu \gamma_5 \varepsilon V_\mu^{(1)} + i\vec{\tau}\varepsilon \vec{X}^{(1)} - [\bar{\varepsilon}\gamma_5 \lambda, \lambda] + \omega\lambda^{(2)}$$

$$\delta V_\mu^{(1)} = -i\bar{\varepsilon}\gamma_5 \mathcal{D}_\mu \lambda^{(1)} + i\bar{\varepsilon}\,\gamma_\mu \lambda^{(2)} - 2\,[\,\bar{\varepsilon}\gamma_5\lambda, V_\mu] + \omega V_\mu^{(2)}$$

$$\delta\vec{X}^{(1)} = \bar{\varepsilon}\,\vec{\tau}\gamma^\mu \mathcal{D}_\mu \lambda^{(1)} - \bar{\varepsilon}\vec{\tau}\gamma_5\,\lambda^{(2)} - [\bar{\varepsilon}\gamma_5\,\lambda, \vec{X}] + i\,[\,\bar{\varepsilon}\,\vec{\tau}\,\lambda^{(1)}, W]$$

$$+\; i\,[\,\bar{\varepsilon}\vec{\tau}\lambda, W^{(1)}] - i\,[\,\bar{\varepsilon}\vec{\tau}\gamma^\mu\lambda, V_\mu]$$

$$+\;\omega\vec{X}^{(2)} \qquad (6.8)$$

We may summarize the field content of the gauge multiplet (superfield) as one scalar W, a Majorana spinor doublet $\lambda_{\alpha i}$, one axial vector $V_\mu$, one vector $A_\mu$ contained in $f_{\mu\nu} = \partial_\mu A_\nu - \partial_\nu A_\mu + i[A_\mu, A_\nu]$, one triplet of scalars $\vec{X}$, and the infinite tower of fields resulting from central charge transformations on these fields.

The reader may have noticed that this is not the field content of either of the known N=2 Yang-Mills theories. This apparent paradox is related to the fact that the gauge multiplet is not an irreducible representation of supersymmetry. Let us turn to this point now.

## 7. CHOICE CONSTRAINTS FOR THE GAUGE MULTIPLET

In this section we will show how the two known formulations (Grimm, Sohnius and Wess, 1978); Sohnius, Stelle and West, 1980 a,b) of the N=2 Yang-Mills theory are contained in the gauge multiplet W and may be extracted by placing two different supersymmetric constraints on W.

Choice I. We impose the constraint

$$\frac{\partial}{\partial z} W(x,z,\theta) = 0 \quad , \qquad (7.1)$$

which implies that the superfield W does not depend on the coordinate z. In terms of the component fields of W it implies the infinite set of constraints

$$\begin{aligned} W^{(1)} &= i\,[A_z, W] \\ \lambda^{(1)} &= i\,[A_z, \lambda] \quad \text{etc.} \end{aligned} \qquad (7.2)$$

The remaining component fields are W, $\lambda$, $A_\mu$, $\vec{X}$ and $V_\mu$ with the caution that now

$$f^{(1)}_{\mu\nu} = \mathcal{D}_\mu V_\nu - \mathcal{D}_\nu V_\mu = i\,[A_z, f_{\mu\nu}] \tag{7.3}$$

and consequently that $V_\mu = -\mathcal{D}_\mu A_z$. The constraint (7.1) is gauge-covariant only if we restrict the gauge parameter to not depend on z either. Then $A_z$ has homogeneous gauge transformation laws.

The component fields of the gauge-multiplet subject to equation (7.1) are a scalar W and a pseudoscalar $A_z$, a Majorana spinor doublet $\lambda_{\alpha i}$, a gauge vector $A_\mu$ and a scalar triplet $\vec{X}$. A supersymmetric action constructed from these fields is

$$I = \int d^4x \; \mathrm{Tr}\,\Big(-\tfrac{1}{4} f_{\mu\nu} f^{\mu\nu} + \tfrac{1}{2}\mathcal{D}_\mu W \mathcal{D}^\mu W + \tfrac{1}{2}\mathcal{D}_\mu A_z \mathcal{D}^\mu A_z - \tfrac{i}{2}\bar{\lambda}^i \gamma^\mu \mathcal{D}_\mu \lambda_i \tag{7.4}$$

$$+ \tfrac{1}{2}\vec{X}^2 - \tfrac{1}{2}\bar{\lambda}^i [(W + \gamma_5 A_z), \lambda_i] + \tfrac{1}{2}[W, A_z]^2 \Big)$$

This is the N-2 Yang-Mills theory found by Grimm, Sohnius and Wess (1978).

In superspace, the action (7.4) is given by

$$I \sim \int d^4x \; \bar{D}^i D^j \bar{D}_i D_j \; \mathrm{Tr}\,(Y^2 + \mathrm{h.c.}) \tag{7.5}$$

where Y is a chiral superfield whose $\theta=0$ component is $W + iA_z$.

<u>Choice II</u>. We impose the constraint

$$\bar{\mathcal{D}}^i \vec{\tau}_i{}^j \mathcal{D}_j W = 0. \tag{7.6}$$

In terms of component fields it implies the following infinite set of constraints

$$0 = \vec{X} \tag{7.7a}$$

$$0 = \lambda^{(1)} - \gamma^\mu \gamma_5 \mathcal{D}_\mu \lambda + i[\gamma_5 \lambda, W] \tag{7.7b}$$

$$0 = W^{(2)} - \mathcal{D}_\mu \mathcal{D}^\mu W + \tfrac{1}{2}\{\bar{\lambda}^i, \lambda_i\} \tag{7.7c}$$

$$0 = V^{(1)}_\mu + \mathcal{D}^\nu f_{\mu\nu} + \tfrac{1}{2}\{\bar{\lambda}^i, \gamma_\mu \lambda_i\} - i[W, \mathcal{D}_\mu W] \tag{7.7d}$$

$$0 = \mathcal{D}^\mu V_\mu + \tfrac{1}{2}\{\bar{\lambda}^i, \gamma_5 \lambda_i\} + i\,[W^{(1)}, W] \tag{7.7e}$$

and all their central charge variations.

Unlike the Choice I constraint, the constraint of equation (7.6) does not imply that W is independent of z. In fact, the Taylor expansion of W in z still has an infinite number of terms, however now the higher coefficients are functions of the lower ones and their space-time derivatives. The structure of the gauge multiplet subject to the Choice II constraint is, for an abelian gauge group,

$$\begin{array}{llll} W & ; W^{(1)} & ; \Box W & ; \Box W^{(1)} \ldots\ldots \\ \lambda_{\alpha i} & ; (\not{\partial}\gamma_5\lambda)_{\alpha i} ; & \Box \lambda_{\alpha i} & \ldots \\ V_\mu & ; -\partial^\nu f_{\mu\nu} & \ldots & \\ f_{\mu\nu} ; & \partial_\mu V_\nu - \partial_\nu V_\mu & \ldots & \end{array} \tag{7.8}$$

subject to

$$\partial^\nu {}^*f_{\mu\nu} = \partial^\mu V_\mu = 0,$$

where each column corresponds to a given power of z in the Taylor expansion of $W(x_\mu, z, \theta_{\alpha i})$.

The field content is then one scalar W, one conserved axial vector $V_\mu$, one Majorana spinor doublet $\lambda_{\alpha i}$, one gauge vector $A_\mu$, and one pseudo-scalar of dimension two, $W^{(1)}$.

A supersymmetric action constructed from these fields, is

$$I = \int d^4x \; \mathrm{Tr} \Big( -\frac{1}{4} f_{\mu\nu} f^{\mu\nu} + \frac{1}{2} \mathcal{D}_\mu W \mathcal{D}^\mu W - \frac{1}{2} V_\mu V^\mu - \frac{1}{2} \bar{\lambda}^i \gamma^\mu \mathcal{D}_\mu \lambda_i$$

$$+ \frac{1}{2} W^{(1)} W^{(1)} + \frac{1}{2} \bar{\lambda}^i [\lambda_i, W] \Big). \tag{7.9}$$

This is the N=2 Yang-Mills theory found by Sohnius, Stelle and West (1980 a,b).

This action can be expressed in superspace as

$$I \sim \int d^4x \; \bar{D}^i \gamma_5 \vec{\tau}_i{}^j D_j \; \mathrm{Tr} \; (\bar{\mathcal{D}}^i W \; \vec{\tau}_i{}^j \gamma_5 \mathcal{D}_j W), \tag{7.10}$$

and can easily be seen to be invariant, if one realizes that $\bar{\mathcal{D}}^i W \vec{\tau}_i{}^j \gamma_5 \mathcal{D}_j W$ is a linear superfield.

## 8. CONSTRAINTS AND SUPERSYMMETRIC LAGRANGE MULTIPLIERS

As has been explained in section four, the method of dimensional reduction by Legendre transformation enables the systematic discovery of off-shell formulations of all supersymmetric theories in four dimensions. There is, however, a serious drawback to this procedure, namely that in theories with a gauge invariance it leads to equations which restrict the space-time dependence of some of the fields in four dimensions. This arises in exactly the same way as it does in the Hamiltonian formulation of electromagnetism where the constraint is

$$\vec{\nabla}\cdot\vec{E} = 0, \tag{8.1}$$

and $\vec{E}$, the electric field, is the canonically conjugate momentum to the vector potential $\vec{A}$. If we construct the N=2 (or 4) supersymmetric abelian gauge theory by dimensional reduction by Legendre transformation from the five-dimensional theory, we find the corresponding constraint

$$\partial^{\mu}V_{\mu} = 0 \tag{8.2}$$

for the field $V_{\mu}$, the canonically conjugate momentum to $A_{\mu}$.

In the abelian cases such constraints do not present a serious problem since they can be solved explicitly, if one is prepared to neglect topological effects. The solution to eq. (8.2) is given in terms of an antisymmetric tensor $A_{\mu\nu}$,

$$V_{\mu} = \partial^{\nu *}A_{\mu\nu} \tag{8.3}$$

The relevant term in the Lagrangian then becomes

$$-\frac{1}{2}(V_{\mu})^2 = -\frac{1}{2}(\partial^{\nu}\,{}^{*}A_{\mu\nu})^2 \tag{8.4}$$

and is known to describe a spin zero particle.

For non-abelian gauge groups, however, the constraints are more complicated. For N=2 it is found, in section four, to be

$$0 = \mathcal{D}^{\mu}V_{\mu} + i\,[W^{(1)},W] + \frac{1}{2}\{\bar{\lambda}^{i},\gamma_5\lambda_i\} \tag{8.5}$$

There is one obvious solution (W. Siegel) to this equation in terms of an antisymmetric tensor $A_{\mu\nu}$,

$$V_{\mu} = \mathcal{D}^{\nu *}A_{\mu\nu} - \frac{i}{2}\mathcal{D}_{\mu}\frac{1}{\mathcal{D}^2}([f^{\rho\sigma},A_{\rho\sigma}] + 2[W^{(1)},W] - i\{\bar{\lambda}^{i},\gamma_5\lambda_i\}) \tag{8.6}$$

which is, however, non-local and so of little interest. The important question to ask is, are all solutions of eq. (8.5) non-local? Although we cannot give a definitive answer at the present, the following observations may be relevant.

Firstly, just because the solution in the abelian case involves an anti-symmetric tensor, it does not necessarily follow that it should in the non-abelian case as well. This situation would imply a singularity of the solution as the Yang-Mills coupling constant goes to zero, and this would pose problems for a perturbation expansion in the coupling constant, but it cannot be categorically ruled out .

Secondly, the corresponding constraint for the Hamiltonian formulation of Yang-Mills theories is

$$0 = \mathcal{D}^{\mu} f_{o\mu} = -\vec{\mathcal{D}}\cdot\vec{E} \tag{8.7}$$

and has a local solution (Goldstone and Jackiw, 1978; Das and Townsend, 1979). However, this solution does not possess manifest gauge and Lorentz symmetry. This consideration suggests that equation (8.5) does have a local solution which may not exhibit the symmetries of the theory in a manifest fashion. However, even supposing we found one such solution, it is not entirely clear that it is a correct procedure to just substitute it into the Lagrangian.

We believe that the best method of approach is not to try to solve the constraint, but to implement it in the Lagrangian by using Lagrange multipliers. For example, we could adopt the Lagrangian

$$L^{Y.M} + \mathrm{Tr}\Big(A(\mathcal{D}^{\mu}V_{\mu} + i\,[W^{(1)}, W] + \frac{1}{2}\{\bar{\lambda}^{i}, \gamma_5\lambda_i\})\Big) \tag{8.8}$$

where $L^{Y.M.}$ is the N=2 Yang-Mills Lagrangian with the term $-\frac{1}{2}V^{\mu}V_{\mu}$ and A is a pseudo-scalar field.

Unfortunately, this is not a supersymmetric procedure and consequently of no interest to us. The trouble arises because, although the supersymmetry variation of the constraint is zero, its enforcement is required for the algebra to close. Consequently, $L^{Y.M.}$ , is no longer invariant under supersymmetry if the constraint is relaxed. Furthermore, the last term in (8.8) is also not invariant, since we cannot demand that the Lagrange multiplier A be an arbitrary function of x and yet inert under supersymmetry.

To impose the troublesome constraint in a supersymmetric fashion, we must impose it by using a supermultiplet's worth of Lagrange multipliers. That is, we must find a supermultiplet of constraints which includes the one of question (8.5) and impose them with the corresponding supermultiplet of Lagrange multipliers in such a way that the entire Lagrangian is supersymmetric. Since in the formulation obtained by dimensional reduction the constraint is inert under supersymmetry, we must add new fields to the theory to create a large multiplet so that the constraint now varies into these new fields and becomes a member of a supermultiplet of constraints. It is here that we begin to make contact with the considerations in the previous section, where the set of Choice II constraints indeed contained our eq. (8.5), see eqs. (7.7a–e).

The supermultiplet of constraints that we wish to impose by Lagrange multiplier is the Choice II constraint. This will be carried out in detail in section eleven, but first we turn to two simpler examples of supersymmetric Lagrange multipliers in theories without gauge invariance.

## 9. AN N=1 EXAMPLE

There is a good example for Lagrange multiplier techniques in N=1 supersymmetry.

The simplest free supersymmetric theory with an on-shell spectrum of two spin-0's and one spin-$\frac{1}{2}$ can be formulated in such a way that one of the spin-0 particles (here the pseudo-scalar) is described by a constrained vector $V_\mu$ (Siegel, 1979). The supersymmetric Lagrangian is

$$L = \frac{1}{2}\,\partial_\mu C \partial^\mu C - \frac{i}{2}\,\bar{\lambda}\,\not{\partial}\lambda - \frac{1}{2}\,V_\mu V^\mu \tag{9.1}$$

where $V_\mu$ is subject to $\partial^\mu V_\mu = 0$. In superspace, the theory is described by a superfield $\phi_1(x_\mu,\theta_\alpha)$, subject to the condition

$$0 = \bar{D}D\,\phi_1 = \bar{D}\,\gamma_5 D\phi_1\ . \tag{9.2}$$

This superfield is called the linear superfield (Ferrara, Wess and Zumino, 1974). The superspace action is given by

$$I \sim \int d^4x(\bar{D}D)^2\phi_1{}^2 \tag{9.3}$$

We now want to impose the constraint $\partial^\mu V_\mu = 0$ by Lagrange multipliers in a supersymmetric way. To do this we must first relax the constraint (9.2) and consider an unconstrained scalar superfield $\phi(x_\mu,\theta_\alpha)$. The component field content of this superfield is

$$\phi = (C,\chi_\alpha,M,N,V_\mu,\lambda_\alpha,D) \tag{9.4}$$

and eq. (9.2) implies the chiral supermultiplet of constraints

$$0 = M = N = \lambda - i\not{\partial}\chi = D - \Box C = \partial^\mu V_\mu \tag{9.5}$$

We now consider the Lagrangian

$$L_B = -\tfrac{1}{2}\,(CD + \bar{\chi}\lambda + M^2 + N^2 + V_\mu V^\mu) \tag{9.6}$$

and impose the constraints of equation (9.5) by adding an $L_H$ to $L_B$ where

$$L_H = A\partial^\mu V_\mu + B(D-\Box C) + \bar{\psi}(\lambda - i\not{\partial}\chi) + FM + GN \tag{9.7}$$

The fields $A, B, \psi, F, G$ are the Lagrange multipliers. We must now verify that the Lagrangian

$$L = L_B + L_H \tag{9.8}$$

is supersymmetric. This is the case if the fields $C, \chi, M, N, V_\mu, \lambda, D$ form a general pseudoscalar superfield (then $-2\, L_B$ is just the D-component of $\phi^2$) and if the fields $A, B, \psi, F, G$ form a chiral superfield (since the constraints form also a chiral superfield $L_H$ takes the form of the F-component of the product of two chiral superfields).

The action of eq. (9.8) gives the action of eq.(9.1) upon elimination of the Lagrange multipliers. One can, however, also eliminate all the fields <u>except</u> the Lagrange multipliers, and then one obtains the Wess-Zumino formulation of the theory (Wess and Zumino, 1974).

We call such a formulation of a supersymmetric theory with Lagrange multipliers a "Grandfather", since it is the ancestor to different formulations of the same theory depending on whether one eliminates the Lagrange multipliers, which occur only in $L_H$ (the "head"), or the other fields, which also occur in $L_B$ (the "body").

## 10. AN N=2 TOY MODEL

The linear multiplet in N=2 supersymmetry has features which are quite similar to those of the gauge multiplet of section six. In particular, it is a good example for studying how choice constraints can be imposed by Lagrange multipliers.

The linear multiplet (superfield) is an SU(2) triplet $K_{ij} = K_{ji}$ and fulfils **the defining condition**

$$\sum_{(ijk)} D_{\alpha i}\, K_{jk} = 0 \quad . \tag{10.1}$$

It has been called "linear multiplet" (Breitenlohner and Sohnius, 1980) because, in the absence of a central charge, it contains a conserved vector and decomposes into a chiral multiplet and a linear multiplet upon restriction to N=1 supersymmetry.

The field component content of the linear multiplet is given by the table

$$\begin{array}{llll} \vec{K} & \vec{K}^{(1)} & \vec{K}^{(2)} & \cdots\cdot \\ \varphi_{\alpha i} & \varphi_{\alpha i}^{(1)} & \varphi_{\alpha i}^{(2)} & \cdots\cdot \\ S & S^{(1)} & \cdots\cdot & \\ P & \partial^\mu V_\mu & \cdots\cdot & \\ V_\mu & V_\mu^{(1)} & \cdots\cdot & \end{array} \tag{10.2}$$

where each column corresponds to a given power of z in the Taylor expansion of $K(x,z,\theta)$ about $z=0$. Consequently, the central charge transformations with parameter $\omega$ are

$$\delta_\omega K = \omega K^{(1)} \;,\; \delta_\omega \varphi = \omega \varphi^{(1)}, \delta_\omega P = \omega \partial^\mu V_\mu \quad \text{etc.} \tag{10.3}$$

The supersymmetry transformations are

$$\begin{aligned} \delta_\varepsilon \vec{K} &= -\bar{\varepsilon}^i \vec{\tau}_i{}^j \varphi_j \\ \delta_\varepsilon \varphi_i &= -\varepsilon_i S - \gamma_5 \varepsilon_i P + \gamma^\mu \varepsilon_i V_\mu + i\gamma_5 \vec{\tau}_i{}^j \varepsilon_j \vec{K}^{(1)} \\ &\quad -i \gamma^\mu \vec{\tau}_i{}^j \varepsilon_j \partial_\mu \vec{K} \\ \delta_\varepsilon S &= -i \bar{\varepsilon}^i \not\partial \varphi_i + i \bar{\varepsilon}^i \gamma_5 \varphi_i^{(1)} \\ \delta_\varepsilon P &= i \bar{\varepsilon}^i \not\partial \gamma_5 \varphi_i \\ \delta_\varepsilon V_\mu &= -\bar{\varepsilon}^i \sigma_{\mu\nu} \partial^\nu \varphi_i + i \bar{\varepsilon}^i \gamma_\mu \gamma_5 \varphi_i^{(1)} \;, \end{aligned} \tag{10.4}$$

plus those for the higher components in the z-expansion, which can be derived from these with the help of $[\delta_\varepsilon, \delta_\omega] = 0$.

The infinite tower of fields (10.2) can again be truncated in two ways :

<u>Choice I</u> implies independence of z:

$$\frac{\partial}{\partial z} K_{ij} = 0 \;. \tag{10.5}$$

In terms of field components this implies the infinite set of constraints

$$0 = \vec{K}^{(1)} = \phi^{(1)} = S^{(1)} = \partial^\mu V_\mu = V_\mu^{(1)} = \vec{K}^{(2)} = \ldots \tag{10.6}$$

As we can see, these constraints involve one of the type we are interested in, namely we get a transversality condition for the vector :

$$\partial^\mu V_\mu = 0 \tag{10.7}$$

The remaining field content is the SU(2) triplet of scalars $\vec{K}$, the Majorana spinor doublet $\phi_{\alpha i}$, and the scalar-pseudoscalar pair S and P. We note the interesting difference to the gauge multiplet: the conserved vector is present in the version <u>without</u> a central charge.

A supersymmetric action for these fields is

$$I = \int d^4x \left(\frac{1}{2}\partial_\mu \vec{K}\partial^\mu \vec{K} - \frac{1}{2} V_\mu V^\mu - \frac{i}{2} \bar{\phi}^i \not\partial \phi_i + \frac{1}{2}(S^2+P^2)\right) \tag{10.8}$$

The corresponding superspace action,

$$I \sim \int d^4x \, \bar{D}^i D^j \bar{D}^k D^\ell \sum_{(ijk\ell)} K_{ij} K_{k\ell}, \tag{10.9}$$

is invariant only in the absence of a central charge (i.e. if equation (10.5) is imposed).

<u>Choice II</u>. Here we impose

$$\bar{D}^i D^j K_{ij} = 0 \tag{10.10}$$

which implies

$$0 = S = \varphi^{(1)} - \not\partial\gamma_5\varphi = \partial_\mu V_\nu - \partial_\nu V_\mu = V_\mu^{(1)} - \partial_\mu P = K^{(2)} - \Box K \tag{10.11}$$

and the central charge variations of this. In particular, we see that $V_\mu = \partial_\mu A$ with $\delta_\varepsilon A = i\bar{\varepsilon}^i \varphi_i$ and $A^{(1)} = P$. The supersymmetric action is

$$I = \int d^4x \left(\frac{1}{2}\partial_\mu \vec{K}\partial^\mu \vec{K} + \frac{1}{2}\partial_\mu A\partial^\mu A - \frac{i}{2}\bar{\varphi}^i \not\partial \varphi_i + \frac{1}{2}(\vec{K}^{(1)}\vec{K}^{(1)} + A^{(1)}A^{(1)})\right) \tag{10.12}$$

and we see that for both choices the theory describes one Majorana spinor doublet $\varphi_i$, a scalar triplet $\vec{K}$, and a spin-zero particle which is represented by the conserved vector $V_\mu$ for choice I and by the scalar field A for choice II.

We now return to choice I. We wish to implement the constraint (10.5) for it, which includes the condition $\partial^\mu V_\mu = 0$, by a Lagrange multiplier technique as discussed in section 8. To do this we must add to a "body" Lagrangian $L_B$ a "head" $L_H$.

$$L_B = \frac{1}{2}\partial_\mu \vec{K}\partial^\mu \vec{K} - \frac{1}{2} V_\mu V^\mu - \frac{i}{2}\bar{\varphi}^i \not\partial \varphi_i + \frac{1}{2}(S^2+P^2+\vec{K}^{(1)}\vec{K}^{(1)}) \tag{10.13}$$

$$L_H = -A\partial^\mu V_\mu + \sum_{n=o}^{\infty} L_H^{(n)} \tag{10.14}$$

where the "head" has infinitely many "hairs",

$$L_H^{(n)} = \vec{X}^{(-n)}\vec{K}^{(n+1)} + i\,\bar{\lambda}^{i(-n)}\gamma_5\varphi_i^{(n+1)} - B^{(-n)}\,S^{(n+1)} + A^{\mu(-n)}V_\mu^{(n+1)} \qquad (10.15)$$

so that the infinite tower of fields in (10.2) can be truncated.

We could have tried to only impose the constraints

$$0 = \vec{K}^{(1)} = \varphi_i^{(1)} = S^{(1)} = \partial^\mu V_\mu = V_\mu^{(1)}$$

by Lagrange multiplier and then to impose

$$\frac{\partial^2}{\partial z^2} K_{ij} = 0$$

by hand. The trouble with this is that it contains again a $\partial^\mu V_\mu^{(1)} = 0$ and we are in the same position as before.

Of course, the construction so far does not guarantee at all that $L = L_B + L_H$ is a supersymmetric Lagrangian. To prove that it is supersymmetric we must find a set of transformations of the fields which leave it invariant and then show that these transformations form a representation of the supersymmetry algebra on all fields.

One can easily verify that L is central charge invariant, if the fields of the original linear multiplet have the same central charge variations as before and if the Lagrange multipliers vary as follows under central charge transformations :

$$\delta_\omega \vec{X}^{(-n)} = -\omega\vec{X}^{(-n+1)}$$
$$\delta_\omega \lambda_{\alpha i}^{(-n)} = -\omega\lambda_{\alpha i}^{(-n+1)}, \text{ etc. for } n\neq 0 \qquad (10.16)$$

and

$$\delta_\omega \vec{X}^{(o)} = \omega(\Box\vec{K} - \vec{K}^{(2)})$$
$$\delta_\omega \lambda_\alpha^{(o)} = \omega(\varphi^{(1)} - \partial\!\!\!/\gamma_5\varphi) \qquad (10.17)$$
$$\delta_\omega A_\mu^{(o)} = \omega(V_\mu - \partial_\mu A)$$
$$\delta_\omega A = \omega P$$
$$\delta_\omega B^{(o)} = \omega S$$

The much stronger requirement of invariance under supersymmetry can be shown by a relatively simple calculation using the following transformation properties for the Lagrange multipliers:

$$\delta_\varepsilon B^{(-n)} = i\bar{\varepsilon}^i\gamma_5\lambda_i^{(-n)}$$

$$\delta_\varepsilon A_\mu^{(-n)} = i\bar{\varepsilon}^i\gamma_\mu\gamma_5\lambda_i^{(-n)} - i\bar{\varepsilon}^i\partial_\mu\lambda_i^{(-n-1)}$$

$$\delta_\varepsilon\lambda_i^{(-n)} = -\gamma^\mu\varepsilon_i A_\mu^{(-n+1)} + \varepsilon_i B^{(-n+1)} + \gamma^\mu\gamma_5\varepsilon_i\partial_\mu B^{(-n)} - i\sigma^{\mu\nu}\gamma_5\varepsilon_i\partial_\mu A_\nu^{(-n)} - i\,\vec{\tau}_i{}^j\gamma_5\varepsilon_j\vec{X}^{(-n)} \qquad (10.18)$$

$$\delta_\varepsilon\vec{X}^{(-n)} = \bar{\varepsilon}^i\,\vec{\tau}_i{}^j\partial\!\!\!/\gamma_5\,\lambda_j^{(-n)} - \bar{\varepsilon}^i\,\vec{\tau}_i{}^j\,\lambda_j^{(-n+1)}$$

for $n\neq 0$ and

$$\delta_\varepsilon A = i\varepsilon^i\varphi_i - i\bar{\varepsilon}^i\lambda_i^{(o)}$$

$$\delta_\varepsilon B^{(o)} = i\bar{\varepsilon}^i\gamma_5\lambda_i^{(o)}$$

$$\delta_\varepsilon A_\mu^{(o)} = i\bar{\varepsilon}^i\gamma_\mu\gamma_5\lambda_i^{(o)} - i\bar{\varepsilon}^i\partial_\mu\lambda_i^{(-1)}$$

$$\delta_\varepsilon\lambda_i^{(o)} = \gamma^\mu\varepsilon_i V_\mu - \varepsilon_i S + \gamma^\mu\gamma_5\varepsilon_i\partial_\mu B^{(o)} - \gamma^\mu\varepsilon_i\partial_\mu A - i\sigma^{\mu\nu}\gamma_5\varepsilon_i\partial_\mu A_\nu^{(o)} - i\vec{\tau}_i{}^j\gamma_5\varepsilon_j\vec{X}^{(o)} \qquad (10.19)$$

$$\delta_\varepsilon\vec{X}^{(o)} = \bar{\varepsilon}^i\vec{\tau}_i{}^j\partial\!\!\!/\gamma_5\lambda_j^{(o)} + \bar{\varepsilon}^i\vec{\tau}_i{}^j\varphi_j^{(1)} - \bar{\varepsilon}^i\,\vec{\tau}_i{}^j\partial\!\!\!/\gamma_5\varphi_j$$

for $n=0$. The transformation laws for the fields of the linear multiplet are unchanged from (10.4). These transformations form a representation of the N=2 supersymmetry algebra, with the central charge transformations (10.16) and (10.17).

If we eliminate the Lagrange multipliers $A, B^{(-n)}$, $\lambda^{(-n)}$, $A_\mu^{(-n)}$ $\vec{X}^{(-n)}$ from the Lagrangian we end up with our choice I situation. We can also obtain the choice II situation by eliminating an alternative set of fields.

## 11. THE GRANDFATHER

We now wish to implement, by Lagrange multipliers, the constraint of the N=2 supersymmetric abelian theory found by Sohnius, Stelle and West (1980b). It is this formulation which occurs when one obtains the theory by dimensional reduction by Legendre transformation from the five dimensional theory. To do this in a supersymmetric manner we have to consider many more fields than occur in the formulation of the theory with the constraint imposed by hand. In fact we must extend the fields considered to include the infinite number of fields contained in the gauge multiplet discussed in section six and then impose by Lagrange multipliers the constraints of equation (7.7) given in superspace by equation (7.6).

The construction proceeds along very similar lines to the N=1 example of section nine and the toy model of the previous section. The final Lagrangian is

$$L = L_B + \sum_{n=0}^{\infty} L_H{}^{(n)} \tag{11.1}$$

where

$$L_B = -\frac{1}{4} f_{\mu\nu}^2 + \frac{1}{2}(\partial_\mu W)^2 - \frac{1}{2} V_\mu^2 - \frac{i}{2} \bar{\lambda}^i \partial\!\!\!/ \lambda_i + \frac{1}{2}(W^{(1)})^2 + \frac{1}{2}\vec{X}^2$$

and

$$L_H{}^{(n)} = \vec{K}^{(-n)} \vec{X}^{(n-1)} + i\, \bar{\chi}^{i(-n)} \gamma_5 \left( \lambda_i{}^{(n)} + \partial\!\!\!/ \gamma_5 \lambda_i{}^{(n-1)} \right)$$

$$+ A^{(-n)} \partial_\mu V^{\mu(n-1)} + B^{(-n)} \left( W^{(n+1)} - \Box W^{(n-1)} \right)$$

$$+ B_\mu^{(-n)} \left( V^{\mu(n+1)} + \partial_\nu f^{\mu\nu(n-1)} \right) . \tag{11.2}$$

This Lagrangian is invariant under supersymmetry transformations, if those of the Lagrange multipliers with n≠0 are given by

$$\delta B^{(-n)} = -i\bar{\varepsilon}^i \gamma_5 \chi_i^{(-n)}$$

$$\delta A^{(-n)} = i\bar{\varepsilon}^i \chi_i^{(-n)} + \Lambda^{(-n)}$$

$$\delta \chi_i^{(-n)} = \Big( -B^{(-n+1)} - B\!\!\!/^{(-n+1)} + \partial\!\!\!/ A^{(-n)} - \partial\!\!\!/ \gamma_5 B^{(-n)}$$

$$- i\gamma_5 \sigma^{\mu\nu} \partial_\mu B_\nu^{(-n)} \Big) \varepsilon_i + i\gamma_5 \vec{\tau}_i{}^j \varepsilon_j \vec{K}^{(-n)}$$

$$\delta B_\mu^{(-n)} = i\,\bar{\varepsilon}^i \gamma_\mu \gamma_5 \chi_i^{(-n)} + \partial_\mu \Lambda^{(-n-1)}$$
$$\delta \vec{K}^{(-n)} = -i\,\bar{\varepsilon}^i \vec{\tau}_i{}^j \partial\!\!\!/ \gamma_5 \chi_j^{(-n)} + \bar{\varepsilon}^i \vec{\tau}_i{}^j \chi_j^{(-n+1)} \tag{11.3}$$

and those for n=0 given by

$$\delta B^{(o)} = -i\,\bar{\varepsilon}^i \gamma_5 \chi_i^{(o)} \quad ; \quad \delta A^{(o)} = i\,\bar{\varepsilon}^i \chi_i^{(o)} - i\,\bar{\varepsilon}^i \lambda_i$$
$$\delta \chi_i^{(o)} = (\,-W^{(1)} + V\!\!\!/ + \partial\!\!\!/ A^{(o)} - \partial\!\!\!/ \gamma_5 B^{(o)} - i\gamma_5 \sigma^{\mu\nu} \partial_\mu B_\nu^{(o)}) \,\varepsilon_i + i\gamma_5 \vec{\tau}_i{}^j \varepsilon_j\, \vec{K}^{(o)} \tag{11.4}$$
$$\delta B_\mu^{(o)} = i\,\bar{\varepsilon}^i \gamma_\mu \gamma_5 \chi_i^{(o)} + \partial_\mu \Lambda^{(-1)} \quad ; \quad \delta \vec{K}^{(o)} = -\bar{\varepsilon}^i \vec{\tau}_i{}^j \partial\!\!\!/ \gamma_5 \chi_j^{(o)} - \bar{\varepsilon}^i \vec{\tau}_i{}^j \lambda_j^{(1)}$$

The parameters $\Lambda^{(-n)}$ are undetermined and correspond to the gauge invariances of $L_B + L_H$. In fact, by a judicious choice of these parameters we can eliminate all mention of $A^{(-n)}$ for n>0 from the theory. The transformations for the fields which belong to the original gauge multiplet are unchanged and given in eq. (6.7). This set of transformations form a realization of the supersymmetry algebra with one central charge.

The Lagrangian $L_H + L_B$ is composed from a large supermultiplet containing the original gauge multiplet and the Lagrange multiplier. The field content of the multiplet is given in the following table:

$$\begin{array}{llllll}
\ldots\; A^{(-2)} & A^{(-1)} & A^{(o)} & & & \\
\ldots\; B^{(-2)} & B^{(-1)} & \left.\begin{array}{l} B^{(o)} \\ W \end{array}\right\} & W^{(1)} & W^{(2)} & \ldots \\
\ldots\; B_\mu^{(-2)} & B_\mu^{(-1)} & \left.\begin{array}{l} B_\mu^{(o)} \\ A_\mu \end{array}\right\} & V_\mu & V_\mu^{(1)} & \ldots \\
\ldots\; \chi_i^{(-2)} & \chi_i^{(-1)} & \left.\begin{array}{l} \chi_i^{(o)} \\ \lambda_i \end{array}\right\} & \lambda_i^{(1)} & \lambda_i^{(2)} & \ldots \\
\ldots\; \vec{K}^{(-2)} & \vec{K}^{(-1)} & \left.\begin{array}{l} \vec{K}^{(o)} \\ \vec{X} \end{array}\right\} & \vec{X}^{(1)} & \vec{X}^{(2)} & \ldots
\end{array} \tag{11.5}$$

The action of the central charge on the fields of the original gauge multiplet is still given by equation (6.5). Its action on the Lagrange multipliers is given for $n\neq 0$ by

$$\delta B^{(-n)} = -\omega B^{(-n+1)}, \quad \delta A^{(-n)} = -\omega A^{(-n+1)}, \quad \delta \chi_i^{(-n)} = -\omega \chi_i^{(-n+1)}$$

$$\delta B_\mu^{(-n)} = -\omega B_\mu^{(-n+1)}, \quad \delta \vec{K}^{(-n)} = -\omega \vec{K}^{(-n+1)}$$

and for $n=0$ by

$$\delta B^{(o)} = -\omega W^{(1)}, \quad \delta A^{(o)} = 0, \quad \delta \chi_i^{(o)} = \omega \lambda_i^{(1)}$$

$$\delta B_\mu^{(o)} = \omega V_\mu, \quad \delta \vec{K}^{(o)} = -\omega \vec{X}^{(1)} \qquad (11.6)$$

That is, the central charge transformation just moves one from left to right in the table of equation (11.5).

One can see from the supersymmetry and central charge transformations of the multiplet of equation (11.5) that it is an extension of the gauge multiplet to contain the Lagrange multipliers. The supermultiplet of table (11.5) is far from being irreducible. One of its most important irreducible parts are the field combinations

$$B^{(o)}+W; \quad A^{(o)}; \quad \chi_i^{(o)} - \lambda_i; \quad B_\mu^{(o)} - A_\mu \text{ and } \vec{K}^{(o)}+\vec{X}, \qquad (11.7)$$

which are central charge inert and transform into themselves under supersymmetry. In fact, they have exactly the transformations of the set of fields out of which the N=2 Abelian theory, found by Grimm, Sohnius and Wess (1978) is constructed.

Hence, we have obtained a supersymmetric Lagrangian which describes the N=2 supersymmetric Abelian theory and contains an infinite set of fields, none of which are constrained. This Lagrangian is constructed from the fields which make up the gauge multiplet as well as an infinite set of Lagrange multipliers.

It follows from the method of construction that the elimination of these Lagrange multipliers by their equations of motion, leads to the formulation of the N=2 Abelian theory which can be obtained from the five dimensional theory by dimensional reduction by Legendre transformation (Sohnius, Stelle and West 1980 a,b). However, if instead of eliminating the Lagrange multipliers we eliminate all fields except the field combinations given in (11.7) we arrive at the formulation of the N=2 supersymmetric Abelian theory found by Grimm, Sohnius and Wess (1978). This is possible since these field combinations by themselves form an irreducible supermultiplet and have the correct supersymmetry transformations for this formulation. Indeed, such a multiplet must be present if we demand that the Grimm, Sohnius, Wess formulation may be obtained by elimination of suitably chosen fields.

This is exactly the same phenomenon as in the N=1 supersymmetry example of section nine. One could also, as in the N=1 case, perform a series of field redefinitions in the Lagrangian, in such a way as to give two disjoint terms, one of which is the N=2 Abelian formulation of Grimm, Sohnius and Wess, the other being a term giving rise to no propagating degrees of freedom.

Consequently, depending on what fields we choose to eliminate we can obtain either of the two N=2 supersymmetric formulations. As a result, we call this formulation the Grandfather formulation and the multiplet of equation (11.5) the Grandfather multiplet.

## 12. CONSTRAINTS AND DIMENSIONAL REDUCTION

We now wish to give a simple derivation of the Grandfather and hopefully throw some light on its origin. Both of the previously known formulations of the N=2 supersymmetric abelian theories can be obtained by dimensional reduction from a five dimensional N=2 abelian theory whose supersymmetric action is given by

$$I_{(5)} = \int d^5x\left(-\frac{1}{4} F_{ab}^2 - \frac{i}{2} \bar{\lambda}^i \partial\!\!\!/ \lambda_i + \frac{1}{2}(\partial_a A)^2 + \frac{1}{2}\vec{X}^2\right)$$

$$a,b = 1 \rightarrow 5 \tag{12.1}$$

Using the most straightforward dimensional reduction technique, that is neglecting the dependence of the fields on the fifth coordinate $x^5$, we find the supersymmetric four dimensional theory of Grimm, Sohnius and Wess (1978), given by the action

$$I^{(4)} = \int d^4x \left\{-\frac{1}{4} F_{\mu\nu}^2 - \frac{i}{2} \bar{\lambda}^i \partial\!\!\!/ \lambda_i + \frac{1}{2}(\partial_\mu A)^2 + \frac{1}{2}(\partial_\mu B)^2 + \frac{1}{2} \bar{X}^2\right\} \tag{12.2}$$

where $B = A_5$.

The other known formulation of N=2 abelian theory can be derived by the technique of dimensional reduction by Legendre transformation from the same five dimensional action of equation (12.1), as explained in section five. The advantage of this technique is that one need not know the auxiliary fields to gain an off-shell formulation in four dimensions. In fact, if one starts with an off-shell formulation, this technique irradicates all knowledge of the auxiliary fields in the higher dimension (in this case $\vec{X}$). As such it is natural to ask if the Grandfather formalism of the previous section can be obtained by some form of dimensional reduction from the five dimensional theory of equation (12.1). Support for this possibility is that the supermultiplet of the Grandfather has a central charge, and since the central charge is just equivalent to a fifth dimension we may view the theory as being constructed from a five dimensional supermultiplet.

Neglecting for the moment the origin of the action, we now focus on the question to which type of five dimensional supermultiplet the component fields of table (11.5) belong. That is, we want to know what dependence on $x_5$ would lead to the properties we have found for these fields; a first step will be to account for their central charge transformations.

Since the Grandfather multiplet is an extension of the gauge multiplet, we now examine the five dimensional formulation of the gauge multiplet. The conditions of equation (6.2) on the superfield $W(x_a,\theta_{\alpha i})$ take on the five dimensional form

$$\bar{\mathcal{D}}^i \vec{\tau}_i{}^j \gamma_a \mathcal{D}_j W = 0 \; ; \qquad a = 1 \to 5. \tag{12.3}$$

The central charge transformations are given by assuming that the five dimensional dependence of W is analytic at z=0 and then expanding W in a Taylor series in z about z=0, viz.

$$W(x^a, \theta_{\alpha i}) = W^{(o)}(x^\mu, \theta) + z\, W^{(1)}(x^\mu, \theta) + \frac{z^2}{2} W^{(2)}(x^\mu, \theta) + \ldots + \frac{z^n}{n!} W^{(n)}(x^\mu, \theta) + \ldots \tag{12.4}$$

where

$$z \equiv x^5$$

Each term in the above expansion corresponds to a column in the field content of table (11.5). The action of the central charge is a shift in z and can be read off from the equation

$$W'(x_a,\theta) - W(x_a,\theta) = \omega \frac{dW}{dz}(x_a,\theta) \quad . \tag{12.5}$$

Consequently, we find the action of the central charge on the component fields to be given by equation (6.5), namely

$$\delta W^{(n)}(x_\mu,\theta) = \omega W^{(n+1)}(x_\mu,\theta) \quad . \tag{12.6}$$

As we have already observed, the Grandfather multiplet is an extension of the gauge multiplet, being extended by the addition of Lagrange multipliers. In fact, the transformations of the Lagrange multipliers are of the same general form as those of the component fields of the gauge multiplet. A closer examination shows that we can represent the Grandfather multiplet by a superfield $G(x_a,\theta_\alpha)$ which is subject to the same conditions (12.3) as the gauge multiplet, namely

$$\bar{\mathcal{D}}^{i}\vec{\tau}_i{}^{j}\gamma_a\mathcal{D}_j G = 0, \quad a = 1 \to 5 \quad . \tag{12.7}$$

As the gauge multiplet was obtained by demanding that it satisfy the condition of equation (12.3) and that it be analytic in $x_5$ , it follows that its extension, the Grandfather multiplet, must be non-analytic in $x_5$.

The simplest possibility from the point of view of mathematical formalism would be to assume that $G(x_a,\theta_{\alpha i})$ possesses a Laurent expansion in $x_5$, viz.

$$G(x^a, \theta_{\alpha i}) = \ldots - \frac{(n-1)!}{(-z)^n} G^{(-n)}(x^\mu, \theta) - \ldots - \frac{1}{-z} G^{(-1)}(x^\mu, \theta)$$

$$+ G^{(o)}(x^\mu, \theta) + z\, G^{(1)}(x^\mu, \theta) + \ldots$$

$$\ldots + \frac{z^n}{n!} G^{(n)}(x^\mu, \theta) + \ldots \tag{12.8}$$

The action of the central charge on the coefficients of this expansion, $G^{(n)}$, for $n \neq -1$ is

$$\delta_\omega G^{(n)}(x^\mu, \theta_{\alpha i}) = \omega\, G^{(n+1)}(x^\mu, \theta_{\alpha i}) \tag{12.9}$$

An important new phenomenon of this type of non-analytic expansion is the appearance of a central charge invariant object, the coefficient of $\frac{1}{z}$, $G^{(-1)}(x_\mu,\theta)$. Although the coefficients of the $z^{-n}(n>o)$ terms correctly mimic the central charge and supersymmetry transformations of the Lagrange multipliers, they do not transform into the analytic parts of the series (the original gauge multiplet). But this ought to happen for the Grandfather multiplet!

The expansion in $x_5$ which does correctly account for the supersymmetry and central charge transformations of the Grandfather multiplet is given by

$$G(x^a, \theta_{\alpha i}) = \ldots - \frac{(n-1)!}{(-z)^n} G^{(-n)} - \ldots - \frac{1}{-z} G^{(-1)} + G_{(o)}$$

$$+ \ell n\, z\; G^{(o)} + z\, \ell n\, z\; G^{(1)} + \ldots$$

$$\ldots + f^{(n)}(z)\; G^{(n)} + \ldots \tag{12.10}$$

where $f^{(n)}(z)$ is recursively defined by

$$f^{(n)}(z) = \int^z d\zeta\, f^{(n-1)}(\zeta) \quad \text{and} \quad f^{(1)} = z\, \ell n\, z \; . \tag{12.11}$$

The correct rotation of the Lagrange multipliers into the gauge multiplet fields under the action of the control charge is provided by the log-term. The higher dependence on z is governed by the simple requirement that the coefficients of $f^n(z)$ shift up one under the action of the central charge.

It is not entirely clear that eq. (12.8) is the only expansion which will give the required behaviour; however, any other expansion will hopefully be related by field redefinitions to this one.

As with the Laurent expansion we have a central charge invariant expression,

$$G^{(o)}(x_\mu,\theta_{\alpha i}) - G_{(o)}(x_\mu,\theta_{\alpha i}) \quad . \tag{12.12}$$

This corresponds to the irreducible supermultiplet contained in the Grandfather and commented on in the previous section in eq. (11.7).

As well as the square of the above expression, there are an infinite number of central charge invariant bilinear expressions which we can form from the coefficients of two expansions of the type of equation (12.10). They are of the form

$$G_1^{(n)} G_2^{(o)} - G_1^{(n-1)} G_2^{(1)} - G_1^{(n+1)} G_2^{(-1)} + G_1^{(n-2)} G_2^{(2)} + G_1^{(n+2)} G_2^{(-2)}$$
$$- \ldots - \ldots + \ldots + \ldots \tag{12.13}$$

The invariant which will interest us is one of the above (n=0) combined with the square of the invariant (12.12):

$$G_{1(o)}G_{2(o)} - G_1^{(o)}G_{2(o)} - G_{1(o)}G_2^{(o)} + G_1^{(-1)}G_2^{(1)} + G_1^{(1)}G_2^{(-1)}$$
$$- G_1^{(-2)}G_2^{(2)} - G_1^{(2)}G_2^{(-2)} + \ldots + \ldots \tag{12.14}$$

To obtain the action for the Grandfather we simply substitute the expression for $G(x_a,\theta_{\alpha i})$ given by equation (12.11) (or the equivalent expression for the component fields) into the five dimensional action of equation (12.1) and then combine the resulting coefficients into an invariant according to equation

(12.14). The resulting expression agrees with that found in equations (11.1) and (11.2).

Hence, we have found a dimensional reduction technique which yields the Grandfather. This technique involves the substitution of a more complicated and non-analytic dependence on $x_5$ into the five dimensional action and the replacement of the integration over $x_5$ by the instruction for combining the coefficients into an invariant contained in equation (12.14). It is interesting to consider this as another way of achieving the same results as the dimensional reduction by Legendre transformation. We perform the above outlined dimensional reduction technique and then eliminate the Lagrange multipliers. The elimination of the highest dimensional Lagrange multipliers just imposes the same five dimensional equations of motion as imposed by the technique of dimensional reduction by Legendre transformation. Presumably the application of this technique to the five dimensional non-abelian theory will produce a non-abelian Grandfather.

The infinite set of fields is, of course just a reflection of the arbitrary dependence, up to analytic structure, of $G(x_a,\theta_{\alpha i})$ on $x_5$. Lagrangians such as the Grandfather take their five dimensional origins very seriously, but in one very important aspect they differ from five dimensional theories: they forsake the Lorentz invariance.

This discussion naturally leads us to wonder what is the precise relation between specifying the analytic structure of $G(x_a,\theta_{\alpha i})$ on $x_5$ and the resulting Grandfather and what theories other choices of analytic structure would lead to. Certainly, an arbitrary dependence on $x_5$ , up to analytic behaviour, does not always lead to an infinite number of propagating fields in four dimensions as one would naively expect.

## 13. OUTLOOK

In this article we have seen how the demands of preservation of representations of supersymmetry in the presence of Yang-Mills fields has led to constraints on the N=2 Yang-Mills field-strength. The full consequences of these constraints were worked out by using the Bianchi identities.

We also saw how we could obtain a formulation of the N=2 Yang-Mills theory by dimensional reduction by Legendre transformation. This formulation, as with any formulation of a gauge theory obtained from this technique, involved fields which were subject to constraints in ordinary space, the solution of these being unknown. The superspace version of this formulation in ordinary space contains constraints on the Yang-Mills field strengths which are stronger than, but include those obtained by the demands of preservation of representations.

This dilemma is resolved by the presence of another formulation of the N=2 supersymmetric Yang-Mills theory (Grimm,Sohnius and Wess 1978). The superspace constraints in this formulation are also stronger than but include those obtained by the demands of preservation of representations. In fact, given the representation preserving constraints, we can make two alternative further choices of constraint which do not put the theory on-shell. These two alternative choices together with the representation preserving constraints lead to the two alternative formulations of the N=2 supersymmetric Yang-Mills theory.

The choice of the further constraint which leads to the formulation found recently by the authors is responsible for the restriction of the ordinary space dependence of one of the component fields found in this theory. The representation preserving constraints lead to no such restriction and there is no such restriction implied by the choice of constraint required to give the Grimm,Sohnius,Wess (1978) formulation.

Since the only known off-shell formulation of the N=4 supersymmetric Yang-Mills theory (Sohnius, Stelle and West 1980a,b) can be obtained by the technique of dimensional reduction by Legendre transformation from a five dimensional theory, it necessarily contains some fields whose dependence on ordinary space is restricted. Therefore we sought methods of implementing these constraints in a supersymmetric fashion by the method of Lagrange multipliers.

This was achieved for the N=2 Abelian theory. The supermultiplet of constraints which were imposed by Lagrange multipliers, are just the choice constraints which are required in addition to the representation preserving constraints to obtain the formulation resulting from dimensional reduction by Legendre transformation of a five dimensional theory. Hence we arrived at what we called the Grandfather formulation, constructed from a supermultiplet which is subject only to the representation preserving constraints. Necessarily, elimination of the Lagrange multipliers by their equations of motion leads to the N=2 Abelian theory found by the authors. However, the elimination of another set of fields leads to the Grimm,Sohnius Wess formulation (1978)of the N=2 supersymmetric Abelian theory. It is interesting to observe that what is the choice constraint for one formulation turns out to be the equation of motion in the other formulation.

In terms of component fields the Grandfather formulation involved an infinite set of Lagrange multipliers. This infinity, which was not found in the N=1 analogous case, reflects the presence of the central charge.

The Grandfather formulation can be obtained by dimensional reduction from the five dimensional Abelian theory. The dependence of the fields on the fifth dimension is then non-analytic and, apart from its exact type of non-analytic behaviour, arbitrary. It is this arbitrariness that gives rise to the infinity of fields in four dimensions. One point of considerable importance is that in order to obtain the Grandfather formulation we must forsake the five dimensional Lorentz invariance of the original five dimensional Abelian theory.

Although we can derive the Grandfather formulation from the five dimensional Abelian theory, the method used for its construction given in sections one to eleven does not utilize the knowledge of the five dimensional theory. It only requires a knowledge of the conditions necessary for preservation of representations and the appropriate action formula. In fact, the Grandfather formulation does not even imply the existence of the five dimensional theory, since this theory has the much stronger requirement of five dimensional Lorentz covariance, a point we will return to later.

We could carry out a similar procedure in the case of the maximal supersymmetric Yang-Mills theory. The first step is to know the consequence of the requirement of the preservation of representations of N=4 supersymmetry. However, these can only be found if we have a knowledge of the representations of N=4 supersymmetry.

We now give one of these representations. It is provided by the superfield $\phi_i(x_\mu, z, \theta_{\alpha i})$, subject to the conditions

$$\sum_{(ij)} D_{\alpha i}\,\phi_j = \sum_{(ij)} \bar{D}_{\dot{\alpha} i}\,\phi_j = 0 \quad . \tag{13.1}$$

These conditions are Lorentz covariant in five dimensions and thus any consequences of these equations must be covariant as well, and consequently they cannot lead to four dimensional equations of motion.

By analogy with N=2, we can call this representation the N=4 hypermultiplet.

It is interesting to observe that the corresponding anti-symmetric conditions

$$\sum_{[ij]} D_{\alpha i}\phi_j = 0 = \sum_{[ij]} \bar{D}_{\dot{\beta} i}\phi_j \tag{13.2}$$

do not lead to a representation because they imply that the super-field is a constant.

In writing down these conditions we assume the presence of a central charge in the algebra. It is a necessary consequence of this that the internal symmetry group of the algebra (not of all theories) be SP(4). Indices can be raised with the anti-symmetric metric of this group.

Promoting the N=4 hypermultiplet to the presence of Yang-Mills fields requires the modified defining conditions

$$\sum_{(ij)} \mathcal{D}_{\alpha i}\phi_j = 0 = \sum_{(ij)} \bar{\mathcal{D}}_{\dot{\beta} i}\phi_j \quad . \tag{13.3}$$

It is a consequence of these conditions that

$$\sum_{(ijk)} \{\mathcal{D}_{\alpha i}, \mathcal{D}_{\beta j}\}\phi_k = i \sum_{(ijk)} F_{\alpha i \beta j}\, \phi_k = 0$$

$$\sum_{(ijk)} \{\mathcal{D}_{\alpha i}, \bar{\mathcal{D}}_{\dot{\beta} j}\}\phi_k = i \sum_{(ijk)} F_{\alpha i \dot{\beta} j}\, \phi_k = 0$$

$$\sum_{(ijk)} \{\bar{\mathcal{D}}_{\dot{\alpha} i}, \bar{\mathcal{D}}_{\dot{\beta} j}\}\phi_k = i \sum_{(ijk)} F_{\dot{\alpha} i \dot{\beta} j}\, \phi_k = 0 \quad , \tag{13.4}$$

which in turn imply that

$$\sum_{(ij)} F_{\alpha i \beta j} = \sum_{(ij)} F_{\alpha i \dot{\beta} j} = \sum_{(ij)} F_{\dot{\alpha} i \dot{\beta} j} = 0 \quad . \tag{13.5}$$

To these constraints we can add some conventional constraints; those for $A_\mu$ and $A_z$ are

$$F_{\alpha i \dot{\beta}}{}^{i} = 0 \quad \text{and} \quad F_{\alpha i}{}^{\alpha i} - F^{\dot{\alpha}}{}_{i \dot{\alpha}}{}^{i} = 0 \quad . \tag{13.6}$$

Although we can not guarantee that the preservation of other representations will not lead to further conditions, past experience has shown that it is sufficient to preserve the most fundamental representation. Certainly, the N=4 chiral representation does not lead to any further conditions.

It is of considerable interest to compare these superspace constraints, equations (13.5) and (13.6), with the superspace constraints for the known N=4 formulation (Sohnius, Stelle and West, 1980a ). Just as in the N=2 case we see that these constraints are stronger than the representation preserving constraints, the difference being the constraints :

$$\sum_{[ij]} (F_{\alpha i \dot{\beta} j} - \text{trace}) = 0 \tag{13.7}$$

This leads one to suspect that the pattern of N=2 is repeated in N=4, and to conjecture that there does exist an alternative formulation of the maximally extended Yang-Mills theory. Hopefully, this formulation will not contain a field whose ordinary space dependence is constrained; it is the constraints of equation

(13.7) which are the five dimensional field equations and thus responsible for the x-space constraints.

Assuming for the sake of argument that an alternative formulation exists, we could find this theory in two ways. Firstly, we could simply search for a constraint other than that of equation (13.7) which when taken with the representation preserving constraints of equation (13.5) and the conventional constraints of equation (13.6) does not put the theory on-mass-shell. If the theory were to possess an SU(4) symmetry this additional constraint would have to set the action of the central charge on the field strengths to zero. Such a constraint would be consistent with the N=2 pattern and is the equation of motion for the N=4 theory with eq. (13.7) imposed.

The second approach would be to build the N=4 Grandfather by implementing the constraints of equation (13.7) by Lagrange multipliers. The first set of such multipliers we hope to be the fields of the alternative formulation, which would appear upon elimination of a suitable set of fields in the Grandfather.

To substantiate the above conjecture would require us to calculate, using the Bianchi identities, the field content of the theory when subject to the representation preserving and conventional constraints of equations (13.5) and (13.6). Application of the choice constraint would then give us the field content of the new formulation. Out of this field content we could hopefully construct the Lagrangian.

An intriguing feature of the representation preserving constraints is the way they are embedded into the ten dimensional Yang-Mills theory. Simple considerations show that these constraints are contained in the ten dimensional constraints

$$\sum_{[abcde]} (\gamma^a\gamma^b\gamma^c\gamma^d\gamma^e C_{(10)})^{\alpha\beta} F_{\alpha\beta} = 0 \qquad \begin{array}{l} a,b,c,d,e = 1,\dots,10 \\ \alpha,\beta = 1,\dots,32 \end{array} \qquad (13.8)$$

The conventional constraints are given by

$$(\gamma^a C_{(10)})^{\alpha\beta} F_{\alpha\beta} = 0 \ . \qquad (13.9)$$

However, as is well known, these constraints, in ten dimensions, put the theory on-shell.

The resolution of this dilemma is that the existence of the ten dimensional theory has one much stronger requirement than the four dimensional theory, namely ten dimensional Lorentz covariance. This requires us to add the further constraints contained in equation (13.8) to those found by the representation preserving arguments in four dimensions. This argument shows that the existence of an off-shell supersymmetric Yang-Mills theory in ten dimensions is a much stronger requirement than the existence of the off-shell N=4 theory in four dimensions.

In this context it is helpful to use the technique of ghost counting (reviewed in Sohnius, Stelle and West, 1980c) to assist us in the determination of the constraint-free version of the theory. Let us demand that this new formulation possess two properties:

(i) The internal symmetry is SU(4) even off-shell.

(ii) All the particle states in the spectrum correspond to their usual field representations, i.e. spin-0 to scalars, spin-1 to vectors and spin-$\frac{1}{2}$ to spinors.

The first property implies that the central charges in the theory vanish for the S-matrix. The second demand is an attempt to ensure that the component fields of the theory do not have constraints upon their dependence on Minkowski space. A theory with these properties would be a straightforward off-shell extension of the on-shell N=4 supersymmetric Yang-Mills theory found by Gliozzi, Scherk and Olive (1978).

Upon insertion of a d'Alembertian operator into such a theory the resulting spectrum falls into two N=4 supersymmetry representations. One possesses the same spectrum as the original theory, the other is a massive ghost representation. The simplest massive representation with no central charge occurs with the following spins and SU(4) representations:

| Spin | 2 | $\frac{3}{2}$ | 1 | $\frac{1}{2}$ | 0 |
|---|---|---|---|---|---|
| SU(4)-content | $\underline{1}$ | $\underline{4}+\underline{4}^*$ | $\underline{6}+\underline{6}+\underline{15}$ | $\underline{4}+\underline{4}^*$ $+\underline{20}+\underline{20}^*$ | $\underline{1}+\underline{1}$ $+\underline{10}+\underline{10}^*+\underline{20}'$ |

(13.10)

The $\underline{20}'$ indicates the real representation of SU(4) with 20 states. The existence of such a supermultiplet can be ascertained by considering a massless representation, made massive by means of Higgs effects. The intriguing fact about the representation (13.10) is that it does not contain a singlet of spin one, and consequently cannot be obtained by insertion of a d'Alembertian from the Yang-Mills theory. One singlet spin-1 must arise from the gauge vector field.

Hence, the massive ghosts must belong to a different representation. The simplest representation which does contain a singlet of spin one is obtained by taking the product of table (13.10) with the representation $\underline{6}$:

| Spin | 2 | $\frac{3}{2}$ | 1 | $\frac{1}{2}$ | 0 |
|---|---|---|---|---|---|
| SU(4)-content | $\underline{6}$ | $\underline{4}+\underline{4}^*$ $+\underline{20}+\underline{20}^*$ | $\underline{1}+\underline{1}+\underline{64}$ $+\underline{10}+\underline{10}^*$ $+\underline{15}+\underline{15}$ $+\underline{20}'+\underline{20}'$ | $\underline{4}+\underline{4}+\underline{4}^*+\underline{4}^*$ $+\underline{20}+\underline{20}$ $+\underline{20}^*+\underline{20}^*$ $+\underline{36}+\underline{36}^*$ $+\underline{60}+\underline{60}^*$ | $\underline{6}+\underline{6}+\underline{6}$ $+\underline{15}+\underline{15}^*$ $+\underline{45}+\underline{45}^*$ $+\underline{50}+\underline{64}$ |

(13.11)

If this were the answer, then the auxiliary fields would have to provide upon insertion of the d'Alembertian a $\underline{6}$ of spin two as well as many other states.

It is useful in this context to consider how a theory having the above two properties could arise out of the SP(4) covariant superspace arguments given earlier in this section. According to the above considerations, the theory would have the SP(4) covariant superspace constraints of equation (13.5) and (13.6), as well as the choice constraint

$$\frac{\partial}{\partial z} F_{\alpha i\beta j} = \frac{\partial}{\partial z} F_{\alpha i\dot{\beta} j} = \frac{\partial}{\partial z} F_{\dot{\alpha} i\dot{\beta} j} = 0 \qquad (13.12)$$

However, to preserve the final constraint, we must restrict the superspace Yang-Mills gauge group to be independent of z, like in the N=2 case. In such circumstances, the potential $A_z$ becomes a covariant field and can be used explicitly in the action, as in the Sohnius-Grimm-Wess formulation for N=2. For N=4, this superfield might contain the component fields required to fill out the SP(4) representations to make up representations of SU(4). Say the five scalars in $F^{\alpha}{}_{i\alpha j}$ fill up to the six scalars expected.

This last section is rather speculative, but it does point into possible directions for further investigations.

REFERENCES

Breitenlohner, P., and M. Sohnius (1980). Superfields, Auxiliary Fields and Tensor Calculus for N=2 Extended Supergravity. Nucl. Phys. B165, p.483

Brink, L., J. Schwarz and J. Scherk (1977). Supersymmetric Yang-Mills Theories. Nucl. Phys. B121, p.77

Das, A., and P. Townsend (1979). Gauge Fixing Ambiguities, Flux Strings and the Unconstrained Yang-Mills Theory. Nucl. Phys. B149, p.109

Cremmer, E., S. Ferrara, K.S. Stelle and P.C. West (1980). Off-shell N=8 Supersymmetry with Central Charges. Phys. Lett. 94B, p.349

Cremmer, E., and B. Julia (1979). The SO(8) Supergravity. Nucl. Phys. B159, p.141

de Wit, B., and J.W. van Holten (1978). Covariant Quantization of Gauge Theories with Open Gauge Algebra. Phys. Lett. 79B, p.389

Fayet, P. (1976). Hypersymmetry. Nucl. Phys. B113, p.135

Gates Jr., S.J., K.S. Stelle and P.C. West (1980). Algebraic Origins of Superspace Constraints in Supergravity. Nucl. Phys. B169, p. 347

Gliozzi, F., J. Scherk and D. Olive (1978). Supersymmetry, Supergravity and the Dual Spinor Model. Nucl. Phys. B122, p.253

Goldstone, J., and R. Jackiw (1978). Unconstrained Temporal Gauge for Yang-Mills Theory. Phys. Lett. B74, p.81

Grimm, R., M. Sohnius and J. Wess (1978). Extended Supersymmetry and Gauge Theories. Nucl. Phys. B133, p.275

Siegel, W. (1979). Gauge Spinor superfield as Scalar Multiplet. Phys. Lett. 85B, p. 333

Sohnius, M.F. (1978). Supersymmetry and Central Charges. Nucl. Phys. B138, p. 109

Sohnius, M.F., K.S. Stelle and P.C. West (1980a). Dimensional Reduction by Legendre Transformation Generates Off-shell Supersymmetric Yang-Mills Theories. Nucl. Phys. B,in press

Sohnius, M., K.S. Stelle and P.C. West (1980b). Off-mass-shell Formulation of Extended Supersymmetric Gauge Theories. Phys.Lett. 92B, p. 123

Sohnius, M.F., K.S. Stelle and P.C. West (1980c). Supersymmetric Yang-Mills Theories. To be published in the proceedings of the Europhysics Conference "Unification of the Fundamental Interactions", Erice, March 1980

Taylor, J.G. (1980). On representations of Extended Supersymmetry Algebras on Superfields. Nucl. Phys. B169, p.484

Wess, J. (1975). Supersymmetry and Internal Symmetry. Acta Phys. Austriaca 41, p. 409

Wess, J., and B. Zumino (1974). Supergauge Transformations in Four Dimensions. Nucl. Phys. B70, p. 39

Ferrara, S., J. Wess and B. Zumino (1974). Supergauge Multiplets and Superfields. Phys. Lett. 51B, p. 239

# GROUP DISINTEGRATIONS

B. Julia (ENS)

Laboratoire de Physique Théorique
de l'Ecole Normale Supérieure, Paris

*dedicated to the memory of Joël Scherk*

## 1 THE PROBLEM

Our ultimate goal is to understand the origin of the remarkable symmetries obtained "experimentally" in the various models of supergravity. These theories exist for various dimensions (we consider $d \geqslant 3$ ) and for various numbers of supersymmetry spinor charges (Salam & Strathdee, 1978 ; Scherk, 1978) ; $N \leqslant 8$ is the number of "real" bispinor charges in the 4 dimensional form of the theories. The deus ex-machina is probably supersymmetry itself; however the theory of superalgebras is still so young and unfamiliar to most people that it seemed to us worthwhile concentrating on the ordinary (Lie) symmetries. The plan of this discussion is to describe the inverse process to dimensional reduction, hereafter called "group disintegration", and also to study the N-reduction. The starting point will be supergravity N = 8 (presumably equivalent to N = 7) in 3 dimensions, (note that it has 16 2 component-spinor charges). Finally we shall present conjectures for the structure of the 11 dimensional model and the 2 dimensional supergravities and sketch the generalization of the previous considerations to superalgebras. But our first task is to set the scene in 11 dimensions and to briefly review dimensional reduction.

## 2 DIMENSIONAL REDUCTION

### 2.1 The Action

As promised all fermionic fields are set equal to zero, the 11 dimensional Lagrangian is (Cremmer, Julia & Scherk, 1978) :

$$K^2 L_{11} = -\frac{1}{4} V R_{MN\cdot B}{}^{A} e^{M}{}_{A} e^{NB} - \frac{1}{48} V F_{MNPQ} F^{MNPQ} + \frac{2\,V}{(12)^4} \varepsilon^{M_1 \dots M_{11}} F_{M_1 \dots M_4} F_{M_5 \dots M_8} A_{M_9 M_{10} M_{11}} \tag{1}$$

with a parallelizable orientable manifold $\mathcal{M}_{11}$ on which we define a moving frame, that is 11 1-forms $e^A = e_M{}^A dx^M$ ; the Lorentz metric $\eta_{AB}=(+1,-1,\ldots,-1)$ induces a pseudo-Riemannian metric $g_{MN} = \eta_{AB}e_M{}^A e_N{}^B$ on $\mathcal{M}_{11}$ ; we are also given a 3-form $A = A_{MNP}\, dx^M{}_\Lambda dx^N{}_\Lambda dx^P$ and its exterior derivative $F = 4dA$ with its dual $^*F=\frac{1}{24}\varepsilon^{M_1..M_4}{}_{M_5\ldots M_{11}} F_{M_1..M_4} dx^{M_5}{}_{\Lambda..\Lambda} dx^{M_{11}}$ (here the Levi-Civita tensor is a true tensor) ; finally there is a 1-form $\omega^A{}_B = \omega_M{}^A{}_B dx^M$ taking its values in the Lie algebra of the Lorentz group SO (1,10) (preserving $\eta_{AB}$ ). The action can be rewritten covariantly

$$K^2 S_{11} = \frac{K^2}{11!}\int L_{11}\, d^{11}x = \int_{\mathcal{M}_{11}} -\frac{1}{2.7!} F_\Lambda {}^*F + \frac{2}{(12)^4} F_\Lambda F_\Lambda A - \frac{1}{8.9!} R^A{}_{B\,\Lambda} e^C{}_\Lambda \cdots{}_\Lambda e^K \varepsilon^B{}_{A.C\ldots K} \qquad (2)$$

$\varepsilon_{AB\ldots K}$ is the totally antisymmetric symbol, $e_M{}^A$ (resp. $e_A{}^M=(e_M{}^A)^{-1}$) is used for contracting "tangent" and "curved" indices (relative respectively to the bases $e^A$ and $dx^M$ ), $V = \det\,(e_M{}^A)$

$$R_{MN}{}^A{}_B = (\partial_M \omega_N{}^A{}_B + \omega_M{}^A{}_C \omega_N{}^C{}_B) - (M \leftrightarrow N)$$

$$R^A{}_B = R_{MN}{}^A{}_B dx^M{}_\Lambda dx^N = 2(d\omega^A{}_B + \omega^A{}_{C\,\Lambda}\omega^C{}_B) \qquad (3)$$

The vector fields $e_A = e_A{}^M \partial/\partial x^M$ define an anholonomic (orthonormal) basis and one defines the anholonomity as $[e_A, e_B] = \Omega_{AB}{}^C e_C$ , the coefficients $\Omega_{AB}{}^C$ or rather $\Omega_{MN}{}^C$ have a simple expression : $\Omega^C = -\frac{1}{2}\Omega_{MN}{}^C dx^M{}_\Lambda dx^N = de^C$ . The Einstein-Hilbert action can be simply rewritten (up to a surface term that we shall <u>neglect as well as other global topological effects</u>) in terms of $\Omega^C$ if we solve the (algebraic) equation of motion of $\omega^A{}_B$ ($de^A + \omega^A{}_{B\Lambda} e^B = 0$ : no torsion)

$$\int -\frac{1}{4}\, RVdx = \frac{1}{16}\int V[\Omega_{AB}{}^C \Omega^{AB}{}_C + 2\Omega_{AB}{}^C \Omega^A{}_C{}^B - 4\Omega_{AB}{}^B \Omega^A{}_C{}^C]dx \qquad (4)$$

in invariant form the last term in (2) reads :

$$K^2 S_{\text{Einstein}} = \frac{1}{12.8!}\int_{\mathcal{M}_{11}} X^{AB}{}_\Lambda {}^*X^{CD}(2\eta_{AD}\,\eta_{BC} - \eta_{AB}\,\eta_{CD}) \qquad (5)$$

where we introduced the 3-forms $X^{AB} = 3\Omega^A{}_\Lambda e^B$ and their duals

$${}^*X^{AB} = \frac{1}{6}\,\varepsilon^{M_1 M_2 M_3}{}_{M_4\ldots M_{11}}\, X^{AB}{}_{M_1 M_2 M_3}\, dx^{M_4}{}_{\Lambda..\Lambda} dx^{M_{11}}$$

such a formula exists in every dimension higher than 2 (3-forms must exist) ; one recalls that Einstein's action is a total divergence in 2 dimensions : $S_E \simeq 0$ .

## 2.2 Killing Vectors

The simplest dimensional reduction (à la Kaluza-Klein) corresponds to the existence of (11-d) commuting space-like Killing vectors $K^M_m \partial/\partial x^M$ (a generalization is discussed by Scherk & Schwarz, 1979). The action (2) in 11 dimensions is by construction coordinate reparametrization invariant but also "locally" Lorentz invariant, (this invariance is hidden in (5) because a change $\delta e^A = \Lambda^A{}_B(x^M)\, e^B$ changes the anholonomy 2-forms inhomogeneously : $\delta\Omega^A = d\Lambda^A{}_B \wedge e^B + \Lambda^A{}_B \wedge \Omega^B$ ).(An invariance is called local if the parameter ( $\Lambda^A{}_B$ say) depends arbitrarily upon the space time point). The manifold $\mathcal{M}_{11}$ acquires a trivial principal fibre bundle structure with fibre the torus $T_{11-d}$ and a d- dimensional basis $\mathcal{M}_d$. Solutions in 11 dimensions with the above symmetry can be obtained from a d-dimensional variational principle; the reduction of (2) to d dimensions by symmetry yields upon variation equations of motion that are necessary conditions for the existence of symmetric solutions in eleven dimensions; they are also sufficient because the symmetry is compatible with the time evolution in 11 dimensions (the equations of motion are hyperbolic).

The action in d dimensions admits an invariance group ($SL(11-d)_{global} \times SO(11-d)_{local}$) if the original action is invariant under general coordinate transformations and if the fields are assumed to have vanishing Lie derivative with respect to the Killing vectors : let us combine these two symmetries, the action $S_{11}$ is invariant under the following infinitesimal changes of fields ( $f^m$ are arbitrary functions, M N... are "curved" indices in 11 dimensions) :

$$T^{MN\cdot\cdot}{}_{PQ\cdot\cdot} \to T'_{(x)} = T_{(x)} + \delta T_{(x)}$$

$$\delta T^{MN\cdot\cdot}{}_{PQ\cdot\cdot} = \partial_p f^m K_m{}^{P'} T^{MN\cdot\cdot}{}_{P'Q\cdot\cdot} - \partial_{M'} f^m K_m{}^{M} T^{M'N\cdot\cdot}{}_{PQ\cdot\cdot} + \text{analogs} \quad (6)$$

In order to obtain symmetries in d-dimensions, two conditions must be satisfied.

2.2.1. The Killing vectors must be preserved by (6), namely for $K_m{}^P = \delta_m{}^P$ ($x^m$ cyclic coordinate) one must require $\partial_m(\partial_p f^n) = 0$ or :

$$f^n = a^n{}_m x^m + g^n(x^\mu) \quad (7)$$

$$(\mu = 1..d)$$

2.2.2 The coordinate transformation we have effected (with parameters $x^M - x'^M = \xi^M = f^m(x)\ K_m{}^M$ ) must preserve the internal volume of the torus (whose global structure is irrelevant apart from this). This condition restricts the invariance group to be SL (11-d) plus (11-d) abelian gauge invariances with gauge fields $A_\mu{}^m$ plus the remnant of the local Lorentz invariance SO(1,10);SO(11-d), using the notations of (Cremmer & Julia, 1979 referred to later as (C.J.))

$$\delta e_\mu{}^a = e_i{}^a \partial_\mu f^i \qquad \mu = 1,\dots,d \quad \text{curved}$$

$$a(i) = d+1,\dots,11 \quad \text{flat (curved) "internal"}$$

suggests $(e_d)^i{}_a = (e_i{}^a)^{-1}$ (with internal indices only)

and $A_\mu{}^i = (e_d)^i{}_a\, e_\mu{}^a$ (8)

In fact the previous invariance group can be considerably enlarged. The first addition is a scale transformation when $d = 4$ (but not for $d = 5$ ); in four dimensions the equations of motion have been shown to have an invariance group $E_7$ (+7) global times $SU(8)_{local}$ which commute with the space-time transformations. In the case of SL(7,ℝ) one sees that a subgroup of the group of diffeomorphisms (changes of coordinates) in 11 dimensions became a group of invariances when combined with the Killing symmetries. It is clearly unsatisfactory to have to go through complicated and unsystematic computations to find such a large symmetry as $E_7$ , and to end up being convinced that there is more symmetry to be discovered. This conviction was formulated as a conjecture in (C.J.) ; because the action for the spin 0 fields takes the very simple form

$$\frac{1}{4} \det(e_\mu{}^\alpha) g^{\mu\nu} \mathrm{Tr}\left[(g^{-1}\partial_\mu g - h_\mu)(g^{-1}\partial_\nu g - h_\nu)\right] \qquad (9)$$

with $g(x^\mu)$ an arbitrary element of $E_7$ (+7) and $h_\mu dx^\mu$ a 1-form taking its values in the maximal (compact) subalgebra SU(8) (in the representation 28 ℂ) ; one is led to conjecture an analogous structure <u>for the complete set</u> of fields (including fermions) : $E_7$ would be replaced by a larger group, a super-group (Berezin & Katz, 1970 ; Kac, 1977) mixing in a non trivial way space-time symmetries and internal symmetries. We even expect the subgroup of transformations that preserve separately the bosonic sector (discussed here) to include the Lorentz group, the "gauge" group SU(8) should be replaced by a larger group containing probably the Lorentz-group as well and the total action should be analogous to (9). We shall discover a group that unifies spin and internal symmetry. It is also important to simplify the theory in its symmetric form (C.J.) because more realistic versions with symmetry breaking are even more cumbersome. In particular compaction on a twisted torus (Scherk & Schwarz, 1979) or on a sphere (Nahm, 1978a) would benefit from a better understanding of the group theory.

## 2.3 The State of the Art

Let us now recall the (partially) proven internal symmetries of the $N=7$ (or 8) supergravity theories between 3 and 10 dimensions : they are listed in the first column of Table 1 (C.J.; Cremmer 1980a). A general pattern emerges : in dimension d the global invariance group is $E_{11-d}$ in its "normal" real form with Cartan character $\chi = (11-d) = r$ equal to the rank. The normal form is the maximally non compact form. A canonical real basis has $(11-d)$ commuting (non-compact) generators (Cartan maximal abelian subalgebra) with a diagonal action on the other generators ; $g-(11-d)$ (where g is the dimension of the group) is always even and if we denote the commuting generators by $C_a$ and the others by $G_{\pm R}$ (corresponding to positive and negative "roots" respectively), one has :

$$\begin{cases} [C_a, C_b] = 0 \quad (a,b = 1,\dots,r) \\ [C_a, G_{\pm} R] = \pm\, (C_R, C_a)\, G_{\pm R} \\ [G_{\pm R}, G_{\pm R'}] = N_{\pm R, \pm R'} G_{\pm R \pm R'} \quad \text{if } R \neq R' \;,\; [G_R, G_{-R}] = C_R \end{cases} \qquad (10)$$

The generators $(G_{+R} - G_{-R})$ form the maximal compact subalgebra $(N_{R,R'} = -N_{-R,-R'})$. The real numbers $(C_R, C_a)$ are the coordinates of $C_R$ a linear form (a root) on the Cartan subalgebra $\mathcal{C}$ ; but the Killing form is diagonal on $\mathcal{C}$ and can be set equal to $(+1)_{ab}$ ; it allows an identification of $\mathcal{C}$ and its dual $\mathcal{C}^*$ (the real linear forms on $\mathcal{C}$ ): $(C_R, C_a)$ can be viewed as the Killing product of two elements of $\mathcal{C}$ $(C_R \in \mathcal{C})$ . Finally Dynkin (1962) has shown that one can choose r "simple positive roots" such that all the roots be linear combinations of them with integer coefficients of the same sign ; they must satisfy certain constraints but once they do, and given their scalar products (or the Cartan matrix of elements $C_{ab}$ $\frac{2(R_a, R_b)}{(R_a, R_b)}$ ), one can reconstruct the full Lie algebra. Cremmer (1980a) noticed that the rank increases with the dimension of the internal torus ; more recently B. Morel and J. Thierry-Mieg (1980) proposed a crystallization of space-time from a general lattice in 11 dimensions to the product of a general lattice in d dimensions times a crystal : the rigid lattice generated by the simple roots which indeed has dimension $(11-d)$ . We note that this discretization of the whole problem is not necessary, the group generated by the $(11-d)$ $C_a$'s of the Cartan subalgebra is a maximal torus in the group $E_{11-d}$ that reminds us of the internal torus of dimensional reduction. Most of our efforts will be devoted to trying to find further explicit examples of this symmetry between space-time and group manifold. The remarks of the previous workers can be generalized by :

Rule 1 : $r+d=4+N$ where r is the rank of the global group of Table 1 for given N and d. Another proposal has been put forward by P. Cvitanovic (1979) who noticed the similarity of Table 1 to a "magic" triangle generalizing the magic square of Tits & Freudenthal (1964). However the analogy is not complete. Let us now proceed to new results.

## 2.4 The $d \leftrightarrow (13 - d)$ Symmetry

The key observation now is that the number of degrees of freedom does not change by dimensional reduction. What is meant here by degrees of freedom is the set of normal modes of the linearized equations of motion (neglecting cubic and higher interactions in the Lagrangian). Considering the Hamiltonian evolution problem, one notices that the initial data must be given on a space-like surface of variable dimension $(d - 1)$ or on a partially light-like surface of dimension $(d - 2)$; however, the fields to be specified have the same tensor character in any dimension. This is most striking when one fixes a light-like momentum $P_M = (1,1,0...0)$ $(x' = t)$ ; its little group contains what we shall call the helicity group $SO(d-2)$ (Nahm 1978b ; Ne'eman 1980) : it acts on the states of the linearized theory and for fixed $P_M$ there are 128 of them labelled by 2 tensors $A_{[\overset{\perp}{M}NP]}$ and $g_{(M \overset{\perp}{} N)}$ (traceless); the only change when $d$ varies is the separation of the indices (taking 9 values) into space-time valued indices ( $d$ values) and internal indices ( $(11 - d)$ values). In particular, one notices that the exchange of $d$ and $d' = 13 - d$ exchanges the roles of both types of indices. For example, the theory in 3 dimensions is thought to have internal symmetry $E_8(+8)$ , consequently the model in 10 dimensions (namely the small slope limit of a closed string dual model) may have a space-time symmetry including the same group $E_8$ ; in fact the fields parametrize the coset space $E_8(+8)/SO(16)$ in an $SL(8,\mathbb{R})$ covariant way (for $d = 11$ it is even $SL(9,\mathbb{R})$ covariant , Freudenthal (1953)). The transformations of $E_8$ act naturally in 11 dimensions ; it remains to be seen whether $E_{d-2}$ is really a space-time symmetry in $d$ dimensions. At this point, we recall that the field $A_{\overset{\perp}{M}NP}$ $(M,N,P = 1,...,8)$ comes from the fermion-fermion sector of the closed string dual model in 10 dimensions (Cremmer et al. 1978 ; Nahm 1978b) and is closely connected with the supersymmetry, which consequently must play a crucial role behind the scene.

We shall now focus on the space of degrees of freedom (128) functions (initial data) and study its symmetries in various dimensions. Although the most complete analysis has been carried out in 4 dimensions, it is easy to guess its generalization to $d = 3$ (where the model is almost unified) : the Lagrangian there must be given by (9) but with $g$ in $E_8(+8)$ and $h_\mu$ representing the Lie algebra of $SO(16)$. We shall thus start from $d = 3$. It is important, however, to remember that if one goes beyond the linearized level (trilinear and higher terms in the action) one must keep the Einstein action (4) or (5) as well as (9).

## 3. GROUP DISINTEGRATION OF THE NORMAL "ISOTOPES" : $E_{n(+n)}$

In Table 1, one can find the dimensions of the internal non-compact invariance groups and of their maximal compact subgroups

$H_d$ ; if one identifies them with mass and proton numbers of nuclei respectively, the symmetric spaces correspond to highly radioactive nuclei (Europium ( Z = 63) has mass ∿ 152 and not 133 ! hence the name "disintegration"). One must work hard to find the $E_8$ symmetry after dimensional reduction. Now we study the converse operation : group disintegration ; there, one must work hard to restore space-time symmetry. Before going from 3 to 4 dimensions for N = 7, let us do it for N = 0.

## 3.1 From 3 to 4 Dimensions (N = 0)

The dimensional reduction has been done by Gibbons & Hawking (1979) in Euclidean space but can be readily transposed to the Minkowski signature : Einstein's action becomes, in 3 dimensions,

$$- \frac{1}{4} e_3 R_3 - \frac{1}{16} e_3 \Delta^2 g_3{}^{\mu\rho} g_3{}^{\nu\sigma} F_{\mu\nu} F_{\rho\sigma} + \frac{1}{8} e_3 g_3{}^{\mu\nu} \frac{\partial_\mu \Delta \partial_\nu \Delta}{\Delta^2}$$

with

$$e_{3\mu}{}^{\alpha} = e_{4\mu}{}^{\alpha}\sqrt{\Delta}, \quad e_{4M}{}^{A} = \begin{pmatrix} e_{4\mu}{}^{\alpha} & A_\mu{}^i e_i{}^a \\ o & e_m{}^a \end{pmatrix} = \begin{pmatrix} e_{4\mu}{}^{\alpha}, & \sqrt{\Delta} A_\mu \\ o \;, & \sqrt{\Delta} \end{pmatrix} \quad \mu,\alpha = 1,2,3 \text{ only} \tag{11}$$

If we now use a "duality transformation", the equations of motion can be rewritten with a more symmetrical appearance (of course their symmetry exists from the start at least for classical fields); the observation is that $\partial_\mu (\Delta^2 e g^{\mu\rho} g^{\nu\sigma} F_{\rho\sigma}) = 0$ can be solved locally by introducing a "magnetic" potential B such that

$$\partial_\mu B = \frac{1}{2} \varepsilon_\mu{}^{\nu\rho} F_{\nu\rho} \Delta^2 \tag{12}$$

Once again we deliberately ignore global effects as a first approximation. One can also find a symmetrical Lagrangian (for the general procedure see (C.J.)), it reads :

$$\tilde{L} = \frac{e}{8} g^{\mu\nu} \left( \frac{\partial_\mu B \partial_\nu B + \partial_\mu \Delta \partial_\nu \Delta}{\Delta^2} \right) = - \frac{e}{2} g^{\mu\nu} \frac{\partial_\mu Z \partial_\nu \bar{Z}}{(Z - \bar{Z})^2} \tag{13}$$

where we defined $Z = B + i\Delta$ and we left aside the curvature term $R_3$ that does not represent any degree of freedom in 3 dimensions. This is nothing but the Lagrangian for the $\sigma$ model $SL(2,\mathbb{R})/SO(2)$ ; the transformations of $SL(2,\mathbb{R})$ are given by

$$Z = \frac{a Z + b}{c Z + d} \qquad ad - bc = 1 \qquad (a,b,c,d \text{ real numbers}) \tag{14}$$

The maximal compact subgroup SO(2) can be chosen as the isotropy group of Z = i namely the transformations $\begin{pmatrix} a & b \\ c & d \end{pmatrix} = \begin{pmatrix} \cos\theta & -\sin\theta \\ \sin\theta & \cos\theta \end{pmatrix} = e^{\theta T_1}$; in the linearized theory $(\Delta = 1 + \varepsilon)$ their action is

$$(B'+i\varepsilon') = \frac{ie^{i\theta} + \cos\theta(B+i\varepsilon)}{e^{i\theta} + \sin\theta(B+i\varepsilon)} - i \simeq e^{-2i\theta}(B + i\varepsilon) \tag{15}$$

These are Ehlers transforms (Gibbons & Hawking, 1979) ; for N = 7 these transformations become the SO(16) subgroup of $E_8(+8)$ and their interpretation will be analogous. The non-compact generators of SL(2 $\mathbb{R}$) are $T_2=\begin{pmatrix}0 & 1\\ 1 & 0\end{pmatrix}$ and $T_3 =\begin{pmatrix}1 & 0\\ 0 & -1\end{pmatrix}$ , they correspond to the changes $Z'=Z + \lambda(1 - Z^2)$ and $Z' = Z(1 + 2\lambda)$ ; the second is a dilation of the fields $A_\mu$ and $\Delta$ with inverse weights (not quite what one obtains from a rescaling of the fourth coordinate). An analogous invariance appears in $E_7(+7)$ and $E_8(+8)$ (for N = 7).

By applying a group disintegration to $SL(2,\mathbb{R})/SO(2)$ one would like to go from three to four dimensions and the local complex structure will just allow us to do so. We shall restrict ourselves to solutions of the linearized equations of motion. The first problem is then to make the spin appear (the "helicity" group SO(d - 2), it may not be an easy task considering the duality transformations that have been effected. In four dimensions B and $\Delta$ must parametrize a symmetric traceless 2 x 2 matrix $(g_{(M\ N)}^{\perp})$ with two transverse dimensions) (Weinberg, 1972) in the harmonic gauge $\partial_M(g^{MN}\sqrt{-\det g_{PQ}})=0$. The harmonic condition splits into two equations as can be readily checked

$$\begin{cases}\partial_\mu\left(g_3^{\mu\nu}\sqrt{|g_3|}\right) = 0\\ g_3^{\mu\nu}\,\partial_\mu A_\nu = 0\end{cases} \qquad \text{because} \tag{16}$$

$$g^{MN} = \begin{pmatrix} g_3^{\mu\nu}\Delta\ , & - A^\nu\\ -A^\mu\ , & -\frac{1}{\Delta} + \Delta A_\rho A_\sigma g_3^{\rho\sigma}\end{pmatrix} \tag{17}$$

Consequently, the polarization tensor $\varepsilon^{\perp}_{MN} = g^{\perp}_{MN} - \eta^{\perp}_{MN}$ (the momentum P is taken lightlike in the (1,2) plane) is canonically (for the linearized theory) :

$$\varepsilon_{\perp}^{MN} = \begin{pmatrix} -\varepsilon & -A^3\\ -A^3 & \varepsilon\end{pmatrix} \tag{18}$$

Using (12) we find $B \simeq -A_3$ (the sign convention is arbitrary) and for the circular polarizations

$$\varepsilon \underset{(-)}{+} iB = \overset{(-)}{z} = \underset{(+)}{-} i\ \overline{(Z - i)} \tag{19}$$

We recognize (with hindsight) the SO(2) subgroup of SL(2, $\mathbb{R}$) as the helicity group in four dimensions : it rotates $z$ and $\bar{z}$ in opposite directions (15).

The theory in 3 dimensions describes the simplest kinetic term for 2 scalar fields $\Delta$ and B which are coordinates on the homogeneous space SL(2,$\mathbb{R}$)/SO(2) ; more precisely an SL(2 $\mathbb{R}$) matrix can be written as $g =\begin{pmatrix}x+y & u+v\\ u-v & x-y\end{pmatrix}$ provided $x^2-y^2-u^2+v^2 = 1$ . Among these the symmetric matrices of signature (++) and determinant

1 describe the coset space $SL(2,\mathbb{R})/SO(2)$, they are parametrized by the $x > 0$ sheet of the hyperboloid $x^2 - y^2 - u^2 = 1$. Actually we would like to express g in terms of $\Delta$ and B rather than the constrained variables $(x,y,u)$ ; we know that the action (9) can be rewritten (C.J.) :

$$S \sim \frac{e}{4} g^{\mu\nu} \mathrm{Tr} \left[ (g^{-1}\partial_\mu g)^\perp \, (g^{-1}\partial_\nu g)^\perp \right] \tag{20}$$

where $\perp$ means the (orthogonal) projection of the Lie algebra valued form $g^{-1}\partial_\mu g dx^\mu$ onto $T_2$ and $T_3$ along the compact direction $T_1$ ; but B appears polynomially in (13); it is natural to conjecture that it does so both in g and $g^{-1}$ and we guess the following solution $(\Delta > 0)$

$$g = \begin{pmatrix} \sqrt{\Delta} & B/\sqrt{\Delta} \\ 0 & 1/\sqrt{\Delta} \end{pmatrix} \tag{21}$$

We conclude that $\varepsilon = \Delta - 1$ is a coordinate along $T_3$, and B (for small B ) is a coordinate along the nilpotent generator $T_2 - T_1$. One can find (21) in several ways : by stereographic projection, or equivalently by introducing $Z = \frac{x_1 + i\, y_1}{x_2 + i\, y_2}$ and considering the matrix $g' = \begin{pmatrix} y_1 & x_1 \\ y_2 & x_2 \end{pmatrix}$ for $y_2 = 0$.

We proposed that the group of Ehlers transformations be what is left of the spin, and we proved it for the linearized solutions of the equations of motion. Let us translate this into equations for the Lie algebra of $SL(2,\mathbb{R})$, using $C_1 = -\frac{1}{\sqrt{2}} T_3$ as a Cartan basis, one finds two roots + and − :

$$G_\pm = \frac{T_2 \pm T_1}{2}$$

($(G_+ - G_-)$ is the compact generator $T_1$ of SO(2))

with

$$[C, G\pm] = \pm\sqrt{2}G\pm \quad , \quad [G_+ , G_-] = \sqrt{2}C_1 \tag{22}$$

The adjoint action of the spin SO(2) on $SL(2,\mathbb{R})$ is :

$$\frac{1}{2} \left[T_1, \frac{1}{\sqrt{2}} C_1\right] = - G_- - \left(\frac{1}{2}T_1\right) \tag{23}$$

$$\frac{1}{2} \left[T_1, G_-\right] = \frac{1}{\sqrt{2}} C_1 \tag{24}$$

The last term in (23) is an artefact of the triangular "gauge fixing" of (21) (see C.J.). Let r be a spin rotation, we are looking at the action of r on the left of g , but $g' = rg$ is defined up to a right multiplication by another element r' of SO(2), so we might as well consider $g'' = rgr^{-1}$ ; the adjoint action of r on g preserves a symmetric "gauge fixing" but not the triangular (Iwasawa) choice of coset representative made in (21). In the neighbourhood of the unit matrix $g \simeq 1 + BG_- - \frac{1}{\sqrt{2}}\varepsilon C_1$ if one makes another gauge multiplication $r'' \sim e^{\theta T_1}$ on the right of g (23) results. Let us now proceed to the largest model known (N = 7) and start studying it in 3 dimensions rather than in 11 dimensions.

## 3.2 The Symmetric Space

We are at present trying to find the rules that would allow us to go up in dimension if possible, so we are going to practise using the results of dimensional reduction for N = 7 as we did for N = 0. First of all, we guessed that the N = 7, d = 4 theory leads to an action similar to (9) (or (20) once we have eliminated $h_\mu$ ) in 3 dimensions. The 70 scalar fields of d = 4 remain scalar ; the 28 vector fields (2 degrees of freedom each) formed the 28 $\mathbb{C}$ representation of SU(8) with the spin group SO(2) acting on each state of helicity + 1 or − 1 ; they become 56 scalars upon dimensional reduction, two more scalars come from the spin two states (as for N = 0) ; altogether they build the 128 dimensional representation of SO(16), namely a chiral spinor and it is reasonable to expect the Lagrangian to take the form (9) for the coset space $E_8(+8)/SO(16)$. We see immediately that conversely starting from 3 dimensions, one must decompose the various representations of SO(16) into representations of SU(8) x SO(2) in order to obtain the (d = 4) multiplets of states. This is done in Table 2 : the decompositions of SO(16) and of $E_8/SO(16)$ are shown for each dimension.

Rule 2 : The "spin" group SO(d−2 ) times the maximal compact subgroup $H_d$ of $E_{11-d}(11-d)$ classifies the states for all dimensions.

Several comments are in order. It was obvious from dimensional reduction that one ought to obtain SO(11−d)xSO(8) the local internal subgroup of the Lorentz group SO(1,10) as a subgroup of $H_d$ ($H_3$) ; Rule 2 implies that actually SO(16) contains SO(9):helicity (d=11). In table 2 there are 2 lines for each dimension d, the top line shows how the various multiplets of dimension (d−1) have split and the second line how d=3 multiplets have split under $H_d$ x SO(d−2). One notices that space-time indices appear usually one at a time : scalars become vectors in the next dimension ; exceptions are due to dualities (see also subsection 3.3).

From left to right the multiplets are labelled as representations R x R' of $H_d$ x SO(d−2) and have the tensor character of the graviton $g^{\perp}_{\mu\nu}$ (symmetric traceless), vectors, tensors $A^{\perp}_{\mu\nu}$ for $d \geqslant 6$, $\mu\nu .. = 3,4...,d$ , tensors $A^{\perp}_{\mu\nu\rho}$ ($d \geqslant 8$) and scalars forming the coset space $E_{11-d}/H_d$ ; these are the physical states but we listed also the auxiliary states on the right. For d = 3 they are just the 120 components of the subgroup SO(16), but one may consider them in higher dimensions as well, in particular if one tries to restore manifest $E_8$ symmetry ; they are listed also as representations of Hd x SO(d−2) from left to right, scalars (Hd itself), vectors $B^{\perp}_{\mu}$, $B^{\perp}_{\mu\nu}$ ($d \geqslant 6$), $B^{\perp}_{\mu\nu\rho}$ ($d \geqslant 8$) and finally SO(d−2) : the helicity group.

The auxiliary degrees of freedom are gauged away by auxiliary gauge fields of $H_3$ ($h_\mu$) namely SO(16) in 3 dimensions ; if one were now to generalize (9) to unify all fields it would be natural to keep SO(16) as an auxiliary local gauge group in higher dimensions : it certainly includes $H_d$ but also the helicity group $SO(d-2)$ which has already an associated (auxiliary) gauge field $\omega_\mu{}^{\alpha}{}_{\overset{\perp}{\beta}}$ . In fact for N = 1 the auxiliary field $A_\mu$ of off-shell supergravity has similar transformation laws as $\omega_\mu{}^3{}_4$ , this fact may be related to a previous work of Ferrara & Van Nieuwenhuizen (1979). One may generalize this to higher values of N and hopefully make some progress towards the long awaited construction of off-shell extended supergravities; it should be kept in mind however that duality transformations modify the auxiliary field structure (Siegel, 1979).

Finally we would like to remark that in the linearized theory the duality operation is just a relabelling of transverse indices. Indeed for fixed momentum $(P_1, P_2 = P_1, 0, ..0)$ we have :

$$\varepsilon^{+\mu\nu..\,-\rho\sigma...} P_+ A^{\perp}_{\mu\nu..} \sim P^- B^{\perp\rho\sigma...}$$

$$(\mu,\nu,\rho,\sigma \;... = 3,4,...d \;;\; P_\pm = P_1 \pm P_2) \qquad (25)$$

We used (25) in (19).

### 3.3 The Disintegration of $E_8$

We discussed the disintegration of the coset space $E_8/SO(16)$ and of SO(16) emphasizing in d-dimensions the subgroup $H_d \times SO(d-2)$ of SO(16); in fact we can exhibit an analogous disintegration of $E_8$ considered as a symmetry group now (C.J. ; Nahm unpublished ; Morel & Thierry-Mieg, 1980). Going up from dimension d to dimension (d+1), one reduces the rank of the internal symmetry group by one in a regular fashion. It is useful to recall the structure of (10), where as indicated in Table 1 one takes the Cartan subalgebra in the non compact part of the real Lie algebra $E_8(+8)$. If we plot the Dynkin diagram (simple roots) of $E_8$ : 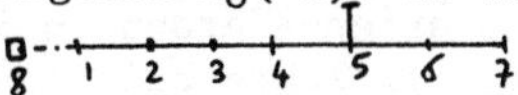

and take away the roots in the indicated order as well as the non simple roots they generate, we find the subgroup $E_{11-d}(11-d)$ ; it is a simple observation now to find :

Rule 3 : This operation is actually a decomposition of $E_8$ into the product of $E_{11-d}$ and $SL(d-2, \mathbb{R})$ ; the latter subgroup contains, as a maximal compact subgroup, the preceding SO(d-2) ("helicity" group). We can also verify that the graviton states for all dimensions are singlets of $H_d$ and transform under $SL(d-2, \mathbb{R})$ as the coset space $SL(d-2,\mathbb{R})/SO(d-2)$ as one would expect.

One can understand why in Table 1 one scalar degree of freedom in the Cartan subalgebra in d dimensions becomes a component of the graviton namely $g_{d+1,d+1}$ in d+1 dimensions : it corresponds to the simple root of $E_{11-d}$ that is given away to $SL((d+1)-2)$.

For N = 0, $E_8$ reduces to SL(2,ℝ). We finally note that SL(9,ℝ) is a maximal subgroup of $E_8(+8)$ which becomes apparent if one introduces the extended Dynkin diagram and the most negative root labelled 8 above (see for example Helgason, 1978).

It is already known that $E_{11-d}$ is a symmetry group of the equations of motion in d dimensions that commutes with the group of coordinate changes. $E_8$ is a symmetry group in 3 dimensions, what is its SL(d-2) subgroup ? The answer is that SL(d-2) is also a symmetry group in d-dimensions, it contains the helicity group SO(d-2) as a subgroup and must be identified with the space-time symmetry group consisting of global affine transformations of the coordinates on $\mathcal{M}_d$. More precisely one considers (in the linearized approximation) only (d-2) coordinates on which the fields do not depend. Starting from $E_8$ in 3 dimensions we have shown that beyond the previously known $E_{11-d} \times H_{d \cdot \text{local}}$ invariance of the theory in d-dimensions, one should also view the symmetry group SL(d-2, ℝ) of linear (transverse) coordinate transformations (perpendicular to a 2-plane containing the wave vector $P_M$ ) as a subgroup of $E_8$. This suggests again that $E_8$ could be a symmetry of the full (non linear) theory in all dimensions.

An objection to this proposal comes from the famous no go theorems. Crudely speaking they forbid symmetry groups that would include the Poincaré group except as a direct factor (under some general assumptions). If $E_8$ or SO(16) were to be such symmetry groups (containing the helicity group SO(d-2)) one would have to avoid these theorems. For example the symmetry could be spontaneously broken so as not to be a symmetry of the one particle states (if the S-matrix is non trivial) ; it would mix states with different numbers of particles. We can see the generators that mix space-time and internal indices in the decomposition of SO(16) into representations of SU(8) x SO(2) (for d = 4) ; we recall that SO(16) is manifestly a local invariance group in 3 dimensions because of the 120 gauge fields of (9). These become $\Omega_\mu{}^A{}_B$ the SU(8) connection in 4 dimensions (C.J.), $\omega_\mu{}^3{}_4$ the "helicity connection" and 56 more vectors one should introduce also to simplify the 4 dimensional theory ; their spin and SU(8) character are given in Table 1: they are of the type $\Omega_\mu^{[AB]}{}_{,3 \pm i4}$ where the indices A, B are indices for the fundamental representation of SU(8) and ($3 \pm i4$) are helicity states. A fully covariant description would require gauge fields $\Omega_{\mu\nu}^{[AB]}$ ; they clearly exchange a pair of indices of SU(8) with a space-time index $\nu$ . They may be needed as auxiliary fields for the closure of the supersymmetry algebra off-shell. The fields in 11 dimensions that would implement a local SO(16) invariance are again the connection for the helicity subgroup SO(9) plus $B_\mu[\nu\rho\sigma]$ , a gauge field separately in the $\mu$ index and in the antisymmetric combination $[\nu\rho\sigma]$ the transverse components of which form the representation 84 of SO(9) (Table 1).

### 3.4 More Symmetries ? Alternative Symmetries ?

The previous discussion cannot apply to the full non linear theory as it stands. The little group is introduced for each Fourier component; one needs a non-linear generalization. The first idea is to reduce the theory to two (or even one) dimension(s) so as to enlarge $E_8$ ; the group $E_9$ is an infinite Lie group, its maximal torus would include the 9 transverse dimensions of the Kaluza torus. One would like also to simplify the spontaneously broken versions of extended supergravities ; the invariance group is apparently smaller there but the above analogy between the Kaluza torus and the maximal torus of the group suggests another invariance when the torus is replaced by a twisted torus (called flat group by Scherk & Schwarz, 1979). Finally it may be possible to find a group of rank 11 that would really put all space-time dimensions on the same footing.

## 4 N-REDUCTION

### 4.1 The Construction of Table 1

For d = 4 the analogs of $E_7(+7)$ global and SU(8) local have been determined in (C.J.) for all values of $N \leqslant 6$. The states are known from the canonical construction (Scherk 1978), they form representations of the groups U(N) for $N \leqslant 6$. In particular, the number of scalars is known, and it permits the discovery of the non compact subgroups of $E_7(+7)$ that have the groups U(N) as maximal compact subgroups (Table 1). One is disintegrating $E_8/SO(16)$according to representations of U(N) ; besides the supergravity (O(N)) multiplet, one finds various matter multiplets; SO(16) contains the subgroups U(N) x SU(8-N) x SO(2) that classify the matter multiplets conveniently. Keeping only the supergravity multiplet, we immediately notice several features of the scalar homogeneous spaces. Firstly, they involve non compact groups (not in their normal form) ; the maximal compact subgroups have rank equal to the rank of the group (this ceases to be true for $d \geqslant 5$). Secondly, the characters vary apparently randomly : from N = 7 to N = 1, we encounter : +7, - 6, -15, -14, - 9, - 1. Thirdly, they have a Kähler structure compatible with SU(N) symmetry (for $N \leqslant 7$) and with the non compact group G (for $N \leqslant 6$). For d = 5, Cremmer (1980b) discussed analogs of the $E_6/USp(8)$ theory; we obtain only the N = 6 model : $SU^*(6)/USp(6)$ by disintegration of the N = 8 model, the others involve matter multiplets (with the possible exception of an N = 5 model, see below).

Again, we can guess that the theories simplify in three dimensions. Starting from $E_8/SO(16)$ the coset space of dimension 128, we would like to obtain a coset space of dimension 64 (another power of 2, that is another bosonic spinor) for N = 6, and a coset space of dimension $2^N$ for generic N. At this point, we notice the resemblance of our embryonic Table 1 to the magic square which is in particular symmetric with respect to the diagonal (Freudenthal 1964) ; the real form can change : SU*(6) becomes SU(5,1) but d = 9 N = 7 looks quite similar to d = 3 N = 1. It is now a simple exercise to fill in the line d = 3 with nested subgroups of $E_8(+8)$ of the appropriate character and containing the d = 4 (non-compact) groups times SL(2,ℝ) with the same N. In 3 dimensions, the number of supersymmetry generators is 2N and they form the fundamental representation of SO(2N), actually the local symmetry group is slightly larger : SO(2N) x SO(2) for N ⩽ 5, SO(2) x SO(3) for N = 6, and SO(16) for the "self-conjugate" N = 8 model.

## 4.2 Ranks and Characters

Let us now generalize the discussion of subsection 2.3. We must now disintegrate $E_7(-5)$ rather than $E_8(+8)$. The symmetric space $E_7(-5)/SO(12) \times SO(3)$ has rank four (this result is due to E. Cartan, see for example Helgason (1978)) : this means that the maximal abelian subalgebra of $E_7(-5)$ formed of non compact generators has dimension four : the rank is also the dimension of the maximal torus (flat totally geodesic manifold). This rank (r') decreases by one under disintegration as in the normal case (N = 7, r' = r) and increases by one under dimensional reduction. By analogy, we can consider the Dynkin diagram of $E_7(-5)$ as 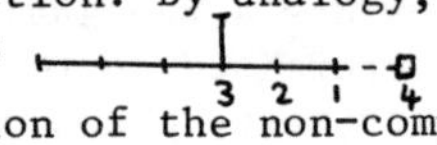
where we exhibited the order of disintegration of the non-compact roots including the smallest root (4) . Although this has not been checked, it seems reasonable because the multiplets of states can be obtained from a scalarless ("N = 6") supergravity in seven dimensions with 3 gauge fields $A_{[MN]}$ , 4 vectors and the graviton ; the box d = 6 N = 6 can accommodate the coset space $\frac{SO(5,1) \times SO(3)}{USp(4) \times SO(3)}$
We note also that the character and the rank of G(r) increase by one under dimensional reduction as well ; for N ⩽ 6, they differ from the rank r'.

The same analogy applied to the case N = 5 suggests the disintegration diagram 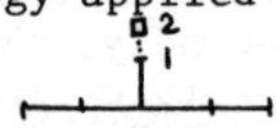 and the existence of a theory in 5 dimensions with one graviton and nine vectors. For N = 4 the diagram might be 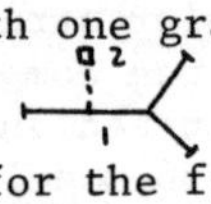 . However there is no simple candidate in 5 dimensions for the fields of pure N = 4 supergravity ; it is known that one can start from 10 dimensions (Scherk, 1978) or 5 dimensions (Cremmer, 1980b) if one adds some matter multiplets. The N = 4 theory is actually even more anomalous; contrary to all the other models going from d = 3 to d = 4 one does not find the factor group SL(2,ℝ) in the tables, yet the graviton is still given by SL(2,ℝ)/SO(2). We hope to return to this problem in the future. Naturally we can disintegrate $E_7/SO(12) \times SO(3)$ for fixed N as we

Table 1 : The Disintegration Triangle

Notation: $r$ = rank $G$; $G/H$ with $\dim G$ (upper) and $\dim H$ (lower); character $\chi$ (upper right); $G/H$ rank $r'$ (lower right).

Cells below are given as: $r$; $G/H$; $\dim G$ / $\dim H$; $\chi$; $r'$.

| $d \backslash N$ | 7 (128) | 6 (64) | 5 (32) | 4 (16) | 3 (8) | 2 (4) | 1 (2) | 0 (2) |
|---|---|---|---|---|---|---|---|---|
| 11 | $g_{MN}, A_{MNP}$ | | | | | | | |
| 10 | 1; $\mathbb{R}$; 1 / 0; 1; 1 | | | | | | | |
| 9 | 2; $SL(2)\times\mathbb{R} / SO(2)$; 4 / 1; 2; 2 | | | | | | | |
| 8 | 3; $SL(3)\times SL(2) / SO(3)\times SO(2)$; 11 / 4; 3; 3 | | | | | | | |
| 7 | 4; $SL(5) / SO(5)$; 24 / 10; 4; 4 | $g_{MN}, 3A_{MN}, 4A_M$ | | | | | | |
| 6 | 5; $SO(5,5) / SO(5)\times SO(5)$; 45 / 20; 5; 5 | 4; $SO(5,1)\times SO(3) / SO(5)\times SO(3)$; 18 / 13; −8; 1 | | | | | | |
| 5 | 6; $E_6 / USp(8)$; 78 / 36; 6; 6 | 5; $SU^*(6) / USp(6)$; 35 / 21; −7; 2 | $g_{MN}, 9A_M$ | ? | | | | |
| 4 | 7; $E_7 / SU(8)$; 133 / 63; 7; 7 | 6; $SO^*(12) / U(6)$; 66 / 36; −6; 3 | 5; $SU(5,1) / U(5)$; 35 / 25; −15; 1 | 4; $SU(4)\times SU(1,1) / SU(4)\times SO(2)$; 18 / 16; −14; 1 | 3; $U(3) / U(3)$; −9; 0 | 2; $U(2) / U(2)$; −4; 0 | 1; $U(1) / U(1)$; −1; 0 | $g_{MN}$ |
| 3 | 8; $E_8 / SO(16)$; 248 / 120; 8; 8 | 7; $E_7 / SO(12)\times SO(3)$; 133 / 69; −5; 4 | 6; $E_6 / SO(10)\times SO(2)$; 78 / 46; −14; 2 | 5; $SO(8,2) / SO(8)\times SO(2)$; 45 / 29; −13; 2 | 4; $SU(4,1) / SO(6)\times SO(2)$; 24 / 16; −8; 1 | 3; $SU(2,1)\times SU(2) / SO(4)\times SO(2)$; 11 / 7; −3; 1 | 2; $SU(1,1)\times SO(2) / SO(2)\times SO(2)$; 4 / 2; 0; 1 | 1; $SL(2) / SO(2)$; 3 / 1; 1; 1 |
| 2 |  $E_9 = E_8^{(1)}$ | 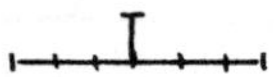 $E_7^{(1)}$ | 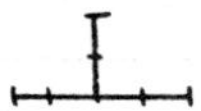 $E_6^{(1)}$ | 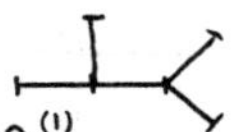 $D_5^{(1)}$ | | | | |

Table 2 : Representations of $H_d$ x SO(d-2)(N = 7)

| d | Physical degrees of freedom | | | Auxiliary fields | | | $H_d$ |
|---|---|---|---|---|---|---|---|
| 3 | (128 Scalars = 120 + 8) x 1 | | | 120 x 1 | | | SO(16) |
| 4 | $1_C$ | $28_{\mathbb{C}}$ (Vectors) | 63 + 7 | 63 | $28_{\mathbb{C}}$ | 1 | SU(8) |
| 5 | 2+2+1 | | 36 + 6 | 36 | 27 + 2 x 27 | 2+1 | USp(8) |
| 5 | 5 | | 42 | 36 | 27 x 3 | 1x3 | |
| 6 | 5+3+1 | | 20 + 5 | (2 x 10) | 16(1+3);(10x3)* | 3+3 | $SO(5)^2$ |
| 6 | 1 x 9 | | 25 x 1 | (2 x 10)x1 | 16 x 4 ; 5 x 6 | 1x(2x3) | |
| 7 | 9+4+1 | | 10 + 4 | 10 | 10(1+4);5(4+6) | 4 + 6 | SO(5) |
| 7 | 1 x 14 | | 14 x 1 | 10 x 1 | 10x5 ; 5 x 10 | 1 x 10 | |
| 8 | 14+5+1 | | 4 + 3 | 4 | 6(1+5);3(5+10);(2x10)* | 5 + 10 | SO(3) x SO(2) |
| 8 | 1 x 20 | | 7 x 1 | 4x1 | 6x6;3x15;1x20 | 1 x 15 | |
| 9 | 20+6+1 | | 1 + 2 | 1 | 3(1+6);2(6+15);20+15 | 6 + 15 | SO(2) |
| 9 | 1 x 27 | | 3 x 1 | 1 | 3x7;2x21;1x35 | 1 x 21 | |
| 10 | 27 + 7 + 1 | | 1 | 1+7;7+21;21+35 | | 7 + 21 | / |
| 10 | 1 x 35 | | 1 | 8 ; 28 ; 56 | | 1 x 28 | |
| 11 | 35 + 8 + 1 | | | 28 + 56 | | 8 + 28 | / |
| 11 | 1x44(graviton) | 84 $(A^{\perp}_{\mu\nu\rho})$ | | 84$(B^{\perp}_{\mu\nu\rho})$ | 1 x 36(helicity) | | |

Same as $2^{nd}$ Column (auxiliary)

did for $E_8/SO(16)$ in Table 2. One can follow the apparition of spin and the decrease of the ranks r and r'; the regular decrease of the character requires a more detailed investigation for some explanation.

## 5 CONCLUSION

The whole discussion has been about the linearized theory and we left aside general covariance ; for example in three dimensions, Einstein's action changes the interactions. As we said in subsection 2.4, a full geometrical understanding of the model requires treating the d space-time dimensions on the same footing. This is particularly important in two dimensions where we cannot apply a Weyl transformation to rewrite Einstein's action as a surface term. By extrapolation of $d \geqslant 3$ results we expect the Kac-Moody algebras $E_8^{(1)}, E_7^{(1)}, E_6^{(1)}, D_5^{(1)}$ to be symmetries of the two-dimensional models N = 7, 6, 5, 4 ; the extra-root in the Dynkin diagrams (Table 1) for d = 2 appears at the location of the smallest root of the extended Dynkin diagrams for d = 3 which coincides with the naïve extrapolation from higher dimensional reductions exhibiting the subgroup SL $(d-2, \mathbb{R})$. By the symmetry $d \leftrightarrow (13-d)$ these algebras ought to appear in the dual string models, and indeed they do. On the other hand they appear in the study of integrable systems in two dimensions. We have just begun to bridge the gap between 3 seemingly disjoint areas of physics ; to make progress it seems important to bring more mathematics (or/and mathematicians) into the game. It is easier to check an invariance after guessing it than to prove it from first principles ( $E_7$ is a good example (for d = 4 N = 7)) ; so we would like to find a group or a coset space of rank 11 that would provide the desired unification for N = 7 supergravities in any dimension.

The N-reduction is intimately connected with supersymmetry ; we observe experimentally some regularity : for example going from N = 7 to N = 6 in any (allowed) dimension factorizes a group SU(2) ; Rule 1 is also rather mysterious (d+r = 4+N). One may increase N beyond 7 ; for N = 8 one just doubles the number of states (to complete the magic triangle). Kac-Moody algebras ought to appear there too. For $N > 8$ one must introduce spins higher than two, in the interacting theory one may find even higher symmetries. At present the only interacting theories that include particles of higher spins are the string models. For N = 9 one may be inspired by d = 1 models ($N \leqslant 7$) because of the symmetry of Table 1 and one may obtain interacting models of some interest.

We also deliberately restricted ourselves to the conventional Lie symmetries, but we expect the superalgebras to clarify the structure of the theory. We conjectured a homogeneous space structure for the full theory (C.J. ; Julia, 1980a) in fact if we look at the 3-dimensional theory, there is a remarkable analogy between bosons and fermions : their number is $2^N$ each and if we look at

the Cartan decomposition of the noncompactalgebra $G_N$ ($E_8$, $E_7$...) we find a common structure : $G_N = H_N \oplus P_N$
where $P_N$ is a (bosonic) spinor of $H_N$, respectively Majorana-Weyl (N = 7), or Weyl ($N \lesssim 6$). The only non trivial commutation rules of $G_N$ are the commutators of two non compact generators of $P_N$, for N = 7 for example ($E_8 = G_7$) it is rather easy to find the commutator by covariance :

$$[\chi\alpha\ , \chi\beta]_- \simeq \Gamma^{[ab]\ L}_{[\alpha\beta]}\ H_{ab} \qquad \alpha,\beta = 1,..,128 \qquad (26)$$

where $\Gamma^{[ab]}_L = \frac{1-\Gamma_{17}}{4}[\Gamma^a,\Gamma^b]$ are the 120 generators of the SO(16) Lie algebra in the (left-handed) Majorana-Weyl representation (they are anti-symmetrical matrices in their indices $\alpha$ and $\beta$ ).
We note that we have a graded Lie algebra $G_N$ ((25) is a commutator). The fermionic extension will include a fermionic spinor $\psi_{\dot{\alpha}}$ whose anticommutator with itself must give a bosonic generator ; the supergroup disintegration must lead to the fermionic spectrum, so (for N = 7) the representation $\psi_{\dot{\alpha}}$ of SO(16) cannot be the same as $\chi_\alpha$ although it has the same dimension ; it must be the right-handed Majorana-Weyl representation. We shall return to this in a later publication but the generalization of the present program to superalgebras, at least at the linearized level, seems rather straightforward, it may explain some of the mysteries of the superfield expansions of Ne'eman, Morel and Thierry-Mieg (1980). In Table 1 all groups G had simple roots of equal length (single line diagrams) and they were nested in the so-called "regular" fashion. The supergroups are expected to have a similar property and will no doubt be discovered before the next Nuffield workshop. It would be interesting to find a general connection between non compact groups (or Cartan decompositions) and supergroups where the grading becomes the "super"-grading ; if one introduces Grassmann parameters, one may talk only about commutators and use results of abstract Lie group theory.

Finally we note that realistic supersymmetric theories are still missing but we hope that a group theoretical understanding of the fully interacting supergravity theories may lead to promising generalizations. Dual models seem important in that respect ; let us indeed remember that one must understand why the slope of Regge trajectories is the same for bosons and fermions. Once again it should be emphasized that we are hunting for interacting theories where one goes from the free theory to the interacting theory as one goes from a Lie algebra to a Lie group (see (9)); in dual models such a connection may exist between the spectrum generating algebras and the vertices (Del Giudice, Di Vecchia & Fubini, 1972).

This article is based on a talk given at Cambridge University, it is a pleasure to thank the D.A.M.T.P. for its congenial hospitality. Since then I benefited from conversations with E. Cremmer,

G. Gibbons, N. Hitchin, J. Schwarz, J. Thierry-Mieg and J. Tits. The talk included a presentation of the generalized Noether theorem, a preliminary account of which can be found in Julia (1980b).

REFERENCES

Berezin F.A. & Katz G.I. (1970). Lie Groups with Anticommuting Parameters. Math. USSR Sbornik 11 p. 311.

Cremmer E., Julia B. & Scherk J. (1978). Supergravity Theory in 11 Dimensions. Phys. Lett. 76B, p. 409.

Cremmer E., Julia B. (noted C.J.) (1979). The SO(8) Supergravity. Nucl. Phys. B159, p. 141.

Cremmer E. (1980a) in Proceedings of the Erice meeting : "Unification of the Fundamental Interactions".

Cremmer E. (1980b), these Proceedings.

Cvitanovic P. (1979) Nordita preprint 79/46 Supersymmetry, Negative Dimensions and the Emergence of $E_7$ Symmetry.

Del Giudice E., Di Vecchia P., Fubini S. (1972). General Properties of the Dual Resonance Model. Ann. of Phys., 70, p. 378.

Dynkin E.B. (1962). The Structure of Semi-Simple Lie Algebras. Translations of the Am. Math. Soc., 1 vol. 9, p. 328.

Ferrara S. & Van Nieuwenhuizen P. (1979). Simplifications of Einstein Supergravity. Phys. Rev., D20, p. 2079.

Freudenthal H. (1953), Sur le Groupe Exceptionnel $E_8$, Indagationes 15, p. 95

Freudenthal H. (1964). Lie Groups in the Foundations of Geometry, Adv. in Math., 1, p. 145

Gibbons G. & Hawking S. (1979). Classification of Gravitational Instanton Symmetries. Comm. in Math. Phys., 66, p. 303.

Helgason S. (1978), Differential Geometry Lie Groups and Symmetric Spaces, (Academic Press).

Julia B. (1980a), Lecture Notes in Physics, 116, (Springer).

Julia B. (1980b), Effective gauge fields (preprint LPTENS 80/14).

Kac V. (1977). Lie Superalgebras. Advances in Mathematics, 26, p.8.

Morel B., Thierry-Mieg J. (1980), Harvard preprint/A 038 : Groups and Supergroups in Exceptional Gravity.

Nahm W. (1978a). An Octonionic Generalization of Yang-Mills. CERN preprint (TH 2489) unpublished.

Nahm W. (1978b). Supersymmetries and their Representations. Nucl. Phys., B135, p. 149.

Ne'eman Y. (1980), noticed a supergroup SU(1/8) that classifies the states.

Salam A. Strathdee J. (1978). Supersymmetry and Superfields. Fortschritte der Physik, 26, p. 57-142.

Scherk J. (1978), Cargèse Lecture in Recent Developments in Gravitation (1979), Plenum P.C.

Scherk J., Schwarz J. (1979). How to Get Masses from Extra-Dimensions. Nucl. Phys., B153, p. 61 (and references therein).

Siegel W. (1979). Gauge Spinor Superfield as Scalar Multiplet. Phys. Lett., 85B, p. 333.

Tits, J. & Freudenthal, H. (1964)

Weinberg, S. (1972), Gravitation and Cosmology, p. 256, (J. Wiley).

# SUPERALGEBRAS IN EXCEPTIONAL GRAVITY

J. Thierry-Mieg & B. Morel

Harvard University

ABSTRACT

The physical spectrum of maximally extended supergravity in dimension D=3-11 is analyzed using Kac's theory of representations of superalgebras. First, we construct explicitly a few of them including for the first time all the atypical representations of SU(1/N). Then we show that the full spectrum corresponds in $D \leq 9$ to a super $\sigma$ model valued in the symmetric space $E_{11-D,11-D}$ divided by its maximal compact subgroup.

## I. REPRESENTATIONS OF LIE SUPERALGEBRAS

### I.1.: Kac-Dynkin Diagrams of Classical Lie Superalgebras

Lie superalgebras have been studied in great detail by Victor Kac (1977 a,b). In this section we shall recall some of his definitions and construct explicitly a few irreps (irreducible representations) relevant to supergravity. The classical Lie superalgebras are the well known SU(M/N) and OSp(M/2N) infinite families and 3 exceptionals, the superalgebra F(4) and G(3) and the one parameter family D(2,1,$\alpha$) (Table 1).

As for Lie algebras one can define a system of r positive simple roots $\alpha_i^+$, of r negative roots $\alpha_i^-$, and a system $h_i$ of r mutually commuting generators spanning a Cartan subalgebra such that $\alpha_i^\pm$, $h_j$ generate by commutation the entire algebra and satisfy the commutation relations:

$$[\alpha_i^+, \alpha_j^-] = \delta_{ij}\, h_j$$

$$[h_i, h_j] = 0$$

$$[h_i, \alpha_j^+] = a_{ij}\alpha_j^+$$

$$[h_i, \alpha_j^-] = - a_{ij}\, \alpha_j^- \qquad i,j = 1,\ldots r$$

r is the rank of the algebra, the matrix $[a_{ij}]$ is the Cartan matrix, it characterizes the Lie superalgebra and is represented by a Kac-Dynkin diagram with r nodes. There are 3 types of nodes

a) The even Bose roots are denoted by a white node. They are normalized and the value of the corresponding diagonal element $a_{ij}$ of the Cartan matrix is 2.

b) Some Fermi odd roots, normalized in the same way, are denoted by a black node. These roots always hide a simple root of the Bose subalgebra of the Lie superalgebra.

c) However, other odd roots, denoted by a crossed node (grey node) are null vectors of the Killing metric. The corresponding diagonal element $a_{ij}$ of the Cartan matrix vanishes. Their normalization is fixed by dialing the next non-zero element (usually $a_{i,i+1}$) to 1.

d) The i and j nodes are not joined if $a_{ij}=a_{ji}=0$. Otherwise they are connected by max $(|a_{ij}|,|a_{ji}|)$ links.

e) There is a certain arbitrariness in the choice of the Cartan subalgebra and several inequivalent Dynkin diagrams may characterize the same super-algebra (see the SU(1/2) case below).

Table 1 (from Kac 1977 b)

| G | Diagram | Bose Subalgebra |
|---|---|---|
| A(m,n) | 1 1 1 1 1<br>o—o— ··· —o—⊗— ···—o | SU(m+1) ⊕ SU(n+1) ⊕ U(1) |
| B(m,n), m > 0 | 2 2 2 2 2 2<br>o—o— ···—o—⊗— ··· —o=>o | SO(2m+1) ⊕ Sp(2n) |
| B(0,n) | 2 2 2 2 2<br>o—o— ··· —o—o=>● | Sp(2n) |
| C(n), n > 2 | 1 2 2 2 1<br>⊗—o— ··· —o—o<=o | U(1) ⊕ Sp(2n) |
| D(m,n) | 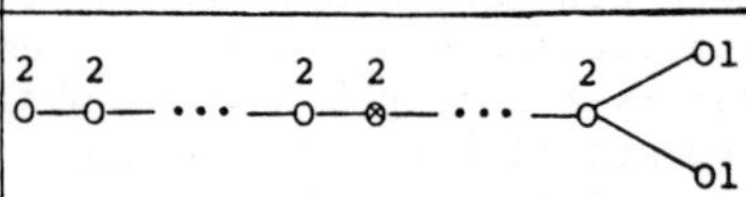 | SO(2m) ⊕ Sp(2n) |
| F(4) | 2 3 2 1<br>⊗—o<=o—o | SO(7) ⊕ SU(2) |
| G(3) | 2 4 2<br>⊗—o<≡o | $G_2$ ⊕ SU(2) |
| D(2,1;α) | 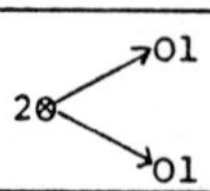 | SU(2) ⊕ SU(2) ⊕ SU(2) |

## I.2.: Irreducible Representations

By proposition 2.3 of Kac finite dimensional irreducible representations $V(\Lambda)$ of a classical Lie superalgebra G are uniquely characterized by their highest weight vector $|\Lambda\rangle$ labeled by its weight system $\lambda_i$:

$$\begin{cases} \alpha_i^+ |\Lambda\rangle = 0 \\ h_i|\Lambda\rangle = \lambda_i|\Lambda\rangle \qquad i = 1,\ldots r \end{cases}$$

The entire representation is then generated by repeated action of the negative simple roots $\alpha_i^-$ on $|\Lambda\rangle$. As for Lie algebras, the weight, referring to Bose roots must take positive integer values for the representation to be finite dimensional. For A(M/N) and C(N) the additional Fermi weight may take arbitrary complex values because, by antisymmetrization, products of more than M x N negative odd roots vanish. However, for the other classical superalgebras, called class II, the odd Fermi root hides a Bose root and the hidden Cartan generator k of the Bose subalgebra must take an integer weight. Reexpressing this generator k in terms of the $h_i$ one finds the condition $k \in Z^+$ given in table 2. For small value of k, Kac found by explicit computation that some additional conditions must be met.

Table 2 (from Kac 1977b)

| G | k | b |
|---|---|---|
| B(0,n) | $\frac{1}{2}a_n$ | 0 |
| B(m,n), m > 0 | $a_n - a_{n+1} - \cdots - a_{m+n-1} - \frac{1}{2}a_{m+n}$ | m |
| D(m,n) | $a_n - a_{n+1} - \cdots - a_{m+n-2} - \frac{1}{2}(a_{m+n-1}+a_{m+n})$ | m |
| D(2,1;α) | $(1+\alpha)^{-1}(2a_1 - a_2 - \alpha a_3)$ | 2 |
| F(4) | $1/3(2a_1 - 3a_2 - 4a_3 - 2a_4)$ | 4 |
| G(3) | $\frac{1}{2}(a_1 - 2a_2 - 3a_3)$ | 3 |

<u>for</u> k < b <u>(in table 2) there are the supplementary conditions:</u>

B(m,n): $a_{n+k+1} = \cdots = a_{m+n} = 0$.

D(m,n): $a_{n+k+1} = \cdots = a_{m+n} = 0$, $k \leq m-2$; $a_{m+n-1} = a_{m+n}$, $k = m-1$.

D(2,1;α): <u>all</u> $a_i = 0$ <u>if</u> $k = 0$; $(a_3+1)\alpha = \pm(a_2+1)$ <u>if</u> $k = 1$.

F(4): <u>all</u> $a_i = 0$ <u>if</u> $k = 0$; $k \neq 1$; $a_2 = a_4 = 0$ <u>if</u> $k = 2$; $a_2 = 2a_4 + 1$ <u>if</u> $k = 3$.

G(3): <u>all</u> $a_i = 0$ <u>if</u> $k = 0$; $k \neq 1$; $a_2 = 0$ <u>if</u> $k = 2$.

I.3.: First Example $A(1/0) \sim SU(2/1)$

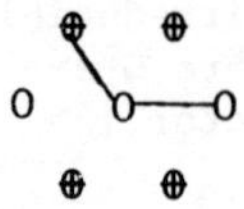

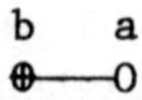


$$[a_{ij}] = \begin{vmatrix} 0 & 1 \\ -1 & 2 \end{vmatrix}$$

SU(2/1), the group of 2/1 superunitary matrices, is the compact form of A(1/0). It is closely related to SU(3) and has been proposed as a framework for a superunified model of the weak and electromagnetic interactions. (Ne'eman 1979, Fairlie 1979, Dondi & Jarvis 1979, Ne'eman & T. M. 1980). Let $\beta^+$ and $\alpha^+$ denote the odd and even positive roots, $h_0$ and $h_1$ the Cartan generators. An irreducible representation is labeled by the weights $|b,a>$ of its highest vector.

$$h_0 |b,a> = b\ |b,a>$$

$$h_1 |b,a> = a\ |b,a>$$

$$\alpha^+ |b,a> = \beta^+ |b,a> = 0$$

With respect to the Bose subgroup SU(2) x U(1) the representation splits into a set of Bose representations, whose highest weights we call the B weights. For SU(2/1) there is a maximum of 4 B-weights.

$$|\phi_1> = |b,a>$$

$$|\phi_2> = \beta^- |\phi_1> = |b,a+1>$$

$$|\phi_3> = a\beta^-\alpha^- - (a+1)\alpha^-\beta^- |\phi_1>$$

$$= |b-1,a-1>$$

$$|\phi_4> = \beta^-\alpha^-\beta^- |\phi_1> = |b-1,a>$$

$$\alpha^+ |\phi_i> = 0$$

However, not all these states necessarily occur as some of the $|\phi_i>$ may vanish. The check is to recover $|\phi_1>$ by repeated action of the positive roots on $|\phi_i>$. By example

$$\beta^+ |\phi_2> = \beta^+\beta^- |b,a> = b\, |b,a>$$

If b vanishes, $|\phi_2>$ must vanish, otherwise $|\phi_2>$ would be the highest vector of a representation not connected to $|\phi_1>$. We then find 3 cases:

a) Typical case

$b \neq 0 \;\; b \neq a+1 \;\; \dim(\text{rep}) = 2a+2/2a+2$

$\{|\phi_i>\} = \{|b,a>/|b,a+1>, |b-1,a-1>/|b-1,a>\}$

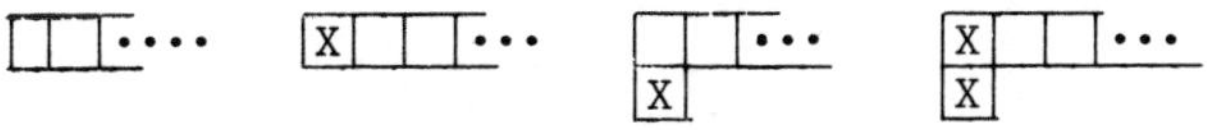

b) Atypical cases

$b=0 \;\; \dim(\text{rep}) = a+1/a$

$\{\phi_i\} = \{|b,a>/|b-1,a-1>\}$

c) $b=a+1 \;\; \dim(\text{rep}) = a+1/a+2$

$\{\phi_i\} = \{|b,a>/|b,a+2>\}$

On this example one already sees the general mechanism at work for SU(N/1). The B weights are obtained by adding to the Young tableau of $|\phi_i>$ up to a full column of new boxes, one by one and wherever this is possible, these B-tableaux occurring once. However, this construction is controlled by the b quantum number. There exist N critical values forbidding respectively the addition of a new box in one of the N lines.

The irreps of SU(2/1) were first calculated by Rittenberg, Nahm and Scheunert (1977). In our notations, their construction is based on the equivalent Kac-Dynkin diagram:

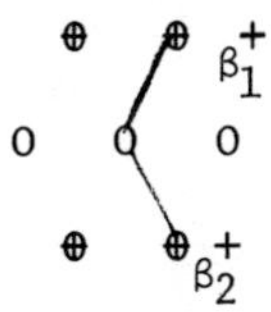


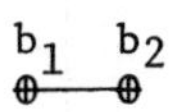

## I.4.: Second Example $A(0,7) \sim SU(1/8)$

$$[a_{ij}] = \left|\begin{matrix} 0 & 1 & & & & \bigcirc \\ -1 & 2 & -1 & & & \\ & -1 & \ddots & \ddots & & \\ & & \ddots & \ddots & \ddots & -1 \\ \bigcirc & & & & -1 & 2 \end{matrix}\right|$$

The group SU(1/8) has also been used in physics. Of course, we can no longer draw the root diagram which would be 8-dimensional. Starting from the highest weight $|\phi_1\rangle = |b,a,\dots a_7\rangle$ the last of the B-weight corresponding to the same Young tableau with a full column added is:

$$|\phi_7\rangle = |b-7,a_1,\dots a_7\rangle$$
$$= \beta^- \alpha_1^- \alpha_2^- \dots \alpha_7^- \quad \beta^- \alpha_1^- \dots \alpha_6^- \dots \beta^- \alpha_1^- \beta^- |\phi_1\rangle$$

However, $|\phi_7\rangle$ occursand the representation is typical only if:

$$\langle\phi_1|\beta^+\alpha_1^+\beta^+ \dots \alpha_7^+ \dots \alpha_1^+\beta^+|\phi_7\rangle \neq 0$$

$$\Longleftrightarrow \quad b \neq 0 \ , \ b \neq a_1+1, \ b \neq a_1+a_2+2$$
$$\dots \ b \neq \sum_{i=1}^{7} a_i + 7$$

In particular, if b=0 it is not allowed to add a new box in the first line of the Young tableau describing $|\phi_1\rangle$. Some representations of this particular type have been found by Ne'eman and Sternberg (1980), and Dondi and Jarvis (1979). By example:

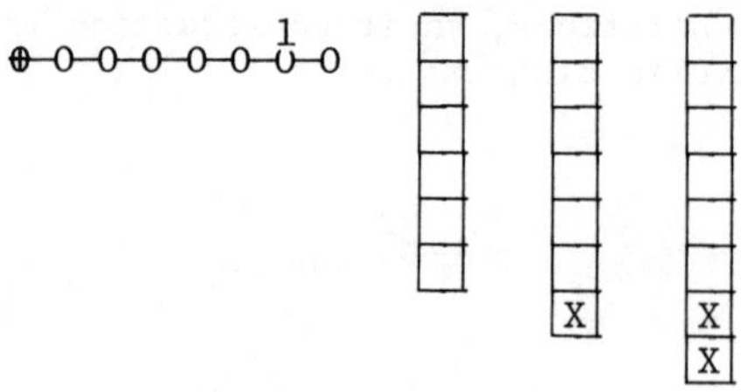


$= \underset{\sim}{\overline{28}}/\underset{\sim}{\overline{8}}/\underset{\sim}{1}$

If $b = \sum_{1}^{7} a_i + 7$, it is not allowed to add a box in the last line.

This case is interesting because the SU(8) content of the set of states with Trace removed considered by Ellis, Gaillard and Zumino (1980) in their attempt towards a phenomenology of supergravity is exactly given by the SU(8/1) atypical representation

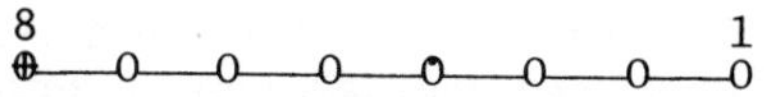


$$\{|\phi_i>\} = \overline{\underset{\sim}{8}} \,/\, \overline{\underset{\sim}{63}} \,/\, \overline{\underset{\sim}{216}} \,/\, \overline{\underset{\sim}{420}} \,/\, \overline{\underset{\sim}{504}} \,/\, \overline{\underset{\sim}{378}} \,/\, \overline{\underset{\sim}{168}} \,/\, \overline{\underset{\sim}{36}}$$

## I.5.: Third Example $B(0,2) \sim OSp(1/4)$

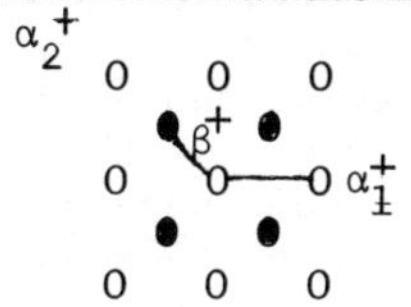


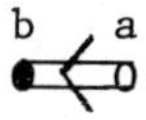


$$[a_{ij}] = \begin{vmatrix} 2 & -2 \\ -1 & 2 \end{vmatrix}$$

B(0,2) is a type II superalgebra. The odd node is black because the odd root is normalizable thanks to the relation

$$\{\beta^+, \beta^+\} = \alpha_2^+$$

$\alpha_2^+$, the second simple root of the Bose subalgebra Sp(4) is hidden in the process, therefore the weight b must be a double integer, $b/2 \in Z^+$. The B-weight system is

$$|\, b,a >$$

$$|\, b-2,\ a+1 > \qquad b \neq 0$$

$$|\, b,\ a-1 > \qquad a \neq 0$$

$$|\, b-2,a > \qquad b \neq 0$$

The adjoint representation is $|\,0,2>$. It is easy to follow the B weights on the root diagram. The OSp(1/8) and OSp(2/4) representations may be constructed in the same way. (We are preparing a general table of representations of Lie algebras labeled by their highest weight and B weights).

# II. EXCEPTIONAL SUPERGRAVITY AS A SUPER σ-MODEL

## II.1.: Hidden supersymmetry

By exceptional supergravity, we mean the Cremmer-Julia (1979) set of maximally extended supergravity models obtained by dimensional reduction from D=11 simple supergravity. In this process, we observe the transmutation of space time dimension into the rank of the Cartan exceptional group $E_{11-D}$. This group is non-linearly realized and appears in its normal (i.e. least compact) form. Its maximal compact subgroup H, together with the helicity group SO(D-2), classifies the physical states. In a preceding work, we (1980) have given the Young tableaux and Dynkin weights of all the relevant representations. As in the σ-model, the scalars are valued in the symmetric space E/H (Cremmer-Julia 1979). Our first remark is that, if one forgets for a moment about the spin, the system of H representations encountered exactly corresponds to systems of B-weights for the irreducible representations listed in table 3 of the superalgebras $\tilde{H}$. In D=3, $\tilde{H}$ is the semi direct super product of SO(16) by its spinorial representation 128 considered as Fermionic. In all cases, $\tilde{H}$ includes H. Moreover, in D=8 and in D=4, the even subalgebra of $\tilde{H}$ is equal to the direct sum of H by the helicity algebra SO(D-2) and the $\tilde{H}$ representations also assign the correct spin to all particles. For this reason, we think that SU(4/2) in D=8 and SU(1/8) in D=4 may define hidden supersymetries of the full theory on shell. This would support the relevance of the atypical representation ofSU(1/8) described in section I to phenomenology.

Table 3 : Symmetry groups in exceptional gravity.

| D | SO(D-2) | S1(D-2) ; $E_{11-D}$ | $H_{11-D}$ | $\tilde{H}$ |
|---|---|---|---|---|
| 10 | SO(8)<br>o-o< (o, o) | S1(8) ; R<br>o-o-o-o-o-o-o<br>x | | |
| 9 | SO(7)<br>o-o⇒o | S1(7); S1(2) ⊕ R<br>o-o-o-o-o-o o<br>x | U(1)<br>x | |
| 8 | SU(4)<br>o-o-o | S1(6) ; S1(3)⊕S1(2)<br>o-o-o-o-o o-o<br>o | SU(2)⊕U(1)<br>o<br>x | SU(4/2)<br>o-⊠-o-o-o<br>4 |
| 7 | Sp(4)<br>o⇒o | S1(5) ; S1(5)<br>o-o-o-o o-o-o<br>o | Sp(4)<br>o⇒o | OSp(2/4)<br>4<br>⊠-o═o |
| 6 | $SU(2)^2$<br>o<br>o | S1(4) ; SO(5,5)<br>o-o-o o-o-o-o<br>o | $Sp(4)^2$<br>o⇒o<br>o⇒o | $OSp(1/4)^2$<br>2●⇐o<br>2●⇐o |
| 5 | SU(2)<br>o | S1(3) ; $E_{6(6)}$<br>o-o o-o-o-o-o<br>o | Sp(8)<br>o-o-o⇐o | OSp(1/8)<br>2<br>●═o-o-o |
| 4 | U(1)<br>x | S1(2) ; $E_{7(7)}$<br>o o-o-o-o-o-o<br>o | SU(8)<br>o-o-o-o<br>o-o-o | SU(8/1)<br>o-o-o<br>8<br>⊠-o-o-o-o |
| 3 | | ∅ ; $E_{8(8)}$<br>o-o-o-o-o-o-o<br>o | SO(16)<br>o-o-o-o< (o, o)<br>o-o | SO(16)⋉$\widetilde{128}_F$ |

## II.2.: The super σ-model

Our second result is that, in D<10, the whole physical spectrum may be recovered by considering a super-σ field $\tilde\sigma(x^\mu,\theta^i_\alpha)$ valued in the symmetic space E/H and depending both on D dimensional space time and on Grassmann coordinates $\theta^i_\alpha$. The $\theta^i_\alpha$ are defined as SO(D-2) spinors, the α index, belonging to the fundamental representation of H, the i index. The expansion is now carried as follows.

| D | H, the E/H tableau; SO(D-2), the graviton tableau. | The $\tilde\sigma$ expansion : φ=scalars, χ=spinors, A=vectors B=2-tensors, C=3-tensor, e=graviton. The indices are symmetrized according to the Young tableaux of column 2. |
|---|---|---|
| 9 | U(1); SO(7), [α / β / γ / δ] | $\tilde\sigma = \phi + \theta_\alpha\chi_\alpha + \theta_\alpha\theta_\beta\cdot(A+B)_{[\alpha\beta]} + \theta_\alpha\theta_\beta\theta_\gamma\cdot(\chi+\psi)_{[\alpha\beta\gamma]}$ $+ \theta_\alpha\theta_\beta\theta_\gamma\theta_\delta\cdot(e+A+C+\phi)_{[\alpha\beta\gamma\delta]}$ +equiv. higher terms. |
| 8 | SU(2)⊕U(1), [i\|j\|k\|l] ; SU(4), [α γ̄ / β δ̄]. | $\tilde\sigma = \frac12\phi^{ijkl} + \theta^i_\alpha\cdot\chi^{jkl}_\alpha + \theta^i_\alpha\theta^j_\beta\cdot A^{kl}_{\alpha\beta} + \frac12\theta^i_\alpha\theta^{\bar j}_\gamma\cdot B^{k\bar l}_{\alpha\bar\gamma}$ $+ \theta^i_\alpha\theta^j_\beta\theta^{\bar k}_\gamma\cdot\psi^l_{\alpha\beta\bar\gamma} + \theta^i_\alpha\theta^j_\beta\theta^k_\gamma\theta^l_\delta\cdot\phi_{\alpha\beta\gamma\delta}$ $+ \frac12\theta^i_\alpha\theta^j_\beta\theta^k_\gamma\theta^{\bar l}_\delta\cdot C_{\alpha\beta\gamma\bar\delta} + \frac12\theta^i_\alpha\theta^j_\beta\theta^{\bar k}_\gamma\theta^{\bar l}_\delta\cdot e_{\alpha\beta\bar\gamma\bar\delta}$ + h.c. |
| 7 | $Sp(4)^i_j$ [⊠⊠ / ⊠⊠] k l<br>$Sp(4)^\alpha_\gamma$ [⊠⊠ / ⊠⊠] β δ | $\tilde\sigma = \phi^{ijkl} + \theta^i_\alpha\chi^{jkl}_\alpha + \theta^i_\alpha\theta^j_\beta\cdot B^{kl}_{\alpha\beta} + \theta^i_\alpha\theta^k_\gamma\cdot A^{jl}_{\alpha\gamma}$ $+ \theta^i_\alpha\theta^j_\beta\theta^k_\gamma\cdot\psi^l_{\alpha\beta\gamma} + \theta^i_\alpha\theta^j_\beta\theta^k_\gamma\theta^l_\delta\cdot e_{\alpha\beta\gamma\delta}$ |
| 6 | $Sp(4)^2$, [⊠ / ⊠] i j<br>$SU(2)^2$, [⊠ / ⊠] k l<br>[α\|β] [γ\|δ] | $\tilde\sigma' = \phi'^{ij} + \theta^i_\alpha\chi'^j_\alpha + \theta^i_\alpha\theta^j_\beta\cdot A'_{\alpha\beta}$<br>$\tilde\sigma = \tilde\sigma'\cdot\tilde\sigma'' = \frac12\phi^{ijkl} + \theta^i_\alpha\cdot\chi^{jkl}_\alpha + \theta^i_\alpha\theta^j_\beta\cdot B^{kl}_{\alpha\beta}$ $+ \frac12\theta^i_\alpha\theta^k_\gamma\cdot A^{jl}_{\alpha\gamma} + \theta^i_\alpha\theta^j_\beta\theta^k_\gamma\cdot\psi^l_{\alpha\beta\gamma} + \frac12\theta^i_\alpha\theta^j_\beta\theta^k_\gamma\theta^l_\delta\cdot e_{\alpha\beta\gamma\delta}$ +h.c. |
| 5 | Sp(8), i j [⊠ / ⊠]<br>SU(2), k l<br>[α\|β\|γ\|δ] | $\tilde\sigma = \phi^{ijkl} + \theta^i_\alpha\chi^{jkl}_\alpha + \theta^i_\alpha\theta^j_\beta\cdot A^{kl}_{\alpha\beta}$ $+ \theta^i_\alpha\theta^j_\beta\theta^k_\gamma\cdot\psi^l_{\alpha\beta\gamma} + \theta^i_\alpha\theta^j_\beta\theta^k_\gamma\theta^l_\delta\cdot e_{\alpha\beta\gamma\delta}$ |
| 4 | SU(8), [i / j / k / l] | $\tilde\sigma = \frac12\phi^{ijkl} + \theta^i\chi^{jkl} + \theta^i\theta^j\cdot A^{kl}$ $+ \theta^i\theta^j\theta^k\cdot\psi^l + \theta^i\theta^j\theta^k\theta^l\cdot e$ + h.c. |
| 3 | SO(16) o-o-o-o-o-o-o< (nodes labelled a, β̄, α) | $\tilde\sigma^\alpha = \phi^\alpha + \theta^a\sigma_{a\alpha\bar\beta}\cdot\chi^{\bar\beta}$ |

a) The σ-superfield, being valued in the E/H H-representation carries a certain number of i (H fundamental representation) indices symmetrized according to a known Young tableau. In the expansion, these indices are progressively moved over, from the component fields, onto the explicit $\theta^i_\alpha$ which appear in the process. The same overall symmetrization is performed and in particular no H indices are contracted between the θ or between the θ and the component fields. In this way, only a few components are generated (6 in D=7) despite the great number of available θ (16 in D=7), and we exactly reconstruct the $\tilde{H}$ representatations of II.1..

b) On the other hand, the spin indices carried by the θ are systematically contracted onto spin indices of the component fields. In this way, because the θ anticommute, these indices automatically undergo the dual symmetrization and correctly induce the spin of all fields. The expansions are given in the table of the preceding page.

A few comments may be added :

-In D=11, there is no scalar and this construction cannot be used.

-In D=10, the construction does not seem to work, probably because there is no H group.

-In D=9 and D=8, E/H is reducible and $\tilde{\sigma}$ is taken respectively as a U(1) singlet and as a $\underset{\sim}{5}$ of SU(2).

-In D=9, antisymmetrized products of SO(7) spinorial representations are reducible and split as shown in the table.

-In D=8, the expansion is carried over both left and right SO(6) spinors, this distinction is irrelevant for lower D.

-In D=6, the helicity group, SO(4)=SU(2)*SU(2), and the H group, Sp(4)*Sp(4), factorize. The spectrum is given by the product of two superfields, each constructed for one of the SU(2)*Sp(4) pair of groups. In this sense, (N=8, D=6) supergravity is the square of (N=4, D=5) supergravity.

-In D=3, there is no helicity group and the expansion degenerates. It however correctly assigns the scalars and spinors to conjugated 128 and $\overline{128}$ representations of SO(16).

In all dimensions, equivalent fields have been reduced to their simplest form by assumed duality transformations.

## II.3.: Group disintegration

The H representations assignments may be checked by working backwards from D=3 to D=11 (Julia, this conference). When the dimension raises, the group $E_{11-D}$ progressively disintegrates and the representations split accordingly. Consider the graviton, it is valued in Sl(D-2)/SO(D-2) in analogy with the scalars valued in E/H. However, the Dynkin diagram of $E_8$ splits, in any dimension, by removing the link between the D-3 and the D-4 node, into the direct sum of the $E_{11-D}$ and Sl(D-2) Dynkin diagrams (table 3). For example, in D=6 :

```
                         o
          E8 : o-o-o-o-o-o-o
                         o
gives Sl(4)⊕SO(5,5) : o-o-o o-o-o-o
```

This splitting explains the quasi exchange of space and internal space when one goes from D to 14-D dimensions, the off the line node however introducing a certain asymmetry. In the limiting case of D=11, the graviton should accordingly be described not by Sl(9)/SO(9), but by $E_8/SO(16)$. The antisymmetric 3-tensor field is thus necessary for purely Bose reasons, independently of supersymmetry; and the would be D=2 (string) theory of the infinite Kac Moody algebra $E_9$ would correspond to the unconstructed ($N=\omega$) D=12 supergravity. An argumentation similar to this section is given by Julia in these proceedings, but he reaches different conclusions.

## CONCLUSION

Exceptional gravity may well be the super-$\sigma$ theory of the $E_8$ group. The various dimensions in which it may be constructed correspond to various splittings of the $E_8$ Dynkin diagram. In dimension 4, the spectrum forms an irreducible representation of the superalgebra SU(1/8) unifying the helicity group to the SU(8) internal symmetry group. We have also constructed all the atypical representations of SU(1/N) and found that the phenomenological set of states "with traces removed" considered by Ellis, Gaillard and Zumino (1980) is irreducible with respect to the hidden supersymmetry SU(1/8).

## ACKNOWLEDGEMENTS

We are very grateful to Victor Kac for his invaluable help and have benefited from conversations with Eugene Cremmer and Bernard Julia.

REFERENCES

Cremmer, E. & Julia, B. (1979), Nuclear Physics B159, p141

Dondi, P.H. & Jarvis, P.D. (1979), Phys. Lett. B84, pp75-78

Ellis, J., Gaillard, M. & Zumino, B. (1980), Phys. Lett. B94, p343

Fairlie, D.B. (1979), Phys. Lett. B82, pp97-100

Kac, V.G. (1977a), Lie superalgebras, Advances in Math. 26,pp 8-96

Kac, V.G. (1977b), Lecture notes in Math. 676, pp596-626

Morel, B. & Thierry-Mieg, J. (1980), Groups & supergroups in exceptional gravity, HUTP-80/B95, to be pub. in Comm.Math.Phys.

Ne'eman, Y. (1979), Phys.Lett. B81, pp190-194

Ne'eman, Y. & Sternberg, S. (1980), Internal symmetry and unifications, P.N.A.S., in press

Ne'eman, Y. & Thierry-Mieg, J. (1980), P.N.A.S. 77, 2 pp720-723

Scheunert, M., Nahm, W. & Rittenberg, V. (1977), J.Math.Phys. 18, 1 pp155-162

# EXTENDED SUPERFIELDS IN LINEARISED SUPERSYMMETRY AND SUPERGRAVITY

J. G. Taylor
King's College, London

## INTRODUCTION

Superfields are fields whose variables include not only the usual space-time vector $x^\mu$ but also a set of N Majorana anti-commuting spinor variables $\theta_{\alpha i}$ ($1 \leq i \leq N$, $1 \leq \alpha \leq 4$). In their unextended form (for N=1) they were introduced originally by Salam and Strathdee (1975) and used there to give an elegant form for supersymmetric Yang-Mills theories as well as elsewhere to describe supergravity. This elegance arises from the possibility of extending the tools of differential geometry to the superspace situation as originally worked out by Wess and Zumino (1977) and independently by the author and S. Downes-Martin (1977). Another benefit of superfields is their automatic inclusion of auxiliary fields and of the tensor calculus, so that various lengthy calculations can be otherwise circumvented.

However the most important reason for wishing to construct extended superfield formulations of supersymmetry and supergravity is that there may be (J.G. Taylor 1979a) ultra-violet divergence cancellations similar to those which have been shown to occur to all orders in unextended supersymmetric Yang-Mills theories (Ferrara and Piguet, 1975). This cancellation is known not to occur (J.G. Taylor 1979b) for unextended supergravity, but may be achieved on going to the largest possible value of N(N=8). We therefore must analyse the nature of extended superfields as to their supersymmetric and particle content before applying what we have learned from this to extended supersymmetric and supergravitational theories.

The plan of the talk is therefore to follow the above prescription and describe various features of extended supersymmetry algebras, in particular the spin content and projectors onto irreducible representations (irreps) on superfields before turning to the application of these ideas.

Some of the results reported here have been worked out in collaboration with Charles Pickup and Victor Rivelles, and I would like to thank them for their help.

THE GLOBAL SUPERSYMMETRY ALGEBRA

We start by considering the global extended supersymmetry algebra in terms of the Majorana spinor generators $S_{\alpha i}$ of supersymmetry transformations with the generators $P_\mu$ and $J_{ab}$ of the Poincaré group:

$$[P_\mu, S_{\alpha i}]_- = 0, \quad [S_{\alpha i}, S_{\beta j}]_+ = (\not{P}\eta)_{\alpha\beta}\delta_{ij},$$

$$[S_{\alpha i}, J_{ab}]_- = i(\sigma_{ab})_\alpha{}^\beta S_\beta \tag{1}$$

and $\eta = -C^{-1}$, C being the Dirac charge conjugation matrix. We use the metric (+---). The total algebra $\mathcal{S}_N$ has generators $\{S_{\alpha i}, P_\mu, J_{ab}\}$. It is possible to include central charges on the r.h.s. of (1) but we will not do so for the initial analysis of $\mathcal{S}_N$; we will later find strong reasons for introducing them in order to discuss SO(4) Yang-Mills and SO(8) supergravity.

It is natural to concern ourselves with the properties of possible representations of (1). These may be obtained from the eigenvalues of the Casimirs of $\mathcal{S}_N$ from which may be deduced projection operators onto these irreps. Since $P_\mu$ commutes with $\mathcal{S}_N$ we can consider eigenstates of $P_\mu{}^2$.

We may proceed more directly in the case of lower N by taking $\gamma_o$ diagonal so that (1) reduces, in the rest frame when $P_\mu{}^2 \neq 0$, to the usual annihilation-creation operator relation between the pair $(S_{1i}, S_{2i})$ and $(S_{3i}, S_{4i})$ with $S_{3i} = S^+_{2i}$, $S_{4i} = -S^+_{1i}$. Given a state $|j\rangle$ of Poincaré spin j with $S^+_1|j\rangle = S^+_2|j\rangle = 0$ for N=1 we may construct a quartet of states $\{|j\rangle, S_1|j\rangle, S_2|j\rangle, S_1S_2|j\rangle\}$, which have spin j, $j-\frac{1}{2}$, $j+\frac{1}{2}$, and j (since $S_1$ decreases $J_3$ by $\frac{1}{2}$ unit, and $S_2$ increases it).

This pattern (1,2,1) for multiplicity of Poincaré spins $j-\frac{1}{2}$, j and $j+\frac{1}{2}$ in an irrep of $\mathcal{S}_1$ can be extended to higher N in terms of the dimensions of the irreps of the algebra Sp(2N). This can be seen from the fact that (Pickup 1980) (now in 2 component notation) the rotation invariant matrices $a_{ij}$ and $b_{ij}$ defined, in the rest frame, by

$$a_{ij} = (\sigma_o)^{\alpha\dot{\beta}}\, S_{\alpha i}\bar{S}_{\dot{\beta} j}, \quad b_{ij} = \varepsilon^{\alpha\beta}S_{\alpha i}\, S_{\beta j} \tag{2}$$

are hermitian and symmetric respectively, so can be used to define the generators of Sp(2N), where the metric of Sp(2N) is taken to be $\begin{pmatrix} 0 & I_N \\ -I_N & 0 \end{pmatrix}$. Hence we expect the antisymmetric representations (due to anticommutation of the S's) of Sp(2N) to be appropriate to the classification of the spin content of irreps of Sp(2N).

Let us consider the case N=2. The irreps for N=2 are classified by Sp(4), so have spin content $[(j-1), (j-\frac{1}{2})^4, (j)^{5+1}, (j+\frac{1}{2})^4, (j+1)]$, where the $\underline{5}+\underline{1}$ is a reducible representation, the extra singlet being required to keep equality between bose and fermi degrees of freedom. This singlet disappears for the case j=0 when the irrep has spin content $[0^5, (\frac{1}{2})^4, 1]$; this is clearly the appropriate multiplet to describe N=2 supersymmetric Yang-Mills theory. The graviton in N=2 supergravity will be described by the irrep with j=1, with spin content $[0, (\frac{1}{2})^4, (1)^6, (\frac{3}{2})^4, 2]$.

For N=4 the general irrep has spin content $[(j-2), (j-\frac{3}{2})^8, (j-1)^{27+1}, (j-\frac{3}{2})^{48+8}, j^{35+35}, (j+\frac{1}{2})^{48+8}, (j+1)^{27+1}, (j+\frac{3}{2})^8, j+2]$. There is thus no irrep for N=4 with maximum spin j=1, so that case must be discussed by extension of $\mathcal{S}_N$ to include central charges. The graviton will be contained in the irrep with j=0, so has spin content $[0^{70}, \frac{1}{2}^{48}, 1^{27}, \frac{3}{2}^{8}, 2]$.

Finally we see that for N=8 there is no irrep with Poincaré spin no larger than 2, so central charges are essential if these higher spins are to be excluded. It is possible to envisage using multiplets with higher spin fields which become auxiliary or can be gauged away, but it does not seem easy to obtain a model embodying such a cancellation. Since it is possible to discuss off mass shell N=8 supergravity by means of N=4 multiplets we will follow that path here.

It must be emphasized here that we have considered only off-mass-shell spin content. This is considerably simpler on-shell, where $P^2=0$ (for massless gauge multiplets). Then for N=1, $W_\mu = \Lambda P_\mu$, $S = i\not{P}S^1$ (following Jarvis 1976) so $\bar{S}i\gamma_5\gamma_\mu S = nP_\mu$, $n = -\frac{1}{2}\bar{S}^1 i\gamma_5\not{P}S^1$, $\bar{n}^2 = \frac{1}{2}$. Then $(\Lambda-\frac{1}{2})$ is a Casimir of $\mathcal{S}_1$ called the super-helicity $\lambda$; the helicity content of a multiplet is thus $[\lambda, \lambda-\frac{1}{2}]$. For higher N this may be shown to be (Taylor 1980a) $[\lambda, \lambda-\frac{1}{2}, ----, \lambda-\frac{1}{2}N]$.

IRREPS ON SUPERFIELDS

An explicit representation for $S_{\alpha i}$ can be given on any superfield $\Phi(x,\theta)$ by the well-known expression $S_{\alpha i} = i(\frac{\partial}{\partial\bar{\theta}^{\alpha}_{i}} + \frac{1}{2}\not{p}_{\alpha}{}^{\beta}\theta_{\beta})$.

We may extend (Taylor 1980a) following Sokatchev (1975) the Pauli Lubanski vector $W_{\mu}$ for the Poincaré group to $\mathcal{S}_N$ as

$$C_{\mu} = \tfrac{1}{2}\varepsilon_{\mu\nu\lambda\rho} \; P^{\nu}J^{\lambda\rho} - \tfrac{1}{4}\sum_{j=1}^{N} \bar{S}_{j}i\gamma_{\mu}\gamma_{5}S_{j} \qquad (3)$$

Then the eigenvalues of $C^{2}_{\mu}$ are $-2P^{2}Y(Y+1)$ where Y is the extended superspin with half-integer or integer values. There are further Casimir operators that are needed to label the irreps of $\mathcal{S}_N$ completely. For N=1 they are obtained from the $\gamma_5$ transformations $S\to e^{\alpha\gamma 5}S$ with generator $\Gamma$ for which $[S_{\alpha},\Gamma]_{-} = (\gamma_{5}S)$; the extra Casimir is (Sokatchev 1975)

$$G = -\bar{S}i\not{P}\gamma_{5}S + 2P^{2} \qquad (4)$$

For higher N we need to include the SO(N) rotation generators $T_{ij}$ with $[\theta_{k},T_{ij}]_{-} = i\delta_{jk}\theta_{i} - i\delta_{ik}\theta_{j}$, the Casimirs being

$$\mathfrak{J}_{\ell m} = T_{\ell m} + (P^{2})^{-1}\,\bar{S}_{\ell}i\not{P}S_{m} \qquad (5)$$

There are further Casimirs arising from the Sp(2N) generators (2), with $S_{\alpha i}$ replaced by $D_{\alpha i}$. Thus for N=2 we have the Casimirs Y, G, $\mathfrak{J}$ and the quadratic Sp(4) Casimirs $C_2$, whilst for N=4 there are the Casimirs Y, G, $\underline{\mathfrak{J}}^{2}$, $\underline{\mathfrak{J}}'^{2}$, $\mathfrak{J}_{3}$, $\mathfrak{J}'_{3}$, where $\underline{\mathfrak{J}}$ and $\underline{\mathfrak{J}}'$, are the generators associated with the 2 SU(2) groups in SO(4) = SU(2) × SU(2), and G is extended from N = 1 as $G = 2P^{2}\Gamma - \sum_{j=1}^{N} \bar{S}_{j}i\not{P}\gamma_{5}S_{j}$. There are numerous identities satisfied by the generators $D_{\alpha i}$ and their products, and these must be taken into account in order to obtain the correct labelling for the irreps by the eigenvalues of the Casimirs in the various cases.

The irreps we are most interested in arise by using the explicit expressions for $S_{\alpha}$ and $D_{\alpha}$ acting on a superfield with external Poincaré spin j, $\Phi_{j}(x,\theta)$. This superfield contains a number of irreps of $\mathcal{S}_N$, such as the chiral one determined by the conditions $D_{\alpha+i}\ \Phi = 0$, where $D_{\alpha+} = [\frac{1}{2}(1+i\gamma_5)\Phi]_\alpha$. $\Phi_j$ gives an irrep of the algebra $\mathcal{S}_N$ extended by addition of the $D_{\alpha i}$; it gives various irreps of $\mathcal{S}_N$ by applying products of $D_{\alpha i}$ to a basic state $|Y\rangle$ with superspin $Y=j$. Since the set $D_{\alpha i}$ have the same commutation relations with $C_j$ as the set $S_{\alpha i}$ with $J_j$ we see that the superspin of the various irreps of $\mathcal{S}_N$ in $\Phi_j$ will be $Y = j+\frac{1}{2}(N-r)$ for $r=0$ to $2N$, with multiplicity given by the irreps of $Sp(2N)$.

Thus for $N=1$ in $\Phi_j$ have $Y = j^2,\ j\pm\frac{1}{2}$

$N=2$ in $\Phi_j$ have $Y = j^6,\ (j\pm\frac{1}{2})^4, (j\pm 1)$

$N=4$ in $\Phi_j$ have $Y = j^{70},\ (j\pm\frac{1}{2})^{56}, (j\pm 1)^{28}, (j\pm\frac{3}{2})^8, j\pm 2$.

The values of the complete set of Casimirs for the scalar and vector superfields ($j$ = 0 and 1) are given in Taylor (1980a) for N=2; similar expressions have been obtained (Pickup and Taylor, 1980; Pickup 1980) for N=4, and also for the case of higher j for N=2 and 4. We have not gone to higher N because of the problem of higher spin fields; the results for N=3 can easily be obtained from those for N=4 by an obvious reduction.

Finally we may construct projectors onto the various irreps by using the operators generalising the N=1 case. For example we may project out from a scalar superfield $\Phi$ the irrep with $Y = G = 0$, $\mathcal{J} = \pm 2$, $C_2 = 2/3$ by means of the projectors onto $A_1 = A_2 = 0$, $J = 3p^4$ (where $A_i = (\bar{D}_i D_i)^2/4p^2$, $J = \frac{1}{4}\{p^2(\bar{D}_1 i\gamma_\mu\gamma_5 D_1)(\bar{D}_2 i\gamma_\mu\gamma_5 D_2) - (\bar{D}_1 i\not{p}\gamma_5 D_1)(\bar{D}_2 i\not{p}\gamma_5 D_2)\}$ and $A_1 = A_2 = 1$, $\bar{D}_1 D_1 \bar{D}_2 D_2 + \bar{D}_1\gamma_5 D_1 \bar{D}_2\gamma_5 D_2 = -8p^2$.

## CENTRAL CHARGES

Before we turn to consider the field component of the above and similar irreps we will remedy our present inability to discuss N=4 Yang-Mills and N=8 supergravity by introducing central charges.

In order to have irreps of the SO(4) supersymmetry algebra on superfields with Poincaré spin at most one it is clear that the spin-reduction property (Sohnius, 1978) of central charges is necessary, where by spin reduction we mean that only half the maximal Poincare spins arise in a given irrep. This will allow certain of the superspin zero irreps to describe the Yang-Mills gauge field. In the case N=2 (for which we may also include a central charge) the SO(2) superalgebra then reduces effectively to an SO(1) superalgebra without central charges and the SO(4) case reduces to SO(2).

Let us consider the supersymmetry algebra (in 4-component form)

$$[S_{\alpha i}, S_{\beta j}]_+ = (\not{p}\eta)_{\alpha\beta}\, \delta_{ij} + i\underline{Z}_{ij}\, \eta_{\alpha\beta} \tag{6}$$

For N=2, $Z_{ij} = Z\epsilon_{ij}$ and for N=4 we may choose $\underline{Z}_{ij} = \alpha_{ij}Z$, where $\alpha$ is one of the antisymmetric 4×4 matrix generators of one of the SO(3)'s in SO(4). We may represent the $S_{\alpha i}$ and associated $D_{\alpha i}$ as differential operators on superfields as before with $Z = id/dz$, where z is an extra bose co-ordinate. We require that $Z^2 = p^2$ for all such superfields.

We write S = V+W (where S denotes the 4N vector with components $S_{\alpha i}$) with $V = (2\not{p})^{-1}(\not{p}-i\underline{Z})S$, $W = (2\not{p})^{-1}(\not{p}+i\underline{Z})S$, so that $[V,V]_+ = [V,W]_+ = 0$ and W satisfies (1). We thus expect to be able to represent V trivially. Since W also satisfies the Dirac equation $(\not{p}-iZ)\,W = 0$ then only half of the components of W are independent; this explains the spin-reduction property analysed by means of rest-frame states by Sohnius (1978). We may make the spin-reduction property even more transparent for N=2 by noting that in the coordinates $\phi = \theta_1 + iZ\not{p}^{-1}\theta_2$, $\psi = \theta_1 - iZ\not{p}^{-1}\theta_2$, $V_{\alpha 1} = i\,\partial/\partial\bar{\psi}^{\alpha}$, $W_{\alpha 1} = i(\partial/\partial\,\bar{\phi}^{\alpha} - \tfrac{1}{2}(\not{p}\phi)_{\alpha})$. For N=4 we introduce the new variables $\phi_1 = \theta_1 + iZ\not{p}^{-1}\theta_2$, $\psi_1 = \theta_1 - iZ\not{p}^{-1}\theta_2$, $\phi_2 = \theta_3 + iZ\not{p}^{-1}\theta_4$, $\psi_2 = \theta_3 - iZ\not{p}^{-1}\theta_4$ (for a suitable choice of $\alpha$) and similar expressions arise. Thus we take only superfields depending on $\phi$ (for N=2) or on $\phi_1$ and $\phi_2$ (for N=4); spin reduction has evidently occurred.

We still have to consider the Casimirs of the reduced algebra of the W's and Z. These may be obtained from the Casimirs of the unreduced algebra and are the Casimirs of the algebra of the W's together with $\mathfrak{M} = [M - i(Z/4)\bar{D}D]$, where $[M,D_i] = \epsilon_{ij}D_j$ for N=2 and $[M,D_i]_- = \alpha_{ij}D_j$ for N=4. The irrep $\phi_i$ for N=2 with $(\mathfrak{M}\Phi)_i = 0$ is Fayet's scalar hypermultiplet (Fayet, 1976) which had already been presented in superfield form by Sohnius (1978) though from a very different point of view. We note that the extension of this multiplet to SO(4) is straightforward by means of the above spin reduction and the SO(2) irreps. This multiplet is an SO(4)-self-dual antisymmetric tensor $\phi_{ij}$ with the chiral properties $D_{\alpha+1}\,\phi_{13} = D_{\alpha+3}\,\phi_{13} = D_{\alpha+2}\,\phi_{14} = D_{\alpha+4}\,\phi_{14} = 0$, $\phi_{12} = 0$ (in a particular representation of $\alpha$) with $A_1 = A_2 = A_3 = A_4 = 1$

on $\phi_{ij}$ (where $A_i = (\bar{D}_i D_i)^2/4p^2$).

FIELD CONTENT

In order to obtain the component content of the irreps on superfields considered above we avoid the complexity produced by using explicit power series expansions of $\Phi(x,\theta)$ when N=2 (with 256 degrees of freedom) and N=4 (with $2^{16}$ = 65,536 degrees of freedom). Thus for N=2 the superfield expanded in powers of $\theta$ has terms up to $\theta^8$ and even in isospin form with an eight-component spinor $\underline{\theta}$ can look very complicated. To avoid this complexity we introduce a suitable set of basis functions in terms of which we may evaluate the component content of any irrep.

For N=1 a set of basis functions have already been introduced by Jarvis (1976). We define these in a $\theta$-independent manner. The basis functions of the $\underline{Y}=\frac{1}{2}$ irrep may be obtained by operating with (1-A) on 1, $(\not\partial D)_\alpha 1$ or $\bar{D}i\gamma_\mu\gamma_5 D1$. We are here denoting by 1 a $\theta$-independent but not x-independent quantity, the resulting basis functions being therefore differential operators in x-space. We have therefore the basis functions of the $Y=\frac{1}{2}$ irrep as

$$e_o = 2(1-A)1, \quad u_\alpha = 4i\,\Box^{-1}(1-A)(\not\partial D)_\alpha 1,$$

$$\omega_\mu = -\,\Box^{-1}(1-A)\bar{D}i\gamma_\mu\gamma_5 D1 \tag{7}$$

The two irreps of Y=0 can be obtained by applying $\bar{D}D_\pm$ to $1, D_{\alpha\pm}1$ or $\bar{D}D_\pm 1$ to give the basis functions

$$e_{\mp} = -\,\Box^{-1}\bar{D}D_\pm\bar{D}D_\mp 1,\quad u_\mp = -\,2\Box^{-1}\bar{D}D_\pm D_{\alpha\mp}1,\quad \omega_\mp = \Box^{-1}\bar{D}D_\pm \tag{8}$$

These basis functions (7) and (8) agree with those of Jarvis (1976) except for $\omega_\mu$, which has expression (after use of the product rules for the $D_\alpha$) $\omega_\mu = \Box^{-1}\bar{D}i\gamma_\mu{}'\gamma_5 D1 = \frac{1}{4}\bar{\theta}i\gamma_\mu{}'\gamma_5\theta$, where $\gamma_\mu{}' = \gamma_\mu - \not\partial\partial_\mu\Box^{-1}$. We note that $\partial_\mu\omega^\mu = 0$, so preserving the invariance of $\omega_\mu A_\mu$ under the variation $\delta A_\mu = \partial_\mu\Lambda$; we do not need to require $\partial_\mu A_\mu = 0$ in any superfield expansion.

For N=2 we may expand a superfield $\Phi$ in the various products of the above basis functions. Suitable projection operators may be used to project these onto the various irreps. As an example of this, and of particular relevance to SO(4)-extended Yang-Mills supersymmetry (J.G. Taylor (1980b)) we may take the eigensubspace $A_1 = A_2 = 0$. The resulting superfield can be expressed in terms of the basis functions as

$$\Phi = e_o(e_oA - u_\alpha\psi^\alpha + \omega_\mu A^\mu) - u_\alpha(e_o\chi^\alpha - u_\beta C^{\alpha\beta} + \omega_\mu\psi^{\alpha\mu})$$

$$+ \omega_\mu(e_oB^\mu - u_\alpha\chi^{\alpha\mu} + \omega_\nu A^{\mu\nu})$$

where $e_oe_o$ denotes $e_o(\theta_1)e_o(\theta_2)$, etc. We may project out the irreps with Y=0 and Y=1 by $\frac{1}{4}(1-\frac{1}{4}\Box^{-1}K)$ and $\frac{3}{4}(1 + (1/_{12})\Box^{-1}K)$. The resulting expression for Y=0 can be written as

$$\Phi|_{Y=0} = (e_oe_o - \Box\omega_\mu\omega_\mu)A' + 2\Box\omega_\mu\omega_\nu A'_{\mu\nu} + (e_o\omega_\mu - \omega_\mu e_o)\varepsilon_{\mu\nu\lambda\sigma}\partial_\nu A'_{\lambda\sigma}$$

$$+ \bar{u}uS + \bar{u}\gamma_5uP + \bar{u}\not{p}uV + \bar{u}\not{p}\gamma_5uT + e_o\bar{u}\,\gamma_5\psi_1 + i\omega_\mu\bar{u}\not{p}\gamma^\mu\psi_1$$

$$+ \bar{u}\,\gamma_5{}^0{}_2e_\psi + \bar{u}\not{p}\gamma^\mu{}_{\mu_2}\omega_\psi$$

where A', S, P, V and T are 5 scalars (neglecting parity), $A'_{\mu\nu}$ is an antisymmetric tensor and $\psi_1$ and $\psi_2$ are 2 Dirac spinors for complex $\Phi$; all the above are real or Majorana respectively for real $\Phi$. The spin content agrees exactly with our earlier analysis. Using the gauge invariance associated with $\omega_\mu$ noted above for N=1 we see that there are 8 bose and 8 fermi degrees of freedom. A similar discussion can be given for the Y=1 irrep.

We may also consider the irrep with $G = \pm 4p^2$, $Y = \mathbf{J} = 0$, $A_1 = A_2 = 1$, $\bar{D}_1D_1\bar{D}_2D_2 = 4p^2$, which is appropriate for the gauge field of SO(2) extended Yang-Mills theory without a central charge (Grimm, Sohnius and Wess, (1978)). By use of the projector $\frac{1}{2}\,[1-(4\Box)^{-1}\bar{D}_1D_1\bar{D}_2D_2\,]$ we may project this out of a superfield expanded in $e_\pm e_\pm$--- basis functions, and obtain, for a real field

$$\Phi_{Y=\mathbf{J}=0,\ \bar{D}_1D_1\bar{D}_2D_2=4p^2} = (e_+e_+ - 4\Box\omega_-\omega_-)A^1 + (e_+\omega_+ + \omega_-e_-)B\ +$$

$$+ [e_+\,\bar{u}^{\alpha+} + \omega_-\,(\bar{u}_-\not{p})^{\alpha+}]\psi_{1\alpha+} + [\bar{u}^{\alpha+}\ e_+ + (\bar{u}_-\not{p})^{\alpha+}\ \omega_-\,]\psi_{2\alpha+}$$

$$+ \text{herm-conj.} + S\,\bar{u}\gamma_5u + i\,\partial_aV_b\bar{u}\sigma_{ab}u$$

where $A^1$, B are complex scalars, S is a real pseudo scalar, $V_\mu$ is a real vector and $\psi_1$and $\psi_2$are two Weyl spinors; there are again 8 fermi and bose degrees of freedom.

Similar discussions may be given for the case of other irreps of the scalar superfield and also for superfields with an external spin index. These latter are of particular interest for supergravity; we will discuss them shortly.

When central charges are present we can obtain the field content of an irrep by doubling (or multiplying) the appropriately spin reduced multiplet. For N=2 with one central charge we take the general solution of

$$Z^2\Phi = P^2\Phi \tag{9}$$

as $$\Phi = \cosh\sqrt{P^2}\, f_o + (\sinh\sqrt{P^2})(\sqrt{P^2})^{-1} f_1 \tag{10}$$

where $f_o$ and $f_1$ are functions of x and $\theta_{\alpha 1}$ only. We have therefore two similar N=1 multiplets $f_o$ and $f_1$. This has the correct degrees of freedom 8(2Y+1) for each irrep of superspin Y, compared to the 16(2Y+1) degrees for SO(2) without such a central charge.

For N=4 there are 6 central charges and 7 for N=8. We find that there are infinite sets of reduced multiplets in either case satisfying the generalisation of (9) to $\underline{Z}$ being a 6 or 7 component vector for N=4 and 8 respectively. To avoid this we need to impose further restrictions. A trivial one is to choose all but one of these components of Z to be zero and so reduce the problem to the form (8); this has been done in the method of dimensional reduction by Legendre transformation (Sohnius, Stelle and West, 1980). It is possible to have alternative non-trivial conditions. Thus for two non-zero components of $\underline{Z}$ we may also choose

$$Z_1^{\ 2} = aP^2,\ Z_2^{\ 2} = (1-a)P^2$$

so $\Phi = \cosh Z_1\sqrt{aP^2}\cosh Z_2\sqrt{(1-a)P^2}\, f_{oo} + \cosh Z_1\sqrt{aP^2}\sin Z_2\sqrt{(1-a)P^2}.$
$[P^2]^{-\frac{1}{2}} f_{o1} + \sinh Z_1\sqrt{aP^2}.[P^2]^{-\frac{1}{2}}\cosh Z_2\sqrt{(1-a)P^2} f_{1o} + \sinh Z_1\sqrt{aP^2}$
$\sinh Z_2\sqrt{(1-a)P^2}[P^2]^{-1} f_{11}.$

We have therefore a quadrupling of the set of states. By similar conditions various multiplets of a fundamental reduced spin multiplet can be obtained; these irreps seem difficult to discuss in terms of Legendre transformations using a multi-time formalism, though this may be possible.

## LINEARISED SUPER SYMMETRIC YANG-MILLS THEORIES

We wish to write down Lagrangians quadratic in the basic superfields using appropriate irreps from those discussed above. We use the basis functions described above, though to preserve locality need to take $(\Box^{-1}e_o, \not\partial^{-1}u, \omega_\mu)$, $(\Box^{-1}e_\pm, \not\partial^{-1}u_\pm, \omega_\pm)$ so as to have the component fields as local functions of the fields in the original superfield from which projection into the irrep occurred.

Let us consider how to write a Lagrangian in the presence of a central charge. For N=2, under a global supersymmetry transformation with parameters $\varepsilon_1$ and $\varepsilon_2$ we have by use of the Dirac equation for $W_1$ and $W_2$ that

$$\delta \int d^4x \int d^4\phi\, (\bar{\varepsilon}_1 W_1 + \bar{\varepsilon}_2 W_2)\, L(x,\phi,z) = \int d^4x \int d^4\phi\, i(\bar{\varepsilon}_2\phi) Z L(x,\phi,z)$$

In order for this to vanish we may take for L the expression $[-(d\Phi/dz)^2 + \Phi\Box\Phi]$ for any superfield $\Phi$ satisfying $d^2\Phi/dz^2 = \Box\Phi$. Thus a supersymmetric Lagrangian density will be

$$L = \int d^4\phi\, [-(d\Phi/dz)^2 + \Phi\Box\Phi]\Big|_{z=o} \tag{11}$$

A similar expression can be used for N=4, though now with $d^4\phi$ replaced by $d^4\phi_1\, d^4\phi_2$ with $\phi_1$ and $\phi_2$ given earlier. Finally we may solve the equation $d^2\Phi/dz^2 = \Box\Phi$ by means of the formal operator power series $f_1 = \cosh z\sqrt{\Box}$ and $f_2 = (\Box)^{-\frac{1}{2}} \sinh z\sqrt{\Box}$. We may multiply either of these by the basis functions of the reduced algebra and suitable projection operators to obtain the basis functions in the presence of the central charge. These basis functions for the Y=0 singlet irrep of SO(2) with spin reducing central charge are thus the set $\{f_1 e_o, f_2 e_o, f_1 u_\alpha, f_2 u_\alpha, f_1\omega_\mu, f_2\omega_\mu\}$. There is thus a doubling of field components, though the physical description of these latter is considerably modified by the fact that $f_1$ and $f_2$ have different dimensions; we will see the details of this in the actual applications to SO(2) and SO(4).

N=2

We apply the above results to SO(2) with a spin-reducing central charge. We use the projection operator method which for super-QED (N=1) has Abelian Lagrangian density $\int d^4\theta\, (\Pi_y^{\frac{1}{2}}\Phi)\Box(\Pi_y^{\frac{1}{2}}\Phi)$, where $\Pi_y^{\frac{1}{2}}$ is the projection operator onto the $Y=\frac{1}{2}$ irrep. For SO(2) a singlet superfield V with $\mathfrak{M}V=0$ must have AV=0, so can only have $Y=\frac{1}{2}$. We may thus write the gauge vector field as $V = \Pi_y^{\frac{1}{2}}\Phi$ with component expansion

$$V = -\Box^{-1}e_o(f_1A - f_2B) + \sqrt{\tfrac{1}{2}}f_1\bar{u}\not\partial^{-1}\psi_1 + \sqrt{\tfrac{1}{2}}f_2\bar{u}\psi_2 + f_1\omega_\mu A_\mu$$
$$+ f_2\omega_\mu B_\mu \quad .$$

The linearised Lagrangian (11) has the value (using the tensors $\int d^4\phi e_{(i)} e_{(j)}$ (which can be evaluated once and for all)

$$L = \tfrac{1}{2}(A^2 + p^2B^2) + \frac{i}{2}\bar{\psi}_\ell \not{\partial}\psi_\ell - \tfrac{1}{2}p^2 A_\mu \bar{\eta}_{\mu\nu} A_\nu - \tfrac{1}{2}B_\mu \bar{\eta}_{\mu\nu} B_\nu$$

where $\bar{\eta}_{\mu\nu} = \eta_{\mu\nu} - p_\mu p_\mu|p^2$. We note that $A_\mu$, $\psi_\ell$ and B are a physical singlet gauge vector, doublet Majorana spinor and scalar. The set of three independent fields $B_\mu$ gives 1 scalar on-shell, as can be seen by introducing the scalar Lagrange multiplier $\chi$ by the additional term $B_\mu \partial_\mu \chi$ to ensure $\partial_\mu B_\mu = 0$. The resulting elimination of $B_\mu = \partial_\mu \chi$ produces the physical scalar term $\tfrac{1}{2}(\partial_\mu x)^2$. We may also extend this discussion to the N=2 matter multiplet; this is more fully discussed in Taylor (1980b).

N=4

We may proceed similarly for the case of SO(4) with a spin-reducing central charge. In that case we must take $A_1 = A_2 = 0$ for the gauge singlet superfield V, so that $\mathfrak{M}.V=0$. We must also have Y=0 only, to give a maximum Poincaré spin of unity, so that we must have the Casimir $\mathfrak{J}$ with values $\mathfrak{J}^2 = 4$. This can be obtained by projecting onto $A_1 = A_2 = 0$, $J = 3p^4$, with $\mathfrak{J} = \frac{1}{p^2}\ \bar{D}_1 i\not{p}\gamma_5 D_2$, $J = \frac{1}{4}\{p^2(\bar{D}_1 i\gamma_\mu\gamma_5 D_1)(\bar{D}_2 i\gamma_\mu\gamma_5 D_2) - (\bar{D}_1 i\not{p}\gamma_5 D_2)(\bar{D}_2 i\not{p}\gamma_5 D_2)\}$. The resulting component expansion of V can be shown, after some algebra, to be

$$V = -\sqrt{2}(\Box)^{-2}\ (e_o e_o - \Box\omega_\mu\omega^\mu)(f_1 A + \Box f_2 B)$$

$$+ \Box^{-1}\sqrt{2}[\omega_\mu\omega_\nu\ (f_1 A_{\mu\nu} + \Box f_2 B_{\mu\nu}) - i(2\Box)^{-1}(e_o\omega_\mu - \omega_\mu e_o)\varepsilon_{\mu\nu\lambda\sigma}p_\nu$$

$$(f_1 A_{\lambda\sigma} + f_2 B_{\lambda\sigma})] - (2\sqrt{2}\Box)^{-1}\ [\bar{u}u(f_1 S + f_2 S') + \bar{u}\gamma_5 u(f_1 P + f_2 P')]$$

$$- (\sqrt{2}\Box)^{-2}\ [\bar{u}\ \not{p}u\ (f_1 V^1 + \Box f_2 V) + \bar{u}\ \not{p}\gamma_5 u\ (f_1 D' + \Box f_2 D)\ ]$$

$$- (\sqrt{2}\Box)^{-1}\ e_o\bar{u}(f_1\ \not{p}^{-1}\gamma_5\psi_1 + f_2\ \gamma_5\psi_2) + i(\sqrt{2})^{-1}\Box^{-1}\omega_\mu\bar{u}(f_1\gamma^{\mu 1}\psi_1$$

$$+ f_2\gamma_\mu\not{p}\psi_2) - (\sqrt{2}\Box)^{-1}\ \bar{u}(f_1 e_o\ \not{p}^{-1}\gamma_5\psi_3 + f_2 e_o\ \gamma_5\psi_4)$$

$$+ i(\sqrt{2})^{-1}\Box^{-1}\bar{u}(f_1\omega^\mu\gamma_\mu\psi_3 + f_2\omega^\mu\gamma_\mu\not{p}\psi_4)$$

where the basis function product $e_o e_o$ is a shorthand for $e_o(\phi_1)$,

$e_o(\phi_2)$(and similarly for the other terms above) with Lagrangian (11) (with an extra box in each term) equal to

$$L = \tfrac{1}{2}(A^2 + p^2B^2) + \tfrac{1}{2}(A_{\mu\nu}A_{\lambda\sigma}\bar{\eta}_{\mu\lambda}\bar{\eta}_{\nu\sigma} + p^2B_{\mu\nu}B_{\lambda\sigma}\bar{\eta}_{\mu\lambda}\bar{\eta}_{\nu\sigma})$$
$$+ \frac{i}{2}\sum_{\ell=1}^{4}\psi_\ell \not\partial\psi_\ell + \tfrac{1}{2}(V'^2 + D'^2 + S'^2 + P'^2)$$
$$+ \tfrac{1}{2}p^2(V^2 + D^2 + S^2 + P^2) \qquad (12)$$

The distinction between physical and auxiliary fields in V is clear from (12) except for the antisymmetric tensor $A_{\mu\nu}$; it represents on-shell a gauge vector, as the additional Lagrangian multiplier term $-2A_{\mu\nu}p_\lambda\chi_\tau\bar{\eta}_{\mu\lambda}\bar{\eta}_{\nu\tau}$ introduced to preserve $\partial_\mu A_{\mu\nu} = 0$ generates the term $-\frac{1}{4}(\partial_{[\mu}\chi_{\nu]})^2$. The other antisymmetric tensor $B_{\mu\nu}$ is well known to describe a scalar. The count of auxiliary and physical fields in V agrees with that reached independently elsewhere (Sohnius, Stelle and West, 1980); the count is two conserved antisymmetric tensors, ten scalars and four Majorana spinors off-shell and 6 scalars, one gauge vector field and four Majorana spinors on-shell.

It is interesting to note that the Lagrangian is obtained by integrating over 8 θ's and not only over 4. Thus the dimension of Φ will be non-canonical and so cause difficulties in making L non-linear. However this does give an example of true integration over the complete set of θ-variables, and so may allow a similar expression in the supergravitational case.

## LINEARISED SUPERGRAVITY THEORIES

The appropriate irreps for the minimal set of auxiliary fields were shown by Ogievetsky and Sokatchev (1977) to be Y=0 and $^3/_2$, though the geometrical approach using the Lagrangian equal to the volume of super-space gives a larger set of irreps, these being at least Y = $^3/_2$, $^1/_2$, $0^2$ (J.G. Taylor, 1979b). In our analysis of extended supergravity we will solely apply the criterion of localisability of the Lagrangian $\int d^{4N}\theta\, \Phi\, \Pi\Box\Phi$. Thus we will search for projectors Π, or combinations thereof, for which □Π has no inverse powers of □. The possibility of relating to a supercurrent should also be considered, but since the supercurrent analysis has not been given in the extended superfield framework we will not discuss it further here.

N=2

By analogy with N=1 we may hope that torsion constraints may be discovered which will reduce the super-geometric objects (multibein and super-connection) to a "pre-pre-potential" which is a scalar superfield V. For then we know (Taylor 1980a) that all irreps will have Poincaré spin limited by 2. There is a unique irrep with Y=1 which contains a spin 2 component field. This latter enters in a basis function expansion in the form $\omega^\mu \omega^\nu h_{\mu\nu}$, and thus is satisfactory to describe the graviton. This is to be compared to other irreps in the vector superfield with Y=1, for which the spin 2 component field is represented by antisymmetric higher rank tensors. (Pickup and Taylor, 1980a).

We can calculate the projector $\Pi_{Y=1}$ to have the value $\Pi_{Y=1} = \frac{3}{4}(1 - A_1)(1 - A_2) - (4\Box^2)^{-1}J$. $\Box\Pi_{Y=1}$ is non-local due to the $\Box A_1 A_2$ and $\Box^{-1}J$ terms. We may remove these terms (uniquely) by use of the projectors $\Pi_+$ onto the irrep with A=2, $R=+4p^4$ and $\Pi_{-+}$ onto A=2, $R=4p^2$, $\sum=-8p^2$, where $\sum = \bar{D}_1 D_1 \bar{D}_2 D_2 + \bar{D}_1\gamma_5 D_1 \bar{D}_2\gamma_5 D_2$ and $R = G_1 G_2$, with the values $\Pi_+ = \frac{1}{2}A_1 A_2[1 + (8\Box^2)^{-1}R]$, $\Pi_{-+} = \frac{1}{4}A_1 A_2 - (16\Box^2)^{-1}R - (16\Box)^{-1}\sum$. Then the combination $(\Pi_{Y=1} - \Pi_+ - \Pi_{-+})$ has the value $\Pi = \frac{3}{4}(1 - A) - (16\Box^2)^{-1}(4J + R) + (16\Box)^{-1}\sum$. Since this combination is local when multiplied by the factor $\Box$, the expression

$$\int d^8\theta \; V\Box \{\tfrac{3}{4}(1 - A) - (16\Box^2)^{-1} (4J + R) + (16\Box)^{-1} \textstyle\sum\} V$$

is a candidate linearised Lagrangian for N=2 extended supergravity. However it involves one Y=1 irrep, one real Y=0 irrep (with projector $\Pi_+$) but also two further Y=0 irreps since $\Pi_+$ projects onto both the irreps with $G = +4p^2$ and $G = -4p^2$. We therefore have at least (48 + 48) bose and fermi degrees of freedom to be compared with the known minimal set of 40 + 40 (de Wit and van Holten 1979, Fradkin and Vasiliev 1979).

We remove one of the Y=0 irreps by projecting $\Pi_+ V$ down with the operator $\Pi_o = [1 - (4\Box)^{-1} \bar{D}_1 D_1 \bar{D}_2 D_2]$; this corresponds to the N=2 chiral irrep of Firth and Jenkins (1975), which breaks G invariance. If we use this operator we find that, with $\Pi'_+ = \Pi_+ \Pi_o$, $\Pi' = \Pi_1 - \Pi_{-+} - 2\Pi'_+$ gives $\Box\Pi'$ is local, with

$$\Box\Pi' = \frac{1}{4}\Box A_1 A_2 - (16\Box)^{-1}R - (16\Box)^{-1} (\bar{D}_1 D_1 \bar{D}_2 D_2 - \bar{D}_1\gamma_5 D_1 \bar{D}_2\gamma_5 D_2).$$

Moreover we may write this expression in the form proportional to

$$\{-(\bar{D}_1 i\gamma^m\gamma_5 D_1 + \bar{D}_2 i\gamma^m\gamma_5 D_2)^2 - 3(\bar{D}_1 D_1 + \bar{D}_2 D_2)^2 + (\bar{D}_1\gamma_5 D_1 + \bar{D}_2\gamma_5 D_2)^2\}$$

suggestive of arising from a six-dimensional vector with components $(\bar{D}_1 i\gamma^m\gamma_5 D_1 + \bar{D}_2 i\gamma^m\gamma_5 D_2)V$, $\sqrt{3}(\bar{D}_1 D_1 + \bar{D}_2 D_2)V$, $(\bar{D}_1 i\gamma_5 D_1 + \bar{D}_2 i\gamma_5 D_2)V$. It would be of interest to obtain this from a six-dimensional approach. These and other items are discussed in Pickup and Taylor (1980b).

N=4

We are presently engaged on analysis of the projectors onto Y=0 irreps discussed in Pickup and Taylor (1980a) in order to carry through an analysis similar to that carried out for N=2. This will allow us to obtain the minimal auxiliary field structure. Central charge spin reduction should then allow us to proceed to N=8.

REFERENCES

de Wit, B. and van Holten, J.W. (1979) Nucl.Phys. B155 pp 530-542.

Downes-Martin, S. and Taylor, J.G. (1977) Nucl.Phys. B126, pp 97-108.

Fayet, P. (1976), Nucl.Phys. B113, pp 135-155.

Ferrara, S. and Piguet, O. (1975), Nucl.Phys. B93, pp 261-302.

Firth, R. and Jenkins J. Nucl.Phys. (1975) B85 pp 525-534.

Fradkin, E.S. and Vasiliev, M.A. (1979) Phys.Lett. 85B pp 47-51.

Grimm, R., Sohnius, M. and Wess, J. (1978), Nucl.Phys. B133, pp 275-284.

Jarvis, P. (1976), Journ.Math.Phys. 17 pp 916-922.

Ogievetsky, V. and Sokatchev, E. (1977) Nucl.Phys. B129, pp 309-316.

Pickup, C. (1980) "Representations of Supersymmetry and their Applications", Univ. of London Ph.D. Thesis (unpublished).

Pickup, C. and Taylor, J.G. (1980a) "Representations of N=4 Supersymmetry on Superfields", King's College preprint.

Pickup, C. and Taylor, J.G. (1980b) "Linearised N=2 Superfield Supergravity", King's College preprint.

Salam, A. and Strathdee, J. (1975) Phys.Rev. D11 pp 1521-1539.

Sohnius, M. (1978), Nucl.Phys. B138, pp 109-121.

Sohnius, M., Stelle, K. and West, P.C. (1980) Phys.Lett 92B, pp 123-127.

Sokatchev, E. (1975), Nucl.Phys. B99, pp 96-108.

Taylor, J.G. (1979a) "The Ultra-Violet Divergences of Superfield Supergravity", In Proc. Int. Conf. on High-Energy Physics, CERN, Geneva, pp 969-973.

Taylor, J.G. (1979b) "Quantisation of Superfield Supergravity", King's College preprint, May 1979 (unpublished).

Taylor, J.G. (1980a) Nucl.Phys. B169, pp 484-500.

Taylor, J.G. (1980b) Phys.Lett. 94B, pp 174-178.

Wess, J. and Zumino, B. (1977) Phys.Lett. 66B, pp 361-364.

# IV N=8 SUPERGRAVITY

# ANTISYMMETRIC TENSORS AND SUPERGRAVITY

M.J. Duff
Physics Department, Imperial College, London.

## 1. INTRODUCTION

The work I am going to describe concerns antisymmetric tensors and is based on a paper written in collaboration with Peter van Nieuwenhuizen (Duff and van Nieuwenhuizen, 1980. See also Duff, 1980a; 1980b; van Nieuwenhuizen, 1980). I shall also include some more recent thoughts and speculations of my own on where I think this avenue of research might lead. Before discussing the applications to supergravity, let me begin by summarizing the main results:

(i) Although the gauge theory of a rank-two antisymmetric tensor potential $A_{\mu\nu}$ describes one degree of freedom, it is not, contrary to naive expectations (see Section 2), equivalent to the theory of a single scalar field A. In particular, the gravitational conformal anomalies are different :

$$T(A_{\mu\nu}) - T(A) = \frac{1}{32\pi^2}(R_{\mu\nu\rho\sigma}R^{\mu\nu\rho\sigma} - 4R_{\mu\nu}R^{\mu\nu} + R^2) \qquad (1.1)$$

where T is the anomalous contribution to the trace of the regularized energy momentum tensor.

(ii) Similarly, the gauge theory of a rank-three antisymmetric tensor potential $A_{\mu\nu\rho}$, which has no degrees of freedom at all, still has a non-vanishing trace anomaly given by

$$T(A_{\mu\nu\rho}) = -\frac{1}{16\pi^2}(R_{\mu\nu\rho\sigma}R^{\mu\nu\rho\sigma} - 4R_{\mu\nu}R^{\mu\nu} + R^2) \qquad (1.2)$$

(iii) Some readers may recognise the RHS of (1.1) and (1.2) to be proportional to the Gauss-Bonnet combination

$$*R_{\mu\nu\rho\sigma}*R^{\mu\nu\rho\sigma} = R_{\mu\nu\rho\sigma}R^{\mu\nu\rho\sigma} - 4R_{\mu\nu}R^{\mu\nu} + R^2, \qquad (1.3)$$

where

$$*R^{\mu\nu\rho\sigma} = \frac{1}{2}\varepsilon^{\mu\nu\alpha\beta}R_{\alpha\beta}{}^{\rho\sigma} \quad , \qquad (1.4)$$

which provides the integrand of a topological invariant, the Euler number $\chi$ ;

$$\chi \equiv \frac{1}{32\pi^2}\int d^4x\,\sqrt{g}*R_{\mu\nu\rho\sigma}*R^{\mu\nu\rho\sigma} = \text{integer} \qquad (1.5)$$

This strongly suggests that these quantum inequivalences may be derived from topological considerations, and this is indeed the case. As we shall see in Section 4, (1.1) and (1.2) will follow very simply as a consequence of the Gauss-Bonnet theorem. In spaces with non-trivial topology, moreover, the Euler number provides a non-vanishing contribution to the one-loop counter-terms $S_1$. For closed loops of A, $A_{\mu\nu}$ and $A_{\mu\nu\rho}$ fields in a gravitational background we find

$$S_1(A_{\mu\nu}) - S_1(A) = \frac{-1}{\varepsilon}\chi \qquad (1.6)$$

and

$$S_1(A_{\mu\nu\rho}) = \frac{2}{\varepsilon}\chi \qquad (1.7)$$

where $\varepsilon = n-4$ is the dimensional regularization parameter, which once again demonstrates the quantum inequivalence between $A_{\mu\nu}$ and A, and between $A_{\mu\nu\rho}$ and nothing. This modifies the previous result of Sezgin and van Nieuwenhuizen (1980) who found equivalent counterterms after discarding total divergences. With hindsight these inequivalences of different field representations at the quantum level are hardly surprising since, as discussed in Section 3, the naive equivalence arguments break down even at the classical level when global considerations are taken into account.

(iv) In addition to the Gauss-Bonnet theorem, there is another well-known topological result concerning antisymmetric tensor representations, namely the Hirzebruch Signature Theorem:

$$n(1,0) - n(0,1) = \tau \qquad (1.8)$$

where n(1,0) and n(0,1) are the number of self-dual and antiself-dual zero eigenvalue modes of the Laplacian acting on second-rank tensors, and

$$\tau \equiv \frac{1}{48\pi^2}\int d^4x\sqrt{g}\ {}^*R_{\mu\nu\rho\sigma}R^{\mu\nu\rho\sigma} \qquad (1.9)$$

is the signature, (= P/3 where P is the Pontryagin number). This theorem bears a strong resemblance to the Atiyah-Singer index theorem

$$n(\tfrac{1}{2},0) - n(0,\tfrac{1}{2}) = -\frac{\tau}{8} \qquad (1.10)$$

where $n(\frac{1}{2},0)$ and $n(0,\frac{1}{2})$ refer to the number of right and left handed zero eigenvalue modes of the Dirac operator acting on spin-$\frac{1}{2}$ fermions. Now just as (1.10) may be regarded as the integrated version of the fermion axial current anomaly, so (1.8) may be regarded as the integrated version of a bosonic "chiral" anomaly for antisymmetric tensors. The gauge theory of $A_{\mu\nu}$ would contribute to such an anomaly by an amount $\tau$, whereas the gauge theory of $A_{\mu\nu\rho}$ (for which the rank-two tensors enter as complex ghosts) would contribute by an amount $-2\tau$. This provides yet another example of the difference between the representations

$A_{\mu\nu}$ and A, and between $A_{\mu\nu\rho}$ and nothing.

(v) Whilst on the subject of anomalies, one might also enquire about possible inequivalence of different fermion representations. For example, one might expect that replacing a spin-$\frac{1}{2}$ fermion $\psi^\alpha$ by an antisymmetric tensor-spinor potential $\psi^\alpha_{\mu\nu}$ (Townsend, 1980) might also lead to different anomalies. Indeed, this could have far-reaching consequences if a change of field representation were to change the Yang-Mills axial anomaly and hence the GIM cancellation mechanism. However, the coupling of such representations to gravity and gauge fields seems to encounter inconsistency problems, which will not be pursued here (See, however, Section 5).

(vi) Yet another feature of the rank-3 potential $A_{\mu\nu\rho}$ coupled to gravity is that the corresponding field strength $F_{\mu\nu\rho\sigma}$ appears in the Einstein equation as

$$G_{\mu\nu} - \frac{\kappa^2 F^2}{2.4!} g_{\mu\nu} = 0 \tag{1.11}$$

But its equation of motion

$$\nabla^\mu F_{\mu\nu\rho\sigma} = 0 \tag{1.12}$$

implies

$$F_{\mu\nu\rho\sigma} = \varepsilon_{\mu\nu\rho\sigma} c \ , \ c = \text{constant} \tag{1.13}$$

Substituting (1.13) into (1.11) therefore yields Einstein's equation with a cosmological constant $\Lambda = -\kappa^2 c^2/2$. This result may also be interpreted as a topological one. See Section 3.

(vii) One might also ask whether this apparent equivalence between pure gravity with a cosmological constant and the combined Einstein-$A_{\mu\nu\rho}$ system also persists at the quantum level. For example, are the one-loop counterterms the same? Fortunately this question was easy to answer because the relevant calculations had already been carried out. We found that the one-loop counterterms for the gravity- $A_{\mu\nu\rho}$ system calculated by Sezgin and van Nieuwenhuizen (1980) yield, upon using (1.11) and (1.13) the same results (except, as one would now expect from (1.7), for the topological term) as those for pure gravity with a cosmological constant calculated by Christensen and Duff (1979b). Thus the cosmological constant may be reformulated as the gauge theory of $A_{\mu\nu\rho}$ !

What are the implications of these results for supergravity? As far as I am aware, there are at least three:

(a) Siegel (private communication) has pointed out to me that in the version of N=8 extended supergravity obtained by dimensional reduction from eleven dimensions (Cremmer and Julia, 1979) the 70 spin-0 fields actually appear as 63A + $7A_{\mu\nu}$+1 $A_{\mu\nu\rho}$. Previously, it was generally assumed that this version was equivalent to the one with 70 A fields. The trace anomaly of this latter version is known (Christensen and Duff, 1978a) to be

$$T = -\frac{5}{32\pi^2} {}^*R_{\mu\nu\rho\sigma} {}^*R^{\mu\nu\rho\sigma} \qquad (1.14)$$

The importance of this, as noted by Siegel, is that when modified by (1.1) and (1.2) the new trace anomaly now vanishes! Thus the antisymmetric tensor version of N=8 supergravity has no anomalous scaling behaviour at the one-loop level and is completely one-loop finite on-shell even when topological Euler number counterterms are taken into account.

I have also checked that there is a cancellation of zero-modes for ALE (asymptotically locally Euclidean) instantons. (Eguchi and Hansen, 1978; Gibbons and Hawking, 1978) in this N=8 theory using the results of Hawking and Pope (1978). This represents an additional cancellation because ALE spaces have boundaries and this means that even the boundary contributions to the one-loop counterterms (Christensen and Duff, 1978b) which are commonly ignored, also cancel. See Section 5.

What about N<8? There does not appear to be an obvious way to truncate this version to lower N. However, Eugene Cremmer has drawn my attention to a version of the N=4 theory in which one of the two scalar fields appears in the Lagrangian only through its derivatives (Cremmer, Scherk and Ferrara, 1978) and is thus a candidate for replacement by a rank-two gauge potential $A_{\mu\nu}$. When modified by (1.1) this theory too would have vanishing trace anomaly (Section 5).

(b) The hitherto unrealized importance of antisymmetric tensors may also help to clear up two mysteries in the axial and conformal anomalies of simple and extended supergravity calculated by Christensen and myself. (Christensen and Duff, 1978a;1979a; Duff, 1979). First, we found that the numerical coefficients appearing in the anomalous trace of the stress tensor and in the anomalous divergence of the axial current were not the same. Thus these supergravity anomalies, along with that of the supercurrent, did not appear to form a supermultiplet as is the case in global supersymmetry (Ferrara and Zumino, 1975) and as is the case for matter multiplets in a classical supergravity background. (For a review, see Grisaru, 1978). This is a rather disturbing result, and one of which Marc Grisaru has constantly reminded me. Secondly, in rederiving these results via the method of Super Index Theorems, we remarked that "the combination of fields appearing in the super index theorems are just those one encounters in physical supermultiplets. The only exception seems to be the appearance of antisymmetric tensor representations (1,0) and (0,1) which are not normally considered as elementary fields in a Lagrangian" (Christensen and Duff, 1979a). Indeed, although for reasons discussed in Section 5 this posed no problem for the trace anomalies, in deriving the corresponding chiral anomalies we found it necessary, at the time, to discard these representations in what seemed a rather unnatural way because there seemed no room for them in the supergravity theories under consideration e.g. simple supergravity.

With hindsight, however, it now seems that these antisymmetric tensors should be taken seriously in the light of the possible bosonic contributions to the chiral anomaly discussed above in (iv). In fact, if the super index theorems are interpreted in the way described in Section 5, they cannot fail to bring the chiral anomaly coefficients in line with those of the trace anomaly! My optimism that this interpretation may be justified physically has been strengthened by conversations at this workshop with Marc Grisaru, Martin Rocek and Warren Siegel. In their superfield Feynman graph programme, (Grisaru, Rocek and Siegel, 1979, See also Grisaru, 1979, 1980; Grisaru and Siegel 1979), the gauge-fixing and ghost procedure is, from start to finish, manifestly invariant under supersymmetry transformations of the background fields. Thus, if applied to the calculation of supergravity anomalies, it would appear to guarantee that they should fall into a supermultiplet. Moreover, the antisymmetric tensor superfield (Siegel, 1979) appears naturally in their formulation. I should add that such bosonic chiral currents have been discussed elsewhere in the literature from a purely mathematical point of view (Dowker, 1978; Nielsen, Romer and Schroer, 1978) but their physical application has always remained obscure. It is remarkable therefore, that supergravity provides the vehicle by which they acquire physical significance.

(c) A third application concerns the cosmological constant. Indeed, it was Ogievetsky and Sokatchev (1980) who first pointed out that their version of the superspace approach to N=1 supergravity naturally yields a cosmological constant in the field equations without a cosmological term in the Lagrangian i.e, in component formalism, without the addition to the Lagrangian of a term linear in the auxiliary field S. (See, however, Section 3). To make contact with result (vi) above, we note that in their formulation the traditional auxiliary fields S and P (Ferrara and van Nieuwenhuizen 1978; Stelle and West, 1978) are replaced by $\partial^\mu S_\mu$ and $\partial^\mu P_\mu$ (Stelle and West, 1978). Setting $A_{\mu\nu\rho} \equiv \varepsilon_{\mu\nu\rho\sigma} S^\sigma$ etc then yields a kinetic $F_{\mu\nu\rho\sigma}F^{\mu\nu\rho\sigma}$ term of the kind described in (vi) and hence a cosmological constant in the Einstein equations. A priori, of course, this $A_{\mu\nu\rho}$-cosmological relation is logically independent of any supergravity considerations. But, as usual, supergravity seems to employ every trick in the book!

Returning now to N=8 supergravity, in a recent paper Aurelia, Nicolai and Townsend (1980) have independently exploited the appearance of $A_{\mu\nu\rho}$ in the Lagrangian obtained by dimensional reduction from 11 dimensions. They observe that N=8 supergravity may now admit of a cosmological constant without gauging the internal symmetry. (Their sign of $\Lambda$ is opposite to that encountered in the gauged models). Their interpretation is rather that this yields a model in which the supersymmetry is spontaneously broken, and 7 of the 28 vector particles acquire a mass. We shall discuss the quantum implications of this model in Section 6.

The remainder of this paper is devoted to filling in the details, and I shall work in a manifold of Euclidean signature which, for the most part, will be taken to be compact without boundary. This will facilitate the introduction of various topological steps leading to the desired results. I wish to emphasize, however, that this topological approach may be viewed in one of two ways. One may regard it, as I do, as providing a fascinating connection between topological ideas on the one hand, and those of quantum gravity and in particular supergravity on the other. Alternatively, it may be viewed merely as a calculational device for fixing the numerical coefficients appearing in the counterterms, trace anomalies, and axial anomalies etc. The chiral anomaly provides the classic illustration of this. One may calculate it using the traditional Minkowski space Feynman rules applied to the triangle graph; or else one may work on a compact Euclidean manifold, invoke the Atiyah-Singer index theorem, and arrive at the same result. Thus it is not necessary to believe that the apparatus of Euclideanization, index theorems, ALE spaces, self-duality, super index theorems etc., are physically significant per se (though I would like to think they might be) because, if nothing else, they provide an efficient and often less laborious route to the desired answers.

I shall adopt the convention that the Euclidean action is, wherever possible, positive definite e.g. the Maxwell Lagrangian is $+\frac{1}{4}\sqrt{g}F_{\mu\nu}F^{\mu\nu}$. Correspondingly, the counterterms have an unconventional sign. The Einstein Lagrangian is

$$\mathcal{L} = -\frac{1}{2\kappa^2}\sqrt{g}(R-2\Lambda) \tag{1.15}$$

where $\kappa^2 = 8\pi G$ and $\Lambda$ is chosen such that de Sitter space, with $S^4$ topology, has $\Lambda>0$.

## 2. NAIVE EQUIVALENCE ARGUMENTS

Consider a Maxwell-type gauge theory but where the vector potential $A_\mu$ is replaced by an antisymmetric tensor $A_{\mu\nu}$ or $A_{\mu\nu\rho}$. It is commonly assumed that these theories are equivalent to a scalar field A, or nothing respectively. The argument begins as follows. The corresponding field strengths, $F_{\mu\nu\rho}$ or $F_{\mu\nu\rho\sigma}$, obey the field equations

$$\nabla^\mu F_{\mu\nu\rho} = 0 \quad ; \quad \nabla^\mu F_{\mu\nu\rho\sigma} = 0 \tag{2.1}$$

or

$$\partial_\mu f_\nu - \partial_\nu f_\mu = 0 \qquad \partial_\mu f = 0 \tag{2.2}$$

where

$$f_\mu \equiv \frac{1}{3!}\varepsilon_{\mu\nu\rho\sigma} F^{\nu\rho\sigma} \quad ; \quad f \equiv \frac{1}{4!}\varepsilon_{\mu\nu\rho\sigma} F^{\mu\nu\rho\sigma} \tag{2.3}$$

Then, so the argument goes, (2.2) imply

$$f_\mu = \partial_\mu A \quad ; \quad f = 0 \tag{2.4}$$

Hence, the $A_{\mu\nu}$ theory is equivalent to $\Box A = 0$ on invoking the Bianchi identity $\nabla^\mu f_\mu \equiv 0$, whereas the $A_{\mu\nu\rho}$ theory is totally vacuous. Similar equivalence arguments are used at the quantum level.

In order to see what is wrong with these proposals, let us remind ourselves of some elementary differential geometry.

## 3. DIFFERENTIAL FORMS

In a coordinate basis a p-form $\omega$ may be written

$$\omega = \frac{1}{p!}\, \omega_{\mu_1 \ldots \mu_p}\, dx^{\mu_1} \wedge \ldots \wedge dx^{\mu_p} \tag{3.1}$$

The exterior derivative d is given by

$$d\omega = \frac{1}{(p+1)!}\, \omega_{\mu_1 \ldots \mu_p;\mu}\, dx^\mu \wedge dx^{\mu_1} \wedge \ldots dx^{\mu_p} \tag{3.2}$$

and its adjoint $\delta = {*d*}$ by

$$\delta\omega = \frac{1}{(p-1)!}\, \nabla^{\mu_1} \omega_{\mu_1 \ldots \mu_p}\, dx^{\mu_2} \wedge \ldots \wedge dx^{\mu_p} \tag{3.3}$$

They obey $d^2 = \delta^2 = 0$. A form $\omega$ is called <u>exact</u> if $\omega = d\alpha$, coexact if $\omega = \delta\beta$, <u>closed</u> if $d\omega = 0$ and coclosed if $\delta\omega = 0$. The Laplacian is defined by

$$\Delta = d\delta + \delta d \tag{3.4}$$

and $\omega$ is called <u>harmonic</u> if $\Delta\omega = 0$. Introducing an inner product

$$(\alpha,\beta) = \frac{1}{p!} \int_M d^4x\sqrt{g}\, \alpha_{\mu_1 \ldots \mu_p}\, \beta^{\mu_1 \ldots \mu_p}\,, \tag{3.5}$$

and using the fact that the manifold M is compact without boundary, we may integrate by parts to show that

$$(\omega,\Delta\omega) = (d\omega,d\omega) + (\delta\omega,\delta\omega) \geqslant 0 \tag{3.6}$$

and hence that $\Delta\omega = 0$ implies $d\omega = \delta\omega = 0$. Note that the Laplacian is not simply the D'Alembertian $-\nabla^\mu\nabla_\mu$ ; on 1-forms for example

$$\Delta A_\mu = -\nabla^\rho\nabla_\rho A_\mu + R_{\mu\nu}A^\nu \tag{3.7}$$

The crucial result for what follows is the Hodge decomposition which states that any p-form $\omega$ may decompose uniquely into three parts: harmonic, exact, and coexact. Thus

$$\omega = \omega_H + d\alpha + \delta\beta \quad ; \quad \Delta\omega_H = 0 \tag{3.8}$$

which means, in particular, that while all exact forms are closed, not all closed forms are exact. For example,

$$\partial^\mu {*F}_{\mu\nu} = 0 \quad \not\Rightarrow \quad F_{\mu\nu} = \partial_\mu A_\nu - \partial_\nu A_\mu \tag{3.9}$$

$$\partial^\mu V_\mu = 0 \quad \not\Rightarrow \quad V_\mu = \partial^\nu H_{[\mu\nu]} \tag{3.10}$$

$$\partial_\mu A_\nu - \partial_\nu A_\mu = 0 \quad \not\Rightarrow \quad A_\mu = \partial_\mu \phi \tag{3.11}$$

In other words, while the equations on the right may be valid locally, they are not valid globally. Similarly, in connection with the field strength $F_{\mu\nu\rho\sigma}$, the integral

$$\int d^4x\sqrt{g}\ \varepsilon^{\mu\nu\rho\sigma} F_{\mu\nu\rho\sigma}$$

vanishes only if F is exact.

It is now clear where the error lies in the naive arguments of section 2. While (2.2) are correct, (2.4) are not because the harmonic parts of $f_\mu$ and f have been overlooked. (It would also be interesting to generalize the Minkowski space equivalence of the non-abelian $A_{\mu\nu}$theory and the non-linear $\sigma$-model (Freedman and Townsend, 1980) to the case of spacetimes and/or gauge fields with non-trivial topology). As we shall see, similar inequivalence will appear at the quantum level in the guise of different anomalies and counterterms. Moreover, the harmonic part of f will also provide the cosmological constant.

At this stage, however, there is an obvious objection. If, as is customary, the field strength is defined to be the curl of a potential, i.e. F=dA, one has automatically ruled out the possibility of a non-vanishing harmonic part to F as is clear from the Hodge decomposition (3.8). Writing the conventional action in first order form does not help, either. To see this, we write the action as

$$S=(F,dA) - \frac{1}{2}(F,F) \tag{3.12}$$

S is invariant under

$$F \to F \quad A \to A + V \quad \text{with} \quad dV = 0 \tag{3.13}$$

(which is somewhat stronger than $A \to A + d\alpha$). Varying with respect to A and F independently yields

$$\delta F = 0 \qquad F = dA \tag{3.14}$$

(The virtue of this compact notation is that we may treat p-forms A of arbitrary rank simultaneously. In the case that A is a vector, for example, S is the conventional Maxwell action in first order form). Now (3.14) imply the equations

$$\delta F = dF = 0 \tag{3.15}$$

but are not implied by them, since (3.15) allow for a harmonic part to F

$$F = \omega + dA \quad ; \quad \Delta\omega = 0 \tag{3.16}$$

(For example, adopting (3.14) rather than (3.15) as the equations of electromagnetism would rule out Dirac magnetic monopoles, $F \sim \sin\theta d\theta \wedge d\phi$, for which F is closed but not exact. One may, of course, force F to be everywhere the curl of a potential but then one pays the price that the potential exhibits a string singularity. However, this is just a physicist's artefact, arising from a failure to recognise that more than one coordinate patch is required to cover the manifold).

The following remedy therefore suggests itself (and I am grateful for discussions on this point with C.J. Isham). Rewrite the action as

$$S = (F,\omega + dA) - \frac{1}{2}(F,F) \tag{3.15}$$

so that, prior to gauge-fixing and ghost additions, the functional integral now reads

$$Z[g,\omega] \sim \int dF\, dA \exp - \left((F, \omega+dA) - \frac{1}{2}(F,F)\right)$$

$$\sim \int dA \exp - \left(\frac{1}{2}(dA,dA) + \frac{1}{2}(\omega,\omega)\right) \tag{3.16}$$

where g is the background gravitational field, and $\omega$ is the most general harmonic p-form (assuming A is (p-1) form) which the background space admits. For a manifold M, the number of independent harmonic p-forms is given by the pth Betti number of M, $n_p = n_{(4-p)}$ (i.e. $n_p$ is the number of zero-eigenvalue modes of $\Delta_p$, and corresponds to the number of independent closed surfaces of dimension p which cannot be continuously shrunk to a point. The Betti numbers are related to the Euler number by $\chi = \sum_{p=0}^{4} (-1)^p n_p$). The most general harmonic p-form will thus be a linear combination of these

$$\omega_p = \sum_{i=1}^{n_p} a_i \omega_i \tag{3.17}$$

When gravity is itself quantized, one must (presumably) integrate over all metrics g and sum over the corresponding $\omega$. Of particular interest is the case where A is our rank-three potential, there is then but one harmonic 4-form

$$\omega_{\mu\nu\rho\sigma} = c\, \varepsilon_{\mu\nu\rho\sigma} \qquad c = \text{constant} \tag{3.18}$$

which yields the promised cosmological constant.

Note that by working in a Euclidean metric, $\Lambda < 0$ in the Einstein equations (1.11) in contrast to Aurelia et. al. (1980) who found $\Lambda > 0$ in a Minkowski metric. The resolution seems to be that $c\varepsilon_{\mu\nu\rho\sigma}$ should be replaced by $ic\,\varepsilon_{\mu\nu\rho\sigma}$ in going from one to the other, if real solutions are desired in both cases. (This is not without precedent: a black-hole with charge Q and angular momentum J yields a real Euclidean solution only if $Q \to iQ$ and $J \to iJ$ under Euclideanization. (Duff and Madore, 1978)) Note, furthermore, that a term similar to our $(F,\omega)$ was also suggested in the case of $A_{\mu\nu\rho}$ by Ogievetsky and Sokatchev (1980) (in analogy with a term linear in the auxiliary field) and by Aurelia et al.(1980) (in order to make the functional integral well-defined). (In these formulations, it is the integral of a total divergence which is non-vanishing because of the boundary conditions. In our formulation, where the space is chosen to be compact without boundary, it is non-vanishing because F, while closed, is not exact). Although not a topological invariant, Aurelia et al.(1980) interpreted this term in analogy with the $\theta$ -vacua term $\theta\varepsilon^{\mu\nu}F_{\mu\nu}$ in the two dimensional Schwinger model. Our formulation suggests an alternative interpretation, however, in which c is treated as a collective coordinate and in which one integrates over c in the functional integral. Integrating over cosmological constant may seem bizarre, but this is just what happens in Hawking's spacetime foam! (Hawking, 1978).

## 4. QUANTUM INEQUIVALENCE

In order to derive the inequivalent trace anomalies and counterterms quoted in the Introduction it is necessary to employ the correct gauge-fixing and ghost procedure for these gauge theories. This is quite subtle, involving as it does the phenomenon of ghosts for ghosts (Siegel, 1980; Thierry-Mieg, 1980; Namazie and Storey, 1979; Curtwright, 1980). Consider first the rank-three potential $A_{\mu\nu\rho}$. It requires two anticommuting rank-two ghosts $B_{\mu\nu}$, three commuting rank-one ghosts $C_\mu$, and four anticommuting scalar ghosts D. The degree of freedom count is thus $1\text{x}4 - 2\text{x}6 + 3\text{x}4 - 4\text{x}1 = 0$.

Functional integration over both physical and ghost fields then yields

$$Z = (\det\Delta_3)^{-\frac{1}{2}}(\det\Delta_2)(\det\Delta_1)^{-3/2}(\det\,\Delta_o)^2 \qquad (4.1)$$

where $\Delta_p$ is the Laplacian on p-forms.(One starts with the action $\frac{1}{2}(dA,dA)$ from (3.16), and then adds a gauge-fixing term $\frac{1}{2}(\delta A,\delta A)$ which from (3.6) yields $\frac{1}{2}(A,\Delta A)$). From the general rules for counterterms and trace anomalies for arbitrary representations (Christensen and Duff, 1979a),it follows that the one-loop counterterm $S_1$ and trace anomaly T are given by

$$S_1 = \frac{-1}{\varepsilon}\,(B_3 - 2B_2 + 3B_1 - 4B_o) \qquad (4.2)$$

$$T = b_3 - 2b_2 + 3b_1 - 4b_o \qquad (4.3)$$

where

$$B_p = \int d^4x \sqrt{g}\, b_p \tag{4.4}$$

is the t-independent term in the asymptotic expansion of Trexp(-Δpt). Explicitly,

$$180(4\pi)^2 b_o = R_{\mu\nu\rho\sigma}R^{\mu\nu\rho\sigma} - R_{\mu\nu}R^{\mu\nu} + \frac{5}{2}R^2 + 6\Box R \tag{4.5}$$

$$180(4\pi)^2 b_1 = -11\, R_{\mu\nu\rho\sigma}R^{\mu\nu\rho\sigma} + 86R_{\mu\nu}R^{\mu\nu} - 20R^2 - 6\Box R \tag{4.6}$$

and

$$b_2 = b_2^+ + b_2^- \tag{4.7}$$

where + and - refer to self-dual and antiself-dual 2-forms respectively, and where

$$180(4\pi)^2 b_2^{\pm} = 33\, R_{\mu\nu\rho\sigma}R^{\mu\nu\rho\sigma} - 93\, R_{\mu\nu}R^{\mu\nu} + \frac{45}{2}R^2 - 12\Box R \pm 30\, R_{\mu\nu\rho\sigma}{}^*R^{\mu\nu\rho\sigma} \tag{4.8}$$

while $b_3$ and $b_4$ are given by the duality relation

$$b_p = b_{(4-p)} \tag{4.9}$$

Note the special combinations

$$\sum_{p=o}^{4} (-1)^p B_p = \chi \tag{4.10}$$

and

$$B_2^+ - B_2^- = \tau \tag{4.11}$$

where $\chi$ and $\tau$ are the Euler and Hirzebruch numbers given by (1.5) and (1.8); (4.10) and (4.11) being the Gauss-Bonnet and Hirzebruch signature theorems, respectively.

The crucial observation, using (4.9), is that is is precisely the Gauss-Bonnet combination of (4.10) which enters (4.2) and (4.3), and we recover the results (1.7) and (1.2) quoted in the Introduction. We learn, furthermore, that the counterterm and trace anomaly are determined only by the zero-modes of the Laplacian. Now the Euler number is known to contribute to part of the integrated trace anomaly (Duff 1977), but for general fields there are non-topological $R^2$ and $R_{\mu\nu}R^{\mu\nu}$ terms as well. For example,

$$T(A) = b_o \tag{4.12}$$

$$T(A_\mu) = b_1 - 2b_o \tag{4.13}$$

In the case of $A_{\mu\nu\rho}$, however, these non-topological terms cancel identically. In other words, after projecting out the zero-modes, the ratio of determinants appearing in (4.1) would be equal to unity.

One may now repeat the exercise for the rank-2 potential A which requires two anticommuting rank-one ghosts $B_\mu$ and three commuting scalar ghosts C. The degrees of freedom count is now $1\times 6-2\times 4+3\times 1 = 1$, and functional integration yields

$$Z = (\det\Delta_2)^{-\frac{1}{2}}(\det\Delta_1)(\det\Delta_o)^{-3/2} \tag{4.14}$$

$$S_{(1)} = \frac{-1}{\varepsilon}(B_2-2B_1+3B_o) \tag{4.15}$$

$$T = b_2-2b_1+3b_o \tag{4.16}$$

which, together with (4.12), again yields the quoted results (1.1) and (1.6).

Once again we see that it is the zero-modes of the Laplacian which are responsible for the quantum inequivalence of different field representations for particles of a given spin and number of degrees of freedom. though it is the harmonic parts of the physical and ghost field potentials that produce the inequivalent counterterms and anomalies, rather than the harmonic part of their field strengths as in Section 3. The harmonic part of F is crucial for the cosmological constant, however, to which we now turn.

So far, we have considered counterterms due to closed loops of antisymmetric tensors in a gravitational background, and we now turn to the quantization of gravity itself. First of all consider pure gravity with a cosmological constant as in (1.15). On dimensional grounds the one-loop counterterms will be a linear combination of $R_{\mu\nu\rho\sigma}R^{\mu\nu\rho\sigma}, R_{\mu\nu}R^{\mu\nu}, R^2$, $R, \Lambda R$ and $\Lambda^2$, but with gauge dependent coefficients. Gauge invariance is achieved by going "on mass shell" by use of the background field equations $R_{\mu\nu}=\Lambda g_{\mu\nu}$. The resulting counterterm may then be written as

$$S_1 = -\frac{1}{\varepsilon}\left(a\chi + b\delta\right), \quad \delta\equiv\int d^4x\sqrt{g}\,\frac{\Lambda^2}{12\pi^2} \tag{4.17}$$

and explicit calculation (Christensen and Duff, 1979b) yields a= 106/45 and b = -87/10. Next consider the combined gravity $-A_{\mu\nu\rho}$ system. Here the counterterms will be a linear combination of $R_{\mu\nu\rho\sigma}R^{\mu\nu\rho\sigma}, R_{\mu\nu}R^{\mu\nu}, R^2$, $R$, $RF_{\mu\nu\rho\sigma}F^{\mu\nu\rho\sigma}$, $(F_{\mu\nu\rho\sigma}F^{\mu\nu\rho\sigma})^2$ and $(\nabla^\mu F_{\mu\nu\rho\sigma})^2$ and the numerical coefficients have been calculated by Sezgin and van Nieuwenhuizen (1980). If one now goes on shell by using (1.11), (1.12) and (1.13), their results yield (4.17) with a = 16/45 and b = -87/10. Thus, apart from the difference in the topological term

$$S_1(\text{gravity}+A_{\mu\nu\rho}) - S_1(\text{gravity}+\Lambda) = \frac{2}{\varepsilon}\chi \tag{4.18}$$

which was to be expected from (1.7), the results are the same.

## 5. SUPERGRAVITY AND SUPER INDEX THEOREMS

As discussed at length elsewhere (Dowker, 1978; Christensen

and Duff 1979a) one way to understand index theorems is to note that the t-independent term in the asymptotic expansion of Tr $\exp{-\Delta t}$, the B coefficient of Section 4, counts the total number of eigenmodes of the differential operator $\Delta$. Thus if two operators $\Delta$ and $\Delta'$ have the same non-zero eigenvalue spectrum, then the difference in the number of zero-modes is given by

$$n-n' = B-B' = \text{integer topological invariant} \qquad (5.1)$$

Normally such theorems relate operators differing by zero or integer spin. In the case that the space is self-dual, however, there can be a mutual cancellation of non-zero modes between boson and fermion operators (Hawking and Pope, 1978). Calculation of the relevant B coefficients then allows one to write down super-index theorems (Christensen and Duff, 1979a, Duff 1979)

$$\begin{array}{lll}
(1) & n[\tfrac{1}{2},o]-2n[o,o] = -\tfrac{1}{24}(\chi+p/_2), & n[o,\tfrac{1}{2}]-2n[o,o] = -\tfrac{1}{24}(\chi-p/_2) \\
(2) & n[\tfrac{1}{2},\tfrac{1}{2}]-2n[o,\tfrac{1}{2}] = -\tfrac{1}{12}(\chi+p/_2), & n[\tfrac{1}{2},\tfrac{1}{2}]-2n[\tfrac{1}{2},o] = -\tfrac{1}{12}(\chi-p/_2) \\
(3) & n[1,o]-2n[\tfrac{1}{2},o] = \tfrac{5}{12}(\chi+p/_2), & n[o,1]-2n[o,\tfrac{1}{2}] = \tfrac{5}{12}(\chi-p/_2) \\
(4) & n[1,\tfrac{1}{2}]-2n[\tfrac{1}{2},\tfrac{1}{2}] = \tfrac{5}{6}(\chi+p/_2), & n[\tfrac{1}{2},1]-2n[\tfrac{1}{2},\tfrac{1}{2}] = \tfrac{5}{12}(\chi-p/_2) \\
(5) & n[1,1]-2n[\tfrac{1}{2},1] = \tfrac{5}{3}(\chi+p/_2), & n[1,1]-2n[1,\tfrac{1}{2}] = \tfrac{5}{3}(\chi-p/_2)
\end{array} \qquad (5.2)$$

In the above $n[A,B]$ denotes the number of zero modes of the operators $\Delta[A,B]$ which act on fields transforming according to the reducible representations of the Lorentz group which, in two-component spinor notation have 2A undotted and 2B dotted indices.

$$\phi_{\alpha_1\alpha_2\ldots\alpha_{2A}\dot\beta_1\dot\beta_2\ldots\dot\beta_{2B}}$$

The precise form of $\Delta[A,B]$ may be found in Christensen and Duff (1979a). These reducible representations are related to the irreducible ones

$$\phi_{(\alpha_1\alpha_2\ldots\alpha_{2A})\,(\dot\beta_1\dot\beta_2\ldots\dot\beta_{2B})} \quad ,$$

which we denote (A,B), in an obvious way. Thus

$$\begin{aligned}
[1,1] &= (1,1) + (1,o) + (o,1) + (o,o) \\
[1,\tfrac{1}{2}] &= (1,\tfrac{1}{2}) + (o,\tfrac{1}{2})
\end{aligned} \qquad (5.3)$$

etc. These in turn are related to the physical fields appearing in supergravity (after Faddeev-Popov ghost subtractions) in the way shown symbolically below

$$e^a_\mu \;:\; (1,1) + (o,o) - 2(\tfrac{1}{2},\tfrac{1}{2})$$

$$\psi^\alpha_\mu \;:\; -\frac{1}{2}\,\{(1,\tfrac{1}{2})+(\tfrac{1}{2},1)-2(\tfrac{1}{2},o)-2(o,\tfrac{1}{2})\}$$

$$A_\mu \;:\; (\tfrac{1}{2},\tfrac{1}{2})-2(o,o)$$

$$\chi^\alpha \;:\; -\frac{1}{2}\,\{(\tfrac{1}{2},o) + (o,\tfrac{1}{2})\}$$

$$A \;:\; (o,o) \tag{5.4}$$

For future reference, we also include the antisymmetric tensors which, from Section 4, read

$$A_{\mu\nu}\!: (1,o) + (o,1) - 2(\tfrac{1}{2},\tfrac{1}{2}) + 3\,(o,o)$$

$$A_{\mu\nu\rho}\!: -2(1,o)-2(o,1)+4(\tfrac{1}{2},\tfrac{1}{2}) -4(o,o) \tag{5.5}$$

The minus signs denote anticommuting fields, and the factor of $\frac{1}{2}$ is for fields obeying first-order equations.

The super-theorems (5.2) may now be used to calculate the supergravity anomalies (Note once again that, for the present purposes at least, no physical significance need be attached to the self-duality condition. The index theorems may be regarded simply as a calculational device. The trace anomaly is given, as in Section 4, only by the zero modes i.e. boson minus one-half fermion, since in supergravity the non-topological terms cancel on mass shell). For example, the representation content of simple supergravity with one graviton and one gravitino is from (5.4) and (5.3)

$$[1,1]-\tfrac{1}{2}[1,\tfrac{1}{2}]-\tfrac{1}{2}[\tfrac{1}{2},1]-[1,o]-[o,1]-2[\tfrac{1}{2},\tfrac{1}{2}]+\tfrac{3}{2}[\tfrac{1}{2},o]+\tfrac{3}{2}[o,\tfrac{1}{2}]+2[o,o] \tag{5.6}$$

and the integrated trace anomaly is simply obtained by <u>adding</u> the left and right hand columns in (5.2) and then taking the appropriate linear combinations. Thus

$$\int T = \tfrac{1}{2}(5)+\tfrac{1}{2}(4)-(3)-\tfrac{1}{2}(1) = \frac{41}{24}\,\chi \tag{5.7}$$

Repeating the exercise for extended supergravity (with the standard representation content) yields the results shown overleaf in Table I. (a is the coefficient appearing in $\int T = a\chi$. The b coefficients calculated by Christensen, Duff, Gibbons and Roček (1980) concern the cosmological constant, see (4.17), and will be discussed shortly).

TABLE I

| N | $e_\mu^a$ | $\psi_\mu^\alpha$ | $A_\mu$ | $\chi$ | A | a | b |
|---|---|---|---|---|---|---|---|
| 1 | 1 | 1 | | | | 41/24 | -77/12 |
| 2 | 1 | 2 | 1 | | | 11/12 | -13/3 |
| 3 | 1 | 3 | 3 | 1 | | 0 | - 5/2 |
| 4 | 1 | 4 | 6 | 4 | 2 | -1 | - 1 |
| 5 | 1 | 5 | 10 | 11 | 10 | -2 | 0 |
| 6 | 1 | 6 | 16 | 26 | 30 | -3 | 0 |
| 7 | 1 | 8 | 28 | 56 | 70 | -5 | 0 |
| 8 | 1 | 8 | 28 | 56 | 70 | -5 | 0 |

Next we examine the chiral anomaly $\nabla^\mu J_\mu^5$. If the anomalies were to form a supermultiplet, one would expect the same coefficients a to determine the axial anomaly

$$\int \nabla^\mu J_\mu^5 = 2(n_+ - n_-) = aP \tag{5.8}$$

and also the spinor current anomaly. For a review, see Grisaru (1978). This is indeed guaranteed by the super index theorems (5.2) if, instead of adding the right and left columns as we did for the trace anomaly, we subtract the left from right and take the same linear combinations of (1)...(5)in both cases. Applying this procedure to matter multiplets in a background gravitational field leads to the standard results for both trace and chiral anomalies. (Christensen and Duff 1978a; Grisaru 1978). Note that the difference $n[\frac{1}{2},0] - n[0,\frac{1}{2}]$ would appear in the vector multiplet with a weight -3 relative to the scalar multiplet, but this is as it should be by supersymmetry (Grisaru, 1978).

The mystery arises if we now apply the same procedure to supergravity. Thus the axial theorem in simple supergravity would read

$$\frac{3}{2}[1,\tfrac{1}{2}] - \frac{3}{2}[\tfrac{1}{2},1] - [1,0] + [0,1] + \frac{3}{2}[\tfrac{1}{2},0] - \frac{3}{2}[0,\tfrac{1}{2}] = \frac{41}{24}\, p/2 \tag{5.9}$$

Aside from the fermion chiral weight assignments, which are perhaps to be expected, the crucial feature is the appearance of an antisymmetric tensor contribution coming from the Hirzebruch signature theorem. Thus, if the anomalies are to form a supermultiplet, even simple supergravity requires antisymmetric tensors! These can only come either from non-traditional auxiliary field representations, as discussed in (c) of the Introduction, or else from Lorentz ghosts. This is presently being investigated, but the superfield approach of Grisaru, Roček, and Siegel would seem to support the latter. There is also the question of gauge invariance of $J_\mu^5$ (Nielsen, Grisaru, Römer and van Nieuwenhuizen, 1978; Nielsen 1979).

According to the super theorems, moreover, the inclusion of antisymmetric tensors as physical fields in the extended models would, when done supersymmetrically, change not only the bosonic contribution to $\nabla^\mu J^5_\mu$ but also the fermionic contribution by virtue of the different chiral weights.

In the meantime, let us return to the trace anomaly where we are on firmer ground (assuming as suggested by (5.6) that Lorentz ghost contributions will be cancelled by the anti-symmetric part of the vierbein). The $a$ coefficients displayed in the table above correspond to the usual representation assignments, but the inclusion of physical fields $A_{\mu\nu}$ or $A_{\mu\nu\rho}$ would alter these numbers since, from (1.1) and (1.2)

$$a(A_{\mu\nu}) - a(A) = 1$$
$$a(A_{\mu\nu\rho}) = -2 \qquad (5.10)$$

Moreover, as Siegel points out, the N=8 version obtained by dimensional reduction (Cremmer and Julia, 1979) replaces the 70A fields by $63A + 7A_{\mu\nu} + 1\,A_{\mu\nu\rho}$. Hence the $a$ coefficient of $-5$ is, from (5.10), replaced by

$$a = -5+7-2 = 0 \qquad (5.11)$$

Since $a$ is an integer for all $N \geq 3$ with the usual representations, one could arrange a vanishing trace anomaly in this range if

$$3-N\,[\psi_\mu] + N\,[A_{\mu\nu}] - 2N\,[A_{\mu\nu\rho}] = 0 \qquad (5.12)$$

but apart from the N=8 version described above and the N=4 version discussed in the Introduction, these assignments do not correspond to any known supergravity theory. (It would be interesting to repeat Grisaru's (1978) comparison of the anomalies in the two N=4 theories taking into account the antisymmetric tensors).

Finally, we may ask what happens if we relax the condition that the space be compact without boundary. In this case, the topological invariants and the index theorems themselves acquire boundary term corrections. Consequently, not only the axial anomalies but only the trace anomalies and counterterms will be modified by surface-integral corrections (the latter are, without much justification in my opinion , almost always ignored in the literature). These have not yet been calculated for arbitrary spacetimes, but we may once more gain some insight by considering a special case, namely ALE spaces, which are self-dual with $\chi = \tau + 1$ (the case $\tau=0$ corresponds to flat space). On such spaces, the number of (normalizable) zero modes are given by (Hawking and Pope, 1978)

$$n(1,0) = \tau,\quad n(1,\tfrac{1}{2}) = 2\tau\,,\quad n(1,1) = 3\tau$$

while $n(A,B) = 0$ for the other relevant representations. This would yield the coefficient $a=(3-N)\tau$ for extended supergravity with the usual representations and hence $a=0$ for all N if condition (5.12) were fulfilled. Once again, the N=4 theory with one A and the N=8 theory with seven $A_{\mu\nu}$ and one $A_{\mu\nu\rho}$ provide concrete realizations of this.

One final puzzle remains. Unless (5.12) is fulfilled, the Gauss-Bonnet combination (1.3) still makes its appearance for N>4. Yet according to De Wit and Ferrara (1979), no supersymmetric extension of $C_{\mu\nu\rho\sigma}C^{\mu\nu\rho\sigma}$ exists for N>4. Are these statements compatible?

## 6. FURTHER PUZZLES

We have seen how, contrary to a popular belief, the gauge theory of $A_{\mu\nu}$ is not equivalent to a scalar A theory and the gauge theory of $A_{\mu\nu\rho}$ is non-trivial. Moreover, antisymmetric tensor representations make their appearance in supergravity either as ghosts and/or auxiliary fields in conventional supergravity or as physical fields in versions of extended supergravity theories which, it now seems, are physically distinct from their more conventional counterparts. In particular, the antisymmetric tensor version of N=8 supergravity seems to have better ultraviolet behaviour at least to one loop. The question naturally arises whether this version is also a better candidate for a theory of super unification. See the work of Ellis, Gaillard, Maiani and Zumino (1980) and Zumino (1980, this volume), a somewhat different approach having been proposed earlier by Curtwright and Freund (1979). Both approaches rely on the observation of Cremmer and Julia that the conventional N=8 theory possesses a hidden local SU(8) and global $E_7$ invariance, which do not seem to be present in the antisymmetric tensor N=8 theory. An understanding of the symmetries of this model would therefore seem crucial.

Equally crucial will be the quantum implications of the spontaneous symmetry breaking of Aurelia, Nicolai and Townsend (1980) induced by the cosmological constant. Indeed, our claims of better ultraviolet behaviour must be qualified by the effects of this cosmological constant. Now the one-loop counterterms of pure gravity (Christensen and Duff 1979b), simple supergravity and gauged extended supergravity (Christensen, Duff, Gibbons and Roček 1980) with a cosmological constant have now been carried out, with the remarkable result that the cosmological counterterm (the b coefficient of Table I) vanishes for N>4, assuming these theories exist, and assuming the scalar couplings to be the natural extension of those in the O(4) model i.e. with a conformal $\sqrt{g}RA^2$ coupling. However, gauge invariance forbids such couplings for the rank-two fields $A_{\mu\nu}$ which yield the same b coefficient (though different a coefficient) as <u>minimally</u> coupled scalars. One cannot jump to the conclusion that N=8 now ceases to be one-loop finite, though, because the scalar effective potential of Aurelia, Nicolai and Townsend contains terms linear in A and so the conventional renormalization arguments cannot even be applied. (c.f. the comments of Christensen, Duff, Gibbons and Roček (1980) on gauged N=4). Since their cosmological constant is opposite in sign to the gauged models (Freedman and Das, 1977), it would also be interesting to know whether a gauged version exists.

The absence of a conformal coupling for $A_{\mu\nu}$ has, as its flat space counterpart, the absence of a new improved stress tensor. This may have a bearing on the off-shell formulation of N=4 Yang-Mills theories of Sohnius, Stelle and West (1980) where one of the six A fields is swapped for an $A_{\mu\nu}$. In a gravitational background, moreover, this theory has a vanishing a coefficient if there are six A fields but not with five A and one $A_{\mu\nu}$. However, I do not know what significance, if any, ought to be attached to this.

Finally, since this flat space N=4 theory has vanishing $\beta$ function , and hence trace anomaly up to 2 loops (or is it 3? see Grisaru, this volume), it is tempting to invoke some kind of Adler-Bardeen theorem on the axial anomaly which, by supersymmetry, might demonstrate finiteness to all orders. One might then repeat the exercise for N=8 supergravity which also now has a vanishing 1-loop trace anomaly and hence, as now seems likely, a vanishing 1-loop axial anomaly as well. Unfortunately, the status of the Adler-Bardeen theorem in supersymmetric theories is notoriously obscure, since renormalization of the trace anomaly takes place in some models. The superfield approach would, on the face of it, appear to keep the anomalies in a supermultiplet, but one would then be forced to conclude that Adler-Bardeen does not apply to these currents. However, to finish on a speculative note, suppose the anomalies always do form a supermultiplet so that a proof of finiteness to all orders devolves upon a proof of vanishing axial anomaly to all orders. Could it be that, as suggested at one loop by the cancellation of zero modes in the index theorems, that the self-conjugate nature of N=4 Yang-Mills and N=8 supergravity forbids any excess of right-handed fermion modes over left handed or self-dual antisymmetric tensors over antiself-dual, and hence forbids the chiral anomaly?

REFERENCES

Aurelia, A., Nicolai, H., and Townsend, P.K. (1980). Hidden constants: The θ parameter of QCD and the cosmological constant of N=8 supergravity. CERN Preprint TH-2875.

Christensen, S.M. and Duff, M.J. (1978a). Axial and conformal anomalies for arbitrary spin in quantum gravity and supergravity. Phys. Lett. 76B, 571.

Christensen, S.M. and Duff, M.J. (1978b). Quantum gravity in 2+ε dimensions. Phys. Lett. 79B, 213.

Christensen, S.M. and Duff, M.J. (1979a). New gravitational index theorems and supertheorems. Nuc. Phys. B154, 301.

Christensen, S.M. and Duff, M.J. (1979b). Quantizing gravity with a cosmological constant. Nuc. Phys. B. (to appear).

Christensen, S.M. , Duff, M.J., Gibbons, G.W. and Roček, M. (1980). Vanishing one-loop β function in gauged n>4 supergravity. Phys. Rev. Lett, 45, 161.

Cremmer, E., Scherk, J. and Ferrara, S. (1978) SU(4) invariant supergravity theory. Phys. Lett. 74B, 61.

Cremmer, E. and Julia, B. (1979). The SO(8) supergravity. Nuc. Phys. B. 159, 141.

Curtwright, T and Freund, P. (1979). SU(8) unification and supergravity. In Supergravity, eds. P van Nieuwenhuizen and D.Z. Freedman, North-Holland.

Curtwright, T. (1980). Generalized gauge fields. Chicago preprint EFI 80/04.

De Wit, B and Ferrara, S. (1979). On higher order invariants in extended supergravity. Phys. Lett. 81B, 317.

Dowker, J. (1978). Another discussion of the axial vector anomaly and the index theorem. J. Phys.A.11, 347.

Duff, M.J. (1977). Observations on conformal anomalies. Nuc. Phys. B125, 334.

Duff, M.J. and Madore, J. (1978). Einstein-Yang-Mills pseudoparticles and electric charge quantization. Phys. Rev. D18, 2788.

Duff, M.J. (1979). Self-duality, helicity and supergravity. In Supergravity, eds. P. van Nieuwenhuizen and D.Z. Freedman North Holland.

Duff, M.J. and van Nieuwenhuizen, P. (1980). Quantum inequivalence of different field representations. Phys. Lett. 94B, 179.

Duff, M.J. (1980a). Quantization of supergravity with a cosmological constant. In Unification of the Fundamental Interactions, proceedings of the Europhysics conference, Erice, Sicily, eds. J. Ellis and S. Ferrara. Plenum Press.

Duff, M.J. (1980b). The cosmological constant in quantum gravity and supergravity. In Quantum Gravity: A Second Oxford Symposium, eds. C.J. Isham, R. Penrose and D.W. Sciama OUP.

Eguchi, T. and Hansen, A. (1978). Asymptotically flat self-dual solutions to euclidean gravity. Phys. Lett. 74B, 249.

Ellis, J., Gaillard, M.K., Maiani, L., and Zumino, B. (1980). In Unification of the Fundamental Interactions, proceedings of the Europhysics conference, Erice, Sicily, eds. J. Ellis and S. Ferrara. Plenum Press.

Ferrara, S., and Zumino, B. (1975). Transformation properties of the supercurrent. Nucl. Phys. B87, 207.

Ferrara, S, and van Nieuwenhuizen, P. (1978). The auxiliary fields for supergravity. Phys. Lett. 74B, 333.

Freedman, D.Z., and Das. A. (1977). Gauge internal symmetry in extended supergravity. Nucl. Phys. B120, 221.

Freedman, D.Z., and Townsend, P.K. (1980). Antisymmetric tensor gauge theories and non-linear σ-models. Stonybrook preprint ITP-SB-80-25.

Gibbons, G.W. and Hawking, S.W. (1978). Gravitational multi-instantons. Phys. Lett. 78B, 430.

Grisaru, M.T. (1978). Anomalies in supersymmetric theories. In Recent Developments in Gravitation, Cargese, eds. S. Deser and M. Levy. Plenum Press.

Grisaru, M.T. (1979). Superloops. In Supergravity, eds. P. van Nieuwenhuizen and D.Z. Freedman, North Holland.

Grisaru, M.T., Rocek, M, and Siegel, W. (1979). Improved methods for supergraphs. Nucl. Phys. B159, 429.

Grisaru, M.T., and Seigel, W. (1979). Supergraphity (To appear).

Grisaru, M.T. (1980). This volume.

Hawking, S.W. (1978). Spacetime foam. Nuc. Phys. B144, 349.

Hawking, S.W., and Pope C.N. (1978). Symmetry breaking by instantons in supergravity. Nuc. Phys. B146, 381.

Namazie, M.A. and Storey, D. (1979). On secondary and higher order ghosts. Nuc. Phys. B157, 170.

Nielsen N.K., Romer, H, Schroer, B. (1978). Anomalous currents in curved space, Nuc. Phys. B136, 475.

Nielsen, N.K., Grisaru, M.T., Römer, H. and van Nieuwenhuizen,P (1978). Approaches to the gravitational spin-3/2 axial anomaly. Nuc. Phys. B140, 477.

Nielsen, N.K. (1979). Background field gauge in supergravity. In Supergravity, eds. P. van Nieuwenhuizen and D.Z. Freedman. North Holland.

Ogievetsky, V and Sokatchev, E. (1980). Equation of motion for the axial gravitational superfield. Dubna preprint E2-80-139.

Sezgin, E. and van Nieuwenhuizen, P. (1980). Renormalizability properties of antisymmetric tensor fields coupled to gravity. Phys. Rev. D22, 301.

Siegel, W. (1979). Gauge spinor superfield as scalar multiplet. Phys. Lett. 85B, 333.

Siegel, W. (1980). Hidden ghosts. Princeton preprint.

Sohnius, M., Stelle, K.S. and West, P.C. (1980). Off-mass-shell formulation of extended supersymmetric gauge theories. Phys. Lett. 92B, 123.

Stelle, K.S. and West, P.C. (1978). Minimal set of auxiliary fields for supergravity. Phys. Lett. 74B, 330.

Theirry-Mieg, J.(1980). BRS Structure of antisymmetric tensor gauge theories. Harvard preprint HUTMP/B86.

Townsend, P.K. (1980). Gauge invariance for spin ½. CERN Preprint TH.2790.

van Nieuwenhuizen, P. (1980). Quantum supergravity. In Quantum Gravity: A Second Oxford Symposium, eds. C.J. Isham R. Penrose and D.W. Sciama, OUP.

# SPONTANEOUS BREAKING OF SUPERSYMMETRY AND THE COSMOLOGICAL CONSTANT IN N = 8 SUPERGRAVITY *)

A. Aurilia**), H. Nicolai and P.K. Townsend

CERN -- Geneva

ABSTRACT

The Abelian gauge field $A_{\mu\nu\rho}$ that appears in one formulation of N = 8 supergravity can be used to construct an analogue of the Fayet-Illiopoulos term for this theory, such that the local supersymmetry is spontaneously broken. The eight Goldstone fermions are eliminated by the super-Higgs effect and a cosmological constant of the expected sign is generated. A gauging of the (spontaneously broken) O(8) symmetry, if possible, would allow a cancellation of the cosmological constant.

## INTRODUCTION: THE "TOPOLOGICAL" GAUGE FIELD

The genesis of the work reported here was a study of the gauge theory for the three index antisymmetric tensor Abelian gauge field $A_{\mu\nu\rho}$ (Aurilia, Nicolai and Townsend (1980)). Although apparently trivial there is one interesting aspect that we shall be able to exploit, so let us first introduce this idea. The generalized Maxwell Lagrangian for $A_{\mu\nu\rho}$ is

$$\mathcal{L} = -\frac{1}{48}\, e\, F_{\mu\nu\rho\sigma}\, F^{\mu\nu\rho\sigma}$$

$$F_{\mu\nu\rho\sigma} = \partial_\mu A_{\nu\rho\sigma} + 3 \text{ terms} \qquad (1.1)$$

---

*) Talk given by P.K. Townsend.

**) Permanent address: Instituto Nazionale di Fisica Nucleare, Trieste, Italy.

$F_{\mu\nu\rho\sigma}$ is the antisymmetric field strength tensor for $A_{\mu\nu\rho}$. We include the space-time volume factor e for general co-ordinate invariance. To analyze the field equations for this system it is convenient to replace (1.1) by an equivalent first order form

$$\mathcal{L} = \frac{1}{4!}\,\rho\,\varepsilon^{\mu\nu\rho\sigma}\,F_{\mu\nu\rho\sigma} - \frac{1}{2}\,e\,\rho^2 \qquad (1.2)$$

From (1.2) we see that the $\rho$ field equation determines $\rho$ to be a function of $A_{\mu\nu\rho}$ and the metric tensor.

$$\rho = \frac{1}{4!}\,e^{-1}\,\varepsilon^{\mu\nu\rho\sigma}\,F_{\mu\nu\rho\sigma} \qquad (1.3)$$

while the $A_{\mu\nu\rho}$ field equation is

$$\partial_\mu \rho = 0 \Rightarrow \rho = a = \text{constant} \qquad (1.4)$$

The dynamics of the $A_{\mu\nu\rho}$ field are therefore trivial: no physical particle is propagated. But there is a static effect, represented by the arbitrary integration constant a (Aurilia, Christodoulou and Legovini (1978)). Comparing (1.4) to (1.3) we see that a represents the strength of a constant background "electric" field $F_{oijk}$. The only effect of such a constant field, in this example, is to provide an additional contribution to the background energy density. In the presence of gravitational interactions this will become an addition to the cosmological constant and, indeed, one sees from (1.2) that there will be a term $\frac{1}{2}\,a^2\,g_{\mu\nu}$ on the right-hand side of Einstein's field equation.

So far we have discussed only field equations. In practice we are more often interested in extracting information via a functional integral. In particular if the Lagrangian (1.1) appears in a functional integral we should be able to eliminate the $A_{\mu\nu\rho}$ field by performing the Gaussian integral over it. Usually this would be equivalent to eliminating $A_{\mu\nu\rho}$ by its field equation, but in this case there is subtlety. Gaussian integration cannot produce an arbitrary constant if it is well defined (which can be checked by putting the system on a lattice), so where does the arbitrary constant come in? The answer to this is that there is the possibility to add to the Lagrangian (1.1) a total derivative term with arbitrary coefficient, a:

$$a\,\varepsilon^{\mu\nu\rho\sigma}\,F_{\mu\nu\rho\sigma} \qquad (1.5)$$

This does make a difference to the functional integral as one can see immediately from the perturbation expansion of the vacuum energy density which includes the following diagram

Fig. 1 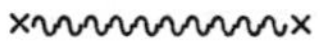 $\sim \quad a^2$

Although the a-dependent source for $A_{\mu\nu\rho}$ has a zero at zero momentum, the $A_{\mu\nu\rho}$ propagator has a pole and the net result is that the above diagram contributes a constant to the energy density.

The addition of (1.5) to the Lagrangian does not affect the field equations but the freedom represented by the constant is already contained in them in the form of an integration constant. Conversely, the integration constant makes no contribution to the functional integral, but the total derivative term does. These are two equivalent ways of explaining the origin of the parameter a. All of this is a straightforward generalization from two dimensions, where the ordinary Maxwell field $A_\mu$ plays the same role as $A_{\mu\nu\rho}$ in four dimensions (Coleman (1976)). Just as the corresponding total derivative $\varepsilon^{\mu\nu}F_{\mu\nu}$ can also be regarded as a topological invariant in some circumstances, so can $\varepsilon^{\mu\nu\rho\sigma}F_{\mu\nu\rho\sigma}$ be regarded as an effective representation of the QCD topological invariant $\varepsilon^{\mu\nu\rho\sigma}F^a_{\mu\nu}F^a_{\rho\sigma}$, in some circumstances. This is the origin of the nomenclature "topological gauge field" for $A_{\mu\nu\rho}$ (Lüscher (1978)).

If we start with the Lagrangian,

$$\mathcal{L} = -\frac{1}{48}\, e\, F_{\mu\nu\rho\sigma}\, F^{\mu\nu\rho\sigma} + a\, \varepsilon^{\mu\nu\rho\sigma}\, F_{\mu\nu\rho\sigma}$$

then, because of the fact that $A_{\mu\nu\rho}$ propagates no physical modes, and because of the cancellation of zeros and poles of $p^2$ in Fig. 1, it is equivalent to treat $F_{\mu\nu\rho\sigma}$ as the independent field instead of $A_{\mu\nu\rho}$. It is clear from (1.6) that the effect of eliminating $F_{\mu\nu\rho\sigma}$ in this case is simply to produce a cosmological term. It will be important to us later that this replacement will not break any symmetries of the action that $A_{\mu\nu\rho}$ finds itself a part of. This criterion is satisfied because:

i) $A_{\mu\nu\rho}$ appears only through $F_{\mu\nu\rho\sigma}$;

ii) $F_{\mu\nu\rho\sigma}$ transforms as a total derivative under any symmetry because by definition it is chosen to transform as ($\partial_\mu A_{\nu\rho\sigma}$ + three terms);

iii) the Bianchi identity for $F_{\mu\nu\rho\sigma}$ ($\partial_\mu F_{\nu\rho\sigma\lambda}$ + four terms) $\equiv 0$

imposes no restriction of $F_{\mu\nu\rho\sigma}$, i.e., it is not required for any symmetry that $F_{\mu\nu\rho\sigma}$ be the curl of $A_{\mu\nu\rho}$.

The applications of this idea are various. Depending on the context one can interpret the constant that is hidden in theories with the $A_{\mu\nu\rho}$ field as (i) the bag constant (Aurilia, Christodoulou and Legovini (1978)), (ii) the $\Theta$-parameter of QCD effective Lagrangians (Aurilia, Nicolai and Townsend (1980)), or (iii) a cosmological constant, as in the above example. It is the third application that is relevant to supergravity theories. In the case of N = 1 supergravity in the superspace formalism of Ogievetsky and Sokatchev it is not possible to add a cosmological constant to the action in the usual way. But, as these authors observed, it does arise in the form of an integration constant in the equation of motion of their field $S_\mu$ (Ogievetsky and Sokatchev (1980); Duff and van Nieuwenhuizen (1980)). The relation between this field and our Abelian gauge field is simply $A_{\mu\nu\rho} = \varepsilon_{\mu\nu\rho\sigma}S^\sigma$, and the gauge transformation of $A_{\mu\nu\rho}$ allows $S_\mu$ to be gauged to the form $\partial_\mu M$. The scalar field M is related to the scalar field S of the minimal set of auxiliary fields (Ferrara and van Nieuwenhuizen (1978); Stelle and West (1978)) as $M = \Box S$. We have recently applied this idea to N = 8 supergravity (Aurilia, Nicolai and Townsend (1980)). The key point here is that the N = 8 supergravity can be obtained from the N = 1 supergravity in 11 dimensions and that this contains the 11-dimensional gauge field $A_{MNP}$. On reduction to four dimensions $A_{MNP}$ produces the wanted $A_{\mu\nu\rho}$ field. This fact is already contained in the work of Cremmer and Julia but they did not appreciate that the elimination of $A_{\mu\nu\rho}$ allows an additional parameter to enter the theory. It is this modification of the N = 8 supergravity Lagrangian that we describe in the following.

## THE N = 8 SUPERGRAVITY

The Lagrangian of 11-dimensional supergravity (Cremmer, Julia and Scherk (1978)) is (with $\kappa = 1$)

$$\begin{aligned}\mathcal{L}_{11} = &- \frac{1}{4} e_{11}\, R(\omega) - \frac{1}{48} e_{11}\, F_{MNPQ}\, F^{MNPQ} \\ &- \frac{i}{2} e_{11}\, \bar{\Psi}_M\, \Gamma^{MNP}\, D_N\left(\frac{\omega+\hat{\omega}}{2}\right)\Psi_P \\ &+ \frac{e_{11}}{192}\left(\bar{\Psi}_M\, \Gamma^{MNWXYZ}\, \Psi_N + 12\, \bar{\Psi}^W\, \Gamma^{YZ}\, \Psi^X\right)\left(F_{WXYZ} + \hat{F}_{WXYZ}\right) \\ &+ \frac{2}{(12)^4}\, \varepsilon^{M_1\cdots M_{11}}\, F_{M_1M_2M_3M_4}\, F_{M_5M_6M_7M_8'}\, A_{M_9M_{10}M_{11}} \end{aligned} \quad (2.1)$$

and the supersymmetry transformations are

$$\delta e_M{}^A = - i \bar{\varepsilon} \Gamma^A \Psi_M$$

$$\delta A_{MNP} = \frac{3}{2} \bar{\varepsilon} \Gamma_{[MN} \Psi_{P]}$$

$$\delta \Psi_M = D_M(\hat{\omega}) \varepsilon + \frac{i}{144} (\Gamma_M{}^{NPQP} - 8 \Gamma^{PQR} \delta_M^N) \varepsilon \hat{F}_{NPQR} \qquad (2.2)$$

The notation and conventions are those of Cremmer and Julia (1979) except for $e_{11}$ which is det $(e_M{}^A)$. In order to dimensionally reduce (2.1) it is most convenient to make a partial gauge choice for the 0(10,1) local Lorentz invariance. We choose (Cremmer and Julia (1979))

$$e_M{}^A = \begin{pmatrix} e_\mu{}^\alpha & B_\mu{}^a \\ 0 & e_m{}^a \end{pmatrix}$$

$$\Rightarrow e_{11} = e \, \Delta^{\frac{1}{2}}; \quad e = \det e_\mu{}^\alpha, \ \Delta^{\frac{1}{2}} = \det e_m{}^a \qquad (2.3)$$

We aim to pick out contributions to the four-dimensional field strength $F_{\mu\nu\rho\sigma}$ coming from the dimensional reduction of $F_{MNPQ}$. Equivalently we look for terms with $F_{\alpha\beta\gamma\delta}$, where

$$F_{\alpha\beta\gamma\delta} = e_\alpha{}^M e_\beta{}^N e_\gamma{}^P e_\delta{}^Q F_{MNPQ}$$

$$= e_{[\alpha}{}^\mu e_\beta{}^\nu e_\gamma{}^\rho e_{\delta]}{}^\sigma (F_{\mu\nu\rho\sigma} - 4 B_\sigma{}^k F'_{\mu\nu\rho k}) \qquad (2.4)$$

$e_A{}^M$ is the inverse of $e_M{}^A$. $B_\sigma{}^k$ is $B_\sigma{}^a e_a{}^k$ with $e_a{}^k$ the inverse of $e_m{}^a$. $F'_{\mu\nu\rho k}$ is $(\partial_\mu A'_{\nu\rho k} + \partial_\rho A'_{\mu\nu k} + \partial_\nu A'_{\rho\mu k})$ with

$$A'_{\mu\nu i} \equiv A_{\mu\nu i} - B_\mu^j A_{j\nu i} - B_\nu^j A_{\mu j i} + B_\mu^j B_\nu^k A_{jki} \qquad (2.5)$$

The reason for the redefinition (2.5) is that the general coordinate transformations in 11 dimensions include the transformation $\delta B_\mu{}^i = \partial_\mu \xi^i$ for $B_\mu{}^i$ when reduced to four dimensions. But $A_{\mu\nu i}$ also transforms under this symmetry. The redefined $A'_{\mu\nu i}$ is invariant with respect to $\xi^i$ transformations, as is $F'_{\mu\nu\rho i}$. Multiplying (2.4) by $\varepsilon^{\alpha\beta\gamma\delta}$, the tangent space four-dimensional Levi-Civita symbol, we have

$$\varepsilon^{\mu\nu\rho\sigma} F_{\mu\nu\rho\sigma} = e\,\varepsilon^{\alpha\beta\gamma\delta} F_{\alpha\beta\gamma\delta} - 4\,\varepsilon^{\mu\nu\rho\sigma} B^i_\sigma F'_{\mu\nu\rho i}$$

$$\varepsilon^{\mu\nu\rho\sigma} = e\, e_\alpha{}^\mu e_\beta{}^\nu e_\gamma{}^\rho e_\delta{}^\sigma \varepsilon^{\alpha\beta\gamma\delta} \qquad (2.6)$$

That is, $\varepsilon^{\mu\nu\rho\sigma} F_{\mu\nu\rho\sigma}$ is a total derivative, and can be expressed in terms of $F_{\alpha\beta\gamma\delta}$ as above.

The terms in (2.1) involving $F_{\alpha\beta\gamma\delta}$ can be written as

$$\mathcal{L}_F = -\frac{1}{48} e\,\Delta^{\frac{1}{2}} F_{\alpha\beta\gamma\delta} F^{\alpha\beta\gamma\delta}$$

$$+ \frac{e}{96} \Delta^{\frac{1}{2}} (\bar{\Psi}_M \Gamma^{MN\alpha\beta\gamma\delta} \Psi_N + 12\, \bar{\Psi}^\alpha \Gamma^{\gamma\delta} \Psi^\beta) F_{\alpha\beta\gamma\delta} \qquad (2.7)$$

The 32 x 32 Dirac matrices may be chosen as

$$\gamma^\alpha \otimes 1_A{}^B\ ,\ \gamma^5 \otimes (\Gamma^a)_A{}^B \qquad (2.8)$$

where $\Gamma^a$ are the seven 8 x 8 matrices of Cremmer and Julia. The charge conjugation matrix of 11 dimensions $C_{11}$ may also be chosen to be $C \otimes 1_A{}^B$ where C is the usual charge conjugation matrix of four dimensions. Using also the notation

$$F_{\alpha\beta\gamma\delta} = -\varepsilon_{\alpha\beta\gamma\delta} f$$

$$f \equiv \frac{1}{4!} \varepsilon^{\alpha\beta\gamma\delta} F_{\alpha\beta\gamma\delta} \qquad (2.9)$$

the four-dimensional equivalent of (2.7) is

$$\mathcal{L}_f = \frac{1}{2} e \, \Delta^{\frac{1}{2}} f^2$$

$$+ \frac{e}{4} \Delta^{\frac{1}{2}} (2\bar{\Psi}_\alpha{}^A \, i\gamma^5 \sigma^{\alpha\beta} \Psi_{\beta A} - \bar{\Psi}_a{}^A \, i\gamma^5 (\Gamma^{ab})_{AB} \Psi_b{}^B) f \qquad (2.10)$$

where

$$\sigma^{\alpha\beta} = \frac{1}{2}\Gamma^{\alpha\beta} = \frac{1}{4}[\Gamma^\alpha, \Gamma^\beta]$$

According to our arguments of the previous section we may now add to (2.10) the total derivative:

$$\mathcal{L}_a = a\,(e\,f - \frac{1}{6} \varepsilon^{\mu\nu\rho\sigma} B_\sigma{}^i F'_{\mu\nu\rho i}) \qquad (2.11)$$

with a an arbitrary constant. Also according to our previous arguments we may replace $A_{\mu\nu\rho}$ by f as the independent variable. Elimination of f then gives the following a-dependent terms in the Lagrangian

$$\mathcal{L}_{new} = -\frac{1}{2} a^2 e \, \Delta^{-\frac{1}{2}} - \frac{a}{6} \varepsilon^{\mu\nu\rho\sigma} B_\sigma{}^i F'_{\mu\nu\rho i}$$

$$+ \frac{a}{4} e \, (\Psi_a{}^A \, i\gamma^5 (\Gamma^{ab})_{AB} \Psi_b{}^B - 2\,\bar{\Psi}_\alpha{}^A \, i\gamma^5 \sigma^{\alpha\beta} \Psi_{BA}) \qquad (2.12)$$

The transformation laws which are f-dependent also pick up new a-dependent terms on elimination of f. These are,

$$\delta_{new} \Psi_\mu{}^A = \frac{a}{3} \Delta^{-\frac{1}{2}} \gamma_\mu \gamma_5 \, \varepsilon^A$$

$$\delta_{new} \Psi_\alpha{}^A = -\frac{a}{6} \Delta^{-\frac{1}{2}} (\Gamma_a)^A{}_B \, \varepsilon^B \qquad (2.13)$$

(One can check directly that the combination of (2.12) and (2.13) does constitute a supersymmetric addition to the N = 8 Lagrangian). However, the fields that appear in (2.12) are not those of the final Lagrangian. Many field redefinitions are necessary both to exhibit the O(8) symmetry and to put the kinetic terms into canonical form. In particular the $\gamma_5$ appearing in (2.12) does not indicate an explicit breaking of parity. It is removed by one of the field redefinitions. Without going into these complications it is still possible to interpret the principal features of (2.12).

One difficulty that we should mention first is the question of how to expand the scalar field function $\Delta$. The problem is that the potential for these fields has no extremum and therefore it is not clear which vacuum values to choose for the scalar fields. From the point of view of perturbation theory we must expand $\Delta$ as $(1 + \ldots)$ or $\Delta^{-1}$ will have no inverse, although there is no other justification for this. This problem is already familiar from gauged N = 4 supergravity (Freedman and Schwarz (1978)), and we shall follow the example of that model by calling the $\Delta^{-\frac{1}{2}}e$ term of (2.12) a "generalized cosmological term". The principal difference between our modification of the N = 8 model and the gauged N = 4 model is that we have no gauging of the O(8) symmetry. Also the sign of the cosmological constant is opposite to that of the N = 4 model and is that ordinarily associated with a spontaneous breakdown of supersymmetry. The validity of this interpretation of the cosmological constant can be seen from (2.13) where the eight fermions $\Delta^{\frac{1}{2}}(\Gamma^a\Psi_a)_A$ can be identified as the Goldstone fermions of eight spontaneously broken supersymmetries. Because the supersymmetry is local these fermion fields can be gauged away, i.e., by choice of the "unitarity gauge", $(\Gamma^a\Psi_a)_A = 0$. One can see from (2.10) that after this gauge choice we will be left with masses for the spin-3/2 fields and remaining spin-$\frac{1}{2}$ fields. Another interesting feature is the $B_\sigma{}^i$ - $A_{\mu\nu\rho i}$ mixing in (2.12) by which the $B_\sigma{}^i$, $A_{\mu\nu\rho i}$ Lagrangian becomes equivalent to seven massive Maxwell Lagrangians (Ogievetsky and Polubarinov (1966); Kalb and Ramond (1974)), i.e., seven of the spin-0 particles in the guise of antisymmetric tensors are eaten by seven of the massless spin-1 particles to become seven massive spin-1 particles.

Unfortunately, the question of the mass spectrum in this model is obscured by the existence of the cosmological term and the problem of how to choose the vacuum values of the scalar fields. This problem would be circumvented if we were able to gauge the O(8) symmetry because this would induce a cosmological term of the opposite sign to cancel (Freedman and Das (1977); Townsend (1977); MacDowell and Mansouri (1977); Deser and Zumino (1977)). Moreover this cancellation would be natural, in the sense that quantum corrections to the cosmological constant in this model are finite, at least at one loop (Christensen, Duff, Gibbons and Roček (1980)).

Finally we would like to draw attention to the similarity between our new mechanism for driving a spontaneous breakdown of supersymmetry with that of Fayet-Illiopoulos in super-QED (Fayet and Illiopoulos(1974)). In both cases, one adds a term to the Lagrangian that is linear in a field that transforms as a total derivative under supersymmetry. In our case, however, the supersymmetry is local. It will be interesting to see if there is any connection between our work and that of Cremmer, Scherk and Schwarz (1979), who obtain a model of N = 8 supergravity with a spontaneous breakdown of supersymmetry by means of a generalized dimensional reduction (Scherk and Schwarz (1979)).

## REFERENCES

Aurilia, A., Christodoulou, D. & Legovini, F. (1978). A classical interpretation of the bag model for hadrons. Phys. Lett. B73, 429.

Aurilia, A., Nicolai, H & Townsend, P.K. (1980). Hidden constants: the Θ-parameter of QCD and the cosmological constant of N = 8 supergravity, CERN preprint TH 2875, to be published in Nuclear Physics B.

Christensen, S.M., Duff, M.J., Gibbons, G.W. & Roček, M. (1980). Vanishing one loop β function in gauged N > 4 supergravity, Phys.Lett. 84B, 411

Coleman, S. (1976). More about the massive Schwinger model. Ann. of Phys. 101, 239.

Cremmer, E. & Julia, B. (1979). The SO(8) supergravity. Nucl. Phys. B 159, 141.

Cremmer, E., Julia, B & Scherk, J. (1978). Supergravity theory in 11 dimensions. Phys. Lett. B 76, 409.

Cremmer, E., Scherk, J & Schwarz, J. (1979). Spontaneously broken N = 8 supergravity. Phys. Lett. B 84, 83.

Deser, S. & Zumino, B. (1977). Broken supersymmetry and supergravity. Phys. Rev. Lett. 38, 1433.

Duff, M. & van Nieuwenhuizen, P. (1980). Quantum Inequivalence of different field representations. Phys. Lett. B 94, 179.

Fayet, P. & Illiopoulos, J. (1974). Spontaneously broken supergauge symmetries and Goldstone spinors. Phys. Lett. B 51, 461.

Ferrara, S. & van Nieuwenhuizen, P. (1978). The auxiliary fields of supergravity. Phys. Lett. B 74, 333.

Freedman, D.Z. & Das, A. (1977). Gauge internal symmetry in extended supergravity. Nucl. Phys. B 120, 221.

Freedman, D.Z. & Schwarz, J. (1978). N = 4 supergravity with local SU(2) x SU(2) invariance. Nucl. Phys. B 137, 333.

Kalb, M. & Ramond, P. (1974). Classical direct interesting action. Phys. Rev. D 9, 2273.

Lüscher, M. (1978). The secret long range force in quantum field theories with instantons. Phys. Lett. B 78, 465.

Macdowell, S.W. & Mansouri, F. (1977). Unified geometric theory of gravity and supergravity. Phys. Rev. Lett. 38, 739.

Ogievetsky, V. & Polubarinov, V. (1966). The Notoph and its possible interactions. (Soviet) Journal of Nuclear Physics 4, 216.

Ogievetsky, V. & Sokatchev, E. (1980). Equation of motion for the axial gravitational superfield. Dubna preprint E2-80-139.

Scherk, J. & Schwarz, J. (1979). Spontaneous breaking of supersymmetry through dimensional reduction. Phys. Lett. B 82, 60.

Stelle, K. & West, P.C. (1978). Minimal auxiliary fields for supergravity. Phys. Lett. B 74, 303.

Townsend, P.K. (1977). Cosmological constant in supergravity. Phys. Rev. D 15, 2802.

COUNTERTERMS FOR EXTENDED SUPERGRAVITY

Paul Howe†
I.T.P. Chalmers, Göteborg

Ulf Lindström
I.T.P. University of Stockholm

ABSTRACT

We construct 3-loop counterterms for N ⩽ 4 supergravities and a 7-loop counterterm for N = 8 supergravity.

INTRODUCTION

One of the main hopes for extended supergravity theories was that they would have better quantum properties than the ordinary Einstein theory. This conjecture has been supported to some extent by the remarkable quantum properties of the maximal (N=4) supersymmetric Yang-Mills theory, (Brink, Schwarz & Scherk 1977) where the β-function has been shown to vanish up to 3-loops so far (Grisaru, Roček & Siegel 1980, Tarasov 1980).

However, ordinary Yang-Mills is already a renormalizable theory, so that this result, although dramatic, does not imply that extended supersymmetry should necessarily rescue a non-renormalizable theory such as quantum gravity. Indeed, it has already been demonstrated that there are problems with N=1 and 2 supergravity, where 3-loop counterterms have been shown to exist, firstly at the linearized level on shell (Deser, Kay & Stelle 1977, Deser & Kay 1978) and subsequently for the full theory off-shell (Ferrara & van Nieuwenhuizen 1978b, de Wit 1980). It would therefore seem that unless there is a new feature of quantum supersymmetry which is not as yet understood, the N=1 and 2 supergravities are unlikely to be finite.

In this contribution, which is a condensed version of a paper in preparation (Howe and Lindström, 1980) we shall show that the 3-loop effect persists at least up to N=4, and that we can also construct a 7-loop counterterm for N=8 (there are 5-loop invariants for N=5, 6 which we refer to Howe & Lindström (1980) for details). We shall employ the superspace technique and present our results explicitly at the linearized on-shell level. We shall then discuss how to extend these results to the full (non-linear) case (even off-shell) and comment on possibilities which may invalidate our method.

LINEARIZED SUPERGRAVITY ($N \leqslant 4$)

All linearized on-shell supergravities in superspace may be obtained by linearizing and truncating the N=8 superspace theory (Brink & Howe, 1979), but for our purposes it is more convenient to consider the $N\leqslant 4$ theories together and N=8 separately (we shall not consider N=5,6). On-shell the $N\leqslant 4$ theories are described by a chiral superfield W of dimension (2-N/2) (mass) which has (4-N) spinor indices and is totally symmetric on all of them. Latin indices are internal indices, Greek denote spinors and vectors have two (one dotted and one undotted) spinor indices.

$$W = \Psi_{\alpha\beta\gamma},\ N = 1$$
$$M_{\alpha\beta},\ N = 2$$
$$\chi_{\alpha},\ N = 3$$
$$V,\ N = 4 \tag{1}$$

$$\bar{D}_{\dot{\alpha}a} W = 0 \tag{2}$$

(For N=1 $\Psi_{\alpha\beta\gamma}$ is just the Wess-Zumino field $W_{\alpha\beta\gamma}$ (Grimm, Wess & Zumino 1978).

W is a singlet under the internal symmetry group which is taken to be SU(N), although we could easily extend it to U(N) by making W totally skew on N internal indices instead. Its components may be obtained by evaluating successive spinor derivatives on it at $\Theta$=0 (Wess & Zumino, 1978), and are all totally antisymmetric on the internal indices and totally symmetric on the spinor indices, i.e.

$$D^{a_1}_{\alpha_1}\ldots D^{a_i}_{\alpha_i} W_{\beta_1\ldots\ldots\beta_{4-N}} = D^{[a_1}_{(\alpha_1}\ldots D^{a_i]} W_{\beta_1\ldots\ldots\beta_{4-N})} \tag{3}$$

We therefore see that W's components are invariant symmetric spinors having 2s indices for spin s.

As an example, let us consider N=4 where the generating field is a scalar V. V's components are V itself (at $\Theta$=0) and

$$\chi^a_\alpha = D^a_\alpha V$$
$$M^{ab}_{\alpha\beta} = D^a_\alpha D^b_\beta V$$

$$\Psi^{abc}_{\alpha\beta\gamma} = D^a_\alpha D^b_\beta D^c_\gamma V$$

$$\varepsilon^{abcd} C_{\alpha\beta\gamma\delta} = D^a_\alpha D^b_\beta D^c_\gamma D^d_\delta V \tag{4}$$

The fields on the left hand side of (4) are superfields whose leading components correspond to 4 spin-½ fields, 6 spin-1 field strengths, 4 spin-3/2 field strengths and the spin-2 Weyl spinor, and there are no further independent components in V (one has D C = 0).

To show that the theory really is on-shell, we may differentiate the identity

$$\varepsilon^{\alpha\beta} D^a_\alpha D^b_\beta V = 0 \tag{5}$$

with a $\overline{D}$ to obtain

$$\partial_{\dot{\alpha}}{}^{\alpha} \chi^a_\alpha = 0 \tag{6}$$

Successively differentiating (6) with D's one readily obtains

$$\partial_{\dot{\alpha}} M^{ab}_{\alpha\beta} = \partial_{\dot{\alpha}}{}^{\alpha} \Psi^{abc}_{\alpha\beta\gamma} = \partial_{\dot{\alpha}}{}^{\alpha} C_{\alpha\beta\gamma\delta} = 0 \tag{7}$$

Finally, differentiating (6) with a $\overline{D}$, one finds

$$\Box V = 0 \tag{8}$$

Equations (6), (7) and (8) are the required on-shell equations of motion and Bianchi identities for the fields of spins 0 to 2 which occur in the theory.

For N<4, one proceeds in a similar fashion, only beginning with the appropriate W (1). For all these theories, we see that the spectrum of different spin fields falls into the familiar binomial coefficient pattern.

3-LOOP COUNTERTERM N $\leqslant$ 4

The form of the 3-loop counterterm action in x-space is given by

$$S^{(3)} = \kappa^4 \int d^4x \, \{ C_{\alpha\beta\gamma\delta} C^{\alpha\beta\gamma\delta} \overline{C}_{\dot{\alpha}\dot{\beta}\dot{\gamma}\dot{\delta}} \overline{C}^{\dot{\alpha}\dot{\beta}\dot{\gamma}\dot{\delta}} + \text{supersymmetric completion} \} \tag{9}$$

where the terms necessary to make the action supersymmetric become progressively more complicated as we increase the number of fields in the theory by increasing N. As we have already mentioned, (9) has been explicitly constructed in x-space for N=1, 2 (Deser, Kay & Stelle 1977, Deser & Kay 1978, Ferrara & van Nieuwenhuizen 1978b, de Wit 1980) and the superspace version of the linearized N=1 has also been given by Ferrara and Zumino 1978. This latter is a special case of the general superspace form valid for all $N \leqslant 4$:

$$S^{(3)} = \kappa^4 \int d^4x \, d^{4N}\Theta \, W^2\bar{W}^2 \tag{10}$$

We observe that although the dimension of the superspace measure increases with N, there is a corresponding decrease in the dimension of W so that (10) always has zero dimension. Since $S^{(3)}$ is an integral over all of superspace of a superfield, it is manifestly supersymmetric, so that if we can establish the spin-2 part of (9) it will automatically follow that the supersymmetric extension exists. For N=4, one has

$$V = \varepsilon^{abcd}\Theta^{\alpha}_{a} \ldots \Theta^{\delta}_{d} \, C_{\alpha\beta\gamma\delta} + \ldots . \tag{11}$$

so that

$$V^2 = \Theta^8 C_{\alpha\beta\gamma\delta}C^{\alpha\beta\gamma\delta} + \ldots . \tag{12}$$

since the product of 8 Θ's is a singlet under SL(2,C) and SU(4). (9) then follows immediately from (12) upon integrating out the Θ's. The analysis for N<4 is similar and we therefore conclude that for $N \leqslant 4$ there exists a 3-loop counterterm whose spin-2 part is the square of the Bel-Robinson tensor.

Although we have used an SU(N) formalism, (10) is manifestly invariant under an additional U(1), so that we really have a U(N) counterterm. For N=1, this U(1) corresponds to the chiral transformation of the gravitino field which arises as a restricted super-Weyl transformation, and has been discussed by Christensen, Deser, Duff & Grisaru (1979).

To obtain the full non-linear result it suffices to generalize (10) by including the superdeterminant of the vielbein, E, so we get

$$S^{(3)} = \kappa^4 \int d^4x \, d^{4N}\Theta \, EW^2\bar{W}^2 \tag{13}$$

(13) can also be considered as a full off-shell invariant (in this case the non-linear generalization of (3) is no longer valid) and for N=1 and 2 where the auxiliary fields are known (Ferrara &

van Nieuwenhuizen 1978a, Stelle & West 1978, de Wit & van Holten 1979, Fradkin & Vasiliev 1979, Breitenlohner & Sohnius 1979) it could in principle be integrated out over the Θ's to obtain the corresponding x-space invariant (cf. Lindström, Karlhede and Roček in this volume for the N=1 case). For N=3 and 4 this can as yet not be done.

There are two cases in which the argument leading to (13) may not be valid off-shell: firstly, for N=4, one or both of the scalars (V is complex) may occur as antisymmetric tensors and this would alter the superspace description, secondly, for N=2 and 4 it is possible that there are versions of the theory which have off-shell central charges (Sohnius, Stelle & West 1980, Cremmer, Ferrara, Stelle & West 1980) and this would invalidate our arguments.

N = 8

The N=8 theory was first constructed at the linearized level by de Wit and Freedman (1977), and generalized to the full non-linear case by Cremmer and Julia (1979) by dimensional reduction of the eleven-dimensional theory (Cremmer, Julia & Scherk 1978). The full theory has local SU(8) and global $E_7$ invariances and the corresponding superspace formulation has also been given (Brink & Howe 1979). By directly linearizing this last version one finds that N=8 supergravity is described by a scalar superfield $V_{abcd}$ transforming as a 70 under the (global) SU(8):

$$V_{abcd} = \frac{1}{4!}\,\varepsilon_{abcdefgh}\bar{V}^{efgh} \tag{14}$$

V satisfies

$$\bar{D}_{\dot{\alpha}a}V_{bcde} = \bar{D}_{\dot{\alpha}[a}V_{bcde]} \tag{15}$$

and its components are

$$D^a_\alpha\, V_{bcde} = 4\,\delta^a_{[b}\,\chi_{cde]\alpha}$$

$$D^a_\alpha\, \chi_{bcd\beta} = 3\,\delta^a_{[b}\,M_{cd]\alpha\beta}$$

$$D^a_\alpha\, M_{bc\beta\gamma} = 2\,\delta^a_{[b}\,\Psi_{c]\alpha\beta\gamma}$$

$$D^a_\alpha\, \Psi_{b\beta\gamma\delta} = \delta^a_b\, C_{\alpha\beta\gamma\delta} \tag{16}$$

and hence one sees the familiar spectrum of N=8 (1 spin-2, 8 spin-3/2, 28 spin-1, 56 spin-½, 70 spin-0).

The equations of motion corresponding to (6), (7) and (8) may be derived by using arguments similar to those we used for N=4, and almost identical to those one can employ to find the equations for N=4 Yang-Mills in superspace (Sohnius, 1978). One finds

$$\partial_{\dot{\alpha}}{}^{\alpha_1} D^{a_1}_{\alpha_1} \ldots\ldots D^{a_i}_{\alpha_i} V_{bcde} = 0 \quad , \quad 4 \geq i \geq 1$$

$$\Box V_{abcd} = 0 \tag{17}$$

so that the theory is on-shell.

From our point of view, the main differences encountered as we pass from N≤4 to N=5 and above are that the basic superfield ceases to be chiral and that the dimension of the 3-loop Lagrangian no longer matches the dimension of the superspace measure. For these reasons, we shall focus on the simplest counter term we can construct which occurs at the 7-loop level:

$$S^{(7)} = \kappa^{12} \int d^4x \, d^{32}\theta \, (V_{abcd} \bar{V}^{abcd})^4 \tag{18}$$

The spin-2 contribution to V is given by

$$V_{abcd} = \theta^{\alpha}_{[a} \ldots \theta^{\delta}_{d]} C_{\alpha\beta\gamma\delta} + \frac{1}{4!} \varepsilon_{abcdefgh} \bar{\theta}^{\dot{\alpha}e} \ldots \bar{\theta}^{\dot{\delta}h} \bar{C}_{\dot{\alpha}\dot{\beta}\dot{\gamma}\dot{\delta}} \tag{19}$$

Taking the 8th power of V and integrating out the Θ's one obtains

$$S^{(7)} = \kappa^{12} \int d^4x \, \{ \, (C_{\alpha\beta\gamma\delta} C^{\alpha\beta\gamma\delta} \bar{C}_{\dot{\alpha}\dot{\beta}\dot{\gamma}\dot{\delta}} \bar{C}^{\dot{\alpha}\dot{\beta}\dot{\gamma}\dot{\delta}})^2 +$$

$$+ \text{ supersymmetric completion} \} \tag{20}$$

so that (18) is non-zero.

To generalize (18) to the non-linear case, we recall that the scalars are described by a matrix V which can be considered as an element of the non-compact group $E_{7(+7)}$ (Cremmer & Julia 1979). However, one may choose an SU(8) gauge such that V has the form (Cremmer & Julia 1979)

$$V = \exp \begin{pmatrix} 0 & V^{abcd} \\ & \\ V_{abcd} & 0 \end{pmatrix} \tag{21}$$

where $V_{abcd}$ transforms as a 70 under the remaining (global) SU(8). Therefore, we can generalize (18) to the full non-linear case merely by an insertion of a determinant as in the $N \leqslant 4$ case, although such a term will not be $E_7$ invariant. The resultant invariant will also be an off-shell invariant (if the off-shell group is SU(8) or SO(8), global or local), provided that we relax the non-linear version of (15).

If the off-shell theory has antisymmetric tensors replacing some of the scalars, or central charges, then as we have discussed previously, our arguments will no longer be applicable.

## CONCLUSIONS

We have demonstrated that for all extended supergravity theories one can construct on-shell non-linear counterterms (for N=5, 6 see Howe & Lindström 1980). This is not an encouraging result since although one may argue that the coefficient of a counterterm may be zero in an actual calculation, one can construct an infinite series of counterterm Lagrangians from a given one. For example, for $N \leqslant 4$, the expression

$$S^{(p)} = \kappa^{4+8p} \int d^4x \; d^{4N}\theta \; W^2\bar{W}^2(C^2\bar{C}^2)^p \tag{22}$$

gives rise to a manifestly supersymmetric series of (3+4p)-loop counterterms. On the face of it, it would therefore seem as if supergravity is only an improvement on quantum gravity in the sense that the divergences start at higher loop order. Although our analysis has not been exhaustive, it is tempting to conjecture that this improvement increases beyond N=4, so that N=5,6 may be finite up to 4 loops and N=8 up to 6 loops. To prove this one would have to rule out all possible 6-loop and lower terms (for N=8) and this would require demonstrating that the only supersymmetric actions that can be constructed from V.... have to be integrals over the whole of superspace.

Another partially optimistic point is that all of the counterterms we have constructed are of the form allowed by Christensen, Deser, Duff & Grisaru, 1979, i.e. they vanish for half-flat spaces, although whether or not half-flatness is compatible with N=8 supergravity is not obvious since C and $\bar{C}$ enter on an equal footing in the basic field V. (This was pointed out to us by M. Duff).

The question remains as to whether or not there are ways around our result. We have already mentioned the fact that, in the off-shell case, either central charges or antisymmetric tensors would invalidate our construction. However, it is the on-shell case that is of paramount importance and here it is not clear whether either of these alternative types of theory would help us. Furthermore, it has been pointed out that the central charge method is not free from difficulties in the Yang-Mills case (Siegel 1980).

Another possibility is that we have failed to take into account all relevant invariances, but on this point, it would appear that all the invariants we have constructed have a U(N) invariance which is the maximal one allowed on-shell by the Haag, Lopuszanski & Sohnius 1975 classification scheme. (For N=8 we can only have SU(8) because of the self-duality condition (14)).

We should mention that the theory we are ultimately interested in will have some mechanism that breaks supersymmetry, such as the one proposed by Scherk & Schwarz, 1979. If this breaking is spontaneous, one would expect that the ultra-violet behaviour should be unchanged so that our result would still stand, whereas if the mechanism is not of this type, it is not at all obvious that the quantum behaviour should be improved.

Finally it might be that we are being too naive in our approach to quantum gravity or perhaps that there are some new features of quantum supersymmetry that are as yet not understood.

ACKNOWLEDGEMENTS

We would like to thank Peter van Nieuwenhuizen for his comments on the manuscript and numerous colleagues at this conference for helpful discussions.

† Current address: CERN, Geneva.

REFERENCES

Breitenlohner, P. & Sohnius, M. (1979).Superfields, Auxiliary Fields, and Tensor Calculus for N=2 Extended Supergravity. To be published in Nucl. Phys. B.

Brink, L. & Howe, P. (1979). The N=8 supergravity in superspace. Phys. Lett. 88B pp. 268-272.

Brink, L., Schwarz, J. & Scherk, J. (1977). Supersymmetric Yang-Mills Theories. Nucl. Phys. B121 pp. 77-92.

Christensen, S.M., Deser, S., Duff, M.J. & Grisaru, M.T. (1979). Chirality, self-duality, and supergravity counterterms. Phys. Lett. 84B pp. 411-415.

Cremmer, E., Ferrara, S., Stelle, K.S. & West, P.C. (1980). Off-shell N=8 supersymmetry with central charges. Phys. Lett. 94B pp. 349-354.

Cremmer, E. & Julia, B. (1979). The SO(8) supergravity. Nucl. Phys. B159 pp. 141-212.

COUNTERTERMS FOR EXTENDED SUPERGRAVITY


Cremmer, E., Julia, B. & Scherk, J. (1978). Supergravity theory in 11 dimensions. Phys Lett. 76B pp. 409-412.

Cremmer, E., Scherk, J. & Schwarz, J.H. (1979). Spontaneously broken N=8 supergravity. Phys. Lett. 84B pp. 83-86.

Deser, S. & Kay, J.H. (1978). Three-loop counterterms for extended supergravity. Phys Lett. 76B pp. 400-403.

Deser, S., Kay, J.H. & Stelle, K.S. (1977). Renormalizability properties of supergravity. Phys. Rev. Lett. 38 pp. 527-530.

de Wit, B. (1980). Private communication (for N=2).

de Wit, B. & Freedman, D.Z. (1977). On SO(8) extended supergravity. Nucl. Phys. B130 pp. 105-113.

de Wit, B. & van Holten, J.W. (1979). Multiplets of linearized SO(2) supergravity. Nucl. Phys. B155 pp. 530-542. (For N=2).

Ferrara, S. & van Nieuwenhuizen, P. (1978a). The auxiliary fields of supergravity. Phys. Lett. 74B pp. 333-335. (For N=1).

Ferrara, S. & van Nieuwenhuizen, P. (1978b). Structure of supergravity. Phys Lett. 78B pp. 573-576. (For N=1).

Ferrara, S. & Zumino, B. (1978). Structure of linearized supergravity and conformal supergravity. Nucl. Phys. B134 pp. 301-326.

Fradkin, E.S. & Vasiliev, M.A. (1979). Minimal set of auxiliary fields in SO(2)-extended supergravity. Phys. Lett. 85B pp. 47-51.

Grimm, R., Wess, J. & Zumino, B. (1978). Consistency checks on the superspace formulation of supergravity. Phys. Lett. 73B pp. 415-417.

Grisaru, M.T., Roček, M. & Siegel, W. (1980). Zero 3-loop β-function in N=4 super Yang-Mills theory. Brandeis preprint submitted to Phys. Rev. Lett.

Haag, R., Lopuszanski, J.T. & Sohnius, M. (1975). All possible generators of supersymmetries of the S-matrix. Nucl. Phys. B88 pp. 257-274.

Howe, P. & Lindström, U. (1980). In preparation.

Scherk, J. & Schwarz, J.H. (1979). Spontaneous breaking of supersymmetry through dimensional reduction. Phys. Lett. 82B pp. 60-64.

Siegel, W. (1980). Off-shell central charges. Princeton preprint.

Sohnius, M. (1978). Bianchi identities for supersymmetric gauge theoris. Nucl. Phys. B136 pp. 461-474.

Sohnius, M., Stelle, K.S. & West, P.C. (1980). Dimensional Reduction by Legendre Transformation generates Off-shell Supersymmetric Yang-Mills theories. Imperial College preprint to appear in Nucl. Phys. B.

Stelle, K.S. & West, P.C. (1978). Minimal auxiliary fields for supergravity. Phys. Lett. 74B pp. 330-332.

Tarasov. (1980). Private communication of A.A. Slavnov.

Wess, J. & Zumino, B. (1978). The component formalism follows from the superspace formulation of supergravity. Phys. Lett. 79B pp. 394-398.

# SUPERGRAVITY AND GRAND UNIFICATION

Bruno Zumino

CERN, Geneva, Switzerland

## GRAND UNIFIED THEORIES

At the present time the (low energy) interactions among elementary particles appear to be successfully described by the so-called extended Glashow-Weinberg-Salam model (or standard model). This is a renormalizable gauge theory based on the low energy group

$$SU(3)_{colour} \times SU(2)_{left} \times U(1) \ . \tag{1}$$

Corresponding to the non-simple nature of this group, the theory has three gauge couplings. The three coupling constants vary with energy according to the renormalization group equations and become approximately equal at an energy of about $10^{14}$-$10^{15}$ GeV, where one can attempt a grand unified theory (GUT) based on a gauge group with only one coupling constant and containing the low energy group (1). As discussed by Georgi and Glashow (1974), the minimal GUT is that based on the group SU(5). In reaching this conclusion they used the fact that (1) has rank four and therefore the GU group must have at least rank four. The 15 left-handed states forming a family (or generation) are in a complex representation of (1) and therefore it is natural to assume that they will be in complex representations of the GU group, which excludes groups without complex representations. Finally, Georgi and Glashow imposed the cancellation of anomalies necessary for a renormalizable gauge theory. With these requirements they were led to the GU group SU(5) with the 15 left-handed states of a family in the reducible representation $\bar{5}$ + 10 (10 is the antisymmetric tensor with two indices). Most other GUTs with larger groups which have been proposed till now go through SU(5) (and it seems that these are the only viable proposals). Larger groups [like SO(10) or $E_6$] achieve the aim of having the fermions of one family in a single irreducible representation but have a left-right symmetry which makes it harder to understand why the neutrinos are massless (if they are). Even larger groups can unify different families in a single representation.

The GUT picture has had some phenomenological successes, which to a large extent do not depend on the particular GU group chosen. The assumptions of a common unification mass scale for the three low energy gauge groups and of essentially no new physics below this scale determine the weak angle which parameterizes neutral current couplings at the value (Georgi, Quinn and Weinberg, 1974; Chanowitz, Ellis and Gaillard, 1977; Buras, Ellis, Gaillard and Nanopoulos, 1978)

$$\sin^2\theta_W = 0.206 \pm 0.006 \, , \tag{2}$$

compatible with values obtained from analyses of the present experimental data. In addition, the minimal GUT relates quark and lepton masses at $10^{15}$ GeV. Evaluation of symmetry breaking effects gives, for example, the ratio (Chanowitz et al., 1977; Buras et al., 1978; Nanopoulos and Ross, 1979)

$$m_b/m_\tau \simeq 3 \, , \tag{3}$$

again compatible with experiment. Other, less successful, mass relations can perhaps also be understood. A crucial test of at least the minimal GUT will soon be forthcoming with searches for nucleon decay with the predicted lifetime

$$\tau_{\text{proton}} \simeq 10^{30\pm2} \text{ years} \, . \tag{4}$$

However, the existing models for grand unification also have shortcomings which strongly suggest that they are incomplete. They contain far too many free parameters ($\gtrsim 20$): most of the fermion masses and weak mixing angles are undetermined, as well as the couplings and even the field content of the Higgs sector. So, in spite of the occurrence of only one gauge coupling, GUTs are not unified. There is a notorious "gauge hierarchy" problem (Gildener, 1976) which is the difference of about 12 orders of magnitude in the mass scales at which the initial breaking

$$\text{GUT} \rightarrow SU(3)_{\text{colour}} \times SU(2)_{\text{left}} \times U(1) \tag{5}$$

and the final breaking

$$SU(2)_{\text{left}} \times U(1) \rightarrow U(1)_{\text{e.m.}} \tag{6}$$

are found to occur. This requires a readjustment of parameters in the Higgs potential to many orders of magnitude in each order in perturbation theory. Finally, perhaps the most obvious defect of grand unification is that it excludes gravity which we know must become important at a mass scale $m_P \simeq 10^{19}$ GeV, not much larger than the GU mass of $10^{14}$-$10^{15}$ GeV.

SUPERGRAVITY AND GRAND UNIFICATION

A natural way to unify gravity with lower spin fields is provided by supergravity (Freedman, van Nieuwenhuizen and Ferrara, 1976; Deser and Zumino, 1976; Ferrara and van Nieuwenhuizen, 1976; Ferrara, Scherk and Zumino, 1977; Freedman, 1977; Das, 1977; Cremmer, Scherk and Ferrara, 1977; Cremmer and Scherk, 1977; de Wit and Freedman, 1977). Supergravity is the supersymmetric extension of Einstein's gravity. In working out this appealing idea one is faced with the difficulty that the largest supergravity theory (largest theory with a supermultiplet with maximum spin not exceeding two) is the N = 8 theory (N is the number of supersymmetries) which has an SO(8) internal symmetry. Now, the group SO(8) is too small, it does not contain the low energy group (1), and the fundamental supermultiplet of the N = 8 theory (which consists of 1 spin 2, 8 spin 3/2, 28 vectors, 56 spin 1/2 and 70 scalars) is not rich enough to accommodate all known quarks, leptons and spin-one gauge fields (Gell-Mann, 1977).

In principle, there are several possible ways out of this difficulty. The first is to observe that the group SO(8) can contain $SU(3)_{colour} \times U(1)_{e.m.}$, which in our present picture is the exact gauge group of nature. The additional $SU(2)_{left}$ could arise in a purely dynamical way and the intermediate bosons of the weak interactions could be bound states of the fundamental fields (the photon and the gluon fields, instead, would be elementary). This requires to elevate the 28 vectors of the basic supermultiplet to gauge vectors of SO(8). Now, it is a general feature of supergravity theories that the vector fields are exactly the right number and in the right representation to become the gauge fields of SO(N). For $N \leq 4$ it has been possible to introduce a gauge coupling g for the vector fields (Freedman and Das, 1977; Freedman and Schwarz, 1978; Cremmer and Scherk, unpublished). Supersymmetry requires it to be accompanied by a mass term $\sim g/\kappa$ for the spin 3/2 fields and by a cosmological term $\sim (g/\kappa^2)^2$. Perhaps this gauging can be achieved also for the N = 8 theory, but the resulting cosmological constant is enormous, if g is of the order unity. Still, it is not impossible that this unwanted cosmological term is cancelled by that generated from the spontaneous breaking of supersymmetry (Deser and Zumino, 1977). Alternatively, as suggested by Hawking, the very large cosmological constant could give rise to a foamy structure of space time. Over distances large compared to the Planck length the foamy structure would not be visible and space time could look smooth and approximately Minkowskian. At the same time the SO(8) internal symmetry would be masked and a larger approximate effective symmetry could result. However interesting this point of view, it is hard to see how one would proceed to implement it. In the following we shall assume that the cosmological constant vanishes, due to cancellation among different contributions to it.

An approach which seems more promising at this time is based on the observation that N = 8 supergravity (without the gauge coupling g) has not only an SO(8) symmetry but also global SU(8) internal symmetry (the same which is an automorphism of the supersymmetry algebra) and even a global $E_7(+7)$ invariance, where $E_7(+7) \supset SU(8) \supset SO(8)$. The situation is analogous to that observed first by Ferrara, Scherk and the author (1977) for the N = 2 and 3 theories and by Cremmer, Scherk and Ferrara (1978) for the N = 4 theory. The complex transformations of the SU(8) group are realized on the spinor fields as chiral transformations and on the spin-one fields as duality rotations which transform the electric into the magnetic field strengths. The transformations of the (non-compact) $E_7$ which are not contained in its (maximal compact) subgroup SU(8), are realized non-linearly, the 70 scalar fields of the theory playing a role analogous to that of the pion in non-linear pion dynamics.

Cremmer and Julia (1978, 1979) have formulated this situation in a way which is similar to that in the $CP^{n-1}$ model (d'Adda, Di Vecchia and Lüscher, 1978, 1979; Witten, 1979). In addition to the 70 physical scalar fields, they introduce 63 additional scalars. The theory becomes invariant under a local SU(8) group (which can be used to transform away the redundant scalars) and under a global non-compact $E_7$. If we call $X^i$ the 63 generators of SU(8) and $Y^a$ the additional 70 non-compact generators of $E_7$, the 63 scalars $a_i$ and the 70 scalars $b_a$ can be combined into a group element of $E_7$

$$\exp\,[a_i X^i + b_a Y^a] \tag{7}$$

(sum over repeated indices). The global $E_7$ operates on the scalars by group multiplication on the right, the local SU(8) by group multiplication on the left. Clearly, one can transform (7) by a local SU(8) transformation to the form with 70 scalars

$$\exp\,[b_a Y^a] \tag{8}$$

(symmetric gauge). If we now operate with $E_7$ on the right, this form is not preserved in general, but we can re-establish it by operating on the left with a local SU(8) transformation having parameters which are functions of the scalars $b_a$. The resulting transformation on the $b_a$ is therefore non-linear and that on the other fields of higher spin is linear in those fields but non-linear in the scalars. For an element of the global SU(8) contained in $E_7$ the transformation is linear. The symmetric gauge, which contains only physical fields, is convenient for the description of the physical spectrum of the theory, but the more general formulation is more flexible and useful for a Lagrangian description. In analogy with the case of the $CP^{n-1}$ model, Cremmer and Julia have introduced non-propagating SU(8) gauge vectors to ensure the SU(8) invariance of the Lagrangian. These gauge vectors have no kinetic term of their own, their equations of motion are algebraic and permit to express them in terms of the other fields. Substituting these expressions into the Lagrangian, they can be completely eliminated, but the SU(8) gauge invariance is preserved. Still, as it is clear from the above discussion, this gauge invariance is artificial, and is due to the introduction of the 63 redundant scalars $a_i$. On the other hand, the global invariances SU(8) and $E_7$ are true symmetries of the equations of motion.

Cremmer and Julia have carried the analogy with the $CP^{n-1}$ model even further. In that model, in two space-time dimensions and in the 1/n expansion, the non-propagating U(1) gauge field becomes dynamical, its propagator acquires a zero mass pole. They suggested that the same may happen to the SU(8) gauge vectors. In other words, these gauge vectors, which are functions of the fundamental fields of the theory, should bind into zero mass particles. It is natural then to assume (Ellis, Gaillard, Maiani and Zumino, 1980; Ellis, 1980; Ellis, Gaillard and Zumino, 1980) that the fields of the fundamental supermultiplet of N = 8 supergravity bind together to form another massless supermultiplet containing spin-one fields in the adjoint representation of SU(8). Just as the group SU(8) is large enough to contain the group of grand unification, the supermultiplet of bound states can be sufficiently rich to contain all known quarks, leptons, vectors and scalars. All fields considered elementary at the present time are taken instead to be composite (for an earlier attempt in this direction see Curtright and Freund, 1979).

In the two-dimensional $CP^{n-1}$ model in the 1/n approximation there occurs another interesting phenomenon. The U(1) vector field, which has acquired a zero mass pole, acts as a binding and confining force among the fundamental scalars of the model. The spectrum consists of bound states of those scalars and one finds (Haber, Hinchliffe and Rabinovici, 1980) that it exhibits an SU(n) symmetry. The fundamental scalars are in the coset space

$$CP^{n-1} = SU(n)/S[U(1) \times U(n-1)] \tag{9}$$

but the spectrum exhibits the symmetry of the larger group SU(n). Now, in N = 8 supergravity the fundamental scalars are in the 70-dimensional coset space

$$E_7/SU(8) \ . \tag{10}$$

By analogy one may expect that the bound states form one (or more) representation of $E_7$. Since this group is non-compact, its unitary representations are infinite dimensional and one would expect the bound states to form an infinite dimensional supermultiplet containing arbitrarily high spins. Observe that $E_7$ is represented non-linearly on the finite dimensional supermultiplet of the fundamental fields. Here we are suggesting that it is represented linearly on the bound states (this possibility arose in discussions with W. Bardeen and E. Rabinovici).

One can imagine the following picture. At an energy comparable with (or higher than) the Planck mass $m_P$, supersymmetry is valid and also the larger internal symmetry SU(8) (and also, perhaps, $E_7$). The dynamics is specified by the gravitational constant $\kappa \sim 1/m_P$. Coming down to the GU mass, all supersymmetries are broken and SU(8) is broken down to the GU group, say SU(5). This symmetry breaking is spontaneous and characterized by the vacuum expectation value $\langle\phi\rangle$ of some scalar field. At the GU mass the interactions are determined by a dimensionless coupling constant $g \sim \kappa\langle\phi\rangle$ which emerges as the GU gauge coupling constant. The GUT appears as a "low energy" effective theory derived from the dynamics which is valid at the Planck mass. That it is a renormalizable gauge theory can be perhaps understood from what we shall call the "zero mass decoupling theorem".

This theorem states that if a large mass (say $m_P$ for definiteness) occurs in a field theory, and if a subset of fields or states remains <u>massless</u> as $m_P \to \infty$, then these states decouple from the rest in the limit and their effective interactions at low energy are renormalizable (with additional non-renormalizable interactions inversely proportional to some power of $m_P$). The zero mass decoupling theorem is physically intuitive and can actually be proven, if the original field theory is a well defined renormalizable theory (for instance a gauge theory) and the zero mass states are those corresponding to some of the fields of the original theory. Presumably, it can also be proven for zero-mass bound states. Indeed, if their effective interactions were not renormalizable, divergences would arise in the computation of vertex functions, for which the only cut-off can be the inverse size of the bound states, which we expect to be $m_P$. This would result in masses of order $m_P$, or a breakdown of perturbation theory for the effective interactions of the bound states (we have learned this argument from M. Veltman). We wish to apply the theorem to the case where the original theory is N = 8 supergravity and the zero mass states are composites of the fundamental fields, admittedly a daring extrapolation. A consequence of the zero mass decoupling theorem is that the zero mass states cannot have spin higher than one, since their interaction would be necessarily non-renormalizable. For the same reason, the spin-one fields must be gauge fields and the zero mass spin one-half fields must form an anomaly free set. Scalar fields can survive in any number. All fields or states which do not belong to this subset with renormalizable interactions must disappear as $m_P \to \infty$, by acquiring very large masses (or perhaps by becoming unbound).

## THE COMPOSITE SUPERMULTIPLET

We would like to discover in which supermultiplet are contained the SU(8) gauge fields. According to Cremmer and Julia, these composite fields are expressions (bilinear plus higher order terms) in the fundamental scalar fields. In the symmetric gauge, where all fundamental fields are physical, they agree (up to higher non-linearities) with the conserved currents of the global SU(8) invariance of the equations of motion, or rather the part of those currents involving the scalar fields. In spite of the differences in higher non-linear terms, one may imagine that these composite operators will describe the same vector bound states. We would like then to find to which supermultiplet belong the currents of global SU(8). Observe that, to make the SU(8) currents members of a supermultiplet, one must add to them terms involving all other fundamental fields of spins up to 3/2. Since the spin 1 fields transform by duality rotations, the full generators of SU(8) are not really obtainable as integrals of time components of local currents, but then neither are the generators of the SO(8) subgroup: this non-locality is really a consequence of the vanishing mass of the fundamental vector fields.

In order to find the supermultiplet of the SU(8) currents, one can look first at supersymmetric theories with smaller N. For a massless supersymmetric system with $N \leq 4$ and maximum spin 1, all currents are local and this supermultiplet is the supercurrent supermultiplet (Ferrara and Zumino, 1975; Sohnius, 1979) which contains also the energy momentum tensor and the N spinor currents. One can put the fields of this supermultiplet on the mass shell and one obtains in this way a massive supermultiplet of supersymmetry with $N \leq 4$ spinor charges. The maximum spin is 2 and is a singlet. States with lower spins are in representations of Sp(2N). As the mass tends to zero the supermultiplet breaks up into a finite number of massless supermultiplets in which the various helicities are in representations of SU(N). It is not difficult to identify which of these massless supermultiplets contains the SU(N) currents in the adjoint representation and one finds that it is of the type given below in (11), (12). This argument is reliable for $N \leq 4$; one can assume that the multiplet has the same form for $N = 8$.

Following this long (and admittedly not very strong) chain of arguments, Ellis, Gaillard, Maiani and the author (1980) have come to the conclusion that the relevant supermultiplet of states is given by (A, B, ... = 1, 2 ... 8)

$$\left(\frac{3}{2}\right)^{A}, \left(1\right)^{A}_{B}, \left(\frac{1}{2}\right)^{A}_{[BC]}, \left(0\right)^{A}_{[BCD]}, \ldots \left(-\frac{5}{2}\right)^{A}, \tag{11}$$

to which one must add the TCP conjugate states

$$\left(\frac{5}{2}\right)_{A}, \left(2\right)_{AB}, \left(\frac{3}{2}\right)_{A[BC]}, \ldots \left(-\frac{3}{2}\right)_{A}, \tag{12}$$

(a set of k antisymmetrized lower indices is equivalent to 8-k antisymmetrized upper indices). It contains spin-one states in the adjoint representation but also other spin-one states. The SU(8) representations contained in (12) are given explicitly in Table 1.

Table 1

Helicity states of the composite multiplet of SU(8) gauge vectors

| Helicity | -3/2 | -1 | -1/2 | 0 | +1/2 | +1 | +3/2 | +2 | +5/2 |
|---|---|---|---|---|---|---|---|---|---|
| SU(8) content | $\overline{8}$ | 1<br>63 | 8<br>216 | 28<br>420 | 56<br>504 | 70<br>$\overline{378}$ | $\overline{56}$<br>$\overline{168}$ | $\overline{28}$<br>$\overline{36}$ | $\overline{8}$ |

Ideally one would like to give masses by super-Higgs effect to all unwanted states, those which do not belong to the surviving zero mass subset with effective renormalizable interactions. However, it is easy to see that this cannot be done in an SU(5) invariant way, and not even in an SU(3) × SU(2) × U(1) invariant way. For instance, in order to give an invariant mass to a spin 5/2 state, one needs all the helicities 5/2, 3/2, 1/2, -1/2, -3/2, -5/2 and they must be in the same representation of the invariance group. It is clear that the supermultiplet (11), (12) does not contain all the necessary helicities. It is also easy to convince oneself that no irreducible supermultiplet and probably no finite set of supermultiplets will do. We find here another hint that one may need an infinite set of states. This is due, of course, to our desire to be left with complex (chiral) massless representations; it is very easy to construct finite sets of massless supermultiplets which combine to massive ones, in a vector-like way, simply by starting from a finite mass supermultiplet and going to the limit as the mass tends to zero when it breaks up into massless ones.

As already remarked, the above discussion suggests that one may need an infinite dimensional set of irreducible supermultiplets This would mean that the dynamics at the Planck mass involves states of arbitrarily high spin most of which become infinitely massive, after the symmetry breaking, as $m_P \sim \langle\phi\rangle \to \infty$. Those higher spin states which, like the graviton, remain massless, will be left with non-renormalizable interactions proportional to reciprocal powers of $m_P$. After the symmetry breaking the situation would resemble that in the dual models interpreted as gravity theories (Scherk and Schwarz, 1974) with a slope of the trajectories related to the Planck mass. As $m_P$ tends to infinity and the slope to zero, the zero mass states decouple from the massive states and their residual interactions have the form of renormalizable Yang-Mills and non-renormalizable gravitational interactions.

In an attempt to extract as many properties of the GUT as possible from the dynamics at the Planck mass, without actually knowing this dynamics (nor the possibly infinite-dimensional multiplet), Ellis, Gaillard and the author (1980) have assumed that all surviving low energy states are already contained in the single supermultiplet (11), (12). We have then looked for the maximal set of states of spin 0, 1/2 and 1 contained in it which can have renormalizable interactions, without however keeping those SU(8) representations which are obtained by saturating in (11), (12) upper with lower indices (traces). With these (admittedly very drastic) simplifications, we found that the breaking of SU(8) to SU(5) [rather than SU(6) or SU(7)] is preferred. For the maximal set of left-handed states we found two solutions, of which only one is vector-like for $SU(3)_{\text{colour}} \times U(1)_{\text{e.m.}}$. This solution has three generations of $\overline{5} + 10$ of SU(5), in agreement with present experimental data (if the top quark exists).

To obtain these results we observe first that the 63 states of spin 1 in Table 1 can be identified with the SU(8) gauge bosons. The left-handed spin 1/2 traceless multiplets decompose under $SU(5) \times SU(3)$ as

$$\begin{aligned} 216 &= (10,\overline{3}) + (\overline{5},\overline{3}) + (\overline{45},1) + (1,3) + (5,1) \\ &\quad + (1,\overline{6}) + (24,3) + (5,8) , \\ \overline{504} &= (45,3) + (40,\overline{3}) + (24,1) + (15,1) + (10,1) \\ &\quad + (10,8) + (\overline{10},6) + (\overline{10},\overline{3}) + (5,3) + (\overline{5},3) . \end{aligned} \tag{13}$$

The first two terms represent obvious candidates for the three generations of light fermions. The scalar 420-plet decomposes according to

$$\begin{aligned} 420 &= (24,\overline{3}) + (5,3) + (5,\overline{6}) + (45,3) + (\overline{40},1) \\ &\quad + (\overline{10},\overline{3}) + (\overline{5},1) + (10,1) + (10,8) + (1,\overline{3}) . \end{aligned} \tag{14}$$

Here we observe the occurrence of three scalar SU(5) singlets. If one of them acquires a non-vanishing vacuum expectation value, it will break SU(8) down to SU(5). [In the analogous reduction of SU(8) to SU(7) or SU(6) there are no singlets.]

From an analysis of SU(N) anomalies we learn the following. The effective SU(8) gauge theory cannot be made anomaly free. Sub-theories based on SU(7) or SU(6) which are anomaly free cannot be made vector-like under $SU(3)_{\text{colour}} \times U(1)_{\text{e.m.}}$ unless they are completely vector-like. On the other hand there is a set of fermions contained in $(216)_{\text{left}} + (\overline{504})_{\text{left}}$ which has no SU(5) anomalies, is chiral, and is vector-like under $SU(3)_{\text{colour}} \times U(1)_{\text{e.m.}}$:

$$(45 + \overline{45})_{\mathrm{left}} + 4(24)_{\mathrm{left}} + 9(10 + \overline{10})_{\mathrm{left}} + \\ + 3(5 + \overline{5})_{\mathrm{left}} + 9(1)_{\mathrm{left}} + 3(\overline{5} + 10)_{\mathrm{left}} . \tag{15}$$

This choice is maximal in the sense that it contains the largest number of anomaly free helicity states which can be found in $(216)_{\mathrm{left}} + (\overline{504})_{\mathrm{left}}$. [There is another subset with the same number of states which is excluded phenomenologically because it is not vector-like under $SU(3)_{\mathrm{colour}}$.] We note that it contains precisely three families of $(\overline{5} + 10)_{\mathrm{left}}$ which is the minimal number compatible with observation, and the preferred number for giving a satisfactory calculation of the b quark mass. The remaining content of (15) is vector-like under SU(5) and can acquire SU(5) invariant (Dirac or Majorana) masses of order $m_P$. [The content of (15) is too large to fit into one of the vector-like anomaly free SU(6) or SU(7) sub-theories of SU(8).]

Irrespective of the validity of this particular set of states, one may ask by what techniques one can obtain more information about the GUT. Now, in our picture, dynamics at the Planck mass involves higher spin states. It seems that no consistent classical local Lagrangian describing the interaction of higher spin fields exists. Therefore the only available technique for describing these interactions is that of on-shell scattering amplitudes (S-matrix). It is not difficult to find the constraints imposed on these amplitudes by supersymmetry (Grisaru, van Nieuwenhuizen and Pendleton, 1977; Grisaru and Pendleton, 1977) and it seems also possible to formulate the process of spontaneous breaking. In this way one can obtain, in principle, not only the group representations of the GUT but also the details of the interaction, the Higgs potential and the Yukawa coupling. Perhaps this approach can give some understanding of the hierarchy problem mentioned earlier.

Having taken the point of view that higher spins are entering the dynamics at (or above) the Planck mass, it may seem natural to relax the restriction to supersymmetries with $N \leq 8$ and admit spins higher than 2 in the fundamental supermultiplet. Our point of view that local Lagrangians are low energy approximations to a theory formulated in terms of on shell amplitudes seems to give us this freedom. Still, there seems to be some virtue in keeping with $N = 8$ supergravity, a theory remarkable for its convergence and its symmetries.

In conclusion, we have given suggestions for how one could bridge the gap between supergravity and Grand Unified Theories. Much work remains to be done, but the situation is not without hope.

ACKNOWLEDGEMENT

This lecture is based on work done in collaboration with J. Ellis, M.K. Gaillard and L. Maiani.

REFERENCES

Buras, A.J., Ellis, J., Gaillard, M.K. & Nanopoulos, D.V. (1978). Nucl. Phys. B, 135, p. 66.

Chanowitz, M.S., Ellis, J. & Gaillard, M.K. (1977). Nucl. Phys. B, 128, p. 506.

Cremmer, E. & Julia, B. (1978). Phys. Lett., 80B, p. 48.

Cremmer, E. & Julia, B. (1979). Nucl. Phys. B, 159, p. 41.

Cremmer, E. & Scherk, J. (1977). Nucl. Phys. B, 127, p. 259.

Cremmer, E., Scherk, J. & Ferrara, S. (1977). Phys. Lett., 68B. p. 234.

Cremmer, E., Scherk, J. & Ferrara, S. (1978). Phys. Lett., 74B, p. 61.

Curtright, T. & Freund, P. (1979). In Supergravity, ed. P. van Nieuweuhuizen and D.Z. Freedman, North Holland, Amsterdam, p. 197.

d'Adda, A., Di Vecchia, P. & Lüscher, M. (1978). Nucl. Phys. B, 146, p. 63.

d'Adda, A., Di Vecchia, P. & Lüscher, M. (1979). Nucl. Phys. B, 152, p. 125.

Das, A. (1977). Phys. Rev. D, 15, p. 2805.

Deser, S. & Zumino, B. (1976). Phys. Lett., 62B, p. 335.

Deser, S. & Zumino, B. (1977). Phys. Rev. Lett., 38, p. 1433.

de Wit, B. & Freedman, D.Z. (1977). Nucl. Phys. B, 130, p. 105.

Ellis, J. (1980). In Proceedings of the First Workshop on Grand Unification, Durham, New Hampshire.

Ellis, J., Gaillard, M.K., Maiani, L. & Zumino, B. (1980). LAPP preprint TH-15/CERN preprint TH-2481, to appear in the Proceedings of the Europhysics Study Conf. on "Unification of Fundamental Interactions", Erice.

Ellis, J., Gaillard, M.K. & Zumino, B. (1980). Phys. Lett., 94B, p. 343.

Ferrara, S. & van Nieuwenhuizen, P. (1976). Phys. Rev. Lett., 37, p. 1669.

Ferrara, S. & Zumino, B. (1975). Nucl. Phys. B, 87, p. 207.

Ferrara, S., Scherk, J. & Zumino, B. (1977a). Phys. Lett., 66B, p. 35.

Ferrara, S., Scherk, J. & Zumino, B. (1977b). Nucl. Phys. B, 121, p. 393.

Freedman, D.Z. (1977). Phys. Rev. Lett., 38, p. 105.

Freedman, D.Z. & Das, A. (1977). Nucl. Phys. B, 120, p. 221.

Freedman, D.Z. & Schwarz, J.H. (1978). Nucl. Phys. B, 137, p. 333.

Freedman, D.Z., van Nieuwenhuizen, P. & Ferrara, S. (1976). Phys. Rev. D, 13, p. 3214.

Gell-Mann, M. (1977). Talk at the Washington Meeting of the American Physical Society.

Georgi, H. & Glashow, S.L. (1974). Phys. Rev. Lett., 32, p. 438.

Georgi, H., Quinn, H.R. & Weinberg, S. (1974). Phys. Rev. Lett., 33, p. 451.

Gildener, E. (1976). Phys. Rev. D, 14, p. 1667.

Grisaru, M.T. & Pendleton, H.N. (1977). Nucl. Phys. B, 124, p. 81.

Grisaru, M.T., van Nieuwenhuizen, P. & Pendleton, H.N. (1977). Phys. Rev. D, 15, p. 996.

Haber, H.E., Hinchliffe, I. & Rabinovici, E. (1980). Berkeley preprint LBL-10519.

Nanopoulos, D.V. & Ross, D.A. (1979). Nucl. Phys. B, 157, p. 273.

Scherk, J. & Schwarz, J.H. (1974). Nucl. Phys. B, 81, p. 118.

Sohnius, M. (1979). Phys. Lett., 81B, p. 8.

Witten, E, (1979). Nucl. Phys. B, 149, p. 285.

# SPIN(8), TRIALITY, $F_4$ AND ALL THAT

J. F. Adams

UNIVERSITY OF CAMBRIDGE

## 1. INTRODUCTION

This piece will be written from the viewpoint of a pure mathematician, because that is what I am. However, if I make an effort I can probably leave out a lot of the sort of detail that applied mathematicians would like to neglect. Let me sketch the material to be covered.

Both pure and applied mathematicians need groups, because symmetry is important to both. Now, a group G can itself show symmetry. More precisely, we can first form the group Aut G of all automorphisms of G. Inside this we can distinguish the subgroup Inn G of inner automorphisms of G, that is, automorphisms $\alpha_h$ given by conjugation in the sense $\alpha_h(g) = ghg^{-1}$ $(g,h \in G)$. We wish to discard the inner automorphisms as trivial, so we check that Inn G is a normal subgroup of Aut G and form the quotient Out G = Aut G/Inn G. This is called the group of outer automorphisms; thus, an "outer automorphism" is a coset of ordinary automorphisms - it is not a particular kind of automorphism.

When we apply this process to the simple Lie groups, we find that one (namely $D_4$) has more symmetry than all the rest, because its outer automorphism group is $\Sigma_3$, the symmetric group on three letters, rather than $\Sigma_2$ or the trivial group $\Sigma_1$ as we find in all other cases.

In this statement, $D_4$ means the compact, connected and simply-connected group Spin(8). (At least, this explanation is good enough for present purposes, even if it involves the judicious neglect of detail which I promised you.)

Here I should pause to remind just a few readers that Spin(n) is the double covering group of the rotation group SO(n). That is, there is a homomorphism of Lie groups from Spin(n) onto SO(n); the kernel of this homomorphism has exactly two elements; and Spin(n) is connected (for $n \geq 2$). Readers whose tastes run to topology will realise that these properties characterise Spin(n), and may indeed be used to construct Spin(n) starting from SO(n). Readers whose tastes run to algebra will probably prefer a direct construction of the spinor groups using Clifford algebras; I advise them to consult Atiyah, Bott & Shapiro (1964), or Porteous (1969), but to take the representation-theory of Clifford algebras from Eckmann (1942).

The word "triality" is applied to the algebraic and geometric aspects of the $\Sigma_3$ symmetry which Spin(8) has. The word was chosen by analogy with the word "duality"; the reader already knows about duality, and if one examines the algebraic and geometric aspects of the $\Sigma_2$ symmetry which SU(n+1) has, one finds that they agree with what the reader knows already. I gather that applied mathematicians are not much interested in the geometric aspects.

All this goes to show that the group $D_4$ is not quite as the other groups $D_n$ are. The group $D_4$ is almost exceptional; there is a close relationship between it and the exceptional group $F_4$. In fact, we can embed Spin(8) in $F_4$, and then the $\Sigma_3$ symmetry of Spin(8) is worked by conjugation with suitable elements of $F_4$.

In order to understand this, we shall need to know more about $F_4$. Two concrete presentations of $F_4$ are helpful: (i) $F_4$ is the group of automorphisms of the exceptional Jordan algebra J; (ii) $F_4$ is the groups of isometries of the Cayley projective plane $\Pi$ (if $\Pi$ is provided with the appropriate Riemannian metric). These two presentations are closely related, since $\Pi$ may be embedded in J so that $\Pi$ inherits its $F_4$-action from J (and similarly inherits its Riemannian metric from that on J). All these matters deserve explanation.

This completes my sketch of my programme.

## 2. THE $\Sigma_3$ SYMMETRY

In this setion I will be more specific about the $\Sigma_3$ symmetry. We need to know what three objects the symmetric group $\Sigma_3$ permutes; and the answer is that it permutes representations.

More precisely, let V be a vector space (over R or C), and let Aut(V) be the general linear group of automorphisms of V. Then a representation in which G acts on V may be viewed as a homomorphism $\theta: G \to \text{Aut}V$. If we are given also an automorphism $\alpha: G \to G$, we can form the composite

$$G \xrightarrow{\alpha} G \xrightarrow{\theta} \text{Aut } V.$$

If $\alpha$ is an inner automorphism, then the representation $\theta\alpha$ is equivalent to $\theta$. Thus the outer automorphism group Out G = Aut G/Inn G permutes the (equivalence classes of) representations of G (of a given degree).

Next we need to recall how many representations there are to be permuted. Let G be a group which comes with a given action on $R^n$ or $C^n$ (as Spin(n) does); then I write $\lambda^i$ for the representation of G of degree $n!/(i!(n-i)!)$ on the $i^{th}$ exterior power of $R^n$ or $C^n$, that is, the $i^{th}$ component of the exterior algebra or Grassmann algebra. In particular, $\lambda^1$ means the originally-given representation on $R^n$ or $C^n$. The group Spin(2r) has two basic half-spinor representations $\Delta^+$, $\Delta^-$ of degree $2^{r-1}$; for the group Spin(8s), these are real representations (of degree $2^{4s-1}$). In particular, the group Spin(8) has just three irreducible representations of degree 8, all real, namely $\lambda^1$, $\Delta^+$ and $\Delta^-$. Since the outer automorphism group of Spin(8) permutes $\lambda^1$, $\Delta^+$ and $\Delta^-$, we get a homomorphism from it to $\Sigma_3$.

Theorem 2.1. This homomorphism is an isomorphism from the outer automorphism group of Spin(8) to $\Sigma_3$.

This result should be compared with the following one for the case of "duality".

Theorem 2.2. The outer automorphism group of SU(n+1) (for $n > 1$) is $\Sigma_2$, acting by permuting $\lambda^1$ and $\lambda^n$.

If you interpret $\lambda^1$ as the space of column vectors, with matrices A acting on the left in the usual way, then $\lambda^n$ may be interpreted as the space of row vectors v, with action by forming $vA^{-1}$. However you interpret it, $\lambda^n$ is the dual space of $\lambda^1$.

I pause to illustrate Theorem 2.1. The group SO(8) has an outer automorphism $\bar{\alpha}$ given by conjugation with a matrix A of determinant -1 in O(8). This automorphism $\bar{\alpha}$ lifts to an automorphism $\alpha$: Spin(8) $\to$ Spin(8). The automorphism $\alpha$ fixes $\lambda^1$ and interchanges $\Delta^+$, $\Delta^-$; this is easily checked using characters.

Given the preceding remarks, the proof of Theorem 2.1 is easy, and the idea is this. A map G $\to$ Spin(8) can be interpreted as giving an 8-dimensional representation of G; but you can also argue in the converse direction if you take care.

The approach sketched in this section is due to Élie Cartan (1938). He had found the $\Sigma_3$ symmetry rather earlier, as soon as he set himself the task of determining the automorphism groups of all the simple Lie groups (Cartan, 1925).

## 3. MOTIVATION

In this section I want to motivate the constructions I will give below.

If one wishes to construct the Cayley projective plane, it is natural to ask how one would construct the projective plane over the quaternions H if one were forbidden to use their associativity. To construct this projective plane one normally begins with the space of column vectors $H^3$ (considered as a right module over H); one then takes the one-dimensional subspaces. With each such subspace S one can associate a linear map $H^3 \to H^3$, namely orthogonal projection on S. In this way the subspaces S correspond to those 3 x 3 matrices over H which are Hermitian, idempotent, and of trace 1. These conditions on matrices never involve multiplying more than two quaternions together, so they do not involve associativity. We can now generalise. The exceptional Jordan algebra J is normally defined as the set of 3 x 3 matrices over the Cayley numbers which are Hermitian. Therefore, we may hope to find the Cayley projective plane $\Pi$ embedded inside J as the set of idempotents of trace 1; and this is exactly what happens.

However, if we adopt this usual construction of J, then the symmetry group which is visible from the beginning is only $\Sigma_3 \times G_2$; $\Sigma_3$ because you can permute the rows of a matrix in any way provided you permute the columns the same way, and $G_2$ because that is the group of automorphisms of the Cayley numbers.[2] It would be nice to have the Spin(8) symmetry visible from the beginning, so I will do it that way. To do so I need a little algebra, and I devote the next section to it.

## 4. THE ALGEBRA OF TRIALITY

In the context of Theorem 2.2 (duality), we have a bilinear map $\lambda^n \otimes \lambda^1 \to C$ which is invariant under SU(n+1); with the usual interpretations, the product of a row vector and a column vector is a scalar. In this section we will look at the corresponding phenomena in the case of triality.

Let $V^1$, $V^2$, $V^3$, be three vector-spaces of dimension 8 over R which afford the three inequivalent representations of G = Spin(8). (I put it this way to emphasise the symmetry of my assumptions.) I assume that $V^1$, $V^2$, $V^3$ come with given inner products ( , ) invariant under G = Spin(8).

<u>Lemma 4.1</u>. There is a trilinear map

$$f: V^1 \otimes V^2 \otimes V^3 \to R$$

which is invariant under G = Spin(8) and not identically zero; it is unique up to a non-zero scalar factor. It may be normalised so that for $x^i \in V^i$ with $(x^i, x^i) = 1$ we have

$$-1 \leq f(x^1 \otimes x^2 \otimes x^3) \leq +1,$$

and these bounds are attained. It is then unique up to sign.

Sketch proof. An easy calculation with characters gives

$$\Delta^+ \otimes \Delta^- \cong \lambda^1 + \lambda^3.$$

Therefore

$$\lambda^1 \otimes \Delta^+ \otimes \Delta^- \cong (\lambda^1 \otimes \lambda^1) + (\lambda^1 \otimes \lambda^3).$$

Now $\lambda^1$ is dual to $\lambda^1$ but $\lambda^3$ is not, so that the right-hand side contains the trivial representation R with multiplicity 1. Now, if you do things properly, the defined meaning of that phrase "with multiplicity 1" is

$$\dim_R \operatorname{Hom}_{RG}(\lambda^1 \otimes \Delta^+ \otimes \Delta^-, R) = 1.$$

The rest is easy.

The map f has various interesting aspects. Let us define the associated product

$$\mu : V^2 \otimes V^3 \to V^1$$

by

$$(x^1, \mu(x^2 \otimes x^3)) = f(x^1 \otimes x^2 \otimes x^3);$$

for example, $\mu$ might be the projection of $\Delta^+ \otimes \Delta^-$ on $\lambda^1$, suitability normalised.

Lemma 4.2. Then

$$||\mu(x^2 \otimes x^3)|| = ||x^2|| \; ||x^3|| \,.$$

Sketch proof. Consider arguments with $||x^2|| = 1$, $||x^3|| = 1$. According to our normalisation of f we have

$$||\mu(x^2 \otimes x^3)|| \leq 1,$$

and this bound is attained for at least one pair $(x^2, x^3)$. It can be shown that the group G = Spin(8) acts transitively on pairs $(x^2, x^3)$ with $||x^2|| = 1$, $||x^3|| = 1$, so we have

$$||\mu(x^2 \otimes x^3)|| = 1$$

for all such pairs.

Lemma 4.2 shows us again that there is an 8-dimensional normed algebra over R. In fact, we have only to choose base-points $e^2$, $e^3$ of norm 1 in $V^2$, $V^3$, and identify $V^2$, $V^3$ with $V^1$ under multiplication by $e^3$, $e^2$ respectively; we get on $V^1$ a normed product with unit. However, our strategy will be precisely to avoid breaking the symmetry by choosing these base-points.

The next lemma tells us that the map f has no more symmetry than we know already.

Lemma 4.3. Suppose given a triple of maps $(\alpha^1, \alpha^2, \alpha^3)$ with $\alpha^i \in SO(V^i)$ and such that

$$f(\alpha^1 x^1 \otimes \alpha^2 x^2 \otimes \alpha^3 x^3) = f(x^1 \otimes x^2 \otimes x^3).$$

Then there exists $g \in G = \mathrm{Spin}(8)$ such that $\alpha^i x^i = g x^i$ for each i.

If one is given only the map $\alpha^1$, there are two solutions for $g \in G = \mathrm{Spin}(8)$; going from one solution to the other changes the signs of both $\alpha^2$ and $\alpha^3$.

Lemma 4.3 is usually stated after identifying $V^1$, $V^2$ and $V^3$ with the Cayley numbers (as indicated above). It is then regarded as one of the standard algebraic expressions of triality.

Sketch proof of (4.3). Without loss of generality we can assume $\alpha^2 = 1$. Choose a base-point $e^2$ in $V^2$. Pass from f to $\mu$ as in (4.2), and define an action of $V^2$ on $V^1$ as follows: given $x^1 \in V^1$, $x^2 \in V^2$, first solve for $x^3$ in

$$\mu(e^2 \otimes x^3) = x^1$$

and then take $\mu(x^2 \otimes x^3) \in V^1$. Proceeding in the standard way, we get a Clifford algebra acting on $V^1$. Granted the dimensions, $V^1$ has to be an irreducible representation of this Clifford algebra, and such that any map $\theta\colon V^1 \to V^1$ of Clifford modules is multiplication by a real scalar. But $\alpha^1$ is such a map, so $\alpha^1$ is multiplication by a real scalar, necessarily $\pm 1$. Similarly for $\alpha^3$.

## 5. CONSTRUCTION OF THE EXCEPTIONAL JORDAN ALGEBRA J

In this section I will construct J.

I begin with the vector space $R^3 \oplus V^1 \oplus V^2 \oplus V^3$, where $V^1$, $V^2$, $V^3$ are as in Section 4 and the standard basis vectors $e^1$, $e^2$, $e^3$ in $R^3$ are in (1-1) correspondence with $V^1$, $V^2$, $V^3$. This gives me a vector-space of dimension 27, which is correct for J; but of course it needs more structure yet.

I had better give some explanation. The vector $\lambda_1 e^1 + \lambda_2 e^2 + \lambda_3 e^3 \in R^3$ in my construction will correspond to the diagonal matrix

$$\begin{bmatrix} \lambda_1 & 0 & 0 \\ 0 & \lambda_2 & 0 \\ 0 & 0 & \lambda_3 \end{bmatrix}$$

in the usual construction. After I have finished explaining the structure of J, I shall be able to state that any element of J can be thrown by an automorphism of J onto an element $\lambda_1 e^1 + \lambda_2 e^2 + \lambda_3 e^3 \in R^3$. Two such elements $\Sigma \lambda_i e^i$, $\Sigma \mu_i e^i$ can be thrown onto one another by automorphisms of J if and only if $\{\mu_1, \mu_2, \mu_3\}$ is a permutation of $\{\lambda_1, \lambda_2, \lambda_3\}$. It follows that the polynomial functions on J which are invariant under automorphisms of J can be generated by three, which I may take to be the power-sums

$$\lambda_1 + \lambda_2 + \lambda_3, \qquad \lambda_1^2 + \lambda_2^2 + \lambda_3^2, \qquad \lambda_1^3 + \lambda_2^3 + \lambda_3^3.$$

I propose to give a direct construction of these crucial invariant functions, or rather the associated symmetric multilinear functions. So I must define a linear function $\ell$, a bilinear function b and a trilinear function t.

First I define $\ell: J \to R$. I set

$$\ell(\sum_i \lambda_i e^i) = \sum_i \lambda_i$$
$$\ell|V^i = 0 \quad \text{for } i = 1, 2, 3.$$

In the usual construction, this corresponds to the <u>trace</u> of a $3 \times 3$ Hermitian matrix over the Cayley numbers.

Secondly I define the inner product $b: J \otimes J \to R$. The summands $R^3$, $V^1$, $V^2$, $V^3$ are to be orthogonal. On $R^3$ you are to take the usual inner product, so that

$$b(e^i \otimes e^i) = \delta_{ij}.$$

On each $V^i$ you are to take twice the originally-given inner product:

$$b(x \otimes y) = 2(x,y).$$

This factor 2 has to come in somewhere, and this is where it causes least trouble.

Thirdly I define the trilinear function $t: J \otimes J \otimes J \to R$. On any summand $V^i \otimes V^j \otimes V^k$ with i, j, k, all distinct I permute the factors to get to $V^1 \otimes V^2 \otimes V^3$ and use the map f of lemma 4.1. On $e^i \otimes y \otimes z$ with $y \in V^j$, $z \in V^j$ I take

$$\begin{cases} 0 & \text{if } i=j \\ (y,z) & \text{if } i \neq j, \end{cases}$$

and similarly if $e^i$ comes in the second or third place. On $e^i \otimes e^j \otimes e^k$ I take

$$\begin{cases} 1 & \text{if } i = j = k \\ 0 & \text{otherwise.} \end{cases}$$

On all summands of $J \otimes J \otimes J$ not yet mentioned I take zero. This completes the construction of the trilinear map t.

Finally I will construct the Jordan product, and so make J into an algebra. The function $t(x \otimes y \otimes z)$ is evidently a linear function of x if I fix y and z, and the inner product b is non-degenerate, so I can write

$$t(x \otimes y \otimes z) = b(x, y \circ z)$$

for a unique $y \circ z$ in J, which clearly depends linearly on y and z. This gives the Jordan product $y \circ z$; since t is symmetric in y and z, so is $y \circ z$. However, we have clearly lost on symmetry by going from t to the Jordan product.

The Jordan product has $e^1 + e^2 + e^3$ as a unit.

## 6. THE GROUP $F_4$

The group $F_4$ can now be constructed as the set of R-linear maps $J \to J$ which preserve $\ell$, b and t, or equivalently, which preserve the Jordan product. It has been visible from the beginnning that all our constructions are invariant under Spin(8), so we get

$$\mathrm{Spin}(8) \subset F_4.$$

Proposition 6.1. (i) The subgroup of $F_4$ which fixed $e^1$, $e^2$ and $e^3$ is exactly Spin(8). (ii) The subgroup of $F_4$ which fixes $e^1$ is isomorphic to Spin(9); more precisely, it is the Spin group of the subspace $V^1 \oplus (\lambda(e^2 - e^3))$, acting so as to fix $e^1$, $e^2 + e^3$ and by a 16-dimensional spin representaion on $V^2 \oplus V^3$. Similarly with $e^1$ replaced by $e^2$ or $e^3$.

Sketch proof. Part (i) comes easily from lemma 4.3. From this it is as easy to see that the subgroup of $F_4$ which fixes $e^1$ is no larger than claimed in part (ii). The essential point is to show that J does have the automorphisms asserted in part (ii). Here I must explain that when I constructed t, I used the components of some functions which I already knew to be invariant under Spin(9). More precisely, Spin(9) has a 16-dimensional real spin-representation $\Delta$, and I can construct invariant maps which involve it just as in the sketch proof of lemma 4.1. When I restrict from Spin(9) to Spin(8), $\Delta$ splits as $\Delta^+ + \Delta^-$, and I can identify the components of my invariant maps. These are the components I already used in constructing t, so t is invariant under Spin(9). Since $\ell$ and b are clearly invariant under Spin(9), I get $\mathrm{Spin}(9) \subset F_4$.

Corollary 6.2. The subgroup of $F_4$ which preserves the set $\{e^1, e^2, e^3\}$ maps onto $\Sigma_3$ with kernel Spin(8), and the quotient $\Sigma_3$ acts on Spin(8) by its outer automorphism group.

Sketch proof. Consider the subgroup Spin(9) of $F_4$ which fixes $e^1$, as in (6.1). In this subgroup we find at least one element which reverses the sign of $e^2 - e^3$, so that it interchanges $e^2$ and $e^3$, and therefore interchanges $V^2$ and $V^3$. Similarly with $e^1$ replaced by $e^2$ or $e^3$.

With this I have carried out my brief and shown you that the $\Sigma_3$ symmetry of Spin(8) is worked by conjugation in $F_4$. It is not my brief to say as much as I would like about the geometry of the Cayley projective plane $\Pi$, but I will define it for you. It is the $F_4$-orbit of the point $e^1 \in J$; equivalently, it is the set of points $j \in J$ where

$$\ell(j) = 1, \qquad b(j \otimes j) = 1, \qquad t(j \otimes j \otimes j) = 1;$$

equivalently, it is the set of points $j \in J$ where

$$\ell(j) = 1, \quad j \circ j = j.$$

The last description reconciles it with the "idempotents of trace 1" in Section 3.

## 7. TRIALITY AND COMPLEX QUADRICS

The oldest phenomenon associated with triality goes back before Élie Cartan. To see it, it will be well to complexify our groups. The group SO(8) is a group of real 8 x 8 matrices which preserve the real quadratic form

$$\sum_1^8 x_i^2;$$

it is embedded in the corresponding group of complex 8 x 8 matrices of determinant 1, which preserve the complex quadratic form

$$\sum_1^8 z_i^2.$$

The injection is a homotopy equivalence, so the complex group has a double cover G which contains the group Spin(8). Let us complexify the vector spaces $V^1$, $V^2$ and $V^3$ of Section 4; our complex group G acts on $V^1 \otimes C$, $V^2 \otimes C$ and $V^3 \otimes C$ so as to preserve a non-singular complex quadratic form on each. It therefore acts on the three projective spaces $P(V^1 \otimes C)$, $P(V^2 \otimes C)$ and $P(V^3 \otimes C)$, which are of dimension 7. It also acts on the quadrics $Q^1$, $Q^2$ and $Q^3$ in these projective spaces which are determined by the vanishing of the quadratic forms. These quadrics are of dimension 6. Now just as a complex quadric of dimension 2 has on it two systems of lines, so a complex quadric of dimension 6 has on it two systems of linear subspaces of dimension 3.

I now introduce an idea of Tits (1956); each complex group G of a suitable kind can be interrogated and made to yield "the" geometric system on which it consents to act as a group of automorphisms. For brevity I will not go into the way in which this is done. However, in our case we get three equally valid realisations of Tit's geometrical system, in terms of the geometries of $Q^1$, $Q^2$ and $Q^3$.

Proposition 7.1. These three geometries correspond as follows.

| | | | | |
|---|---|---|---|---|
| (Points on $Q^1$) | $\leftrightarrow$ | (3-spaces of one system on $Q^2$) | $\leftrightarrow$ | (3-spaces of one system on $Q^3$) |
| (3-spaces of one system on $Q^1$) | $\leftrightarrow$ | (Points on $Q^2$) | $\leftrightarrow$ | (3-spaces of the other system on $Q^3$) |
| (3-spaces of the other system on $Q^1$) | $\leftrightarrow$ | (3-spaces of the other system on $Q^2$) | $\leftrightarrow$ | (Points on $Q^3$) |
| (Lines on $Q^1$) | $\leftrightarrow$ | (Lines on $Q^2$) | $\leftrightarrow$ | (Lines on $Q^3$). |

Moreover, these (1-1) correspondences preserve incidence.

Here two linear subspaces on $Q^i$ are "incident" if one is contained in the other, or, when both are of dimension 3, if their intersection is of dimension 2.

I will briefly indicate the first correspondence in the proposition. Take a point of $Q^1$, represented by a non-zero vector $x \in V^1 \otimes C$. Consider the set of vectors $y \in V^2 \otimes C$ such that

$$f(x \otimes y \otimes z) = 0 \quad \text{for all} \quad z \in V^3 \otimes C.$$

Such vectors y form a vector subspace of dimension 4 in $V^2 \otimes C$, yielding a 3-space on $Q^2$.

REFERENCES

M. F. Atiyah, R. Bott & A. Shapiro, (1964), Clifford Modules, Topology 3, Suppl. 1, pp. 1-38.

E. Cartan, (1925), Le principe de dualité et la théorie des groupes simples et semisimples, Bull. des Sciences Mathématiques, 49 pp. 367-374.

E. Cartan, (1938), Leçons sur la théorie des spineurs II Hermann et Cie, Act. Sci.et Ind., 701, pp. 53-54.

B. Eckmann, (1942), Gruppentheoretische Beweis des Satzes von Hurwitz-Radon über die Komposition quadratischer Formen, Comment. Math. Helv., 15 pp. 358-366.

I. Porteous, (1969) Topological Geometry, Van Nostrand Reinhold , especially Chapter 13.

J. Tits, (1956), Les groupes de Lie exceptionnels et leur interprétation géometrique, Bull. Soc. Math. Belgique, 8, pp. 48-81.

(1959)"Correspondence", Ann. Math., 69, pp. 247-251.

# V KÄHLER SPACES AND SUPERSYMMETRY

# REMARKS ON SUPERSYMMETRY AND KAHLER GEOMETRY

Daniel Z. Freedman †

Institute for Theoretical Physics
State University of New York
Stony Brook NY 11794

G.W. Gibbons

Department of Applied Mathematics and Theoretical Physics
University of Cambridge
Silver Street Cambridge CB3 9EW

The geometric nature of the bosonic nonlinear σ-model is evident from the action

$$I_B = \frac{1}{2}\int d^dx g_{ij}(\phi)\partial_\mu\phi^i(x)\partial_\mu\phi^j(x) \tag{1}$$

where the fields $\phi^i(x)$, $i = 1,\cdots,n$, are maps from a flat space-time of d dimensions to an n-dimensional Riemannian manifold with metric $g_{ij}(\phi)$. In the case of $d = 2$ an $N = 1$ supersymmetric extension can be given (Freedman and Townsend, 1980) by defining fields $\phi^i(x,\theta)$ over a superspace of 2 bose coordinates $x^\mu$ and two real fermi coordinates $\theta_\alpha$. The action is

$$I_S = \frac{1}{4i}\int d^2x d^2\theta g_{ij}(\Phi)\bar{D}_\alpha\Phi^i D_\alpha\Phi^j \tag{2}$$

where $D_\alpha$ is the covariant derivative of global supersymmetry.

Kahler geometry enters the subject in connection with σ-models on complex Hermitian manifolds, where it has been shown (Zumino, 1979) that a supersymmetric extension with N = 1 in four dimensions exists if and only if the manifold is Kahler. This connection between extended supersymmetry and Kahler geometry has been applied to elucidate the ultraviolet divergence structure of supersymmetric nonlinear models (Alvarez-Gaumé and Freedman, 1980a) in two space-time dimensions. A one-loop counter term involving the Ricci tensor $R_{ij}$ is present. However, primitive two-loop divergences are absent in all supersymmetric models although allowed by power counting and although present in all bosonic models. The geometrical argument which implied two-loop ultraviolet finiteness can probably be extended to higher order although some non-trivial complications must be handled.

The ultraviolet structure was discussed at the Workshop, but since the technical argument will soon appear in the standard literature we will discuss only one general point here. It concerns a possible analogy between the two dimensional σ-model and the special four-dimensional theories such as the N = 4 super-Yang-Mills theory where we think the β function vanishes and supergravity theories where we hope that the coefficients of higher loop counter terms vanish. These phenomena are presently regarded as "miracles", and to explain them it seems that we will need arguments which incorporate more than the standard symmetries of the theories. We would simply point out that in the two-dimensional supersymmetric σ-models we do have an argument for the ultraviolet properties which goes beyond symmetry. The most striking way to state the situation was suggested to us by C. Isham and M. Duff. Consider the supersymmetric O(n) model (Di Vecchia and Ferrara, 1977; Witten, 1977). Using symmetry alone one would conclude that ultraviolet counter terms proportional to the classical action (2) are permitted at all orders in perturbation theory. Yet by bringing in the full geometric structure of supersymmetric σ-models one can show that the counter term is permitted at one-loop order but prohibited at two-loop level and probably absent in higher order also. At present we see no way to extend the argument to four dimensions but suggest that it may be fruitful to study the question.

Supersymmetric σ-models on Ricci flat manifolds seem to have especially favorable ultraviolet properties. During the Workshop some new results were obtained on Ricci-Flat Kahler manifolds which we now present.

Kahler manifolds are a particularly beautiful branch of differential geometry and we refer readers to a recent pedagogical introduction to the subject prepared by and for physicists. (Alvarez-Gaumé and Freedman 1980b). A Kahler manifold has complex coordinates $z^\alpha$ and conjugates $\bar{z}^\beta$ $(\alpha,\beta = 1,\cdots,n)$ and real line element

$$ds^2 = 2g_{\alpha\bar{\beta}}(z,\bar{z})dz^\alpha d\bar{z}^\beta \tag{3}$$

with a metric tensor which satisfies $\bar{g}_{\alpha\bar{\beta}} = g_{\beta\bar{\alpha}}$ and $\partial_\alpha g_{\gamma\bar{\beta}} = \partial_\gamma g_{\alpha\bar{\beta}}$. These conditions imply that locally there always exists a potential function $F(z,\bar{z})$ such that

$$g_{\alpha\bar{\beta}} = \partial_\alpha\partial_{\bar{\beta}}F \tag{4}$$

Furthermore the Ricci tensor can be written as

$$R_{\alpha\bar{\beta}} = \partial_\alpha\partial_{\bar{\beta}} \log \det g_{\gamma\bar{\delta}}. \tag{5}$$

Therefore any solution of the second order non-linear partial differential equation

$$\det\partial_\alpha\partial_{\bar{\beta}}F = 1 \tag{6}$$

locally determines a Ricci-flat Kahler metric. This equation is known as the Monge-Ampére equation and has been widely studied by complex geometers. It is surprising that the following simple procedure for obtaining solutions apparently escaped the notice of the mathematicians.

In the n-dimensional case it is natural to consider the U(n) invariant Ansatz that F depends only on the single variable $x = z^{\alpha}\bar{z}^{\alpha}$. Under these conditions (6) becomes an ordinary differential equation namely

$$(F')^{n-1}[F' + xF''] = 1 \tag{7}$$

the general solution is

$$F(x) = \int_{c}^{x} \frac{dx'}{x'} (a^{n} + x'^{n})^{1/n} \tag{8}$$

where c is an irrelevant integration constant and a is a real scale parameter. This gives the line element

$$\begin{aligned} ds^{2} &= 2\, F' dz^{\alpha} d\bar{z}^{\alpha} + 2F''\, \bar{z}^{\alpha} dz^{\alpha}\, z^{\beta} d\bar{z}^{\beta} \\ &= 2(a^{n}+x^{n})^{1/n}\, x^{-1}\{dz^{\alpha}d\bar{z}^{\alpha} - x^{-1}(a^{n}+x^{n})^{-1}\bar{z}^{\alpha}dz^{\alpha}z^{\beta}d\bar{z}^{\beta}\} \end{aligned} \tag{9}$$

This local metric has an apparent singularity at x = 0. To show that this is only a coordinate singularity and to clarify the global structure of the manifold, we introduce new coordinates by

$$\begin{aligned} z^{i} &= \lambda w^{i} \quad i = 1,\cdots,n-1 \\ z^{n} &= \lambda \end{aligned} \tag{10}$$

which brings the metric to the form

$$\begin{aligned} ds^{2} = 2(a^{n}+(\lambda\bar{\lambda}\Omega)^{n})^{1/n}\{&(\lambda\bar{\lambda}\Omega)^{n-1}(a^{n}+(\lambda\bar{\lambda}\Omega)^{n})^{-1}(\Omega d\lambda d\bar{\lambda} + \bar{\lambda}d\lambda w^{i}d\bar{w}^{i} \\ &+ \bar{w}^{i}dw^{i}\lambda d\bar{\lambda}) \\ &+\Omega^{-1}dw^{i}d\bar{w}^{i} - \Omega^{-2}(a^{n}+(\lambda\bar{\lambda}\Omega)^{n})^{-1}\bar{w}^{i}dw^{i}w^{j}d\bar{w}^{j}\} \end{aligned} \tag{11}$$

with $\Omega = 1 + w^{i}\bar{w}^{i}$. There is still a singularity at $\lambda=0$ but this is readily removed by the change of variable $\mu=\lambda^{n}$ which makes the metric regular. The submanifold $\mu=0$ is then seen to be $CP^{n-1}$ with its standard Fubini-Study metric. The variable $\mu$ ranges over the entire complex plane, so that the topology of the manifold is that of a complex line bundle over $CP^{n-1}$. From (9) we see that the line element is locally asymptotically flat, but the relation $\mu=\lambda^{n}$ requires us to identify points $z^{\alpha}$ and $\exp(2\pi ik/n)z^{\alpha}$ with $k=1,\cdots,n-1$. The topology of the asymptotic boundary is thus $S^{2n-1}/Z_n$.

The case n=2 was studied previously (Gibbons and Pope, 1978) and shown to give a Kähler parameterization of the gravitational instanton of Eguchi and Hansen (1978). For general n this series of solutions appears to coincide with those obtained by a rather more abstract construction due to Calabi (1979). Another series of manifolds was recently obtained (Curtright and Freedman, 1980) in connection with N=4 supersymmetric σ models and were shown to be Ricci-flat Kahler metrics of even complex dimension n=2m, m=1,2,..., (Alvarez-Gaumé and Freedman, 1980c). In general they are distinct from the series given above since they have different isometry group. However, the case m=1 was also argued to be an alternate Kahler description of the Eguchi-Hansen metric, although the coordinate transformation to the Gibbons-Pope form (which coincides with (9) for n=2) was not given. We now give the explicit coordinate transformation.

The σ-model manifolds were described by Curtright and Freedman as surfaces embedded in $C^{2m+2}$ / U(1) where U(1) is the group of overall phase transformations of complex coordinates $\phi_1^\alpha$ and $\phi_2^\alpha$ where α=1,2,···,m+1. The Lagrangian which describes the metric of the quotient space is

$$\mathcal{L} = \partial_\mu\bar{\phi}^\alpha_i\partial_\mu\phi^\alpha_i + \frac{1}{4\bar{\phi}^\alpha_i\phi^\alpha_i}(\bar{\phi}^\beta_j \overleftrightarrow{\partial}_\mu \phi^\beta_j)^2 \tag{12}$$

and the constraints which describe the embedded surface are

$$\bar{\phi}^\alpha_i\tau^I_{ij}\phi^\alpha_j = b^I \tag{13}$$

where the $\tau^I$ are the 3 standard Pauli matrices and $b^I$ is a fixed isovector parameter whose direction is irrelevant.

The coordinate transformation for m=1 can now be obtianed by fairly straightforward reasoning. SU(2) acting on the α indices of $\phi_i^\alpha$ is an obvious invariance of (12) and (13) whereas SU(2) acting on $z^\alpha$ is an isometry of (9). Thus $\phi_i^\alpha$ and $z^\alpha$ should be simply related. This relation is further specified by U(1) invariance which is the subgroup of the SU(2) acting on the i indices of $\phi_i^\alpha$ which leaves the vector $b^I$ invariant, and which we identify with the overall U(1) phase of the $z^\alpha$. For $b^I$ = (0,0,2) this leads us to the Ansatz

$$\phi_1^{\alpha} = g(x) z^{\alpha}$$
$$\phi_2^{\alpha} = f(x) i\sigma_2^{\alpha\beta} \bar{z}^{\beta} \qquad (14)$$

where $x = z^{\alpha}\bar{z}^{\alpha}$ and $\sigma_2$ is the standard antisymmetric Pauli matrix. This satisfies the constraints if $g^2 - f^2 = 2x^{-1}$. We then insert (14) in (12) and find that the result is identical with (9) if $g^2 = (1 + \sqrt{1+x^2})x^{-1}$. The identification with the Eguchi-Hansen metric was previously made (Alvarez-Gaumé and Freedman, 1980c) on the circumstantial evidence of common isometry group and boundary topology. The construction of the coordinate transformation makes this identification certain.

It has not escaped our notice that the subjects treated here are rather mathematical and that there is no physical application in immediate view. Nevertheless we find the relation between supersymmetry and differential geometry rather rich, and we pursue the subject with some hope that it will reveal new physical aspects of supersymmetry.

We are happy to thank Nigel Hitchin, Don Page and Chris Pope for useful discussion during the Workshop.

REFERENCES

Alvarez-Gaumé, L. & Freedman, D.Z. (1980a). Kahler Geometry and the Renormalization of Supersymmetric Models, Phys. Rev. D22. pp. 846-853.

Alvarez-Gaumé, L. & Freedman, D.Z. (1980b). A Simple Introduction to Complex Manifolds, to appear in Proceedings of the Erice Study Conference on Fundamental Interactions σ-March 1979- ed. J. Ellis and P. van Nieuwenhuizen.

Alvarez-Gaumé, L. & Freedman, D.Z. (1980c). Ricci-Flat Kahler Manifolds and Supersymmetry, Physics Letters. 94B 171-173.

Calabi, E. (1979). Metriques Kahleriennes et fibrés holomorphes, Ann. Sc. de l'E.N.S. 12 pp. 266-292.

Curtright, T.L. & Freedman, D.Z. (1980 ). Nonlinear σ Models with Extended Supersymmetry in Four Dimensions, Physics Letters 90B, pp. 71-74.

di Vecchia,P & Ferrara,S. (1977). Classical Solutions in Two-Dimensional Supersymmetric Field Theories, Nucl. Phys. B130 pp. 93-104.

Eguchi,T. & Hansen,A.S. (1978). Asymptotically Flat Self-Dual Soutions to Euclidean Gravity, Phys. Lett. 74B, pp. 249-251.

Freedman,D.Z. & Townsend,P.K. (1980). Antisymmetric Tensor Gauge Theories and Non-Linear σ-Models, Stony Brook preprint ITP-SB-80-25.

Gibbons,G.W. & Pope, C.N. (1978). The Positive Action Conjecture and Asymptotically Euclidean Metrics in Quantum Gravity, Commun. Math. Phys. 66 pp. 267-290.

Witten,E. (1977). Supersymmetric Form of Nonlinear σ Model in Two Dimensions, Phys. Rev. D16, pp. 2991-2994.

Zumino,B. (1979). Supersymmetry and Kahler Manifolds, Phys. Lett. 87B pp. 203-206.

† Address after 1 September 1980
Department of Mathematics, Rm. 2-381
Massachusetts Institute of Technology
Cambridge, Mass. 02139

# EIGENFUNCTION RELATIONS IN KÄHLER-EINSTEIN SPACES

C. N. Pope

University of Cambridge

## 1. INTRODUCTION

In the functional integral approach to supergravity, the result at the one loop level reduces simply to a ratio of determinants of the operators describing the second variation of the action with respect to the various fields in the theory (including ghost fields arising from the introduction of gauge-breaking terms in the action). In general the boson determinants occur in the denominator and the fermions in the numerator. Each determinant by itself suffers from three orders of divergence; quartic, quadratic and logarithmic (e.g. Hawking 1979). These occur because the spectrum of eigenvalues of the operator in question is unbounded above, and so the product of the eigenvalues diverges. However supersymmetry implies that in the ratios of determinants arising in any of the O(N) theories of supergravity, there is an overall cancellation of the quartic and quadratic divergences; essentially the boson divergences cancel out against those of the fermions.

In an earlier paper it was shown that if the quantization is performed around a background gravitational instanton whose curvature is self-dual or anti-self-dual ($R_{abcd} = \pm{}^*R_{abcd}$), then there is a complete and exact cancellation of all the non-zero eigenvalues (Hawking & Pope 1978 ). This means that the one loop functional integral is then determined simply by the (finite number of) zero modes of the wave operators. In particular this means that the effective action scales as some integer power of k under constant metric rescalings $g_{ab} \to k^2 g_{ab}$, and so the integrated trace anomaly is an integer.

The essential property of self-dual spaces that leads to this exact cancellation of eigenvalues is that, being half-flat, there exists a pair of covariantly constant spinors in the flat half of the space, and these act as generators of global supersymmetry transformations. They may be used to relate the eigenfunctions of the wave operators for the various supergravity fields. The spectrum of eigenvalues turns out to be identical for each spin, and the multiplicities are such as to give an exact mode by mode cancellation in the one loop effective action.

Unfortunately subsequent work has suggested that self-dual metrics are probably not of great physical interest; the non-compact ones all satisfy rather peculiar "twisted" boundary conditions at infinity, and this leaves only the compact self-dual metric K3 -- see for example Pope (1980a). One might therefore ask whether there are any other classes of gravitational instantons for which a cancellation of eigenvalues occurs in supergravity, and which are of greater physical interest. This article does not supply a definitive answer, since further work is needed in order to check the details. However there are indications that a similar cancellation may occur in any Einstein-Kähler background (i.e. a Kähler metric satisfying the Einstein equations $R_{ab} = \Lambda g_{ab}$ ). It is certainly the case that in any such background there are two electrically charged gauge covariantly constant spinors, and that these may be used to generate global supersymmetry transformations and to relate the eigenfunctions of the various wave operators. Thus the spectra of eigenvalues of the operators are related to one another, but it is not yet known whether they coincide. Before going on to describe this recent work we first briefly review the situation when the background geometry is self-dual.

## 2. SUPERGRAVITY IN SELF-DUAL BACKGROUNDS

In the Riemannian regime (i.e. when the metric has positive definite signature) the curvature is given in spinor notation (e.g. Penrose & MacCallum 1972) by the self-dual (or right-handed) Weyl spinor $\Psi_{ABCD}$ , the anti-self-dual (or left-handed) Weyl spinor $\tilde{\Psi}_{A'B'C'D'}$ , the tracefree Ricci spinor $\Phi_{ABA'B'}$ and its trace R (which equals $4\Lambda$ ). (For further details of spinors in the Riemannian regime, and of supergravity in a self-dual background, see Hawking & Pope (1978 )). A self-dual metric automatically satisfies $R_{ab} = 0$ , and in addition $\tilde{\Psi}_{A'B'C'D'} = 0$ . Thus all the curvature is contained in the right-handed Weyl spinor $\Psi_{ABCD}$ . In particular this means that the left hand half of the space is flat; i.e. the connection for primed spinors, which transform under the left hand SU(2) in the local isomorphism between SO(4) and SU(2)×SU(2) , is flat. Therefore one can choose a left hand dyad basis $(\tilde{o}_{A'}, \tilde{\iota}_{A'})$ which is covariantly constant, $\nabla_{AA'}\tilde{o}_{B'} = 0 = \nabla_{AA'}\tilde{\iota}_{B'}$ . These may be normalized so that $\tilde{\iota}^{A'}\tilde{o}_{A'} = 1$ , and $\tilde{\iota}_{A'} = \bar{\tilde{o}}_{A'}$.

Starting from a scalar eigenfunction $\phi$ satisfying

$$-\nabla_a\nabla^a\phi = \lambda\,\phi \tag{1}$$

one can construct spinors with arbitrary numbers of primed and unprimed indices, where each primed index is added by multiplying $\phi$ by $\tilde{o}_{A'}$ or $\tilde{\iota}_{A'}$, and each unprimed index is added by operating with $\tilde{o}^{A'}\nabla_{AA'}$ or $\tilde{\iota}^{A'}\nabla_{AA'}$. By performing the appropriate symmetrizations on primed and unprimed indices one can thus obtain fields that carry irreducible representations of the rotation group SO(4) of any desired spin. Remarkably, these fields all turn out to be eigenfunctions of the second order wave operator of the appropriate spin, in each case with the same eigenvalue $\lambda$ as the original scalar eigenfunction in equation (1). That this should be the case for purely primed spinors is not surprising, since the covariantly constant spinors $\tilde{o}_{A'}$ or $\tilde{\iota}_{A'}$ can be taken straight through the differential operator $\nabla_a\nabla^a$. What is perhaps more surprising is that when the gradient operators needed to add unprimed indices are commuted through $\nabla_a\nabla^a$, they conspire to produce just the right curvature terms involving $\Psi_{ABCD}$ so that the field satisfies the correct eigenvalue equation.

The fact that for each index added one can use either $\tilde{o}_{A'}$ or $\tilde{\iota}_{A'}$ means that a given scalar eigenfunction generates two or more independent eigenfunctions of higher spin. For example starting from $(\phi,\lambda)$ satisfying (1) one obtains 2 right-handed and 2 left-handed spin 1/2 eigenmodes, 4 vectors, 8 right-handed and 8 left-handed spin 3/2 and 9 tracefree spin 2 modes, all with eigenvalue $\lambda$.

## 3. KÄHLER-EINSTEIN METRICS

In a general Einstein metric ($R_{ab} = \Lambda g_{ab}$) the tracefree Ricci spinor $\Phi_{ABA'B'}$ vanishes but the curvature is otherwise unrestricted; $\Psi_{ABCD}$, $\tilde{\Psi}_{A'B'C'D'}$ and $\Lambda$ are all non-zero. The connection for spinors of either handedness therefore has non-zero curvature and so there cannot be any covariantly constant spinors, which means that the construction described in the previous section cannot be used. However if the metric is in addition a Kähler metric then there is a rich additional structure which enables one to find gauge covariantly constant charged spinors. We begin by recalling some properties of Kähler metrics; a detailed discussion may be found in Flaherty (1976).

A real four-dimensional metric is Kähler if it may be written as a complex two-dimensional metric

$$ds^2 = 2\, g_{\alpha\bar{\beta}}\, dz^\alpha\, d\bar{z}^{\bar{\beta}} \quad , \quad \alpha,\ \beta = 1,2 \quad , \qquad (2)$$

and if in addition the 2-form

$$J = i\, g_{\alpha\bar{\beta}}\, dz^{\alpha} \wedge d\bar{z}^{\bar{\beta}} \tag{3}$$

is closed (which in fact means that it will also be covariantly constant). The metric tensor $g_{\alpha\bar{\beta}}$ may be expressed locally as

$$g_{\alpha\bar{\beta}} = \partial/\partial z^{\alpha}\ \partial/\partial \bar{z}^{\bar{\beta}}\ K\ . \tag{4}$$

The 2-form J is the Kähler form, and K is the Kähler function. Transcribed back to four real dimensions, this means that in a Kähler metric there always exists a covariantly constant 2-form of a definite duality, which without loss of generality we may take to be self-dual ($J = {*J}$).

Because J is covariantly constant, any constant multiple of J automatically satisfies Maxwell's equations in the curved background geometry, and because it is self-dual it has vanishing energy-momentum tensor. Thus a Kähler-Einstein metric together with the Kähler form constitute an exact solution of the Einstein-Maxwell equations. One can now look at charged spinors which are minimally coupled to this electromagnetic field, and ask whether there exist any gauge covariantly constant spinors $u_A$ or $\tilde{u}_{A'}$ , satisfying $\tilde{\nabla}_{AA'}\, u_B = 0$ or $\tilde{\nabla}_{AA'}\, \tilde{u}_{A'} = 0$ , where

$$\tilde{\nabla}_{AA'} = \nabla_{AA'} - ieA_{AA'}\ , \tag{5}$$

and A is a potential whose exterior derivative is the electromagnetic field F, which is proportional to J. The answer is that there is no such spinor $\tilde{u}_{A'}$ , but there does exist a gauge covariantly constant spinor $u_A$ (Hitchin, 1974, and private communication).

It is easy to see that no left-handed covariantly constant spinor can exist by considering the commutator $[\tilde{\nabla}_{AA'}, \tilde{\nabla}_{BB'}]\tilde{u}_{C'}$ , which would then have to vanish, and bearing in mind that F is self-dual and so is represented by an unprimed spinor $F_{AB}$ ($F_{ab} \leftrightarrow 1/2\, F_{AB}\, \varepsilon_{A'B'}$). On the other hand, the existence of a gauge covariantly constant right-handed spinor $u_A$ can at least be seen to be plausible by applying the following argument:

By evaluating the commutator $[\tilde{\nabla}_{AA'}, \tilde{\nabla}_{BB'}]u_C$ , which by supposition has to vanish, one can deduce that

$$\Psi_{ABCD}\, u^D - \Lambda/6\, (u_A\, \varepsilon_{BC} + u_B\, \varepsilon_{AC}) - ie/2\, F_{AB}\, u_C = 0\ . \tag{6}$$

The Euclidean conjugate spinor $\bar{u}_A$ is also gauge covariantly constant, with the opposite sign of the electric charge, and hence $\bar{u}^A u_A$ is an ordinary constant, which by appropriate normalization of $u_A$ may be set equal to unity, $\bar{u}^A u_A = 1$.

Since $\bar{u}^A u_A$ is non-zero, the pair $(u_A, \bar{u}_A)$ spans the space of right-handed spinors, and it follows from (6) that

$$F_{AB} = -i\Lambda/e\ (u_A \bar{u}_B + u_B \bar{u}_A) \quad , \tag{7}$$

$$\Psi_{ABCD} = -2\Lambda\ u_{(A} u_B \bar{u}_C \bar{u}_{D)} \quad . \tag{8}$$

Thus the right-handed Weyl spinor is of type D, but the left-handed Weyl spinor is unrestricted.

Now consider the two topological invariants $\chi$, the Euler number, and $\tau$, the Hirzebruch signature, which in a compact Einstein manifold are given by

$$\chi = 1/32\pi^2 \int (C_{abcd}C^{abcd} + 8\Lambda^2/3)\sqrt{g}\,d^4x \quad , \tag{9}$$

$$\tau = 1/48\pi^2 \int C_{abcd}{}^*C^{abcd}\sqrt{g}\,d^4x \quad , \tag{10}$$

where $C_{abcd}$ is the Weyl tensor. Eliminating the anti-self-dual Weyl tensor between these two expressions, and substituting for the self-dual Weyl tensor from equation (8) implies that

$$2\chi + 3\tau = \Lambda^2 V/2\pi^2 \quad , \tag{11}$$

where V is the volume of the space.

To summarize, we have seen that <u>if</u> a right-handed gauge covariantly constant spinor were to exist, then equation (11) follows. However, (11) is just precisely the condition satisfied by a Kähler-Einstein metric (with $\Lambda^2 V/2\pi^2 = c_1^2$, the first Chern number). Thus although this argument does not <u>prove</u> that a gauge covariantly constant spinor must exist in a Kähler-Einstein metric, it is at least consistent with the existence of such a spinor. In fact, such a spinor $u_A$ <u>does</u> always exist in a Kähler-Einstein metric (Hitchin 1974).

The spinor pair $(u_A, \bar{u}_A)$, with $\bar{u}^A u_A = 1$, may be used as a dyad basis for right-handed spinors, and in the standard way one may define the antisymmetric spinor $\epsilon_{AB} = u_A\bar{u}_B - u_B\bar{u}_A$ which lowers unprimed indices, $\xi^A \epsilon_{AB} = \xi_B$, etc. Because $\epsilon_{AB}$ involves products of a spinor of charge +e with a spinor of charge -e it is itself uncharged. Adopting the convention that $\tilde{\nabla}_a$ denotes the gauge covariant derivative for charge appropriate to whatever field it acts on, we see that $\nabla_a(u_A \bar{u}_B) = (\tilde{\nabla}_a u_A)\bar{u}_B + u_A \tilde{\nabla}_a \bar{u}_B = 0 + 0 = 0$, and so $\epsilon_{AB}$ is covariantly constant.

The Kähler form $J_{ab}$ is just a constant multiple of the Maxwell field $F_{ab}$, and since $J_{ab}$ is normalized so that $J_{ab}\,J^b{}_c = -g_{ac}$, and

$J_{ab}$ has spinor equivalent 1/2 $J_{AB}\,\varepsilon_{A'B'}$ , we have

$$J_{AB} = -2i(u_A\bar{u}_B + u_B\,\bar{u}_A) = 2e/\Lambda\;F_{AB} \quad . \tag{12}$$

Clearly $J_{AB}$ is uncharged, and covariantly constant as it should be. The gauge covariantly constant right-handed spinors $u_A$ and $\bar{u}_A$ are the analogues in Kähler-Einstein metrics with self-dual Kähler forms of the covariantly constant left-handed spinors $\tilde{o}_{A'}$ and $\tilde{\iota}_{A'}$ ($=\bar{\tilde{o}}_{A'}$). The difference in handedness is simply because in a self-dual metric (which is automatically a Kähler metric) the Kähler form is anti-self-dual. Note that in a self-dual metric there is a whole 2-sphere's worth of complex structures (i.e. of covariantly constant unit norm anti-self-dual 2-forms), since $\tilde{o}_{A'}\tilde{o}_{B'}$ , $\tilde{\iota}_{A'}\tilde{\iota}_{B'}$ and $\tilde{o}_{(A'}\tilde{\iota}_{B')}$ span the space of anti-self-dual 2-forms. In the general Kähler-Einstein case on the other hand, there is just the one complex structure, given by (12). One can still construct the self-dual 2-forms whose spinor analogues are

$$K_{AB} = u_A\,u_B \quad , \quad L_{AB} = \bar{u}_A\,\bar{u}_B \quad . \tag{13}$$

These are charged ( +2e and -2e respectively), and gauge covariantly constant.

## 4. RELATIONS BETWEEN EIGENMODES IN KÄHLER-EINSTEIN METRICS

One can now proceed by analogy with section 2, and construct higher spin eigenfunctions from scalar eigenfunctions. The only real difference from before is that because $u_A$ and $\bar{u}_A$ are charged, it may be necessary to give the scalar eigenfunction a charge in order to end up with a spinor carrying the desired charge. The original scalar eigenfunction $\phi_p$, of charge pe, must therefore satisfy

$$-\tilde{\nabla}_a\tilde{\nabla}^a\,\phi_p = \lambda_p\,\phi_p \quad , \tag{14}$$

where $\lambda_p = \lambda_{-p}$ .

The idea now is to add unprimed indices by multiplying by $u_A$ or $\bar{u}_A$, and primed indices by operating with $u^A\tilde{\nabla}_{AA'}$ or $\bar{u}^A\tilde{\nabla}_{AA'}$ , symmetrizing where appropriate. For example there are four classes of vectors, corresponding to taking $V_{AA'}$ to be

$$\nabla_{AA'}\phi_0 \;,\; J_{AB}\nabla^B{}_{A'}\phi_0 \;,\; K_{AB}\tilde{\nabla}^B{}_{A'}\phi_{-2} \text{ or } L_{AB}\tilde{\nabla}^B{}_{A'}\phi_2 \;, \tag{15}$$

where K and L are defined in equation (13). In order to obtain neutral vectors in each case, the scalar eigenfunctions have been given charges of 0, 0, -2e and +2e respectively.

## KÄHLER-EINSTEIN SPACES

Consider now the (uncharged) boson operators for vectors (i.e. the Hodge-de Rham operator on 1-forms), the Hodge-de Rham operators for self-dual and anti-self-dual 2-forms, and the Lichnerowicz operator for symmetric tracefree 2 index tensors (i.e. metric perturbations). The spinor transcriptions of these operators in an Einstein background are, respectively,

$$\Delta_1 V_{AA'} = -\nabla_a \nabla^a V_{AA'} + \Lambda V_{AA'} \quad , \tag{16}$$

$$\Delta^+ \psi_{AB} = -\nabla_a \nabla^a \psi_{AB} - 2\Psi_{ABCD}\psi^{CD} + 4\Lambda/3\ \psi_{AB} \quad , \tag{17}$$

$$\Delta^- \tilde{\psi}_{A'B'} = -\nabla_a \nabla^a \tilde{\psi}_{A'B'} - 2\tilde{\Psi}_{A'B'C'D'}\tilde{\psi}^{C'D'} + 4\Lambda/3\ \tilde{\psi}_{A'B'} \quad , \tag{18}$$

$$\Delta_L h_{ABA'B'} = -\nabla_a \nabla^a h_{ABA'B'} - 2\Psi_{ABCD} h^{CD}{}_{A'B'} - 2\tilde{\Psi}_{A'B'C'D'} h_{AB}{}^{C'D'} + 8\Lambda/3\ h_{ABA'B'} \quad . \tag{19}$$

The construction of vector eigenmodes has already been given, in equation (15). There are 3 self-dual eigenmodes of (17), given by $\psi_{AB}$ =

$$J_{AB}\phi_0 \ , \ K_{AB}\phi_{-2} \text{ or } L_{AB}\phi_2 \quad . \tag{20}$$

The 3 anti-self-dual eigenmodes of (18) are given by $\tilde{\psi}_{A'B'}$=

$$J_{AB}\nabla^A{}_{A'}\nabla^B{}_{B'}\phi_0 \ , \ K_{AB}\tilde{\nabla}^A{}_{A'}\tilde{\nabla}^B{}_{B'}\phi_{-2} \text{ or } L_{AB}\tilde{\nabla}^A{}_{A'}\tilde{\nabla}^B{}_{B'}\phi_2 \quad . \tag{21}$$

There are 9 tracefree eigenmodes of (19),

$$h_{ABA'B'} = M_{AB}N_{CD}\tilde{\nabla}^C{}_{(A'}\tilde{\nabla}^D{}_{B')}\phi \quad , \tag{22}$$

where M and N can each be chosen to be one of J, K or L. In each case the charge of $\phi$ must be chosen so as to make $h_{ABA'B'}$ neutral.

The fact that the fields defined by equations (15), (20), (21) and (22) turn out to be eigenfunctions of $\Delta_1$, $\Delta^+$, $\Delta^-$ and $\Delta_L$ is a non-trivial result, particularly in the cases of $\Delta^-$ and $\Delta_L$, which involve $\tilde{\Psi}_{A'B'C'D'}$ terms. The significance of this remark is that a term like, for example, $\Psi_{ABCD}\psi^{CD}$ in (17) could have been put in "by hand" if necessary, since $J_{AB}$, $K_{AB}$ and $L_{AB}$ are eigenvectors of $\Psi_{ABCD}$ with known eigenvalues (see equation (8)). On the other hand, in order to obtain terms involving $\tilde{\Psi}_{A'B'C'D'}$ in (18) and (19) one is dependent upon just such a term being generated when the derivatives in (21) or (22) are commuted through the Laplacian in (14). This is similar to the case of the self-dual metrics in section 2.

It turns out that for all four of the operators (16-19), eigenmodes constructed from uncharged scalar eigenmodes ($\phi_0$, $\lambda_0$) of equation (14) have eigenvalue $\lambda_0$. Those constructed from ($\phi_2$, $\lambda_2$)

or $(\phi_{-2}, \lambda_2)$ have eigenvalue $(\lambda_2 + 2\Lambda)$ and those obtained from $(\phi_4, \lambda_4)$ or $(\phi_{-4}, \lambda_4)$ have eigenvalue $(\lambda_4 + 8\Lambda)$. Thus $\Delta_1$ has 2 eigenvalues $\lambda_0$ and 2 eigenvalues $(\lambda_2 + 2\Lambda)$. $\Delta^+$ and $\Delta^-$ each have $\lambda_0$ once and $(\lambda_2 + 2\Lambda)$ twice. $\Delta_L$ has 3 eigenvalues $\lambda_0$, 4 eigenvalues $(\lambda_2 + 2\Lambda)$ and 2 eigenvalues $(\lambda_4 + 8\Lambda)$.

It should be emphasized that this procedure generates all the eigenfunctions of these operators. To see this, suppose that one had an eigenfunction of one of the operators which did not come from a scalar eigenfunction by the method described above. One could then apply a converse procedure to remove all its indices; the resulting scalar turns out to be an eigenfunction of (14). One could then regenerate the higher spin eigenfunction in the manner set out above, contradicting the supposition that it could not be obtained in this way.

Finally in this section we construct spin 1/2 eigenfunctions of charge e minimally coupled to the electromagnetic field $F_{ab}$. The relevant second order operators are

$$\Delta^+{}_{1/2}\, \phi_A = -\tilde{\nabla}_a \tilde{\nabla}^a \phi_A + \Lambda\, \phi_A + ieF_{AB}\phi^B \quad , \tag{23}$$

$$\Delta^-{}_{1/2}\, \tilde{\phi}_{A'} = -\tilde{\nabla}_a \tilde{\nabla}^a \tilde{\phi}_{A'} + \Lambda \tilde{\phi}_{A'} \quad . \tag{24}$$

There is no explicit Maxwell term in (24) because $F_{ab}$ is self-dual. There are of course still the Maxwell couplings in the derivative term. There are two types of right-handed eigenfunction, $\phi_A$ =

$$u_A \phi_0 \text{ or } \bar{u}_A \phi_2 \quad . \tag{25}$$

The two left-handed eigenfunctions are given by $\tilde{\phi}_{A'}$ =

$$u_A \nabla^A{}_{A'} \phi_0 \text{ or } \bar{u}_A \tilde{\nabla}^A{}_{A'} \phi_2 \quad . \tag{26}$$

In each case it turns out that, just as with the boson operators, the eigenvalues of the operators (23) or (24) corresponding to uncharged scalar eigenfunctions are $\lambda_0$, and the eigenvalues corresponding to charge 2e scalars are $(\lambda_2 + 2\Lambda)$.

To summarize these results, we have seen that the eigenmodes of the Hodge-de Rham operators on 1-forms and 2-forms, of the Lichnerowicz operator on symmetric tracefree metric perturbations, and of the minimally coupled charge e spin 1/2 operators may all be constructed from charged scalar eigenmodes of charges 0, ±2e and ±4e. Furthermore the spectra of those eigenvalues of all of these operators corresponding to a particular charge of scalar eigenfunction are all the same.

Except that one now has several distinct sets of eigenvalues originating from scalars of different charges, the results described above are rather similar to those described in section 2 for self-dual metrics. The results for Kähler-Einstein metrics are much more general, containing the self-dual metrics as a special case. However, the operators considered in this section are not actually the ones which are relevant to, for example, the gauged extended N = 2 supergravity model (Freedman & Das, 1977). These operators are discussed briefly in next section.

## 5. GAUGED EXTENDED SUPERGRAVITY

The operators involved in the gauged extended N = 2 supergravity model will be more complicated than those discussed in section 4 for several reasons. Firstly, because one is quantizing around a non-trivial Einstein-Maxwell background, even the pure Einstein-Maxwell operators will be more complicated. This is because the second variation of the action now involves cross terms containing one variation of the metric, $\delta g_{ab}$, and one variation of the Maxwell potential, $\delta A_a$, in addition to the usual diagonal terms (see Deser & van Nieuwenhuizen 1974). Secondly, the couplings of the spin 3/2 fields to the Maxwell field involve non-minimal terms (Freedman & Das 1977). (The spin 3/2 field was omitted from the discussion in the previous section because a minimal coupling of spin 3/2 to the Maxwell field is inconsistent). Finally, there will also be operators arising from the ghost fields, and these too have to be examined in detail.

At the time of writing, these complications have not been completely investigated. However, it would seem that the higher spin fields constructed in the manner described in the previous section will still be eigenfunctions of the modified operators. The reason for this is that none of the modifications involve the Weyl tensor, but only the Maxwell field $F_{AB}$, or mass terms, etc. Any possible modification involving only $F_{AB}$ or mass terms, no matter what the precise form it takes is, presents no difficulties, since $u_A$ and $\bar{u}_A$ are eigenvectors of $F_{AB}$ (see equation (7)). The only possible modification to the operators which might have created difficulties would have been one involving $\tilde{\Psi}_{A'B'C'D'}$, but in fact these terms are the same in the actual N = 2 supergravity operators as they are in the operators discussed in section 4. Thus in the N = 2 model the eigenfunction construction will still work. What remains to be determined is whether the eigenvalues of the various operators are identical, or whether they are shifted relative to one another, thereby preventing an exact cancellation in the effective action. The extension of these ideas to the N > 2 extended models also remains to be investigated. Presumably one would quantize around a U(1) subgroup of the spin 1 gauge group, taking F as the background field.

It is possible that this work may tie in with the results

obtained recently by Christensen et al. (1980), who found that the one loop term in N > 4 gauge extended supergravity is simply given by an integer multiple of the Euler number - and so the one loop effective action scales as an integer power of k under constant scalings of the metric, $g_{ab} \to k^2 g_{ab}$ (see also the article by Duff in this volume). This suggests the possibility that there may be some "super theorem", with the non-zero modes cancelling exactly. Of course the situation we are considering here is rather different, in that there is a non-trivial background electromagnetic field, and we are considering also O(N) models with N less than 5. There is as yet no agreement on how the results of Christensen et al. are modified by the presence of a non-zero electromagnetic background field.

## 6. CONCLUSION

The procedure described above enables one to express all higher spin eigenfunctions in terms of charged scalar eigenfunctions in an arbitrary Kähler-Einstein background, with electromagnetic field proportional to the Kähler form. Two Kähler-Einstein instantons with positive $\Lambda$ are known; $S^2 \times S^2$ and $CP^2$. There are known to be infinitely many Kähler-Einstein solutions with negative $\Lambda$; for example hypersurfaces of degree >4 in $CP^3$ (see for example Christensen et al. 1980). These are all compact instantons, and are of interest in the spacetime foam picture of the gravitational vacuum (Hawking 1979). It remains to be seen whether the eigenvalues of the operators in supergravity around Kähler-Einstein backgrounds are identical, in which case the effective action is determined just by the zero-modes, or whether the eigenvalues are related but different, in which case no exact cancellation of eigenvalues would take place.

Throughout this article we have been concerned with background geometries which are solutions of the Einstein equations. In fact the essential property of possessing gauge covariantly constant spinors is not confined to Kähler metrics which are also Einstein. It is in fact the case that any Kähler metric has a pair of gauge covariantly constant spinors (Hitchin 1974 and private communication). In this general case the electromagnetic field to which the charged spinors are minimally coupled is proportional not to the Kähler form J, but instead to the Ricci form. (In a Kähler-Einstein metric these two 2-forms are proportional to one another). This will be discussed in detail in a future publication.

Finally, it is worth noting that the background manifold does not have to admit a spin structure in order for this construction to work. One can use the Kähler form as electromagnetic field in order to enable one to define a $spin^c$ structure, in which case one is forced to work with fermion fields carrying charges which are an odd multiple of e and boson fields whose charges are an even multiple of e (Hawking & Pope 1978b). This is discussed in detail for the case of $CP^2$ in Pope (1980b).

REFERENCES

Christensen, S.M., Duff, M.J., Gibbons, G.W. & Roček, M. (1980). One Loop Effects in Supergravity with a Cosmological Constant. In preparation

Deser, S. & van Nieuwenhuizen, P. (1974). One-loop Divergences of Quantized Einstein-Maxwell Fields. Phys. Rev. D, 10, 401-410.

Flaherty, E.J. (1976). Hermitian and Kählerian Geometry in Relativity. Springer 1976.

Freedman, D.Z. & Das, A. (1977). Gauge Internal Symmetry in Extended Supergravity. Nucl. Phys. B, 120, 221-230.

Hawking, S.W. (1979). The Path Integral Approach to Quantum Gravity. In General Relativity: An Einstein Centenary Survey, C.U.P. 1979.

Hawking, S.W. & Pope, C.N. (1978a). Symmetry Breaking by Instantons in Supergravity. Nucl. Phys. B, 73, 42-46.

Hitchin, N.J. (1974). Harmonic Spinors. Adv. Math.. 14, 1-55.

Penrose, R. & MacCallum, M.A.H. (1972). Twistor Theory: An Approach to the Quantization of Fields and Space-time. Phys. Reps. 6, 241-316.

Pope, C.N. (1980a). The Role of Instantons in Quantum Gravity. To appear in Proceedings of the Second Oxford Conference on Quantum Gravity.

Pope, C.N. (1980b). Eigenfunctions and $Spin^c$ Structures in $CP^2$. D.A.M.T.P. preprint.

# V1 OTHER ASPECTS OF SUPERGRAVITY

# IDENTITIES FOR BIANCHI-IDENTITIES

Martin F. Sohnius

The Blackett Laboratory
Imperial College
London SW7 2BZ

## INTRODUCTION

The purpose of this talk is to introduce a rather neat but technical result of differential geometry to those who are most likely to have use for it, the practitioners of supersymmetry and supergravity. The result relates Bianchi-identities to each other and makes it possible to quickly determine which ones are truly independent for a given theory.

Supersymmetric gauge theories and supergravity necessarily have constraints imposed on the geometric quantities torsion and curvature. This is so for a variety of reasons which eventually boil down to the fact that the unconstrained superfield is not an irreducible representation of supersymmetry. Constraints are usually formulated in terms of spinor derivatives, classically denoted by $D_\alpha$ and $\bar{D}_{\dot{\alpha}}$ . The very appearance of the time derivative in their algebra,

$$\{D_\alpha, \bar{D}_{\dot{\beta}}\} = -\, 2i\, \sigma^\mu{}_{\alpha\dot{\beta}}\, \partial_\mu \qquad , \qquad (1)$$

already suggests one of the fundamental dangers of imposing constraints: they may imply equations of motion for the fields (or even set them constant, a very special "equation of motion" indeed). On the other hand, under-constrained fields do not allow Lagrangians, have higher spins than desired, etc. A useful help in walking the narrow ridge between over- and under-constraining a theory are the Bianchi-identities.

If in a supersymmetric Yang-Mills theory or in supergravity the equations of motion do not follow from a given set of constraints via the Bianchi-identities, then these constraints are at least not too strong (they may be too weak, or an inconvenient choice, but that is another matter). Checking the identities is actually a lot easier and quicker than solving the constraints for the connections. This is so because one avoids dealing with pure-gauge degrees of freedom; one has acquired advantages like if working in a Wess-Zumino gauge, but without having given up supersymmetry.

However, all but the simplest theories have dozens of Bianchi-identities. Anyone who has ever tried to check the consistency of constraints will remember a lot of stupid algebra with ever-recurring zeros at the end of the pages. Very quickly then the suspicion arises that some of those zeros are actually consequences of each other and not really independent checks. Indeed, as I will show here, it is exactly the structure constant $\sigma^{\mu}{}_{\alpha\dot\beta}$ on the right-hand side of eq. (1) which has the effect of making many identities algebraically dependent on derivatives of other identities.

On the example of N=1 supersymmetric Yang-Mills theory, we can see that the structure constant makes its appearance not only in the definition of some of the curvatures (field strengths) themselves, e.g. in

$$F_{\alpha\dot\beta} \equiv D_\alpha \mathcal{A}_{\dot\beta} + \bar{D}_{\dot\beta} \mathcal{A}_\alpha + \{\mathcal{A}_\alpha, \mathcal{A}_{\dot\beta}\} + 2i\sigma^{\mu}{}_{\alpha\dot\beta}\mathcal{A}_\mu \quad , \qquad (2)$$

but also in the identities, e.g. in

$$\mathcal{D}_\alpha F_{\dot\beta\mu} + \mathcal{D}_\mu F_{\alpha\dot\beta} - \mathcal{D}_{\dot\beta} F_{\mu\alpha} + 2i\sigma^{\nu}{}_{\alpha\dot\beta}F_{\nu\mu} = 0 \quad . \qquad (3)$$

Eq. (3) gives, for instance, $F_{\mu\nu}$ completely in terms of other curvature components, and it is then not surprising that the identity for $\mathcal{D}_\rho F_{\mu\nu}$ (the "homogeneous Maxwell equation"),

$$\mathcal{D}_\rho F_{\mu\nu} + \mathcal{D}_\nu F_{\rho\mu} + \mathcal{D}_\mu F_{\nu\rho} = 0, \qquad (4)$$

is a consequence of lower dimensional Bianchi-identities (with spinor indices) and the Ricci-identities.

$$\begin{aligned} \{\mathcal{D}_\alpha, \bar{\mathcal{D}}_{\dot\beta}\} &= -2i\, \sigma^{\mu}{}_{\alpha\dot\beta}\mathcal{D}_\mu + F_{\alpha\dot\beta} \\ \{\mathcal{D}_\alpha, \mathcal{D}_\beta\} &= F_{\alpha\beta} \quad , \quad \text{etc.} \end{aligned} \qquad (5)$$

Some time ago (Sohnius, 1978), I have systematically explored which ones of the identities are truly independent in the case of extended supersymmetric Yang-Mills theories without central charges. Another important result has been obtained by N. Dragon (1979), who showed that for supergravity the generalized

Riemann tensor is completely given in terms of the torsions, and that the identities for it can be derived from those for the torsions.

In this talk, I will tie up a few loose ends, and show where these results originate in the structure of gauge theories. The set of quite simple equations which I arrive at can easily be exploited for arbitrary cases: with central charges, with gauged internal symmetries, etc.

In the remaining sections, I will first introduce my notation for differential forms. This is heavily based on J. Wess's canonical "Salamanca lectures" (1978), although I have changed a few definitions to avoid some Grassmann parity factors. I then proceed to derive my results for the most general case, and illustrate their power on a number of examples.

## DIFFERENTIAL FORMS

We define a differential *p-form* as

$$\phi_p \equiv \frac{1}{p!} \quad dx^{M_p} \quad \ldots \quad dx^{M_1} \qquad \phi_{M_1 \ldots M_p}(x) \tag{6}$$

where the coefficient function $\phi_{M_1 \ldots M_p}(x)$ is (graded) antisymmetric in its indices, due to the (graded) anticommutativity of the differentials.

$$0 = dx^M dx^N + (-1)^{\sigma(M)\sigma(N)} \, dx^N \, dx^M . \tag{7}$$

The *wedge-product* of two forms is defined to be

$$\phi_p \wedge \psi_q \equiv \frac{1}{p!q!} \, dx^{N_q} \ldots dx^{N_1} \, dx^{M_p} \ldots dx^{M_1}$$

$$\phi_{M_1 \ldots M_p} \quad \psi_{N_1 \ldots N_q} \quad , \tag{8}$$

and the *exterior derivative* is the wedge-product with the 1-form $d \equiv dx^M \, \partial_M$ :

$$d\phi_p \equiv \frac{1}{p!} \, dx^{N_p} \ldots dx^{N_1} \, dx^M \, \partial_M \, \phi_{N_1 \ldots N_p} . \tag{9}$$

As a consequence of eqs. (8-9), we have the product rule

$$d(\phi_p \wedge \psi_q) = d\phi_p \wedge \psi_q + (-1)^p \phi_p \wedge d\psi_q \ . \tag{10}$$

Under an element $\lambda$ of the *local structure group* (the gauge group), a *covariant p-form* transforms as

$$\phi \xrightarrow{\lambda} e^{-\Lambda} \phi \ . \tag{11}$$

$e^{\Lambda}$ is a matrix representation of $\lambda$, and $\Lambda$ takes its values in a representation of the Lie-algebra of the structure group. It is in general a function over the base manifold, $\Lambda = \Lambda(x)$, i.e. superspace dependent.

The *connection* 1-form $\Omega$ takes its values in the same representation as $\Lambda$ and transforms according to

$$\Omega \xrightarrow{\lambda} e^{-\Lambda} (\Omega + d) \, e^{\Lambda} \ , \tag{12}$$

so that we can define a *covariant exterior derivative*

$$\mathcal{D}\phi \equiv d\phi + \Omega \wedge \phi \tag{13}$$

which transforms just like $\phi$ itself did. Note that $\mathcal{D}$ is not covariantized with respect to the manifold indices on $\phi_{M_1 \ldots M_p}$. Due to its curl-structure, $\mathcal{D}\phi$ is nevertheless a covariant object under general coordinate transformations.

The *curvature* 2-form R,

$$R \equiv d\Omega + \Omega \wedge \Omega , \tag{14}$$

is also Lie-algebra valued and transforms covariantly, under the adjoint representation of the gauge-group,

$$R \xrightarrow{\lambda} e^{-\Lambda} R \, e^{\Lambda} \ . \tag{15}$$

The *vielbein* 1-forms are non-degenerate linear combinations of the differentials,

$$E^A \equiv dx^M \, E_M{}^A(x) , \tag{16}$$

and their wedge-products form a basis in the space of differential forms. They transform into each other under a representation of the gauge-group,

$$E^A \xrightarrow{\lambda} E^B \, (e^\Lambda)_B{}^A \,, \tag{17}$$

for which I will always explicitly write the indices. In extended supergravity these refer to the Lorentz-group and the gauged part of that internal symmetry group which acts non-trivially on the spinor charges.

The covariant derivative of the vielbein is the *torsion* 2-form,

$$T^A \equiv \mathcal{D}E^A = dE^A + E^B \wedge \Omega_B{}^A \,. \tag{18}$$

## THE IDENTITIES

We can use the Poincaré-lemma

$$dd = 0 \tag{19}$$

either on an arbitrary covariant form $\phi$, or on $E^A$, or on $\Omega$ , to give us, respectively, the Ricci-, the first, and the second Bianchi-identity (Wess and Zumino, 1977):

$$\mathcal{D}\mathcal{D}\phi = R \wedge \phi \tag{20}$$

$$\mathcal{D}T^A + E^B \wedge R_B{}^A = 0 \tag{21}$$

$$\mathcal{D}R = dR - R \wedge \Omega + \Omega \wedge R = 0 \tag{22}$$

If we write all forms in the vielbein basis,

$$d \equiv E^A d_A \quad ; \quad \Omega \equiv E^A \, \Omega_A \quad ; \quad \mathcal{D} \equiv E^A \mathcal{D}_A$$
$$R \equiv -\frac{1}{2} E^B \wedge E^A \, R_{AB} \quad ; \quad T^A \equiv -\frac{1}{2} E^C \wedge E^B \, T_{BC}{}^A \,, \tag{23}$$

we can express the 0-forms $R_{AB}$ and $T_{AB}{}^C$ through the covariantized curls of connection and vielbein :

$$R_{AB} = \sum_{[AB\}} (d_A \Omega_B + \Omega_A \Omega_B + \Omega_{AB}{}^C \Omega_C) + T_{AB}{}^C \Omega_C \tag{24}$$

$$T_{AB}{}^{C} = -\sum_{[AB\}} (d_A\, E_B{}^M\, E_M{}^C + \Omega_{AB}{}^C)$$

$$= -\sum_{[AB\}} \mathcal{D}_A\, E_B{}^M\, E_M{}^C \qquad (25)$$

The symbol $\sum_{[AB\}}$ denotes the graded antisymmetrized sum. $E_A{}^M$ is the inverse vielbein.

The identities now take the more familiar førm:

$$[\mathcal{D}_A, \mathcal{D}_B\}\, \phi = -\, T_{AB}{}^{C}\, \mathcal{D}_C \phi + R_{AB}\phi \qquad (26)$$

$$\sum_{[ABC\}} (\mathcal{D}_A\, T_{BC}{}^{D} + T_{AB}{}^{E}\, T_{EC}{}^{D} + R_{ABC}{}^{D}) = 0 \qquad (27)$$

$$\sum_{[ABC\}} (\mathcal{D}_A\, R_{BC} + T_{AB}{}^{D}\, R_{DC}) = 0. \qquad (28)$$

They have picked up additional torsion terms, as compared to eqs. (20-22), because differentiations hàd to be pulled through vielbeins.

It is very convenient to introduce the "names"

$$I^{(1)}{}_{ABC}{}^{D} \quad \text{and} \quad I^{(2)}{}_{ABC}$$

for the left-hand sides of the first and second Bianchi-identities, eqs. (27-28).

Let us quickly review the meaning of these identities in the various cases. In global supersymmetry the Riemann curvatures vanish and the torsions are constants, the structure constants of the super-algebra. The first Bianchi-identity is then just the Jacobi-identity for the algebra and ensures that the structure constants form a representation of that algebra, the adjoint representation. The second Bianchi-identity is the usual set for the Yang-Mills field strengths.

In the general case eq. (27) determines, in terms of the torsion, only that part of $R_{ABC}{}^{D}$ which is totally antisymmetric in ABC.

IDENTITIES FOR BIANCHI-IDENTITIES

For N-extended supergravity, however, the gauge group which acts on the vielbein is restricted to be SL(2,c) × G, where G ⊂ U(N) is that part of the internal symmetry group which rotates the spinor charges. The four-vector and spinor components of the vielbein form, $E^a$, $E^{\alpha i}$, $E_i^{\dot\alpha}$ ($a = 0,\ldots,3$; $\alpha,\dot\alpha = 1,2$; $i=1,\ldots,N$) are not mixed by gauge transformations. This implies severe restrictions on the generalized Riemann tensor $R_{ABC}{}^D$ (Wess and Zumino, 1977). Those components, e.g., where the indices C and D refer to different representations of the SL(2,c) must vanish. Furthermore,

$$R_{AB\alpha i}{}^{\beta j} = -\frac{i}{4}(\sigma_{cd})_\alpha{}^\beta \delta_i{}^j R_{AB}{}^{cd} - i\delta_\alpha{}^\beta \vec{t}_i{}^j \vec{R}_{AB}$$

$$R_{AB\dot\alpha j}{}^{i\dot\beta} = \frac{i}{4}(\bar\sigma_{cd})^{\dot\beta}{}_{\dot\alpha} \delta_j^i R_{AB}{}^{cd} + i\delta_{\dot\alpha}^{\dot\beta} \vec{t}_j{}^i \vec{R}_{AB}, \tag{29}$$

since the spinors do not transform independently of each other and the vectors. The antihermitian matrices $i\vec{t}$ generate a representation of G. The highly restrictive structure of $R_{ABC}{}^D$ makes it possible to express all its components in terms of torsions (Dragon, 1979), via eq. (27). Let me collect which components of R can be determined by which of the identities (I suppress internal symmetry indices):

$$I^{(1)\dot\delta}_{\alpha\beta\dot\gamma} \to R_{\alpha\beta}{}^{cd}, \vec{R}_{\alpha\beta}$$

$$I^{(1)\ \delta}_{\dot\alpha\dot\beta\gamma} \to R_{\dot\alpha\dot\beta}{}^{cd}, \vec{R}_{\dot\alpha\dot\beta}$$

$$\left.\begin{matrix} I^{(1)\ d}_{\alpha\dot\beta c} \\ I^{(1)\ \dot\delta}_{\alpha\dot\beta\dot\gamma} \end{matrix}\right\} \to R_{\alpha\dot\beta}{}^{cd}, \vec{R}_{\alpha\dot\beta} \tag{30}$$

$$I^{(1)\dot\gamma}_{\alpha b\dot\beta} \to R_{\alpha b}{}^{cd}, \vec{R}_{\alpha b}$$

$$I^{(1)\ \gamma}_{\dot\alpha b\beta} \to R_{\dot\alpha b}{}^{cd}, \vec{R}_{\dot\alpha b}$$

$$I^{(1)\beta}_{ab\alpha} \to R_{ab}{}^{cd}, \vec{R}_{ab}$$

IDENTITIES BETWEEN IDENTITIES

If we take one more covariant derivative of the first Bianchi-identity in the form of eq. (21), we get with the help of eq. (20):

$$
\begin{aligned}
0 &= \mathcal{D}(\mathcal{D}T^A + E^B \wedge R_B{}^A) \\
&= -T^B \wedge R_B{}^A + T^B \wedge R_B{}^A - E^B \wedge \mathcal{D}R_B{}^A \qquad (31) \\
&= -E^B \wedge \mathcal{D}R_B{}^A
\end{aligned}
$$

Note that $T^A$ transforms under the inverse transpose of the representation generated by $\Lambda_A{}^B$, and thus $R \wedge T^A = -T^B \wedge R_B{}^A$. The meaning of eq. (31) and its derivation is that

$$0 = \sum_{\{ABCD\}} (\mathcal{D}_A R_{BCD}{}^E + T_{AB}{}^F R_{FCD}{}^E) \qquad (32)$$

is a consequence already of the Ricci- and the first Bianchi-identity. It differs from the second Bianchi-identity, eq. (28), taken for the Riemann tensor, only in the antisymmetrization over four instead of only three indices. Again, for supergravity, the restrictions on the form of $R_{ABC}{}^D$ imply already eq. (28) from eq. (32). This is Dragon's result (1979).

It should be pointed out that the fact that $R_{ABC}{}^D$ can be constructed entirely from torsions is not necessarily sufficient to ensure the Bianchi-identities for it: in general, the identities do not determine the entire superfields to which the curvatures or torsions belong, and $R_{ABC}{}^D$ may quite well have been restricted further.

If we take a covariant derivative on the component form of the first Bianchi-identity, $I^{(1)}_{ABC}{}^E$, instead of on eq. (21), we get (suppressing Grassmann factors):

$$
\begin{aligned}
0 &= \sum_{\{ABCD\}} \mathcal{D}_D(\mathcal{D}_A T_{BC}{}^E + T_{AB}{}^F T_{FC}{}^E + R_{ABC}{}^E) \\
&= \sum_{\{ABCD\}} \Big( -\tfrac{1}{2} T_{DA}{}^F \mathcal{D}_F T_{BC}{}^E + R_{DAB}{}^F T_{FC}{}^E - \tfrac{1}{2} T_{BC}{}^F R_{DAF}{}^E \\
&\qquad + \mathcal{D}_D T_{AB}{}^F T_{FC}{}^E + T_{AB}{}^F \mathcal{D}_D T_{FC}{}^E + \mathcal{D}_D R_{ABC}{}^E \Big)
\end{aligned}
$$

We now use (27) on the fourth term and (32) on the last to get

$$0 = \sum_{[ABCD\}} \Big( -\tfrac{1}{2}\, T_{DA}{}^{F}\, \mathcal{D}_F\, T_{BC}{}^{E} + T_{AB}{}^{F}\, \mathcal{D}_D\, T_{FC}{}^{E}$$

$$+ R_{DAB}{}^{F}\, T_{FC}{}^{E} - \tfrac{1}{2}\, T_{BC}{}^{F}\, R_{DAF}{}^{E}$$

$$- T_{DA}{}^{G}\, T_{GB}{}^{F}\, T_{FC}{}^{E} - R_{DAB}{}^{F}\, T_{FC}{}^{E} - T_{DA}{}^{F}\, R_{FBC}{}^{E} \Big)$$

$$= \frac{1}{2} \sum_{[ABCD\}} T_{AB}{}^{F}\, I^{(1)E}_{FCD}$$

The importance of the result

$$0 = \sum_{[ABCD\}} T_{AB}{}^{F}\, I^{(1)\,E}_{\;\;FCD} \tag{33}$$

lies in the fact that the identities we had to use to derive it involve only the indices ABCDE, but not F. Therefore, given certain invertibility conditions for the torsions, new Bianchi-identities which involve the index F have been derived.

Analogously, we get

$$0 = \sum_{[ABCD\}} T_{AB}{}^{F}\, I^{(2)}_{FCD} \tag{34}$$

if only $I^{(2)}$ holds for any of the indices ABCD.

Equations (33-34) can be widely used in both supersymmetry and supergravity to determine interdependences of Bianchi identities.

For N=1 supersymmetric Yang-Mills theories, e.g., we expect second Bianchi-identities with one vector index and the following SL(2,c) properties :

$$I^{(2)}_{\mu\alpha\beta} \;:\; (\tfrac{3}{2},\tfrac{1}{2}) + (\tfrac{1}{2},\tfrac{1}{2})$$

$$I^{(2)}_{\mu\alpha\dot\beta} \;:\; (1,1) + (1,0) + (0,1) + (0,0) \tag{35}$$

$$I^{(2)}_{\mu\dot\alpha\dot\beta} \;:\; (\tfrac{1}{2},\tfrac{3}{2}) + (\tfrac{1}{2},\tfrac{1}{2})$$

But we now know, that the following combinations will already be a consequence of the lower dimensional identities with only spinor indices:

$$\sum_{\{\alpha\gamma\delta\}} \sigma^{\mu}{}_{\alpha\dot\beta} \; I^{(2)}_{\mu\gamma\delta} \quad : \quad \left(\tfrac{3}{2}, \tfrac{1}{2}\right)$$

$$\sum_{\{\dot\beta\dot\gamma\dot\delta\}} \sigma^{\mu}{}_{\alpha\dot\beta} \; I^{(2)}_{\mu\dot\gamma\dot\delta} \quad : \quad \left(\tfrac{1}{2}, \tfrac{3}{2}\right) \tag{36}$$

$$\sum_{\{\alpha\gamma\}\{\dot\beta\dot\delta\}} \sigma^{\mu}{}_{\alpha\dot\beta} \; I^{(2)}_{\mu\gamma\dot\delta} \quad : \quad (1, 1) \; :$$

This leaves us with only the identities (1,0) + (0,1) + (0,0) + $2(\frac{1}{2},\frac{1}{2})$, which are known to be the following conditions on the spinorial superfield $W^{\alpha}$ which contains the field strengths :

$$\begin{aligned} F_{\mu\nu} &= -\frac{1}{32}\, \sigma_{\mu\nu\,\alpha}{}^{\beta}\, \mathcal{D}_{\beta} W^{\alpha} - \frac{1}{32}\, \bar\sigma_{\mu\nu}{}^{\dot\alpha}{}_{\dot\beta}\, \bar{\mathcal{D}}_{\dot\alpha}\, \bar W^{\dot\beta} \\ 0 &= \mathcal{D}_{\alpha} W^{\alpha} - \bar{\mathcal{D}}_{\dot\alpha} \bar W^{\dot\alpha} \\ 0 &= \mathcal{D}_{\alpha} \bar W^{\dot\beta} = \bar{\mathcal{D}}_{\dot\alpha} W^{\beta} \end{aligned} \tag{37}$$

In general, the higher the identities go in dimension, the more likely are they to become dependent. This is due to the fact that $\sigma^{\mu}{}_{\alpha\dot\beta}$ is the only non-vanishing torsion component in supersymmetry (except for cases with central charges), and with more vector-indices it becomes easier and easier to invert eq. (34).

In supergravity the situation is usually more complicated. There the choice of constraints itself influences the usability of the identities (33-34). It turns out that they are most useful if the constraints imply the vanishing of spinor-spinor-spinor torsions, and

$$T_{\alpha\dot\beta}{}^{a} = 2i\,\sigma^{a}{}_{\alpha\dot\beta} \;\; ; \;\; T_{\alpha\beta}{}^{a} = T_{\dot\alpha\dot\beta}{}^{a} = 0 \,. \tag{38}$$

The steps up in dimension are then like in the flat-space case.

## CONCLUSION

The relations (32-34), which constitute the result of this analysis, have been derived without any assumptions on the geometric structure considered. Their usefulness, however, does depend on such assumptions, most heavily on that of the possibility of a covariant separation of indices into spinor and tangent-space vector indices. This is the case for supersymmetry and supergravity and the relations can be widely used to eliminate unnecessary checks of dependent Bianchi-identities.

A final word on possible central charges: in the presence of central charges, $E^A$ has additional Lorentz scalar components which do not transform under the structure group. The analysis for the Riemann tensor goes straight through to $R_{ABC}{}^D$ with AB possibly also referring to the scalar index types. Eqs. (33-34) still hold, but additional torsion terms will be non-zero and need to be taken into account.

## REFERENCES

Dragon, N. (1979). Torsion and Curvature in Extended Supergravity. _Z. Physik C_, 2, pp.29-32.

Sohnius, M. (1978). Bianchi Identities for Supersymmetric Gauge Theories. _Nucl. Phys. B_136, pp.461 - 474.

Wess, J. (1978). Supersymmetry-Supergravity. _Lecture Notes in Physics_ 77, pp. 81-

Wess, J. and B. Zumino (1977). Superspace Formulation of Supergravity. _Phys. Lett._ 66B, pp.361-364.

# INSTABILITIES IN GRAVITY AND SUPERGRAVITY

Malcolm J. Perry

Department of Physics
Princeton University

A lot of work recently has been devoted to defining quantum gravity using the techniques of Euclidean field theory. Associated with this, there has been a considerable amount of work exploring the properties of gravitational instantons (Hawking, 1979; Perry, 1980a). A gravitational instanton is a solution of the Euclidean Einstein field equations $R_{ab} = \Lambda g_{ab}$ which is geodesically complete and non-singular. However, the physical interpretation of these field configurations is still a little unclear. The purpose of this article is to discuss the stability of these objects, firstly in general; and then to illustrate the instability by a particular physically simple example.

In Euclidean gravity, one constructs a functional integral over the space of all metrics subject to certain boundary conditions. These boundary conditions dictate the problem being solved. Only three types appear to be of physical interest. Asymptotically Euclidean (AE): Here the metric must tend to the flat metric on $R^4$ outside some compact region. This corresponds to the zero temperature vacuum. Asymptotically Flat (AF): Here the metric must tend to the flat metric on $R^3 \times S^1$ outside some compact region. This corresponds to the canonical ensemble at a temperature $T = 1/\beta$ where $\beta$ is the periodicity induced on the $S^1$. Both AE and AF instantons have $\Lambda = 0$. Compact: Here the instantons are simply-connected compact spaces. In this case $\Lambda$ is usually non-zero. These are related to spacetime foam (Hawking, 1978).

The first sign of difficulty is manifested by the fact that the Euclidean action for gravity can be unboundedly negative (Gibbons, Hawking & Perry, 1978). For a metric with Ricci scalar R, the action is

$$I\{g\} = \frac{-1}{16\pi}\int (R-2\Lambda)\, d(\mathrm{Vol}) + \text{boundary terms} \qquad (1)$$

A conformal transformation of the metric $g_{ab} \to \tilde{g}_{ab} = \Omega^2 g_{ab}$ leads to the rescaled action

$$I\{g\} = \frac{-1}{16\pi} \int \Omega^2 R + 6\nabla\Omega\cdot\nabla\Omega - 2\Lambda\Omega^4 \, d(\mathrm{Vol}) + \text{boundary terms} \tag{2}$$

By choosing an $\Omega$ which is a rapidly varying function of position, it is clear that for some $g_{ab}$, $I\{\tilde{g}\} < I_o$ where $I_o$ is any negative number. Furthermore, it is clear that any theory of gravity which has a term linear in R in its action is going to have similar troubles. Thus supergravity also falls into this category.

One method of performing functional integrals is the saddle point method. Thus, one expands the action about a solution of Einstein's equations, $\hat{g}_{ab}$. One assumes that the metric is now $g_{ab} = \hat{g}_{ab} + h_{ab}$, $h_{ab}$ is a metric operator. One then deals with $h_{ab}$ according to the usual rules of quantum field theory. This approximation to the full functional integral includes only small variations about the metric $\hat{g}_{ab}$. We now expand the action in terms of $h_{ab}$.

$$I\{g+h\} = I\{\hat{g}\} + \frac{1}{16\pi} \int \tfrac{1}{4}\bar{h}\Delta h - \tfrac{1}{2}(\delta\bar{h})^2 - \tfrac{1}{4}\Lambda\bar{h}h \, d(\mathrm{Vol}) + O(h^3) \tag{3}$$

where h is a shorthand for $h_{ab}$, $\bar{h}$ for $h_{ab} - \tfrac{1}{2}\hat{g}_{ab}h^c_c$, $\delta$ for $-\nabla_a$, and $\Delta$ for the Lichnerowicz Laplacian

$$\Delta = -\hat{g}_{ac}\hat{g}_{bd}\Box - 2R_{abcd} + \hat{g}_{ac}R_{bd} + \hat{g}_{ad}R_{bc} \tag{4}$$

and $\bar{h}\Delta h = \bar{h}^{ab}(\Delta)_{abcd}h^{cd}$. All operations are carried out with respect to the metric $\hat{g}_{ab}$. Since this system admits gauge freedom in the sense of co-ordinate transformations, we must add a gauge fixing term and a Faddeev-Popov ghost field in the usual way. I shall choose as a gauge

$$\nabla_a \bar{h}^{ab} = \Phi^b(x) \tag{5}$$

where $\Phi^b$ is some vector field defined on $\hat{g}_{ab}$. The resultant effective quadratic part of the action is

$$I\{\hat{g},h,\eta,\bar{\eta}\} = \frac{-1}{16\pi} \int -\tfrac{1}{4}\bar{h}\Delta h + \tfrac{1}{2}\Lambda\bar{h}h + \bar{\eta}C\eta \, d(\mathrm{Vol}) \tag{6}$$

$\eta$, $\bar{\eta}$ are the ghosts, and C is the operator

$$C = (-\Box-\Lambda)\hat{g}_{ab} \tag{7}$$

We now wish to see whether there are infinitesimal physical deformations of $\hat{g}_{ab}$ which decrease the action. This is best done using the following decomposition theorem (Gibbons & Perry, 1978)

$$h_{ab} = h^{TT}_{ab} + \tfrac{1}{4}\hat{g}_{ab}h + (\nabla_a\xi_b + \nabla_b\xi_a - \tfrac{1}{2}\hat{g}_{ab}\nabla_c\xi^c) \tag{8}$$

$h^{TT}_{ab}$ is the transverse ($\nabla_a h^{ab} = 0$) and tracefree part of $h_{ab}$, $h$ is the trace, and $\xi^a$ is the longitudinal tracefree part of $h_{ab}$. Further, we decompose the two vector fields $\xi$ and $\eta$ using the Hodge-de Rham decomposition

$$\eta_a = \eta^c_a + \nabla_a\chi + \eta^H_a \tag{9}$$

$$\xi_a = \xi^c_a + \nabla_a\Phi + \xi^H_a$$

$\eta^c_a$, $\xi^c_a$ are the coexact parts of $\eta_a$, $\xi_a$. Thus they are both divergence-free. $\chi$ , $\Phi$ are the exact parts, hence curl-free. $\eta^H_a$, $\xi^H_a$ are the harmonic pieces. Since we are usually dealing with simply connected spaces, $\pi^1(M)$ is trivial, hence the first Betti number is zero and there are no harmonic vectors. We will henceforth ignore them. (Their inclusion would complicate things but not materially alter anything which follows.) The effective action now becomes

$$I\{g,h^{TT},h,\eta^c,\chi,\xi^c,\Phi\} = \frac{-1}{16\pi}\int (-\tfrac{1}{4}h^{TT}Gh^{TT} + \tfrac{1}{2}hFh$$

$$-4(2\nabla_a\nabla_b\Phi-\tfrac{1}{2}\hat{g}_{ab}\Box\Phi)(2\nabla_a\nabla_b - \tfrac{1}{2}\hat{g}_{ab}\Box)F\Phi - 2\nabla_{(a}\xi^c_{b)}\cdot\nabla^{(a}C\xi^{b)}$$

$$+ \bar{\eta}^c C\eta^c + \nabla_a\chi\cdot\nabla^a F\chi) \quad d(\mathrm{Vol}) \tag{10}$$

where

$$\begin{aligned} G &= \Delta - 2\Lambda \\ C &= (-\Box - 2\Lambda)\hat{g}_{ab} \\ F &= \tfrac{1}{8}(-\Box - 2\Lambda) \end{aligned} \tag{11}$$

If we express the canonical partition function in terms of these operators

$$Z \sim (\det G)^{-\frac{1}{2}}(\det F)^{-\frac{1}{2}}(\det F)^{-\frac{1}{2}}(\det C)^{-\frac{1}{2}}\det C\,\det F$$

$$\sim (\det G)^{-\frac{1}{2}}(\det C)^{\frac{1}{2}} \tag{12}$$

Thus we can see that F drops out of this expression. In fact, F always drops out since the degrees of freedom which couple to it are pure gauge. In fact, we could have picked a gauge, the 'special' gauge, in which the analogue of F is in fact a constant (Perry, 1980b). This is fortunate since if we examine (10), F occurs three times. When it occurs coupled to h its contribution to the quadratic part of the action is such as to generally decrease the action. The other two times it acts so as to increase the action. This is, in fact, the remnant of the conformal problem described earlier. In perturbation theory, this problem thus disappears. We need not consider the operator F any further. However, it may be that there are other deformations which decrease the action.

The operator C acting on divergence-free objects is in fact positive semi-definite, its kernel being Killing vectors. This follows from the following identity in which d is the exterior derivative operator, (Ric) means the Ricci tensor, and V is any vector field

$$\int (\delta V)^2 - \tfrac{1}{2}(\nabla_a V_b + \nabla_b V_a)^2 + V(d\delta + \delta d - 2(\mathrm{Ric}))V \, d(\mathrm{Vol}) = 0 \tag{13}$$

The operator G however fails to be positive semi-definite. The kernel of G corresponds to deformations of the instanton which map solutions of the field equations to new solutions with the same action. For example, the Schwarzschild solution has three zero modes corresponding to translations of the central object. The Eguchi-Hanson (1978) instanton has a zero mode corresponding to scale transformations.

There are two circumstances under which G is known not to have any negative eigenvalues. If the instanton has a self-dual Riemann tensor, then there are no negative eigenvalues since then eigenvalues of the Laplacian on scalars determine the eigenvalues of G (Hawking & Pope, 1978). Since the Laplacian is positive definite, it follows that G is positive semi-definite. If the instanton is compact, and admits a Kähler structure and has $\Lambda<0$, it then follows that G is also positive semi-definite (Friedan, 1979). All other circumstances at present require treatment on a case to case basis. In Table 1, a list of known instantons appears with an indication of the number of negative modes associated with each. A question mark indicates that the answer is not currently known.

Table 1

| Instanton | Number of Negative Modes |
|---|---|
| Compact Kähler | Zero |
| Compact Non-Kähler | ? |
| K3 | Zero |
| $\mathbb{C}P^2$ | Zero |
| $S^2 \times S^2$ | One |
| $S^4$ | Zero |
| $\mathbb{C}P^2 \# \overline{\mathbb{C}P}^2$ | ? |
| Flat AE | Zero |
| $Z_k$ ALE | Zero |
| Flat AF | Zero |
| Multi-Taub-NUT (ALF) | Zero |
| Schwarzschild-Kerr | One |
| Kerr-Taub-Bolt | ? |

We will now move on to treat a specific example, namely the Schwarzschild instanton. It has a topology of $R^2 \times S^2$, and is static. We write the background metric in $(t,r,\theta,\Phi)$ coordinates

$$ds = (1-\frac{2m}{r})dt^2 + (1-\frac{2m}{r})^{-1}dr^2 + r^2 d\theta^2 + r^2\sin^2\theta d\Phi^2 \qquad (14)$$

$\theta$ and $\Phi$ are polar co-ordinates on $S^2$. $r$ is a circumferential co-ordinate which runs from $2m$ to $\infty$, and $t$ is the Euclidean time co-ordinate. For this to be non-singular, $t$ must be identified with period $\beta = 8\pi m$. Its action is $\beta^2/16\pi$, and satisfies AF boundary conditions. We now need to examine the spectrum of G. To do this, we decompose the transverse trace-free metric perturbations into odd and even perturbations in the sense of Regge and Wheeler (1957). Referred to the $(t,r,\theta,\Phi)$ co-ordinates, the odd perturbations are

$$h_{ab} = \begin{pmatrix} 0 & 0 & -h_0(r)\csc\theta\partial_\Phi & h_0(r)\sin\theta\partial_\theta \\ \cdot & 0 & -h_1(r)\csc\theta\partial_\Phi & h_1(r)\sin\theta\partial_\theta \\ \cdot & \cdot & h_2(r)(\csc\theta\partial_\theta\partial_\Phi - \cot\theta\csc\theta\partial_\Phi) & \frac{1}{2}h_2(r)(\csc\theta\partial_\Phi^2 + \cos\theta\partial_\theta - \sin\theta\partial_\theta^2) \\ \cdot & \cdot & \cdot & -h_2(r)(\sin\partial_\theta\partial_\Phi - \cos\theta\partial_\Phi) \end{pmatrix}$$

$$\times \exp(i\omega t)\, Y_{\ell m}(\theta,\Phi) \qquad (15)$$

and the even perturbations are

$$h_{ab} = \begin{pmatrix} H_0(r) & H_1(r) & K_0(r)\partial_\theta & K_0(r)\partial_\Phi \\ \cdot & H_2(r) & K_1(r)\partial_\theta & K_1(r)\partial_\Phi \\ \cdot & \cdot & r^2(G_1(r)+G_2(r)\partial_\theta^2) & r^2G_2(r)(\partial_\theta\partial_\Phi-\cot\theta\partial_\Phi) \\ \cdot & \cdot & \cdot & r^2(G_1(r)\sin^2\theta + G_2(r)(\partial_\Phi^2+\sin\theta\cos\theta\partial_\theta)) \end{pmatrix}$$

$$\times \exp(i\omega t)\, Y_{\ell m}(\theta,\phi) \tag{16}$$

$h_{ab}$ is taken to be symmetric, $Y_{\ell m}(\theta,\Phi)$ are the usual spherical harmonics. It can be seen that $G>0$ for $\ell\geq 2$ by standard Sturm-Liouville techniques. For $\ell=1$, $G\geq 0$ and the zero modes are

$$h_{ab}^{(i)} = \nabla_a\nabla_b\Phi^{(i)} \tag{17}$$

where

$$\Phi^{(i)} = (r-m)Y_{1i}(\theta,\Phi) \tag{18}$$

For $\ell=0$, G is not positive definite. There is only one negative mode which has $\omega=0$ and $\ell=0$. This mode cannot be found analytically, but was discovered as a result of numerical studies. It is most convenient to describe it by referring the mode to an orthonormal tetrad by writing

$$h_{ab}^{(-)} = \mathrm{diag}\left(\left(1-\frac{2m}{r}\right)\tilde{H}_o(r),\ \left(1-\frac{2m}{r}\right)^{-1}\tilde{H}_1(r),\ -\tfrac{1}{2}r^2(\tilde{H}_o(r)+\tilde{H}_1(r)),\ -\tfrac{1}{2}r^2\sin^2\theta(\tilde{H}_o(r)+\tilde{H}_1(r))\right) \tag{19}$$

$\tilde{H}_o(r)$ and $\tilde{H}_1(r)$ obey the following differential equations

$$\frac{r-2m}{r}\frac{d^2}{dr^2}\tilde{H}_1(r) + \frac{4r^2-22mr+24m^2}{r^2(r-3m)}\frac{d}{dr}\tilde{H}_1(r) - \frac{8m}{r^2(r-3m)}\tilde{H}_1(r) = -\lambda\tilde{H}_1(r) \tag{20}$$

$$\tilde{H}_o(r) = \frac{-r(r-2m)}{r-3m}\frac{d}{dr}\tilde{H}_1(r) - \frac{3r-5m}{r-3m}\tilde{H}_1(r) \tag{21}$$

The eigenvalue of G corresponding to this mode is $\lambda^{(-)}$ and is approximately $-0.19\ m^{-2}$. We will assume that $h_{ab}^{(-)}$ is normalized to unity. To understand its physical significance we examine the metric

$$g_{ab} = g_{ab}^{(\mathrm{Schwarzschild})} + \varepsilon h_{ab}^{(-)} \tag{22}$$

This metric is AF and static. At infinity, its mass as measured by the trace of the second fundamental form at infinity is still m. However, another measurement of mass is the mass of the "bolt" (Gibbons & Hawking, 1979), M at r = 2m. The bolt is the fixed point set of the Killing vector $\partial/\partial t$. The area of the bolt is A, and its mass M is

$$M = \left(\frac{A}{16\pi}\right)^{\frac{1}{2}} \qquad (23)$$

For the Schwarzschild solution m = M. For the field configuration (22), we must take that metric and redefine a new radial co-ordinate $\rho$ to be circumferential. This is unambiguously possible since the perturbation is spherically symmetric. A can then be read off by seeing for what value of $\rho$ the coefficient of $dt^2$ vanishes. A is then just $4\pi\rho^2$. It is trivial to verify that this point is still a non-singular bolt. We find that

$$M \simeq m + 0.0094\ \varepsilon m^{-1} \qquad (24)$$

Furthermore, the action I of the new configuration is

$$I \simeq 4\pi m^2 - \frac{0.000945\varepsilon^2}{m^2} \qquad (25)$$

The physical meaning of this perturbation is clear. It moves matter with an effective Euclidean energy away from the bolt. This matter has positive or negative energy depending on whether $\varepsilon$ is negative or positive respectively. We see that in the space of varying M, the action of the Schwarzschild instanton is a maximum.

To illustrate the physical significance of these results, we compute the canonical ensemble at a temperature $T = 1/\beta$ and angular velocity zero confined to a 3-volume V for pure gravity. This means that functional integrals must be taken over all AF metrics. The only known AF instantons are periodically-identified flat space and the Schwarzschild instanton. We in fact know that there are no instantons which contain two Schwarzschild-like topological singularities. If we examine such a field configuration, we discover that there is a wire singularity joining the objects (Gibbons & Perry, 1980). However, the further these are apart, the closer the configuration comes to being a solution. (This is very analogous to what happens in Yang-Mills theory (Callan, Dashen & Gross, 1978)).

We now compute the canonical partition function for this system. Using the methods of Gibbons & Perry (1978) we find that

$$Z \simeq \exp\left(\frac{V\pi^2}{45\beta^3}\right) \{1 + \tfrac{1}{2}i\left(\frac{m}{2\pi\beta}\right)^{3/2} V(\mu\beta)^{212/45} \exp\left(-\frac{\beta^2}{16\pi}\right) + \ldots\} \tag{26}$$

Each term in this expression needs some comment. The exponential factor outside the square brackets represents the effect of thermal gravitons on this system. Its contribution to the free energy is precisely that expected from a system of free, massless, spin-2 particles at a temperature $T = 1/\beta$. The first term in the square brackets comes from the flat space instanton. Its action is zero, and there are no zero modes to contribute further factors. The second term is comprised of four pieces, and comes from the Schwarzschild instanton. The action for this is $e^{-\beta^2/16\pi}$. The factor of $(\mu\beta)^{212/45}$ comes from the anomalous scaling behaviour. This contains a factor of $\mu$. Since the theory we are dealing with is unrenormalizable $\mu$ is essentially a cut-off parameter. It is a mass parameter and its value is usually thought of as roughly the Planck mass. The factors of $(m/2\pi\beta)^{3/2}V$ emerge from the zero modes and correspond to the one particle partition function of an object of mass m at temperature $1/\beta$ in a box of volume V. This clearly corresponds to a black hole constrained to move in such a volume. The factor of $\tfrac{1}{2}i$ comes from the presence of the negative mode in the spectrum of G.

We imagine drawing a diagram of I against M to show how there are two critical points in our system. There is a minimum at I = M = 0. The only other known extremum is that due to the Schwarzschild solution, which is a maximum. It is a maximum because of the presence of the negative mode of G. There is no potential barrier between the two. This is typical in general relativity where topology change is not inhibited by such barriers.

Given the comments concerning multi-black hole configurations, it seems reasonable that the full partition function is derived from the effect of many black holes distributed around the box, and that in the standard way, these exponentiate giving as a reasonable estimate of Z

$$Z \simeq \exp\left(\frac{V\pi^2}{45\beta^3}\right) \exp\left(\tfrac{1}{2}i\left(\frac{m}{2\pi\beta}\right)^{3/2} V\,(\mu\beta)^{212/45} \exp\left(-\frac{\beta^2}{16\pi}\right)\right) \tag{27}$$

In the canonical ensemble Z is to be interpreted as the exp(-F/T). For systems that contain unstable phases, the imaginary part of F is generically non-zero (see Langer, 1967; Coleman, 1977; Callan & Coleman, 1977; Banks, Bender & Wu, 1973; Banks & Bender, 1973).

The imaginary part of F essentially controls the rate at which thermal and quantum fluctuations are able to excite transitions into the unstable phase. In fact, one can relate this to the nucleation rate of regions of instability. Let S be the number of nucleations of unstable regions per unit time, then

$$\frac{S}{V} = \frac{|\lambda^{(-)}|\beta}{\pi} \, \mathrm{ImF} \tag{28}$$

Hence

$$\frac{S}{V} \sim \exp(-\frac{\beta^2}{16\pi}) \tag{29}$$

Of course, the constant of proportionality is unknown owing to the unrenormalizability of the theory.

After an instability has been nucleated we must ask how does it evolve. It is difficult to go into this given the limited amount of space. However, roughly the probability is ½ that it will decrease its mass, and ½ that it will increase. This unstable phase is presumably a black hole, the Lorentzian analytic continuation of the Schwarzschild instanton. Its behaviour will be roughly given by the semi-classical approximation. Thus, if its mass decreases

$$\dot{m} \sim -\pi^{-2}\, m^{-2}\, 9.5^{-1}\, 2^{-12} \tag{30}$$

or if its mass increases

$$\dot{m} \sim \pi^3\, m^2\, \beta^{-4}\, 3\; 10^{-1} \tag{31}$$

If the mass decreases, it will presumably go to zero, to the stable equilibrium. If it increases, since there are no further equilibria, it appears to increase without bound.

It is interesting to note that if we performed this calculation at zero temperature, there would only be a single instanton, which is flat space. There would be no instability, therefore we conclude that gravity undergoes a phase transition at zero temperature.

It is easy to relate this to what one would expect to happen on the basis of semi-classical intuitive notions. The canonical ensemble corresponds to a region of spacetime, enclosed by a box of volume V whose walls are at a fixed temperature T. One would expect such a region to be filled with thermal gravitons. However, every now and again, a large fluctuation would take place which causes a black hole to form. Such a black hole would be in unstable equilibrium (Perry, 1978). It could fluctuate a little so that its mass decreases. It would then be hotter than its surroundings, and would rapidly radiate itself away. Alternatively, it could fluctuate a little so that its mass increases. It would then be cooler than its surroundings and would accrete matter more and more rapidly.

Since in the canonical ensemble the amount of energy in a system is unrestricted, this process will continue until the box has been engulfed by the black hole. This is generally thought to mean that the canonical ensemble for gravity is badly defined, as has been confirmed by this explicit field theory computation.

Further details of this work can be found in Gross, Perry & Yaffe (1980). This research has been supported by the National Science Foundation under grant PHY78-01221.

REFERENCES

Banks, T., Bender, C. & Wu, T.T. (1973). Coupled Anharmonic Oscillators, I: Equal Mass Case. Phys.Rev.D, 8, pp.3346-3366.

Banks, T. & Bender, C. (1973). Coupled Anharmonic Oscillators II: Unequal Mass Case. Phys. Rev. D, 8, pp. 3366-3378.

Callan, C. & Coleman, S. (1977). Fate of the False Vacuum II: First Quantum Corrections. Phys. Rev. D, 16, pp.1762-1768.

Callan, C., Dashen, R. & Gross, D.J. (1978). Towards a theory of the Strong Interactions. Phys. Rev. D, 17, pp. 2717-2748.

Coleman, S. (1977). Fate of the False Vacuum: Semi-Classical Theory. Phys. Rev. D, 15, pp. 2929-2936.

Eguchi, T. & Hanson, A.J. (1978). Asymptotically Flat Self-Dual Solutions to Quantum Gravity. Phys. Lett., 74B, pp. 249-251.

Friedan, D. (1979). Ph.D. thesis, University of California at Berkeley, unpublished.

Gibbons, G.W. & Hawking, S.W. (1979). The Classification of Gravitational Instantons. Comm. Math. Phys., 66, pp. 291-310.

Gibbons, G.W., Hawking, S.W. & Perry, M.J. (1978). Path Integrals and the Indefiniteness of the Gravitational Action. Nucl. Phys., B138, pp. 141-150.

Gibbons, G.W. & Perry, M.J. (1978). Quantizing Gravitational Instantons. Nucl. Phys., B146, pp. 90-108.

Gibbons, G.W. & Perry, M.J. (1980). New Gravitational Instantons and their Interactions. Phys. Rev. D, in press.

Gross, D.J., Perry, M.J. & Yaffe, L. (1980). On the Instability of Quantum Gravity. Princeton preprint.

Hawking, S.W. (1978). Spacetime Foam. Nucl. Phys., B144, pp. 349-362.

Hawking, S.W. (1979). The Path Integral Approach to Quantum Gravity. In General Relativity: An Einstein Centenary Survey, eds. S.W. Hawking and W. Israel. Cambridge University Press.

Hawking, S.W. & Pope, C.N. (1978). Symmetry Breaking by Instantons in Supergravity. Nucl. Phys., B146, pp. 381-392.

Langer, J.S. (1967). Statistical Theory of the Decay of Metastable States. Ann. Phys. (N.Y.), 54, pp. 258-275.

Perry, M.J. (1978). Black Holes and Quantum Mechanics. Ph.D. dissertation, University of Cambridge, unpublished.

Perry, M.J. (1980a). Gravitational Instantons. In Proceedings of the Institute for Advanced Study Symposium on Differential Geometry, ed. S.-T. Yau, in press.

Perry, M.J. (1980b). The Decomposition Theorem and Feynman Diagrams in Quantum Gravity. In preparation.

Regge, T. & Wheeler, J.A. (1957). The Stability of a Schwarzschild Singularity. Phys. Rev., 108, pp. 1063-1069.

# STILL MORE COMPONENTS OF SUPERSPACE

U.Lindström and A.Karlhede
University of Stockholm

M.Roček
University of Cambridge

This is a report on a programme started by U.L.and M.R. (U. Lindström & M.Roček 1978,1979b,c) and its recent development in collaboration with A.K.The basic idea is to write supergravity in a superspace gauge where the relation to the component approach is particularly simple.

Superspace is convenient because supersymmetry is manifest, the notation is compact and operations straight-forward. Tensor calculus, e.g., just corresponds to multiplication of superfields. One can also, using the powerful differential geometry tools, immediately write down objects, such as covariant derivatives of superfields, which in the supergravity case have not yet been constructed in components. This is somewhat formal though; to make contact with physics as expressed in terms of fields over space-time, one has to make a gauge choice to remove the many extra degrees of freedom in superspace and interpret the x-space content. Thus the ideal situation, as we see it, is to go to a Wess-Zumino gauge where the component interpretation is immediate and then to stay in that gauge, still having access to all the superspace tools. This, however, requires us to work out a number of objects in our gauge and this is a report on how far that programme has proceeded.

Our conventions are as follows (for further details see Lindström and Roček (1978, 1979b)): Letters from the beginning of the alphabet are tangent space indices; letters from the middle or end are curved space indices. Gothic letters are superspace indices. Capital latin letters are spinor indices, primed or unprimed for left or right chirality, respectively; a pair of primed and unprimed spinor indices represents a single vector index.

The choice of gauge in superspace in which to identify the physical fields can be made in many ways and it is important to find a convenient one where expressions become simple. We use a Wess-Zumino gauge where the field content is exactly that of the component approach i.e: the graviton $e_{AA'}{}^{MM'}$, the gravitino $\psi_{AMM'}$, $\psi_{A'MM'}$ and the auxiliary fields $A_{MM'}$ and $\mathcal{S} \equiv S + iP$. We work in chiral gauges.In terms of coordinates these correspond to $(z,\theta)$ and $(z^{\dagger},\theta')$, where the complex coordinates z and $z^{\dagger}$ are related by a coordinate shift $z^{\dagger} = z + \Delta z$.We further require that a chiral multiplet $(A,B,\chi,F,G)$ coupled to supergravity takes on the form

$$\mathcal{A} + 2\theta_A\chi^A + \theta^2\mathcal{F} \tag{1}$$

and transforms into

$$\delta\mathcal{A} + 2\theta_A\,\delta\chi^A + \theta^2\delta\mathcal{F} \tag{2}$$

under the allowed coordinate transformations.Here $\mathcal{A}=A+iB$, $\mathcal{F}=F+iG$, and $\delta\mathcal{A}$,$\delta\chi$,$\delta\mathcal{F}$ are the familiar component field variations.This fixes our gauge completely,except for the tangent space Lorentz part. There are two particularly simple choices of Lorentz gauge:The diagonal gauge where the inverse vielbein obeys

$$E_{A'}^{\mathcal{M}} = \phi\,\delta_{A'}^{\mathcal{M}} \tag{3}$$

$\phi$ being a certain superfield,and the"N-gauge" where the connection coefficients $\Omega_{A'BC}$ vanish.

We have found it useful to adopt various short-hand notations for objects frequently encountered.In order to define them and at the same time present some background to the more recent development we here introduce them in the original setting.

In the chiral gauge of global supersymmetry the chiral projector,which is used to obtain a chiral superfield from a general scalar superfield,has the simple form:

$$\Pi = \frac{1}{4}\frac{\partial}{\partial\theta}{}^{A'}\frac{\partial}{\partial\theta_{A'}} \tag{4}$$

For the local case the chiral projector is in general;

$$\Pi = \frac{1}{4}(\mathcal{D}^{A'}\mathcal{D}_{A'} + \frac{1}{3}\mathcal{R}) \tag{5}$$

$\mathcal{R}$ being the Ricci scalar multiplet.In the chiral gauge this projector is as simple as possible in the local theory:

$$\Pi = \frac{1}{4}\frac{\partial}{\partial\theta}{}^{A'}\frac{\partial}{\partial\theta_{A'}}\phi^2 \tag{6}$$

The superfield $\phi$ which enters into this expression,as well as in the definition of the diagonal gauge,is one of the objects we find use for over and over again:

$$\phi^2 = 1 + \sqrt{2}\theta_A\psi_{A'}{}^{AA'} - \tfrac{1}{2}\theta^2(\psi_{A'MM'}\psi^{A'MM'} + \tfrac{4}{3}\mathcal{S}^\dagger)$$

$$+ \sqrt{2}\theta_A\theta_{A'}\rho^{AA'} - \theta^2\theta_{A'}\mu^{A'} + \tfrac{1}{3}\theta'^2\mathcal{R}$$

where

$$\rho^{MM'} = \omega^M{}_{B'}{}^{B'M'} + \tfrac{i}{3}A^{MM'} - \sqrt{2}\,\psi^{BMB'}\psi^{M'}{}_{BB'}$$

$$\mathcal{R} = \mathcal{S} + 2\theta_A\eta^A + \theta^2 r$$

$$\eta^A = \tfrac{\sqrt{2}}{2}\{R^{A'A}{}_{A'} - \psi^{A'A}{}_{A'}\mathcal{S} - \tfrac{i\sqrt{2}}{2}A_{BB'}\psi^{ABB'}\}$$

$$r = \tfrac{1}{2}R - \tfrac{2}{3}(\mathcal{S}\mathcal{S}^\dagger - \tfrac{1}{2}A_{BB'}A^{BB'}) + ie^{-1}\partial_{MM'}(eA^{MM'})$$

$$+ \sqrt{2}\psi^{B'}{}_{BB'}\eta^B - \tfrac{5}{4}\psi_{A'BB'}R^{A'BB'} + \tfrac{1}{4}\psi_{ABB'}R^{ABB'}$$

$$+ (\tfrac{1}{2}\psi_{A'BB'}\psi^{A'BB'} - \psi_{B'BA'}\psi^{A'BB'})\mathcal{S}$$

and

$$\mu^{A'} = -\tfrac{\sqrt{2}}{6}R_A{}^{AA'} - \tfrac{i}{3}A_{BB'}\psi^{B'BA'} + D^{BB'}\psi^{A'}{}_{BB'}$$

$$- \omega_{DD'B'}{}^{A'}\psi^{B'DD'} - \tfrac{\sqrt{2}}{2}\psi^{A'BB'}\psi_{B'}{}^{CC'}\psi_{BCC'} \tag{7}$$

Here $\omega^{\cdots\cdot}$ is the spin connection, $R^{\cdots}$ the spin 3/2 curvature, R the spin 2 curvature scalar and $D^{\cdot\cdot}$ the covariant derivative.

A further operator which is needed is the shift operator \$ corresponding to a complex coordinate transformation from antichiral to chiral coordinates $z^\dagger = z + \Delta z$. This operation is probably most easily understood in terms of the Ogievetsky-Sokatchev (1979) or Siegel-Gates (1979) pictures.

$$\$ = 1 + (2\sqrt{2}\theta^B\theta^{B'} - 2\theta^2\theta_{A'}\psi^{B'BA'} + 2\theta'^2\theta_A\psi^{BAB'}$$

$$- \theta^4 L^{\dagger BB'})\partial_{BB'} - \theta^4\,\partial^{MM'}\partial_{MM'}$$

where

$$- L^{\dagger BB'} = \omega^B{}_{A'}{}^{A'B'} + \Gamma^{BB'} + iA^{BB'} -$$

$$- \sqrt{2}(\psi^{A'}{}_{AA'}\psi^{BAB'} - \psi^{B'}{}_{AA'}\psi^{ABA'} - \tfrac{1}{2}\psi^{B'}{}_{AA'}\psi^{BAA'})$$

and $\Gamma^{MM'} = -\Gamma_{\nu}{}^{\nu\mu}$ (8)

$\Gamma$ being the space-time connection. In the Siegel-Gates formulation this operator reads exp(2iU), in the Ogievetsky-Sokatchev approach it is closely related to the axial field.

To further elucidate our notation we give the expression for the supergravity Lagrangian:

$$\mathcal{L}_{SG} = -\tfrac{1}{2}R + \tfrac{1}{2}(\psi_{A'BB'}R^{A'BB'} + \psi_{ABB'}R^{ABB'}) - \tfrac{1}{3}A^{2} - \tfrac{1}{3}\mathcal{S}\mathcal{S}^{\dagger} \quad (9)$$

The chiral density $\mathcal{E}$, first introduced in superspace by Siegel (Siegel & Gates 1979), subsequently in components by Ferrara and van Nieuwenhuizen (1978), the two approaches being tied together by Wess and Zumino (1978), in our notation reads:

$$e^{-1}\mathcal{E} = 1 - \sqrt{2}\theta_{A}\psi_{A'}{}^{AA'} - \tfrac{1}{2}\theta^{2}(\psi^{A'}{}_{AA'}\psi_{B'}{}^{AB'} - \psi_{A'BB'}\psi^{A'BB'} - 2\mathcal{S}^{\dagger}) \quad (10)$$

(note that the chirality is obvious since $\mathcal{D}_{A'} = \phi\frac{\partial}{\partial\theta^{A'}}$ in this gauge)

Knowing that $\phi^{2}\mathcal{E} = E$ (the graded determinant of the vielbein) holds in the chiral gauge, we derived (1979b):

$$e^{-1}E = 1 + \tfrac{1}{3}\theta^{2}\mathcal{S}^{\dagger} + \sqrt{2}\theta_{A}\theta_{A'}\rho^{AA'} + \tfrac{1}{3}\theta'^{2}\mathcal{S} + \theta^{2}\theta_{A'}(\rho^{AA'}\psi^{B'}{}_{AB'} -$$
$$-\mu^{A'}) - \theta'^{2}\theta_{A}(\tfrac{\sqrt{2}}{\sqrt{3}}\psi_{A'}{}^{AA'}\mathcal{S} - \tfrac{2}{3}\eta^{A}) + \theta^{4}\{-\tfrac{1}{3}\mathcal{L}_{SG} -$$
$$- \tfrac{1}{4}\psi_{A'BB'}R^{A'BB'} + \tfrac{1}{4}\psi_{ABB'}R^{ABB'} + \tfrac{i}{3}e^{-1}\partial_{MM'}(eA^{MM'})\} \quad (11)$$

The appearance of $\mathcal{L}_{SG}$ plus total divergence terms at the $\theta^{4}$ level makes the relation manifest between the superspace and the x-space Lagrangians.

Before discussing the construction of further superspace tools we briefly exemplify the usefulness of the above results.

Mattercouplings are easily constructed in superspace. One rapidly writes down expressions such as:

$$\mathcal{L} \sim \int d^4\theta E(e^{2gV}\Phi_- \$ \Phi_-^\dagger + e^{-2gV}\Phi_+ \$ \Phi_+^\dagger)$$

with $\Phi_\pm = \Phi_1 \pm i\Phi_2$ ($\Phi_i$ independent chiral fields) (12)

and then spends considerably more time abstracting the component content, but the important thing is that it can be done using the hitherto presented formulae. The field V is a general scalar multiplet in chiral gauge and has the ordinary component fields appearing as coefficients in the $\theta$-expansion (Lindström & Roček, 1979b). The power of the superspace method is particularly evident in evaluating expressions such as exp(V).

Another calculation where we have had use for our tools was when coupling a non linear realisation of supersymmetry to supergravity, which involved solving the constraints $\Phi\Phi = 0$, $\Phi = \Phi\Pi\$\Phi^\dagger$ for a chiral field $\Phi$ (Lindström & Roček 1979a).

The material presented so far has been published elsewhere but is needed to interpret the following recent development.

To be able to fully utilize the advantages of superspace we need some more objects, namely the inverse vielbein and some of the connection coefficients. In the diagonal gauge some components of the inverse vielbein look simple e.g.

$$E_{A'}{}^{U} = E_{A'}{}^{MM'} = 0 \,; \quad E_{A'}{}^{U'} = \phi\, \delta_{A'}^{U'} \tag{13}$$

$\phi$ is easily derivable from $\phi^2$ and is

$$\phi = 1 + \frac{\sqrt{2}}{2}\theta_A\psi_{A'}{}^{AA'} + \frac{\sqrt{2}}{2}\theta_A\theta_{A'}\rho^{AA'} + \theta^2\tau + \frac{1}{6}\theta'^2 \mathcal{S}$$
$$+ \theta^2\theta_{A'}\sigma^{A'} + \theta'^2\theta_A\beta^A + \theta^4 N$$

where

$$\tau = -\frac{1}{4}\psi_{A'}{}^{MM'}\psi^{A'}{}_{MM'} - \frac{1}{8}\psi_{A'}{}^{AA'}\psi_{B'A}{}^{B'} - \frac{1}{3}\mathcal{S}^\dagger$$

$$\sigma^{A'} = -\frac{1}{2}\mu^{A'} + \frac{1}{4}\psi_{B'}{}^{BB'}\rho_B{}^{A'}$$

$$\beta^A = \frac{1}{3}\eta^A - \frac{\sqrt{2}}{12}\psi_{B'}{}^{AB'}\mathcal{S}$$

and

$$N = \frac{1}{6}\mathcal{R} + \frac{\sqrt{2}}{12}\psi_{A'B}{}^{A'}{}_{B}\eta + \frac{1}{16}\rho_{BB'}\rho^{BB'} + (\frac{1}{24}\psi_{A'MM'}\psi^{A'MM'}$$
$$- \frac{1}{16}\psi_{A'B}{}^{A'}\psi_{B'}{}^{BB'} + \frac{1}{18}\mathcal{S}^{\dagger}\mathcal{S} \qquad (14)$$

Complex conjugating $E^{A'U'}$ and performing the coordinate transformation back to chiral gauge (shift) we obtain

$$E_A^{U'} = 0 \; ; \; E_A^{U} = \$\phi^{\dagger}\delta_A^U \quad ; \quad E_A^{MM'} = \$\{\phi^{\dagger}\frac{\partial}{\partial\theta}{}_{A}\Delta z^{\dagger MM'}\} \qquad (15)$$

Here

$$\$\phi^{\dagger} = \phi^{\dagger} - \theta'^2\theta_A\partial^{AA'}\psi_B{}^B{}_{A'} + \frac{1}{2}\theta^4\{\partial^{AA'}\rho^{\dagger}{}_{AA'} - \sqrt{2}(\partial^{AA'}\psi_C{}^{CC'})\psi_{A'AC'}\} \qquad (16)$$

and $E^{AMM'}$ is a bit more complicated.We will not give it here since we now want to switch to N-gauge,where the connection coefficients with indices A'BC are zero,and present our results there.The reason for doing this is as follows: Having derived the expressions (13) and (15) for some of the components of the inverse vielbein, the next thing to turn to is to derive $E_{AA'}$ from the formula

$$\{\mathcal{D}_{A'},\mathcal{D}_A\}\Lambda = -2\sqrt{2}\mathcal{D}_{AA'}\Lambda \qquad (17)$$

where the covariant derivatives are acting on a scalar field.This expression will, however, contain certain connection coefficients apart from the vielbeins.N-gauge is defined by the requirement that precisely those coefficients vanish.In that gauge we can therefore derive the missing components of the vielbein from

$$\tilde{E}_{AA'}{}^{\mathcal{M}}\partial_{\mathcal{M}} = -\frac{1}{2\sqrt{2}}\{\tilde{E}_A{}^{\mathcal{M}}\partial_{\mathcal{M}},\tilde{E}_{A'}{}^{\mathcal{N}}\partial_{\mathcal{N}}\} \; . \qquad (18)$$

The key to all this, of course, is to find the Lorentz rotation $N_A{}^B$ taking us from diagonal to N-gauge.There are systematic ways of doing this, but the route we took is a nonsystematic short-cut.We used the existence,in component form, of the field-strength multiplet constructed by Ferrara, Grisaru and van Nieuwenhuizen (1978) and equated its superspace form in our chiral gauge to the projector on the covariant derivative of a general scalar superfield.

$$W_A = \Pi\mathcal{D}_A V \qquad (19)$$

This equation we interpreted in N-gauge which means no connections on either side. Denoting the Lorentz transformation from diagonal to N-gauge by $N_{AB}$ we have (~ denotes N-gauge quantities):

$$\tilde{W}_A = \pi\tilde{\mathcal{D}}_A V = \pi N_A{}^B \mathcal{D}_B V = \pi N_A{}^B (E_B{}^U \partial_U - E_B{}^{MM'} \partial_{MM'})V \qquad (20)$$

Since $W^A$ was only known for the fields of conformal supergravity all expressions in (20) have $\mathcal{S} = 0$. The only unknown quantity in this equation is $N^{\cdot\cdot}$ and it can be evaluated order by order in $\theta$. As a check on our algebra we used the fact that $N^{\cdot\cdot}$ has to have determinant equal to one. With $N_{AB} = \varepsilon_{AB} + X_{AB}$ this implies

$$X_A{}^A = -\tfrac{1}{2}X_{AB}X^{AB} \qquad (21)$$

a fact which we have verified to all orders in $\theta$. The expression for $N^{\cdot\cdot}$ that we find can, with a lot of pain and hard work, be written as

$$N_{AB} = (e^{-1}E\$\phi^{+})^{-1}\{\varepsilon_{AB} + \theta_{C'}h_{AB}{}^{C'}\} \qquad (22)$$

Here we have introduced the chiral field

$$\begin{aligned} h_{AB}{}^{C'} = \sqrt{2}\{&-\psi_{BA}{}^{C'} + \theta_C(\omega^{C'}{}_{AB}{}^{C} - iA_B{}^{C'}\delta_A^C - \varepsilon_{AB}\omega^C{}_{D'}{}^{D'C'} \\ &- \sqrt{2}\psi_{D'D}{}^{C'}\psi_B{}^{DD'}\delta_A^C - \sqrt{2}\varepsilon_{AB}\psi^{C'}{}_{DD'}\psi^{DCD'}) \\ &+ \theta^2(-2D_{AD'}\psi^{C'}{}_B{}^{D'} + 2iA_{AD'}\psi^{C'}{}_B{}^{D'} + iA_{BD'}\psi^{C'}{}_A{}^{D'} \\ &+ \omega_{DD'AB}\psi^{D'DC'} + \varepsilon_{AB}\omega_{DE'}{}^{C'}{}_{D'}\psi^{E'DD'} \\ &- 4S_A{}^{D'}{}_{BE'}{}^{DE'}\psi^{C'}{}_{DD'})\} \end{aligned} \qquad (23)$$

where the torsion is

$$S_{MM'NN'}{}^{AA'} = \frac{\sqrt{2}}{4}\{\psi^A{}_{MM'}\psi^{A'}{}_{NN'} + \psi^{A'}{}_{MM'}\psi^A{}_{NN'}\} \qquad (24)$$

We can now write down the inverse vielbein to all orders in $\theta$ in N-gauge (or in diagonal gauge using $N_{\bar{A}\bar{B}}^{-1} = -N_{(AB)} + N_{[AB]}$ ).Since we have worked in a conformal gauge we have restored the $\mathcal{S}$-dependence at the end using the appropriate Howe-Tucker transformations (P.Howe & R.Tucker 1978,M.Roček & U.Lindström 1979b).The non zero components are (p denotes the Poincaré entities):

$$\overset{p}{\tilde{E}}{}_A{}^U = \exp\{ \tfrac{1}{6}\theta^2\mathcal{S}^\dagger - \tfrac{1}{3}\theta'^2\mathcal{S} \}(eE^{-1})\{\delta_A^U + \theta_{C'}h_A{}^{UC'}\} \tag{25}$$

$$\overset{p}{\tilde{E}}{}_A{}^{MM'} = \exp\{ \tfrac{1}{6}\theta^2\mathcal{S}^\dagger - \tfrac{1}{3}\theta'^2\mathcal{S} \}2\sqrt{2}(eE^{-1})\times$$

$$\times\theta^{B'}(e_{AB'}{}^{MM'} + \sqrt{2}\theta_A\psi^{M'M}{}_{B'}) \tag{26}$$

$$\overset{p}{\tilde{E}}{}_{A'}{}^{U'} = \$(\overset{p}{\tilde{E}}{}_A{}^U)^\dagger =$$

$$= \exp\{ \tfrac{1}{6}\theta'^2\mathcal{S} - \tfrac{1}{3}\theta^2\mathcal{S}^\dagger\}(eE^{\dagger-1})\times$$

$$\times[\delta_{A'}^{U'} + \theta_C h_{A'}{}^{U'C} - 2\theta^2\theta_{D'}\partial^{DD'}\psi^{U'}{}_{DA'}$$

$$- \theta^4\{-\sqrt{2}(\partial_{DD'}\psi_{B'AA'})\psi^{DAD'} + \partial_{DD'}(\omega^D{}_{A'}{}^{U'D'}$$

$$+ iA^{DU'}\delta_{A'}^{D'} - \tfrac{i}{3}\delta_{A'}^{U'}A^{DD'} - \sqrt{2}\psi^{EDE'}\psi^{U'}{}_{EE'}\delta_{A'}^{D'})] \tag{27}$$

$$\overset{p}{\tilde{E}}{}_{AA'}{}^U = -\tfrac{1}{2\sqrt{2}}\exp\{-\tfrac{1}{6}\theta^2\mathcal{S}^\dagger - \tfrac{1}{6}\theta'^2\mathcal{S}\}(eE^{-1})\tilde{E}_{A'}{}^{U'}\times$$

$$\times\{[-\sqrt{2}\theta_A\rho^A{}_{U'} + \tfrac{4}{3}\theta_{U'}\theta_A\eta^A - \theta^2(\mu_{U'} - \psi_{B'}{}^{BB'}\rho_{BU'})$$

$$+ \tfrac{2}{3}\theta_{U'}\mathcal{S} + 4\theta^2\theta_{U'}(2N - \tfrac{1}{6}\mathcal{R})]\times[\delta_A^U + \theta_{C'}h_A{}^{UC'}]$$

$$- h_A{}^U{}_{U'}\} \tag{28}$$

$$\overset{p}{\tilde{E}}{}_{AA'}{}^{U'} = -\tfrac{1}{2\sqrt{2}}(\overset{p}{\tilde{E}}{}_A{}^U\partial_U\overset{p}{\tilde{E}}{}_{A'}{}^{U'} - \overset{p}{\tilde{E}}{}_A{}^{MM'}\partial_{MM'}\overset{p}{\tilde{E}}{}_{A'}{}^{U'}) \tag{29}$$

$$\overset{p}{\tilde{E}}_{AA'}{}^{MM'} = - \exp\{\tfrac{1}{6}\mathcal{S}\theta'^2 - \tfrac{1}{6}\mathcal{S}^\dagger\theta^2\}(eE^{-1})\tilde{E}_{A'}{}^{U'} \times$$
$$\times\{\delta_{U'}^{D'} + \sqrt{2}\theta^D\theta^{D'}\rho_{DU'} + \tfrac{2}{3}\theta'^2\theta_D\delta_{U'}^{D'}\eta^D$$
$$- \theta^2\theta^{D'}(\psi_{B'}{}^{BB'}\rho_{BU'} - \mu_{U'}) + 2\theta^4\delta_{U'}^{D'}(2N - \tfrac{1}{6}\mathcal{h})\}\times$$
$$\times(e_{AD'}{}^{MM'} + \sqrt{2}\theta_A\psi^{M'M}{}_{D'}) \qquad (30)$$

If one would like to have everything multiplied by powers of E (no $E^\dagger$) the following formula can be applied:

$$\frac{E}{E^\dagger} = 1 + \sqrt{2}\theta_A\theta_{A'}(\rho^{\dagger AA'} + \rho^{AA'})$$
$$+ \theta^2\theta_{A'}(- \tfrac{2}{3}\eta^{A'} + \rho^{AA'}\psi^{B'}{}_{AB'} - \mu^{A'})$$
$$+ \theta'^2\theta_A(+ \tfrac{2}{3}\eta^A - \rho^{\dagger AA'}\psi^B{}_{BA'} + \mu^A)$$
$$+ \tfrac{1}{2}\theta^4(\psi_{AMM'}R^{AMM'} - \psi_{A'MM'}R^{A'MM'} + \tfrac{4}{3}ie^{-1}\partial_{MM'}(eA^{MM'})$$
$$- \rho_{AA'}\rho^{\dagger AA'} - \rho^\dagger{}_{AA'}\rho^{\dagger AA'}) \qquad (31)$$

We have also calculated some important connection coefficients (in diagonal gauge):

$$\Omega_{A'B'C'} = 4\varepsilon_{A'(B'}\partial_{C')}\phi =$$
$$= 4\varepsilon_{A'(\underline{B}'}\{\tfrac{\sqrt{2}}{2}\theta_D\rho^D{}_{\underline{C}')} - \tfrac{1}{3}\theta_{\underline{C}')}\mathcal{S} + 2\theta_D\theta_{\underline{C}')}(\tfrac{1}{3}\eta^D + \tfrac{\sqrt{2}}{12}\psi_{D'}{}^{DD'}\mathcal{S})$$
$$+ \theta^2(\tfrac{1}{2}\mu_{\underline{C}')} - \tfrac{1}{4}\psi_{D'}{}^{DD'}\rho_{D\underline{C}')}) + \theta^2\theta_{\underline{C}')}[\tfrac{1}{3}\mathcal{L}_{SG} -$$
$$-\tfrac{1}{4}\psi_{EDD'}R^{EDD'} + \tfrac{1}{4}\psi_{E'DD'}R^{E'DD'} - \tfrac{i}{3}e^{-1}\partial_{MM'}(eA^{MM'})$$
$$+ \tfrac{\sqrt{2}}{6}\psi_{D'D}{}^{D'}\eta^D - \tfrac{1}{8}\rho_{DD'}\rho^{DD'} - \tfrac{1}{12}\psi_{E'DD'}\psi^{E'DD'}\mathcal{S} - \tfrac{7}{24}\psi_{D'D}{}^{D'}\psi_{E'}{}^{DE'}\mathcal{S}$$
$$]\} \qquad (32)$$

$$\Omega_{B'BC} = 2(e^{-1}E\$\phi^{\dagger})^{-1}\phi\ N^{A}{}_{(\underline{B}}h_{A\underline{C})B'} \tag{33}$$

These connection coefficients and their shifted complex conjugations allow us to construct, in diagonal gauge, $\mathcal{D}_A.\mathcal{D}_B$ V where V is a general scalar superfield. The transformations are given by the coordinate transformations $\delta\theta^{\cdot}$, $\$(\delta\theta^{\cdot})^{\dagger}$, $\delta z$ (see Lindström and Roček 1979b) and the Lorentz transformation:

$$-\partial_{(A'}\delta\theta^{B')} + \$(c.c) \tag{34}$$

In particular, we can construct counterterms explicitly by forming $\mathcal{D}_A W_{BCD}$, W... being the Weyl multiplet, and taking it to some power.

The relative simplicity of the inverse vielbein (25)-(30) indicates that the N-gauge is a good choice. We also feel that $E.^{\cdot}$ can be further simplified once we understand its structure, notably the rôle of h..., in more detail.

REFERENCES

Ferrara, S., Grisaru, M.& van Nieuwenhuizen, P. (1978). Poincaré and conformal supergravity models with closed algebras. Nucl. Phys.B 138, pp 430-44.

Ferrara, S. & van Nieuwenhuizen, P. (1978). Tensor calculus for supergravity. Phys.Lett. 76B, pp 404-8.

Howe, P. & Tucker, R. (1978). Scale invariance in superspace. Phys. Lett. 80B, pp 138-40.

Lindström, U. & Roček, M. (1978). Components of superspace, Phys.Lett. 79B, 217-8; (1979a). Constrained local superfields. Phys.Rev.D:8, pp 2300-3; (1979b). More components of superspace. Phys.Lett. 83B, pp 179-84; (1979c). In Supergravity North Holland, Amsterdam.

Ogievetsky, V.& Sokatchev, E. (1979). The simplest group of Einstein supergravity. Dubna preprint E2-12469.

Siegel, W. & Gates, J. (1979). Superfield supergravity. Nucl.Phys. B 147, pp77-104.

Wess, J. & Zumino, B. (1978). The component formalism follows from the superspace formulation of supergravity. Phys.Lett. 79B, pp 394-8.

# LATTICES, GRAVITY, AND SUPERGRAVITY

Michio Kaku

City College of the
City University of New York

In this review article, we will summarize some early work on lattice gravity (Das, Kaku & Townsend, 1979; Smolin, 1979) as well as new work in progress on the lattice gauge theory of supergravity.

Yang-Mills theory can be viewed as the gauge theory of the unitary groups. In the same fashion, Einstein gravity can be viewed as the gauge theory of the Poincaré group, Weyl gravity can be viewed as the gauge theory of the conformal group SU(2,2) (Kaku, Townsend & van Nieuwenhuizen, 1977a), supergravity as the gauge theory of the graded Poincaré group, and superconformal gravity as the gauge theory of the graded conformal group (Kaku, Townsend & van Nieuwenhuizen, 1977b). In principle, all these theories should have a lattice gauge formulation much like Yang-Mills theory.

There are several physical motivations for constructing the lattice gauge theory for gravity:

(1) it provides a convenient way of regularizing the theory which is gauge invariant. Because the theory is gauge invariant independent of the lattice spacing, we can choose the inverse of the lattice spacing to be the cut-off for the theory. In the limit of infinite cut-off, we get back the original continuum theory.

(2) it provides a new approach to the long-standing problem of renormalization. Because we can now sum an entirely new perturbation theory, we expect to see the non-renormalizability of gravity emerge in a new way.

(3) because encouraging numerical results have been obtained recently for Yang-Mills theory using Monte Carlo methods, we expect that similar numerical techniques may also be used for gravity as well.

(4) the techniques developed for Yang-Mills theory on a lattice may eventually be carried over to gravity, esp. the recursion

relations of Migdal (1975a & b) and analogies between D dimensional gauge systems and D/2 dimensional nearest neighbor systems.

(5) because the goal of supergravity is to provide a coherent framework for all particle interactions, and since the electromagnetic, weak, and strong interactions can be expressed in terms of lattice gauge theory, we suspect that gravity, too, must fit into this unified picture.

We must also mention that this formulation of lattice gravity, as opposed to the Regge theory, loses the original geometric picture which motivated Einstein's beautiful work. General covariance, instead, is now reduced to a gauge symmetry on the fields rather than a co-ordinate transformation. Also, in this article we will not attempt to clarify certain ambiguities that are carried over from the usual lattice theory, e.g. how to define spinors on lattices.

DE SITTER GRAVITY

The original problem of constructing a lattice gauge theory of gravity stemmed from difficulties in interpreting the analogue of the Yang-Mills string element:

$$U(a,b) = T \exp \{ ig \int_a^b A_\mu^{\ i} G_i \, dx^\mu \}$$

The problem is that even if the group element $G_i$ is a generator of the Poincaré group or the de Sitter group, the theory still will not be invariant under the local extentions of translations. Local translations may eventually be replaced by general co-ordinate transformations, but this still does not gauge the local translations as part of the lattice gauge group.

The solution to the problem can be found by re-formulating gravity in a new way due to West (1978), which is to write down an action which preserves the de Sitter group in a linear fashion, at the expense of producing a theory non-linear in the curvatures. The West action is ideally suited to putting the theory on a lattice.

We start by writing down the 10 generators of the de Sitter group, with 6 Lorentz rotations and 4 translations. Notice that in the limit of infinite de Sitter radius, this algebra reduces to the usual Poincaré algebra.

$$\{M_{ab}, M_{cd}\}_- = i ( \eta_{bc} M_{ad} + \dots )$$

$$\{M_{ab}, P_c\}_- = i ( \eta_{bc} P_a - \eta_{ac} P_b )$$

$$\{P_a, P_b\}_- = (-i/\lambda) M_{ab}$$

Now we may construct the string elements of the theory in the usual way, where we multiply group generators by gauge fields and where "a" represents the lattice spacing:

$$U_{n,\mu} = \exp i \{e^n_{a\mu} P^a + \tfrac{1}{2} \omega^n_{ab\mu} M^{ab}\} a$$

The gauge potentials are labelled by n, the lattice site. This gives an Sp(4) covariant string element:

$$U_{n,\mu} \rightarrow V_n U_{n,\mu} V^{-1}_{n+\mu} \quad ; \qquad V_n \in Sp(4)$$

where $n + \mu$ indicates the site adjacent to n in the $\mu$ direction. We can now construct the usual quantity:

$$U_{n,\mu\nu} = U_{n,\mu} U_{n+\mu,\nu} U^{-1}_{n+\nu,\mu} U^{-1}_{n,\nu} - 1$$

(no summations), which transforms as:

$$U_{n,\mu\nu} \rightarrow V_n U_{n,\mu\nu} V^{-1}_n$$

At this point, there are an infinite number of actions which are invariant under the linear realization of Sp(4). Our only guide is that they must reduce to the West action in the limit of zero lattice spacing. We are then let to construct:

$$U_n = \sum_{\mu\nu\rho\sigma} U_{n,\mu\nu} U_{n,\rho\sigma} \varepsilon^{\mu\nu\rho\sigma}$$

Since we want the correct parity, we square this object and then take its square root:

$$\tau_n^2 = (U_n - \tfrac{1}{4} \operatorname{Tr} U_n)^2$$

$$I \sim \sum_n (\operatorname{Tr} \tau_n^{\,2})^{\frac{1}{2}}$$

This is our final result. In order to show that this action indeed reduces to the West action and finally to the Einstein action, we now take the continuum limit. In the usual way, we find that our string elements can be expressed eventually in terms of curvatures:

$$U_{n,\mu\nu} \rightarrow ia^2 R_{\mu\nu}$$

$$U_n \rightarrow -a^4 R_{\mu\nu} R^{\rho\sigma} \varepsilon^{\mu\nu\rho\sigma}$$

where:

$$R_{\mu\nu a}(P) = \partial_\nu e_{a\mu} - \partial_\mu e_{a\nu} + \omega_{\nu ab} e_{b\mu} - \omega_{\mu ab} e_{b\nu}$$

$$R_{\mu\nu ab}(M) = \partial_\nu \omega_{\mu ab} - \omega_{\nu ac} \omega_\mu{}^c{}_b + e_{a\mu} e_{b\nu}/\lambda - (\mu \leftrightarrow \nu)$$

Substituting we arrive at the continuum action:

$$I \sim \int d^4x \; \{ (1/16) (\varepsilon^{\mu\nu\rho\sigma} \varepsilon^{abcd} R_{\mu\nu ab}(M) R_{\rho\sigma cd}(M))^2 + (1/\lambda) (\varepsilon^{\mu\nu\rho\sigma} R_{\mu\nu a}(P) R_{\rho\sigma bc}(M) \varepsilon^{abcd})^2 \}$$

which is the West action. Now let us break local translations by imposing four gauge constraints:

$$G_d = R_{\mu\nu a}(P) R_{\rho\sigma bc}(M) \varepsilon^{\mu\nu\rho\sigma} \varepsilon^{abcd} = 0$$

This in turn breaks down our original action to the usual one:

$$I \sim \int d^4x \; \varepsilon^{\mu\nu\rho\sigma} \varepsilon^{abcd} R_{\mu\nu ab}(M) R_{\rho\sigma cd}(M)$$

Smolin (1979) takes the above expression (due to Macdowell-Mansouri) which is locally Lorentz invariant but not invariant under gauge translations and writes down a lattice gauge action as:

$$I \sim \sum_n \mathrm{Tr} \{\gamma_5 U_{n,\mu\nu} U_{n,\rho\sigma} \varepsilon^{\mu\nu\rho\sigma} \}$$

We still have one step left, and that is to show that the four gauge constraints which we imposed to break local translations are compatible with the 24 constraints that one usually imposes to eliminate the connection fields. To show the quantum equivalence, we start with the functional integral for gravity:

$$Z = \int \prod_{\substack{a>b \\ \mu}} D\omega_{\mu ab} \prod_{\substack{a \\ \nu}} De_{a\nu} \; \delta(\chi) \prod_{d} \delta(G_d) \; \Delta_{FP}$$

$$\times \quad \exp \{ i \int d^4x \; (\lambda/16) \; R_{\mu\nu ab}(M) \; R_{\rho\sigma cd}(M) \; \varepsilon^{\mu\nu\rho\sigma} \; \varepsilon^{abcd} \}$$

where $\chi$ are the Lorentz and general co-ordinate gauge fixing conditions and $\Delta_{FP}$ is the Faddeev-Popov determinant. If we separate out the connection field into two parts, one dependent on the classical part and the other the quantum correction, then we can systematically eliminate this latter quantum correction. We write:

$$\omega_{\mu ab} = \omega^{(o)}_{\mu ab}(e) + \rho_{\mu ab}$$

where $\omega^{(o)}_{\mu ab}$ is the classical value of the connection field determined by $R_{\mu\nu a}(P)=0$ We re-express the four gauge constraints breaking local translations by a Lagrange multiplier $f_d$:

$$I \rightarrow I_{\text{Einstein}} + i \int d^4x \; (\rho_{\mu ab} \; \rho^{a\mu b}$$

$$+ f_d \; R_{\mu\nu ab}(M) \; \rho_{\beta c\alpha} \; \varepsilon^{\mu\nu\alpha\beta} \; \varepsilon^{abcd} \; )$$

Now, because terms proportional to $\delta^4(0)$ may be eliminated in dimensional regularization, we can eliminate the determinant:

$$\Delta_{FP}(G_d) = \det \| \; R_{\mu\nu ab} \; R_{\alpha\beta ce} \; \varepsilon^{\mu\nu\alpha\beta} \; \varepsilon^{abcd} \; \delta^4(x-y)$$

$$- \; (2/\lambda) \; R_{\mu\nu a} \; R_{\alpha\beta c} \; \varepsilon^{\mu\nu\alpha\beta} \; \varepsilon^{abcd} \; \delta^4(x-y) \quad \|_{e,d;x,y}$$

We can now eliminate all the unwanted quantum corrections due to these extra fields. If we eliminate the field $\rho_{\mu ab}$, then we are left with an action quadratic in $f_d$. And the equation of motion for this field is just $f_d = 0$ because there are no linear terms. Thus, we have eliminated all unwanted fields, and we are left with the usual Einstein action.

## SUPERGRAVITY

The situation for supergravity is more involved. We can, of course, easily extend the algebra to include spinorial transformations, and likewise define a new string element:

$$U_n = \exp ia \{ e^n_{a\mu} P^a + \tfrac{1}{2} \omega^n_{ab\mu} M^{ab} + \bar{\psi}^{n\alpha}_\mu Q_\alpha \}$$

$$Q_\alpha = \left| \begin{array}{c|c} & C_{\alpha i} \\ \hline \delta_{\alpha i} & \end{array} \right|$$

The problem occurs when one tries to use the previous results to write down the action for supergravity. Again, there are an infinite number of actions one can write down which transform linearly according to the graded Poincare group. The problem is that apparently none of them reduce down to the usual supergravity theory, i.e. none linearizes down to the usual supergravity action when one imposes four gauge conditions to break local translations.

At this point, there are no fully satisfactory solutions to this problem, but one way out of the puzzle is to use auxiliary fields. In West's original work, he also introduced a new field $c(x)$ as follows:

$$C(x) = \gamma_5 P(x) + \gamma_5 \gamma_a A^a(x)$$

Notice that $c(x)$ transforms nicely under Sp(4) when one uses Dirac matrices to represent the 10 generators of the group. The following action then is Sp(4) invariant.

$$I \sim \int d^4x \; \mathrm{Tr} \{ C(x) R_{\mu\nu ab}(M) R_{\rho\sigma cd}(M) \varepsilon^{\mu\nu\rho\sigma} \varepsilon^{abcd}$$

$$+ \rho(C(x)^2 - a^2) \} \qquad a = \text{const.}$$

Eliminating these auxiliary fields in any order yields the original square-root action. Or, one may impose four constraints on the theory $(A_d = 0)$ and reduce down to the broken theory. Now, defining this auxiliary field at every point in the lattice space, we can now write down an equivalent lattice action:

$$I \sim \sum_n \mathrm{Tr} \{ C_n U_{n,\mu\nu} U_{n,\rho\sigma} \varepsilon^{\mu\nu\rho\sigma} \varepsilon^{abcd}$$

$$+ \rho_n (C_n^2 - a^2) \}$$

In the same fashion, one may now introduce an auxillary field defined on the expanded algebra:

$$C(x) = \left| \begin{array}{c|c} \gamma_5 P(x) + \gamma_5 \gamma_a A^a(x) + 1 S(x) & \bar{\phi}^\alpha C_{\alpha i} \\ \hline \bar{\phi}^\alpha \delta_{\alpha i} & -4 S(x) \end{array} \right|$$

West and Chemseddine (1977) then write down actions like:

$$I \sim \Sigma \, \mathrm{Tr} \{ C_n U_{n,\mu\nu} U_{n,\rho\sigma} + A C_n U_{n,\mu\nu} C_n^2 U_{n,\rho\sigma} \} \varepsilon^{\mu\nu\rho\sigma}$$

Let us now try to eliminate the redundant fields. First, by additional constraints on the trace of the square and the cube c(x) matrix, one can eliminate S and P=const. Finally, we now impose four constraints which break local translations: $A_a = 0$ $\phi^\alpha = 0$ and four constraints which break supersymmetry We are now left with the usual supersymmetric action if we now set the value of A to be the proper constant.

The problem with this approach is that we imposed four fermionic constraints which broke supersymmetry, but now we find that supersymmetry re-emerges if we impose the right value of A. Apparently, our original action was actually invariant under eight fermionic symmetries. Macdowell has conjectured that our original was not only linearly invariant under ordinary supersymmetric transformations, but was actually non-linearly invariant under the complete superconformal group, which would account for the extra four invariances. Imposing various trace conditions, then, would amount to breaking scale invariance, chiral invariance, etc. This is still an open question.

RENORMALIZATION AND RECURSION RELATIONS

One of the most elegant results from lattice gauge theories is the equivalence between various two dimensional gauge theories and one dimensional nearest-neighbor interactions, and the analogy between four dimensions and two dimensions. For example, it is not hard to show the exact equivalence between:

| One dimensional nearest-neighbor | Two dimensional gauge |
|---|---|
| Ising model | $Z_2$ gauge theory |
| O(2) Heisenberg model (XY model) | U(1) gauge theory |
| O(4) Heisenberg model | SU(2) gauge theory |

The obvious question is whether gravity and supergravity, as gauge theory of the Poincaré and graded Poincaré groups, also have corresponding relationships between D and D/2 dimensional theories which are exact when D = 2 but only approximate when D = 4. This would be especially helpful when trying to analyze qualitative features of the theory, like phase transitions.

It is easy to show the equivalence between a two dimensional gauge theory and one dimensional nearest-neighbor theory. Consider trace over one plaquette in lattice gauge theory:

$$\mathrm{Tr}\,\{\, U_{n,\mu}\, U_{n+\mu,\nu}\, U^{-1}_{n+\nu,\mu}\, U^{-1}_{n,\nu} \,\}$$

where we eventually sum over all lattice sites. In two dimensions, we can always choose a gauge such that the gauge field vanishes in one space-time direction, so that our original trace now reads:

$$\mathrm{Tr}'\,\{\, U_{n,\mu}\, U^{-1}_{n+\nu,\mu} \,\}$$

For SU(2), for example, we may re-define our variables to read:

$$U = S_0 + S_i\,\tau_i, \qquad U\,U^{\dagger} = 1 \qquad S_0^{\,2} + S_i^{\,2} \doteq 1$$

$$\Sigma\,\mathrm{Tr}\,\{\, U_{n,\mu}\, U^{-1}_{n+\nu,\mu} \,\} = \sum_i \vec{S}_i\, \vec{S}_{i+1}$$

where we now have the usual O(4) nearest-neighbor Heisenberg model in just one dimension (actually an infinite series of just one dimensional Heisenberg models).

We may apply the same techniques to gravity in two dimensions (which reduces to a trivial surface term in the limit of infinite de Sitter radius). Consider the Poincaré group and its curvatures in just two dimensions:

$$\{M_{01},\, P_1\}_- = i\,P_0, \qquad \{M_{01},\, P_0\}_- = -i\,P_1$$

$$\{P_0,\, P_1\}_- = -i\,M_{01}/\lambda$$

$$R^a_{01}(P) = \partial_1 e^a{}_0 - \partial_0 e^a{}_1 + \omega^{ab}_0 e^b{}_1 - \omega^{ab}_1 e^b{}_0$$

$$R^{01}_{01}(M) = \partial_1 \omega^{01}_0 + \lambda^{-1} e^0{}_0 e^1{}_1 - \partial_0 \omega^{01}_1 - \lambda^{-1} e^0{}_1 e^1{}_0$$

$$U_{n,o1} \rightarrow ia^2 R_{01}$$

We can write down a two dimensional action as follows:

$$I \sim \sum_n \mathrm{Tr}\, \{U_{n,\mu\nu}\, \varepsilon^{\mu\nu}\}^{\frac{1}{2}}$$

In the continuum limit, we have:

$$I \sim \int d^4x \;\; \{\mathrm{Tr}\;\; (R_{\mu\nu}\, \varepsilon^{\mu\nu})^2\, \}^{\frac{1}{2}}$$

Now we can impose two gauge conditions which eliminate the connection in terms of the usual gauge fields:

$$R^a_{o1}(P) = 0 \rightarrow \omega^{o1}_\mu(e)$$

Notice that now we have precisely the right number of constraints to eliminate all connections. Our continuum action now reduces to:

$$I \sim \int d^4x \;\; R^{o1}_{o1}(M)$$

which in the limit of infinite de Sitter radius becomes a total derivative.

Now, instead impose 3 new conditions on the original lattice action which eliminates one entire direction of space-time:

$$e^a{}_o = 0, \;\; \omega^{o1}_o = 0$$

Using the same techniques as before, we can now show the equivalence of gravity with the following nearest-neighbor action:

$$I \sim \sum_i (S_i\; S_{i+1})^{\frac{1}{2}}$$

Unfortunately, the exact equivalence relations between gauge theories in D dimensions and nearest-neighbor theories in D/2 dimensions for D = 2 do not carry over exactly to the case of D = 4. Migdal (1975 a & b) has shown that there exist certain "analogies" between four dimensional gauge systems and two dimensional nearest-neighbor interactions, but the precise nature of this approximation is not known, nor is it possible to write down a well-defined perturbation which takes us from one theory in four dimensions and gives us the corresponding theory in two dimensions. Nevertheless, certain qualitative features in four dimensions (e.g. asymptotic freedom for SU(2) theories) seem to correspond nicely to certain features in two dimensions (zero temperature singularities for the O(4) Heisenberg model).

The basis for writing down these analogies are the recursion relations, which can most simply be derived by the process of decimations. In the ordinary Ising model, for example, we can change the distance between points by simply summing over one of the field variables:

$$\sum_{\sigma_2} e^{k\sigma_1\sigma_2} e^{k\sigma_2\sigma_3} = e^{k'\sigma_1\sigma_3} \qquad \tanh k' = \tanh^2 k$$

In this way, we have established a relationship between the coupling constant with one spacing and the coupling constant defined with another spacing. This decimation process, however, does not apply exactly to the four dimensional case, because one link in four dimensions appears in three separate plaquettes, so we have non-trivial interactions between plaquettes as we decimate. The solution to this problem is to simply shift plaquettes around the lattice one at a time, keeping certain constraints fixed; decimations can be applied in all four directions. It's not known how precise these approximations are, though they are probably reliable in the limit of weak and strong coupling.

Previously, we saw in the two dimensional case that a simple gauge choice was enough to prove an exact equivalence relation with a one-dimensional system. We may ask another question: how exact are these "analogies" between four and two dimensional systems? Unfortunately, comparing the four dimensional partition function with the two dimensional case, we find that a simple gauge choice is not sufficient to prove an equivalence. However, if we make decimations in all four directions in both the gauge system in four dimensions and the nearest-neighbor system in two dimensions, then we find that they have identical recursion relations.

Unfortunately, the situation for gravity is much more complex. If a simple two dimensional nearest-neighbor analogy exists for four dimensional gravity, then phase transitions and other qualitative features may be studied in the strong and weak coupling limit, where we expect the approximations to be good. The problem is that we cannot decimate over lattice sites in gravity. The process of functionally integrating out intermediate gauge fields does not leave the action with the same over-all structure. New terms are generated, of a higher order, as we integrate over intermediate gauge fields, leading us to suspect that gravity is not renormalizable.

This is disappointing, because it means that the recursion relations for gravity are not very reliable because of higher order corrections.

The reason for this problem is not hard to find. It stems from the fact that our action contains not just the simple trace over string elements in the xy, xz, xt, etc. planes, but instead traces over dual plaquettes, i.e. traces over two loops in the xy and zt planes, in the xz and yt planes, and xt and yz planes. This is due to the characteristics of the fully anti-symmetric tensor.

In the usual gauge theory for unitary groups, we simply decimate over interior variables to obtain new plaquettes with double the original spacing (see fig. 1). But in the gravitational case, we must now decimate over dual plaquettes, which leave behind combinations of plaquettes which are <u>not</u> part of the original action, which is characteristic of non-renormalizable theories (see fig. 2).

Fig. 1

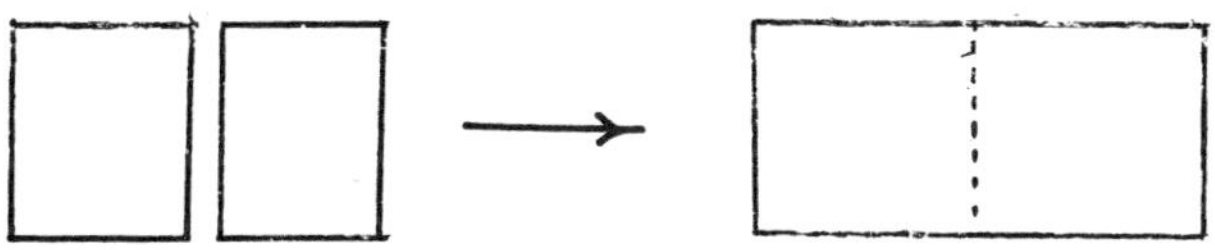

Fig. 2

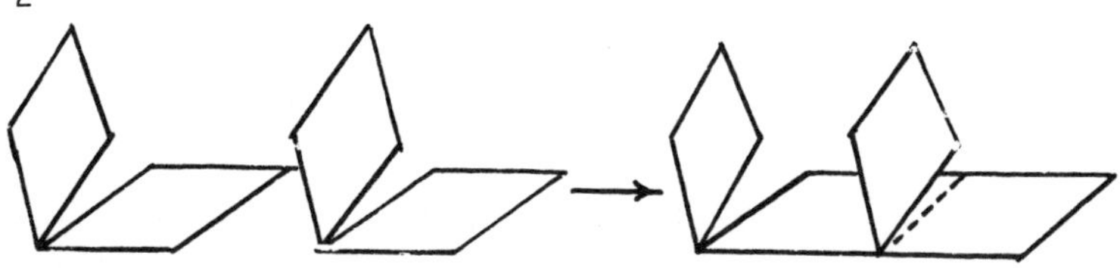

Fig. 3

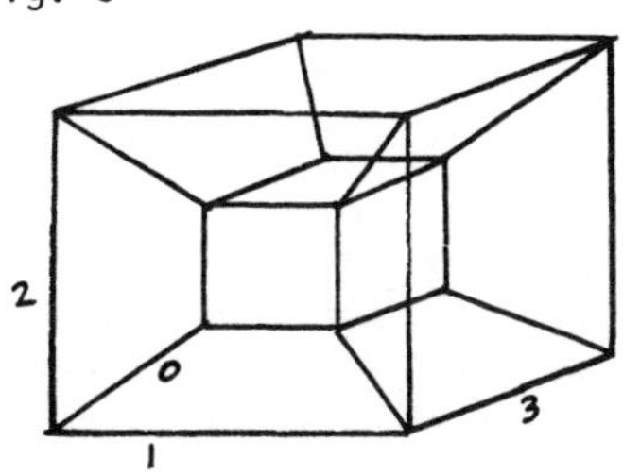


Notice that if we neglect links along the direction 0 and 1, then the links along 2 and 3 form nearest neighbor interactions in the 0-1 plane (actually, an infinite number of copies of the 0-1 plane)

Given the problems with decimations, is it possible to make any conclusion between four dimensional lattice gravity and a two dimensional nearest-neighbor system? Let us make the following speculation. In Migdal's original approximation, his recursion relations contained a hidden appoximation, i.e. the shifting around of certain plaquettes so that all decimations in all four directions can be done. Unfortunately, no one knows how good this approximation is, or even how to estimate how bad the approximation is. But there is another approximation which also takes us from four dimensions down to two dimensions. Simply make two gauge choices which make the gauge field vanish in two space-time directions. Clearly, such gauge choices are not allowed. Gauge invariance only allows us to fix $A_0 = 0$. But if we treat setting $A_1$ to be a small object, so we may perturb around small $A_1$, we find that a four dimensional gauge lattice easily splits up into an infinite sequence of two dimensional nearest-neighbor interactions. But the price we paid for this reduction was high, namely setting $A_1$ equal to zero. Viewed in this way, we see that the correspondence between a four dimensional gauge system and a two dimensional nearest-neighbor interaction is a poor one indeed.

For the moment, let us neglect the higher order terms leftover from decimating links in the lattice gravity theory (or equivalently, making an illegal choice of setting two sets of gauge fields to zero, or perturbing about one gauge field in the approximation that it can be considered small). Making these rather large and unwieldy assumptions, we can show that lattice gravity reduces to the following action in two dimensions:

$$I \sim \Sigma \,\mathrm{Tr}\, \{\gamma_5\, \sigma_{i,j}\, \sigma^{-1}_{i,j+1}\, \theta_{i,j}\, \theta^{-1}_{i+1,j} + \gamma_5\, \sigma_{i,j}\, \sigma^{-1}_{i,j+1}\, \theta_{i,j}\, \theta^{-1}_{i,j+1} \}$$

$$\sigma_{i,j} = U(A_2(x_0 = i,\ x_1 = j)\ ); \quad \theta_{i,j} \doteq U(A_3(x_0 = i,\ x_1 = j)\ )$$

where we have made the drastic approximations $A_0 = 0$ and $A_1$ small enough to perturb around. We are left with links only in the 2nd and 3rd directions, which we designate as $\sigma_{i,j}$ and $\theta_{i,j}$ respectively. (See Fig. 3)

## SUMMARY

Using lattice techniques, we have obtained a new gauge invariant cut-off for gravity and supergravity. The most immediate problem is trying to understand why the theory is not renormalizable, which is inherently linked to the fact that we have a theory expressed in terms of dual plaquettes rather than single plaquettes. Simple decimations, which are useful in deriving recursion relations, give rise to higher order terms not present in our original action. Thus, we are led to suspect that the theory is not renormalizable in the usual sense, but instead might possibly be finite order by order in the coupling constant. Certain statements relating gravity in D dimensions to nearest-neighbor theories in D/2 dimensions can be made, but how reliable they are for D = 4 is a matter of speculation.

## REFERENCES

Chemseddine, A.H. (1977) Massive supergravity from non-linear realizations of orthosymplectic gauge symmetry and coupling to spin 1/2 - spin 1 multiplet. Nuclear Physics B, 131, pp. 494-506

Das, A., Kaku, M., Townsend, P.K. (1979) Lattice Formulation of General Relativity. Phys. Lett. B, 81, pp. 11-14

Kaku, M., Townsend, P.K., van Nieuwenhuizen, P. (1977a) Gauge theory of the conformal and superconformal group. Phys. Lett. B, 69, pp. 304-308.

Kaku, M., Townsend, P.K., van Nieuwenhuizen, P. (1977b) Properties of conformal supergravity. Physical Review D, 17, pp. 3179-87.

Migdal, A. (1975a and b) Phase Transitions in gauge and spin-lattice systems. Zh. Eksp. Teor. Fiz. 69, p. 810 and p. 1457. Also, see Sov. Phys. JETP, 42, p. 413 and p. 743 (1976)

Smolin, L. (1979) Quantum gravity on a lattice. Nuclear Physics B, 148, pp. 333-372.

West, P.C. (1978) Physics Lett. B, 76, p. 569.

West, P.C. & A.H. Chemseddine, (1977) Supergravity as a Gauge Theory of Supersymmetry. Nucl. Phys. B. 129, p. 39.

# GAUGE VIERBEIN SPIN - 5/2, 7/2 AND THEIR COUPLING TO GRAVITY

C. Aragone

Depto. de Fisica, Universidad Simon Bolîvar
Apartado 80659, Caracas 1080 A - Venezuela.

## A INTRODUCTION

The assumption that the maximum helicity must be 2, was shown by Gell-Mann & Nee'man (1977) to lead, in the context of local extended super-symmetry, to very stringent bounds on the grand unified structure of the matter.

Cremmer & Julia (1979) discovered that SO(8) supergravity contains a concealed SU(8) local gauge invariance, rich enough to provide room, when broken at a mass scale of about $10^{15}$ GeV to the generally accepted low energy matter symmetry $SU(3)_{colour} \otimes SU(2)_{left} \otimes U(1)$. Recent attempts at superunification by Ellis, Gaillard, Maiani and Zumino (1980) have indicated SU(5) as the only remaining possibility for the grand unified matter symmetry, in the context of local supersymmetry.

However, the issue of which is the best GUT-group has not yet been settled. Perhaps other groups, like SO(4n+2) proposed by Gell-Mann, Ramond & Slansky (1979); SU(3n+2) ($n \geq 3$) supported by Georgi (1979) or scalarless technicoloured GUT'as conjectured by Dimopoulous & Susskind (1979), could be better candidates than SU(5).

In such a case, extended supergravities will have a spectrum with some spins higher than 2 (Curtright 1979a), it is worth re-analysing the well known problems of coupling higher spins to gravity.

Apart from supergravity, the simplest situation arises when considering the (2, 5/2) supermultiplet.

Recently the action for massless spin-5/2 on flat space (Schwinger 1970) based on a fully symmetric Majorana spinor was carefully examined by Fang & Fronsdal (1978), Berends, van Holten, van Nieuwenhuizen & de Wit (1979a, b) and Aragone & Deser (1980a).
When gravity is coupled to this (symmetric) spin - 5/2 matter, which is what we call hypergravity, the presence of terms proportional to the Rieman tensor, which might preclude the vanishing of curved matter gauge invariance condition, was pointed out by

Berends, van Holten, van Nieuwenhuizen & de Wit (1980). A detailed calculation by Aragone & Deser (1979a) revealed that, due to the $\gamma$-tracelessness of the matter local gauge parameter, either conformally flat backgrounds or half-flat systems (i.e. states with half-flat gravity and matching pure helicity matter) might be consistently coupled in the Euclidean regime. This agreed with similar results of Christensen & Duff (1979) based on arguments of a non-gauge character requiring consistency of a generalized curved d' Alembertian for arbitrary spin.

We addressed ourselves the questions of whether there might exist alternative formulations of massless spin-5/2, presenting a stronger gauge character than the symmetric one, and what should be its behaviour when coupled to gravity.

The answer to the first question turned out to be affirmative. We found (Aragone & Deser 1979b, 1980b) two different, vierbein-like actions equivalent on flat space, one of which is the smooth generalization of the Rarita-Schwinger action for spin 3/2, and the other having a construction requiring a fermionic field of helicity strictly higher than 3/2. Regarding their coupling to gravity they do not constitute a way out: both actions lead to consistency conditions which are proportional in their highest spin gravity-matter part.

After looking at massless spin-5/2, it is difficult not to investigate the fate of the massless fermions of higher (than 5/2) spin. Fang & Fronsdal (1978) were able to generalize the s = 5/2 results to any half integer spin. They found an action for the totally symmetric $\phi_{\mu_1 \ldots \mu_s}$ Majorana spinor that has the right physical content (only helicity $\pm(s + 1/2)$ propagating modes, no lower helicities or ghosts) provided $\phi''' \equiv \gamma\gamma\gamma\phi$ vanishes. This is a discontinuity in the symmetric formulation which starts at s = 7/2 and goes on. This conceptual jump is also present in the generalized-Christoffel symbols approach of Freedman & de Wit (1980). If one emphasizes the gauge structure, a vierbein spin s + 1/2 action is found (Aragone & Deser, 1980b), which is the uniform generalization of the spin - 5/2 situation. This is the reason why, in the presence of a uniform formulation for higher fermionic spins, spin 7/2 has interest by its own, in the same way as for bosons where spin-4 is the lowest case where the parallel condition of double tracelessnees becomes meaningful.

We present several new properties of the vierbein formulation mainly for spin 5/2, 7/2, using the simplest of the two actions. The global supersymmetric transformations and the conserved current of the (2,5/2) supermultiplet will be given. They provide a hint to the possible locally supersymmetric (ss) 1/2-graded transformations of the coupled (2,5/2) system which are clearly different from the possible locally ss, 3/2-graded ones suggested by the curved gauge invariance of the spin-5/2 theory.

The question of whether a locally ss-1/2-graded (2,5/2)

consistent system exists is an interesting one. In fact, for global supersymmetry Curtright (1979b) has shown that a good (massless) spin-s flat action plus a supersymmetric transformation linking it to an associated ss- partner of spin-s±1/2 implies the good (i.e. pure spin - s± 1/2) content of the partner. Nevertheless conservation of the stress tensor still requires the curved gauge invariance associated with a spin-3/2 gauge parameter (Aragone & Deser 1980c).

## B REVIEW AND NOTATION

In the vierbein formulation spin-5/2 is represented by a non symmetric Majorana tensor-spinor $\psi_{\mu\tilde{a}}$, $\gamma$-transverse in the vierbein index $\gamma^{a}\psi_{\mu\tilde{a}}=o$. (We use Greek letters for world indices, Latin for vierbein indices and tilded Latin to indicate $\gamma$-transverse projections).

On flat space $\psi_{\mu\tilde{a}}$ transforms according to the Abelian law

$$\delta_{\zeta}\psi_{\mu\tilde{a}} \equiv \partial_{\mu}\zeta_{\tilde{a}} \quad , \quad \gamma^{a}\zeta_{\tilde{a}} = o. \tag{1}$$

The spin- 5/2 gauge invariant fermionic field strength is given by $f_{\mu\nu\tilde{a}} \equiv \partial_{\mu}\psi_{\nu\tilde{a}} - \partial_{\nu}\psi_{\mu\tilde{a}}$.

It has been shown (Aragone & Deser 1980b), that there are two different, gauge invariant actions for any spin higher than 3/2. For spin 5/2 we have

$$I_{o}^{5/2}(\psi) \equiv 2^{-1}i\int\bar{\psi}_{\rho\tilde{a}}\,\gamma^{\rho\mu\nu}f_{\mu\nu\tilde{a}}(d^4x) \tag{2}$$

where $\psi\equiv\eta^{\mu a}\psi_{\mu\tilde{a}}$ , $\eta_{\mu\nu}$ has signature $(-+++)$, $\gamma$-matrices are real: $\varepsilon\rho\mu\nu\kappa$ : $\varepsilon^{o123} = +1$ , $2^{-1}\{\gamma_a,\gamma_b\}= \eta_{ab}$ , $\sigma_{ab} \equiv 2^{-2}\left[\gamma_a,\gamma_b\right]$ , $\gamma_5 \equiv \gamma_o\,\gamma_1\,\gamma_2\,\gamma_3$ , $\gamma^{\rho\mu\nu} \equiv \varepsilon^{\rho\mu\nu\kappa}\,(-\gamma_5)\,\gamma_{\kappa}$ .

Varying the 12 independent components $\psi_{\mu\tilde{a}}$ yields

$$S_{o\mu\tilde{a}}(\psi) \equiv 2\,\gamma_5\,\gamma_b\,{}^{*}f_{\mu b\tilde{a}} - \gamma_5\gamma_a\,{}^{*}f_{\mu b\tilde{b}} = 0 \ , \tag{3}$$

where ${}^{*}f_{\mu\nu\tilde{a}} \equiv 2^{-1}\,\varepsilon_{\mu\nu}^{\ \ \kappa\lambda}\,f_{\kappa\lambda\tilde{a}}$ is the dual field-strength and $f_b \equiv f_{ba\tilde{a}}$.

Going through the reduction process it can be shown that the two actions are both proportional to the same, pure helicity - 5/2 unconstrained action (which is also the same for the symmetric formulation)

$$i\int\bar{\psi}^{TT\tau}_{i_1i_2}\ \not{\partial}\ \psi^{TT\tau}_{i_1i_2} \tag{4}$$

Here $\psi^{TT\tau}_{i_1i_2}$ is the symmetric, 3-dimensional, transverse and $\gamma$-transverse gauge invariant pure helicity-5/2 dynamical variable whose dynamics is determined by the ordinary Dirac equation.

To obtain the massless spin -7/2 theory in its vierbein form one has to add throughout our spin-5/2 formulation one more vierbein index.

Spin- 7/2 is represented by $\psi_{\mu\tilde{a}\tilde{b}} = \psi_{\mu\tilde{b}\tilde{a}}, \gamma^{a} \psi_{\mu\tilde{a}\tilde{b}} = o$, whose gauge transformation is $\delta_{\zeta}\psi \equiv \partial_{\mu}\zeta_{\tilde{a}\tilde{b}}$ , with $\zeta_{\tilde{a}\tilde{b}}$ symmetric and γ-transverse. Action (2) becomes

$$I_{o}^{7/2}(\psi) \equiv i/2 \int \bar{\psi}_{\rho\tilde{a}\tilde{b}} \gamma^{\rho\mu\nu} f_{\mu\nu\tilde{a}\tilde{b}} (d^{4}x) \qquad (5)$$

Notice that $\psi_{\mu ab}$ is automatically triple γ - traceless: $\psi''' \equiv \gamma^{a} \gamma^{b} \gamma^{\mu} \psi_{\mu\tilde{a}\tilde{b}} = o$.

The (24) independent untwisted spin - 7/2 field equations generalizing eq. (5) are given in terms of the dual fermionic field strength

$$2^{-1}.3 S_{o\mu\tilde{a}\tilde{b}} \equiv 3\gamma_{5}\gamma_{b} {}^{*}f_{\mu c\tilde{a}\tilde{b}} - \gamma_{5} \gamma_{(a} {}^{*}f_{\mu c\tilde{c}\tilde{b})} = o \cdot \qquad (6)$$

Six Bianchi identities occur now. They allow to reduce the dynamical variables down to, at most 6. After performing the explicit reduction process 5 of them disappear and one is left with the massless spin-7/2 action alike to Dirac's for s= 1/2.

## C INDEPENDENCE OF THE LORENTZ PART

In vierbein linearized gravity, the basic field $h_{\mu a}$ transforms as $\delta_{\xi}h_{\mu a} = \partial_{\mu}\xi_{a}$ . Neither its symmetric nor its antisymmetric part $\ell_{ab} \equiv 2^{-1}(h_{ab} - h_{ba})$ is gauge invariant. However the second order field equations $R^{L}_{\mu a}(h) = 0$ are, automatically, symmetric and independent of the 6 variables $\ell_{ab}$.

We shall see that for s = 5/2; 7/2 fermions the situation is alike.

Let us begin with spin-5/2. What is the Lorentz part of $\psi_{\mu a}$?. It turns out that $\psi_{\mu\tilde{a}}$ can be written in the form:

$$\psi_{\mu\tilde{a}} \equiv \psi_{\tilde{\mu}\tilde{a}} + \chi_{\tilde{\mu}\tilde{a}} + 3^{-1} \gamma_{\mu}\psi_{\tilde{a}} - 6^{-1}\gamma_{a}\psi_{\tilde{\mu}} + 4^{-1}\eta_{\mu a}\psi + 6^{-1}\sigma_{\mu a}\psi \qquad (7)$$

where $\psi_{\tilde{a}} \equiv \gamma^{\mu}\psi_{\mu\tilde{a}} - 2^{-1}\gamma_{a}\psi$ is the spin 3/2 part which has been used to locally reparametrize the gauge group.

The first term of (7) is symmetric and γ-transverse, its 6 variables carry the spin-5/2. As ψ, is one Majorana variable , and $\psi_{\tilde{a}}$ involves 3 independent variables, $\chi_{\tilde{\mu}\tilde{a}}$ is left with 2 degrees of freedom. Actually $\chi_{\tilde{\mu}\tilde{a}}$ is anti-symmetric and γ-transverse, corresponds to a $(3,0')\oplus(0,3')$ irreducible representation of SL(2,C) and is, by definition, the Lorentz part of the basic vierbein field.

At this point, we claim that $S_{o\mu\tilde{a}}(\chi)\equiv 0$.

To see this, it is convenient to represent $\chi_{\tilde{\mu}\tilde{a}}$ in terms of a 3 dimensional $\gamma$-transverse spinor $\chi_{\tilde{i}}$; $\gamma^i\chi_{\tilde{i}} = o$:

$$\chi_{\tilde{i}\tilde{j}} = \gamma_i\chi_{\tilde{j}} - \gamma_j\chi_{\tilde{i}}\ , \quad \chi_{oi} = \gamma^o\chi_{\tilde{i}}\ . \tag{8}$$

Then, substituting into $S_{o\mu\tilde{a}}$ this value of $\chi_{\tilde{\mu}\tilde{a}}$ one obtains the total cancellation we stated before, showing the independence of $S_{o\mu\tilde{a}}$ of the Lorentz-like variables $\chi$.

For spin-7/2, eq. (7) is replaced by:

$$\psi_{\mu\tilde{a}\tilde{b}} = \psi_{\tilde{\mu}\tilde{a}\tilde{b}} + \chi_{\tilde{\mu}(\tilde{a}\tilde{b})} + \Sigma c\Gamma\psi_{\tilde{a}\tilde{b}} + \Sigma c\Gamma\psi_{\tilde{a}\tilde{b}} + \Sigma d\Gamma\psi_{\tilde{a}} \tag{9}$$

where $\psi_a \equiv \eta^{\mu a}\ \psi_{\mu\tilde{a}\tilde{b}}$ , $\psi_{\tilde{a}\tilde{b}}$ is the fully symmetric and $\gamma$-transverse projection of $\gamma^{\mu}\psi_{\mu\tilde{a}\tilde{b}}$ and $\chi_{\tilde{\mu}\tilde{a}\tilde{b}}$ is fully $\gamma$-transverse, antisymmetric in its first two indices and satisfies a cyclic identity $\chi_{(\tilde{\mu}\tilde{a}\tilde{b})} = 0$.

We call $\chi_{\tilde{\mu}\tilde{a}\tilde{b}}$ the Lorentz part of the basic spin-7/2 vierbein field. The spatial variables $\chi_{ij\ell}$ are not independent

$$\sigma_{ij}\ \chi_{ij\ell} = 0 = \gamma^i\chi_{i\ell\ell}\quad . \tag{10}$$

These restrictions can be solved in terms of a traceless symmetric tensor $\chi_{\ell m}$ . Introducing the explicit values of $\chi_{\tilde{\mu}\tilde{a}\tilde{b}}$ into $S_{o\mu\tilde{a}\tilde{b}}$ we found a full cancellation, $S_{o\mu\tilde{a}\tilde{b}}\ (\chi)\equiv 0$, the $s = 7/2$ vierbein operator is independent of the Lorentz variables.

On the other hand, straightforward calculation shows that, neither $\chi_{\tilde{\mu}\tilde{a}}$ nor $\chi_{\tilde{\mu}\tilde{a}\tilde{b}}$ are gauge invariant quantities. Even if they essentially are kinematical variables they cannot be eliminated from the formulation. For any spin -s+ 3/2 their number is always one less than the gauge degrees of freedom.

## D FLAT QUANTIZATION AND SOURCES

Gauge invariance (2) of the action implies conservation of the source $j_{\mu\tilde{a}}$

$$\partial_\mu j_{\mu\tilde{a}} = 0. \tag{11}$$

In addition, the independence of $S_{\mu\tilde{a}}$ of the Lorentzian part of $\psi_{\mu\tilde{a}}$ induces an additional (algebraic) symmetry on the external source

$$\chi(j_{\mu\tilde{a}}) \equiv 2^{-1}(j_{\mu\tilde{a}} - j_{a\tilde{\mu}}) - 2^{-2}(\gamma_\mu j_a - \gamma_a j_\mu) + 3^{-1}\sigma_{\mu a} j = 0 \qquad (12)$$

($j \equiv \eta^{\mu a} j_{\mu\tilde{a}}$ and $j_a \equiv \gamma^\mu j_{\mu\tilde{a}}$ respectively).

BRS transformations for the generating functional are given by

$$\delta\psi_{\mu\tilde{a}} = \partial_\mu C_{\tilde{a}}\lambda \ , \ \delta C_{\tilde{a}} = 0 \ , \ \delta C^*_a = - \xi^{-1}(\not\partial\psi_{\tilde{a}})_{\tilde{a}}\,\lambda \ . \qquad (13)$$

If we then go to the effective action we find that in the ξ-harmonic gauge, spin-5/2 and ghost propagators must obey the identity:

$$k_\lambda \langle C_a \overline{C}^*_b \rangle + \xi^{-1} \langle \hat{\psi}_{\lambda\tilde{a}}, \overline{\hat{\psi}}_{\mu\tilde{c}} \rangle_\xi \ \gamma^\mu \not k (\eta_{cb} - 2^{-2}\gamma_c\gamma_b) \equiv 0 \ . \qquad (14)$$

## E CONSISTENCY REQUIREMENTS AND STRESS TENSOR CONSERVATION OF GRAVITY PLUS SPIN-5/2 MATTER

On flat space, Grisaru & Pendleton (1977) and Grisaru, Pendleton & van Nieuwenhuizen (1977) using S-matrix arguments have set very stringent bounds to the coupling of gravitons with massless- spin-5/2 fermions.

Let us minimally couple, for instance, $I_o^{5/2}(\psi)$ to gravity without torsion. It becomes

$$I_o^{5/2} \equiv i/2 \int \overline{\psi}_{\rho a}\, \gamma^{\rho\mu\nu} f_{\mu\nu\tilde{a}}(\psi)(d^4x) \qquad (15)$$

where $\gamma^{\rho\mu\nu} \equiv \varepsilon^{\rho\mu\nu\lambda}\gamma_b\gamma_5 e_\lambda{}^b$ is a density, world and Lorentz indices are related by the vierbein $e_\mu^a$ and its inverse $e_b^\nu$ and $f_{\mu\nu\tilde{a}}(\psi) \equiv D_{[\mu}\psi_{\nu]\tilde{a}}$ where $D_\mu$ is the covariant derivative.

Consistency means that the interacting system composed by gravity and the action (15) will still be gauge invariant under the generalized matter-fermionic gauge transformation $\delta_\zeta\psi_{\mu\tilde{a}} = D_\mu\zeta_{\tilde{a}}$ .

Under this transformation it has been shown that

$$\delta_\zeta I_o^{5/2} = i\int(d^4x)\{\overline{\zeta}_{\tilde{a}} R_{ab\mu\nu}\, \gamma^{\mu\nu\rho}\psi_{\rho\tilde{b}} - \overline{\zeta}_{\tilde{a}} G_{\mu\nu}\gamma_\mu\psi_{\nu\tilde{a}}\} \equiv i\int\overline{\zeta}_{\tilde{a}} C_a \qquad (16)$$

which is a measure of the inconsistency of the established coupling. (For spin 3/2 one gets $\delta_\zeta I^{3/2} \sim \overline{\zeta}\, G_{\mu\nu}\gamma_\mu\psi_\nu$).

**The success of supergravity stems from two facts. The first that the fermion being a spin -3/2 particle, there is no room in** $\delta_\zeta I^{3/2}(\psi)$ for terms like the first of (21), proportional to the Riemann tensor. The second, and this is supergravity, that there exists a coupling scheme such that, when $G_{\mu\nu} \sim T_{\mu\nu}(\psi)$ is substituted into (21) one gets zero.

Actually $C_a$ which has an apparent pure matter origin would preclude stress-tensor conservation if it does not vanish (Aragone

& Deser 1980c).

To illustrate what happens here, let us consider a small coordinate transformation $\xi^{\mu}(x)$. Under this transformation the two vierbein fields transform according to

$$\delta_{\xi} e_{\mu}{}^{a} = \xi^{\lambda} \partial_{\lambda} e_{\mu}{}^{a} + e_{\lambda}{}^{a} \partial_{\mu} \xi^{\lambda} \;, \quad \delta_{\xi} \psi_{\mu\tilde{a}} = \xi^{\lambda} \partial_{\lambda} \psi_{\mu\tilde{a}} + (\partial_{\mu} \xi^{\lambda}) \psi_{\lambda\tilde{a}} \tag{17}$$

and the variation (vanishing because of its explicit covariance) of the full action of the system (up to 4-divergences) has the value

$$\delta_{\xi} I^{2+5/2} = \int \{ -\kappa^{2} \xi^{\lambda} (D_{\mu} T^{\mu}_{\lambda} + \bar{\psi}_{\lambda\tilde{a}} C_{\tilde{a}}) + \xi^{\lambda} \omega_{\lambda\mu\nu} (G_{\mu\nu} + \kappa^{2} T_{\mu\nu}) + i\kappa^{2} \xi^{\lambda} e (\bar{f} S_{\mu\tilde{a}} + \omega\psi S) \} \equiv 0 \tag{18}$$

**It is immediately observed that, on shell for both fields,** one can only state that $D_{\mu} T^{\mu}_{\lambda} + \bar{\psi}_{\lambda\tilde{a}} C_{\tilde{a}} = 0$. Conservation of the stress tensor requires matter-consistency and conversely.

Coming back to $C_{\tilde{a}}$ defined by eq. (16) $R\Gamma'\psi + G\Gamma''\psi$, it is convenient to rewrite it in terms of the self-dual and antiself-dual parts of the conformal Weyl tensor $C^{\pm}_{\tilde{a}\lambda\mu\nu} \equiv 2^{-1} (C \pm i\, C^{*})_{a\lambda\mu\nu}$, $1/2 \varepsilon^{\rho\sigma}_{\mu\nu} C \equiv C^{*}$ and to decompose spinors into their left and right handed parts, $\psi^{L,R} \equiv 2^{-1}(1 \pm i\gamma_{5})\psi$.

Then $\delta I^{5/2}_{o}$ takes the form:

$$\delta_{\zeta} I^{5/2}_{o} = 2i \int S_{a} \{ C^{+}_{a\lambda\mu\nu} \gamma_{\nu} \psi^{R}_{\tilde{\lambda}\tilde{\mu}} + 2^{-1} C^{+}_{a\lambda\mu\nu} \sigma_{\mu\nu} \psi^{L}_{\tilde{\lambda}} + \text{c.c.} - G_{\mu\nu} \gamma_{\nu} \psi_{\tilde{\mu}\tilde{a}} + \\ + O(G, \psi_{\tilde{a}}, \psi) \} \tag{19}$$

where it is shown the full dependence on the Weyl tensor and on $\psi_{\tilde{\mu}\tilde{a}}$, the symmetric- $\gamma$-transverse carrier of spin-5/2.

Notice that the non central, spin-3/2 Lorentz antisymmetric components $\chi_{\tilde{\mu}\tilde{a}}$ do not appear in (19), showing again their kinematical character, as it must be, since $\delta_{\zeta} I^{5/2}_{o} \sim S_{\mu\tilde{a}}$ and it has to be evaluated for $\delta_{\zeta} \psi_{\mu\tilde{a}} = D_{\mu} \zeta_{\tilde{a}}$. Since $S_{\mu\tilde{a}}(\psi)$ are the curved formulation of flat equation (5) they will not contain $\chi_{\tilde{\mu}\tilde{a}}$.

Spin -5/2 is the first case where the coupling to gravity leads in a natural and compulsory way to the appearance of the Weyl -tensor in the inconsistency. Any other coupling scheme (with or without torsion) one may imagine, will always lead to the same algebraic structure in C and $\psi_{\mu\tilde{a}}$ (Aragone & Deser 1979a, 1980b).

Since we are now familiar with supergravity, one can be optimistic and say that actually there might be a better chance of obtaining consistency if one can get rid of the Weyl dependent terms, because perhaps mechanisms similar to the one of supergravity could take care of these (good)- bad terms $\sim G_{\mu\nu}\Gamma\psi_{\lambda a}$. And at least one can explicitly show whether a certain coupling is or is not consistent (which is not ensured "a priori" by the geometrical framework).

In Lorentz space one can set $C_{\mu\nu\rho\sigma} = 0$ and try to see whether there still remains a more or less wide category of spaces which can admit massless spin-5/2. As this assumption is dynamically too strong, it is worth seeing what happens in the Euclidean sector, (Hawking & Pope 1978). In the Euclidean regime, the Dirac adjoint $\widetilde{\psi}_{\mu\tilde{a}}$ and $\psi_{\mu\tilde{a}}$ are respectively represented by

$$\overline{\psi}_{\mu\tilde{a}} \equiv (\psi_{\mu\tilde{a}A}, \widetilde{\psi}_{\mu\tilde{a}}{}^{B'}) , \qquad \psi_{\mu\tilde{a}} \equiv \begin{pmatrix} \psi_{\mu\tilde{a}}{}^{A} \\ \psi_{\mu\tilde{a}B'} \end{pmatrix} , \quad \varepsilon_{12} = +1. \tag{20}$$

and the action (15), quadratic in $\psi$ in the Minkowski sector, becomes bilinear in $\widetilde{\psi}$ and $\psi$ (Aragone 1980).

$$I_o^{5/2}(\widetilde{\psi},\psi) = -2i\int \widetilde{\psi}_{\rho\tilde{a}}\overline{\alpha}^{\rho\mu\nu} D_\mu \psi_{\nu\tilde{a}} , \quad (\overline{\alpha}^{\rho\mu\nu})_{B'A} \equiv \varepsilon^{\rho\mu\nu\kappa}\overline{\alpha}_{bB'A} e_\kappa{}^b . \tag{21}$$

Should the theory be massive this property will be lost. The massless spin-5/2 stress tensor $T^\mu_a(\psi) \sim \delta I_o^{5/2}/\delta e^a_\mu$ has the same structure bilinear in the Euclidean helicities. In particular $T^\mu_a(\psi)$ will vanish for pure helicity (Euclidean) states, either $\psi = 0$ or $\widetilde{\psi} = 0$ and, then "a fortiori" also $G_{\mu\nu} \sim T^\mu_a$ will cancel out for these kinds of matter states.

Looking at the expression (24) we see that for pure helicity matter states one is only left with Weyl-terms (where now $\psi^{L,R}$ have to be read $\widetilde{\psi},\psi$). And then if one goes to half-flat spaces (i.e. $C^+C^- = 0$) the whole expression vanishes and the system becomes consistent for these generalized half-flat (gravity-matter) systems.

Other inequivalent formulations for spin-5/2 will not help to improve the situation, taking into account the results of Christensen & Duff (1979).

## F THE GLOBAL SUPERSYMMETRIC VIERBEIN (2,5/2) MULTIPLET

Massless spin-2 is represented by linearized first order vierbein gravity:

$$I^2(h,\omega) = \int\{h\partial_\nu\omega_\nu - h^{a\mu}\partial_\mu\omega_a + h^{a\mu}\partial_\nu\omega_{\mu\nu a} + \omega_{\mu ab}\omega_{ba\mu} - \omega_b^2\}(d^4x) \tag{22}$$

where the vierbein has been replaced by the weak aproximation $e_{\mu a} \equiv \eta_{\mu a} + \kappa h_{\mu a}$, $e^{\mu a} = \eta^{\mu a} - \kappa h^{a\mu}$ , $\partial_\mu \xi_a(x)$ is the spin-2 gauge transformation, the linearized connection $\omega_{\mu ab} = -\omega_{\mu ba}$ is antisymmetric in its last two indices and $\omega_{aab}$ is abbreviated by $\omega_b$.

GAUGE VIERBEIN SPIN - 5/2, 7/2

Spin - 5/2 is represented by the untwisted action (15). On the field eqs. determined by the action (22), the supersymmetric 1/2 graded transformation is given by:

$$\delta_\zeta h_{\mu a} = 2i\bar{\zeta}\psi_{\mu\tilde{a}} \ , \ \delta_\zeta \psi_{\mu\tilde{a}} = 2^{-1}\omega_{\mu bc}\sigma_{bc}\gamma_a \zeta \ . \tag{23}$$

It is also possible to derive the supercurrent which is conserved by virtue of the global supersymmetry:

$$j^{\nu}_{(2,5/2)} = \frac{i}{2}\ \omega_{\rho bc}\ \gamma_a \sigma^{bc}\ \gamma^{\mu\nu\rho}\ \psi_{\mu\tilde{a}} \tag{24}$$

G LOCAL 1/2 AND 3/2 GRADED SUPER SYMMETRIES AND CONSISTENCY

The existence of higher gradings, $P_{j+1/2}$, of the Poincaré algebra has been explicitly shown by Konopel'chenko (1977). This means that in principle it is possible to write down a square-root of the Poincaré algebra having spinorial charges of spin-3/2, $q_{\tilde{a}}$, whose anticommutator is proportional to the bosonic translations $P_a$. As the gauge field naturally associated to this charge is a vierbein spin-5/2 $\psi_{\mu\tilde{a}}$ one can regard the problem of obtaining a consistent (2,5/2) coupling as being equivalent to gauging $P_{3/2}$. However it has been shown, that on flat space, $P_{3/2}$ can only be trivially represented (Berends, van Holten, van Nieuwenhuizen & de Wit 1979c) and Aragone & Deser 1979b). It is worth asking whether there could be established a consistent, locally 1/2-graded coupling for the supermultiplet (2,5/2). Since we have global transformations for the linearized system, it is immediate to state what must be the transformation ensuring local $P_{1/2}$ supersymmetry:

$$\delta_\beta \psi_{\mu\tilde{a}} = D_\mu \gamma_a \beta \ . \tag{25}$$

Under transformation (25) the curved action $I^{5/2}$ changes by

$$\delta_\beta I^{5/2}(\psi) = \int (d^4x)\{-\bar{\beta}\ \gamma^\mu \gamma^\nu \chi_{\tilde{a}\tilde{b}} C_{\mu\nu ab} + \bar{\beta}\ \gamma^\mu \gamma^\rho \psi_{\rho\tilde{b}} G_{b\mu} + \frac{2}{3}\ \bar{\beta}\psi R\} \tag{26}$$

where the Weyl term is now only coupled to the spin 3/2 -Lorentz part of the spin-5/2 field.

Actually full local supersymmetry of the system requires vanishing of $\delta_\beta I^2 + \delta_\beta I^{5/2}$ but it is clear that $\delta_\beta I^2 \sim \bar{\beta} G \psi_{\mu\tilde{a}}$ will not be the source of additional Weyl terms.

Since there are 2 Majorana degrees of freedom in $\chi_{\tilde{a}\tilde{b}}$, and as $\chi$ is a gauge covariant object one can pick up the covariant gauge $\chi_{\tilde{a}\tilde{b}} = \psi = 0$ and check whether the system is locally supersymmetric or not.

But even if one is lucky enough to obtain a locally ss - 1/2 graded coupling for the supermultiplet (2,5/2), that will not erase the existence of the additional consistency requirement (19) of matter gauge invariance, which is compulsorily linked to the conservation of the stress-tensor.

Notice that the reasoning made does not depend on the assumption of asymptotic flatness, it is still valid even in the presence of a cosmological term. That extends the initial conceptual bounds established the by S- matrix arguments of Grisaru & Pendleton (1977) and Grisaru, Pendleton & van Nieuwenhuizen against the coupling of the supermultiplet (2,5/2) in the Lorentzian region.

## ACKNOWLEDGMENTS

This work was finished during the Nuffield Supergravity Workshop whose attendance was made possible by the CONICIT grant S3-137.

## REFERENCES

Aragone C. & Deser S. (1980a) Hamiltonian form for massless higher spin fermions Phys. Rev. D, 21, pp.352-7.

Aragone C. & Deser S. (1979a) Consistency Problems of Hypergravity Phys. Lett B., 86. pp. 161-3.

Aragone C. & Deser S. (1979b) Consistency Requirements in Hypergravity in Supergravity, ed D.Z. Freedman & P. van Nieuwenhuizen, North Holland.

Aragone C. & Deser S. (1980b) Higher spin vierbein gauge fermions and hypergravities. Nucl. Phys. B, to be published.

Aragone C. & Deser S. (1980c) Consistency problems of spin 2-gravity coupling. N. Cimento, to be published.

Aragone C. (1980) Lorentz variables and vierbein gauge fermions. In preparation.

Berends F. A., van Holten J. W., van Nieuwenhuizen P. & de Wit B. (1980), On spin-5/2 gauge fields. J.Phys.A, 13, 1643-9.

Berends F.A., van Holten J.W., van Nieuwenhuizen P. & de Wit B. (1979a) On Spin-5/2 gauge fields. Phys. Lett. B, 83, pp. 188-90.

Berends F.A., van Holten J.W., van Nieuwenhuizen P. & de Wit B. (1979b) On field theory for massive and massless spin-5/2 particles. Nucl. Phys. B, 154, pp. 261-282.

Christensen S.M. & Duff M.J. (1979) New gravitational index theorems and supertheorems. Nucl. Phys. B., 154, pp. 301-42.

Curtright T. (1979a) report EFI 79/18 unpublished.

Curtright T. (1979b) Massless field supermultiplets with arbitrary spin. Phys. Lett. B., 85, 219-24.

Cremmer E., & Julia, B. (1979) The SO(8) Supergravity. Nucl. Phys. B., 159, 141-212.

Dimopoulos S. & Susskind L., Nucl. Phys. B., (1979) Mass without scalars. 155, pp. 237-252.

Ellis, J. Gaillard M.K., Maiani L. & Zumino B. (1980) Attempts at superunification. CERN report th 2841 Lapp report th 15.

Fang J. & Fronsdal C. (1978) Massless fields with integer spin. Phys. Rev. D., 18, pp. 3630-3633.

Freedman D.Z. & de Wit B (1980) Systematics of higher spin gauge fields. Phys. Rev. D., 21, 358-367.

Gell-Mann, M. & Nee'man, Y. (1977), unpublished.

Gell-Mann, M., Ramond P. & Slansky R. (1979) Complex spinors and unified theories in Supergravity ed. D.Z. Freedman & P. van Nieuwenhuizen, North Holland.

Georgi H., (1979) Toward a Grand Unified Theory of Flavour. Nucl. Phys. B., 156, pp. 126-134.

Grisaru M. & Pendleton H. (1977) Soft spin 3/2 fermions require gravity and supersymmetry. Phys. Lett. B., 67, pp. 323-6.

Grisaru M., Pendleton H. & van Nieuwenhuizen, P. (1977) Supergravity and the S-matrix. Phys. Rev. D., 15, pp. 996-1006.

Hawking S. W. & Pope C.N. (1978) Symmetry breaking by instantons in supergravity. Nucl. Phys. B. 146, pp. 381-92.

Konopel'chenko B. G. (1977) Symmetry groups in quantum field theory. Sov. J. Part. Nucl., 8, pp. 57-72.

Schwinger J. (1970) Particles, Sources and Fields, Addison Wesley, Reading, Mass.